T0093261

AUTOMATION AND COMPUTATION

The proceedings of the International Conference on Automation and Computation 2022 (AUTOCOM-22) consist of complete research articles that were presented at the conference. Each of the research articles was double-blind reviewed by the experts of the corresponding domain. The book contains a blend of problems and respective solutions related to computer-based automation & computation to highlight the recent technological developments in computer-based automation. It serves as an environment for researchers to showcase the latest research results on Data Science & Engineering, Computing Technologies, Computational Intelligence, Communication & Networking, Signal & Image Processing, Intelligent Control Systems & Optimization, Robotics and Automation, Power, Energy & Power Electronics, Healthcare & Computation, AI for human interaction, etc. It aims to give deep insight into the current trends of research in science and technology and shall introduce the reader to the new problems and respective approaches toward the solution and shall enlighten the researchers, students and academicians about the research being carried out in the field.

PROCEEDINGS OF THE INTERNATIONAL CONFERENCE ON AUTOMATION AND COMPUTATION, (AUTOCOM 2022), DEHRADUN, INDIA, 20–22 DECEMBER 2022

Automation and Computation

Edited by

Satvik Vats and Vikrant Sharma
*Computer Science and Engineering, Graphic Era Hill University,
Clement town, Dehradun*

Karan Singh
Computer and System Sciences, Jawaharlal Nehru University, New Delhi

Amit Gupta, Dibyahash Bordoloi and Navin Garg
*Computer Science and Engineering, Graphic Era Hill University,
Clement town, Dehradun*

CRC Press
Taylor & Francis Group
Boca Raton London New York Leiden

CRC Press is an imprint of the
Taylor & Francis Group, an **informa** business

A BALKEMA BOOK

First published 2023
by CRC Press/Balkema
4 Park Square, Milton Park, Abingdon, Oxon, OX14 4RN
e-mail: enquiries@taylorandfrancis.com
www.routledge.com – www.taylorandfrancis.com

CRC Press/Balkema is an imprint of the Taylor & Francis Group, an informa business

Library of Congress Cataloging-in-Publication Data
A catalog record has been requested for this book

ISBN: 978-1-032-36723-1 (hbk)
ISBN: 978-1-032-36724-8 (pbk)
ISBN: 978-1-003-33350-0 (ebk)

DOI: 10.1201/9781003333500

Typeset in Times New Roman
by MPS Limited, Chennai, India

Table of contents

Automation and Computation – Vats et al. (Eds)
© 2023 the Editor(s), ISBN 978-1-032-36723-1

Preface

The International Conference on Automation and Computation (AUTOCOM-22) took place at Graphic Era Hill University, Dehradun on December 20–22. The purpose of the conference was to establish a platform to promote interaction among the knowledge holders belonging to industry, academia, and various sections of society.

The theme of the conference was chosen to cater to all the wings of society which are influenced by the advancement in automation and computation taking place these days. The theme allowed the participants to identify and present the best research ideas. Major of the research tracks include Data Science & Engineering, Computing Technologies, Computational Intelligence, Communication & Networking, Signal & Image Processing, Intelligent Control Systems and Optimization, Robotics and Automation, Power, Energy and Power Electronics, Agriculture & Computation, Healthcare & Computation, AI for human interaction, etc.

More than hundred articles from various domains were received from various researchers in India and across the world. Each article went through a double-blind peer review process by the expert reviewers of the corresponding domain. About 40% of the, which were found relevant in terms of quality innovation and novelty were selected and the authors were invited to present their work in the conference. Apart from this, several renowned speakers (Experts of specific domain) delivered keynote sessions in the conference.

It is believed that this conference will prove to be beneficial for all the participants in terms of knowledge sharing, vision enlightening and motivating the budding researchers to think with the pace of the world.

Prof. Divyahash Bordoloi – Prof. Naveen Garg
Organizing Committee (AUTOCOM-22)
Conference Chair
Graphic Era Hill University Deharadun

Automation and Computation – Vats et al. (Eds)
© 2023 the Editor(s), ISBN 978-1-032-36723-1

Committee Members

ORGANIZING COMMITTEE

Chief Patron(s)

- **Prof. (Dr.) Kamal Ghanshala** *Chancellor, Graphic Era Hill University, Dehradun, India*
- **Dr. B.K. Das** *Director General (ECS), DRDO*

Patron(s)

- **Dr. Ajay Kumar** *Director, IRDE (DRDO), Dehradun*
- **Prof. (Dr.) R. Gowri** *Vice Chancellor, Graphic Era Hill University, Dehradun*
- **Prof. (Dr.) J. Kumar** *Pro Chancellor, Graphic Era Hill University, Dehradun*
- **Prof. (Dr.) Anita Rawat** *Director, Uttarakhand Science Education & Research Center (USERC), Dehradun*

General Chairs

- **Prof. (Dr.) S.N. Singh** *Director, IIITM, Gwalior*
- **Prof. (Dr.) Sanjay Jasola** *Vice Chancellor, Graphic Era (Deemed to be University), Dehradun*
- **Prof. (Dr.) D.K. Lobiyal** *Professor, School of Computer & System Sciences, JNU, New Delhi*
- **Prof. (Dr.) D.P. Kothari** *Vice Chancellor, VIT, Vellore*
- **Dr. Karan Singh** *School of Computer & System Sciences, JNU, New Delhi*
- **Dr. B.B. Sagar** *Computer Science & Engineering, BIT, Mesra*

Organising Chairs

- **Prof. Dibyahash Bordoloi** *HOD, Computer Science & Engineering, Graphic Era Hill University, Dehradun*
- **Prof. Navin Garg** *Computer Science & Engineering, Graphic Era Hill University, Dehradun*
- **Dr. Mahesh Manchanda** *Computer Science & Engineering, Graphic Era Hill University, Dehradun*

Conveners

- **Dr. Satvik Vats,** *Computer Science & Engineering, Graphic Era Hill University, Dehradun*
- **Dr. Vikrant Sharma** *Computer Science & Engineering, Graphic Era Hill University, Dehradun*
- **Prof. Amit Gupta** *Computer Science & Engineering, Graphic Era Hill University, Dehradun*

Organizing Secretaries

- **Mr. Arvind Dhar** *Registrar, Graphic Era Hill University, Dehradun*
- **Prof. (Dr.) Kamlesh Singh,** *Graphic Era Hill University, Dehradun*
- **Prof. (Dr.) A.K. Sahoo** *Computer Science & Engineering, Graphic Era Hill University, Dehradun*

- **Prof. (Dr.) Vrince Vimal,** *Computer Science & Engineering, Graphic Era Hill University, Dehradun*

SCIENTIFIC COMMITTEE

- **Dr. Devender Arora,** *Purdue University, USA*
- **Prof. (Dr.) Om Prakash Kaiwartya,** *Nottingham Trent University, England*
- **Prof. (Dr.) Nur Izura Udzir,** *Dean Academics, University Putra Malaysia, Malaysia*
- **Prof. (Dr.) Shiva kumar Mathapathi,** *Northeastern University and UC- San Diego, California*
- **Dr. Ahmed Aziz,** *Benha University, Benha, Egypt*
- **Prof. (Dr.) Shiva kumar Mathapathi,** *UC- San Diego, California, USA*
- **Dr. Rajeev K Shakya,** *ADAWA Science & Technology University, Ethiopia*
- **Dr. Ahmed Aziz,** *Tashkent State University of Economics Uzbekistan*
- **Prof. (Dr.) R.C. Joshi,** *Chancellor, Graphic Era (Deemed to be University), Dehradun*
- **Prof. (Dr.) H.N. Nagaraja,** *Director General, Graphic Era (Deemed to be University), Dehradun*
- **Prof. (Dr.) D.R. Gangodkar,** *Dean, International Affairs, Graphic Era (Deemed to be University), Dehradun*
- **Prof. (Dr.) S.N. Singh,** *Director, IIITM, Gwalior*
- **Prof. (Dr.) Sanjay Jasola,** *Vice Chancellor, Graphic Era (Deemed to be University), Dehradun*
- **Prof. (Dr.) D.K. Lobiyal,** *Professor, School of Computer & System Sciences, JNU, New Delhi*
- **Prof. (Dr.) R.B. Patel,** *Professor & Head, CCET, Chandigarh*
- **Dr. Karan Singh**, *School of Computer & System Sciences, JNU, New Delhi*
- **Dr. B.B. Sagar**, *Computer Science & Engineering, BIT, Mesra*
- **Prof. (Dr.) Sanjay Jasola,** *Vice Chancellor, Graphic Era (Deemed to be University), Dehradun*
- **Prof. (Dr.) Durgaprasad Gangodkar**, *Dean International Affairs, Graphic Era (Deemed to be University), Dehradun*
- **Prof. (Dr.) Chandra Mani,** *ISC,Bengaluru*
- **Prof. (Dr.) Anurag Jain,***Indra Prastha University, New Delhi*
- **Prof. (Dr.) Devendra Prasad,** *Dean Outreach, PIET, Panipat*
- **Dr. B.B. Sagar,** *Computer Science & Engineering, BIT, Mesra*
- **Dr. Williamjeet Singh,** *Computer Science & Engineering, Punjabi University, Patiala*
- **Prof. (Dr.) Nanhe Singh,** *Professor & Head, CSE, NSUT, New Delhi*
- **Prof. (Dr.) R.S. Rao,** *Registrar, NSUT, New Delhi*
- **Prof. (Dr.) Manoj Kumar Gupta,** *Shri Mata Vaishno Devi University, Katra*
- **Dr. Manisha Manjul,** *G.B. Pant College, New Delhi*
- **Prof. Manjeet Singh,** *G.B. Pant College, New Delhi*
- **Dr. Nisha Chandran,** *School of Computing,Graphic Era Hill University*
- **Dr. Jyoti Parsola,** School of Computing,*Graphic Era Hill University*
- **Dr. Indrajeet Kumar,** *Graphic Era Hill University, Dehradun*
- **Dr. Prateek Srivastava,** *Graphic Era Hill University, Dehradun*
- **Dr. Rakesh Patra,** *Graphic Era Hill University, Dehradun*
- **Dr. Ved Prakash Dubey,** *Graphic Era Hill University, Dehradun*
- **Prof. Saumitra Chattopadhyay,** *Graphic Era Hill University, Dehradun*
- **Prof. Amit Kumar Mishra,** *Graphic Era Hill University, Dehradum*
- **Dr. Chandrakala Arya,** *Graphic Era Hill University, Dehradun*
- **Prof. Himani Sivaraman,** *Graphic Era Hill University, Dehradun*
- **Prof. Divya Kapil,** *Graphic Era Hill University, Dehradun*
- **Prof. Poonam Verma,** *Graphic Era Hill University, Dehradun*

- **Prof. Anupriya,** *Graphic Era Hill University, Dehradun*
- **Dr. Pradeep Singh**, *Graphic Era Hill University, Dehradun*
- **Dr. Gunjan Chhabra,** *Graphic Era Hill University, Dehradun*
- **Dr. Devesh Tiwari,** *Graphic Era Hill University, Dehradun*
- **Prof. Umang Garg**, *Graphic Era Hill University, Dehradun*
- **Prof. Rahul Chauhan,** *Graphic Era Hill University, Dehradun*
- **Prof. Sushant Chamoli,** *Graphic Era Hill University, Dehradun*
- **Prof. Samir Rana**, *Graphic Era Hill University, Dehradun*
- **Prof. Rishika Yadav,** *Graphic Era Hill University, Dehradun*
- **Prof. Richa Gupta,** *Graphic Era Hill University, Dehradun*
- **Prof. Manika Manwal,** *Graphic Era Hill University, Dehradun*
- **Prof. Sonali Gupta,** *Graphic Era Hill University, Dehradun*
- **Prof. Manisha Aeri,** *Graphic Era Hill University, Dehradun*
- **Prof. Preeti Chaudhary,** *Graphic Era Hill University, Dehradun*
- **Prof. Purushotam Das,** *Graphic Era Hill University, Dehradun*
- **Prof. Lisa Gopal,** *Graphic Era Hill University, Dehradun*
- **Prof. Ayushi Jain,** *Graphic Era Hill University, Dehradun*
- **Prof. Aditya Verma,** *Graphic Era Hill University, Dehradun*
- **Prof. Akash Chauhan,** *Graphic Era Hill University, Dehradun*
- **Prof. Chandradeep Bhatt,** *Graphic Era Hill University, Dehradun*
- **Dr. Anil Desai,** *Graphic Era Hill University, Dehradun*
- **Prof. Annirudha Prabhu,** *Graphic Era Hill University, Dehradun*
- **Prof. Anmol Kaundaliya,** *Graphic Era Hill University, Dehradun*
- **Dr. Pankaj Kumar,** *Graphic Era Hill University, Dehradun*
- **Dr. Devesh Tiwari,** *Graphic Era Hill University, Dehradun*
- **Dr. Susheela,** *Graphic Era Hill University, Dehradun*
- **Prof. Sumeshwar Singh,***Graphic Era Hill University, Dehradun*
- **Prof. Resham Taruja,** *Graphic Era Hill University, Dehradun*

Automation and Computation – Vats et al. (Eds)
© 2023 the Author(s), ISBN 978-1-032-36723-1

Heart disease detection using feature optimization and classification

Purushottam Das* & Dinesh C. Dobhal*
Department of Computer Science & Engineering, Graphic Era Deemed to be University, Dehradun, India

Manika Manwal*
Department of Computer Science & Engineering, Graphic Era Hill University, Dehradun, India

ABSTRACT: Classification is mechanism of grouping a set on the basis of their characteristics. These characteristics may be explicitly defined in form of class labels namely in supervised methods and can be used implicitly in un-supervised methods. Image classification categorizes images in separate classes. Categorization of images has been significant area of study for researchers, scientists, engineers etc. In our proposed model an evolutionary approach is used for dimensionality reduction. A dataset of Heart failure clinical records is taken from UCI repository for attaining results. A reduced feature set is obtained after implementing hybrid model. Hybrid model comprises of an evolutionary approach and a classification strategy, i.e., neural network. Results achieved are observed and exploratory research is performed.

Keywords: Classification, evolutionary approaches, dimensionality reduction, reduced feature set, neural network

1 INTRODUCTION

Classification has been a prominent area of research for researchers and scientists. It has been a vital area for every field such as medical, engineering, real life problems, prediction etc. Classification is a mechanism of categorizing groups on the basis of some characteristics present in the data. The data can be in various forms such as images, videos, features etc. [1]. Classification is basically categorized in the two types: supervised and un-supervised. Supervised classification deals with categorization in presence of labels assigned to groups. These labels work as a supervisor for categorizing the data. So, training is performed using labeled data. While in un-supervised classification un-labeled data is used for grouping. In un-supervised classification, the training is done on the basis of clustering, i.e., grouping the data on the basis of characteristics of input data. Once the clusters are formed then classification is become easy. We can say that un-supervised classification in converted in the supervised one after training and formation of clusters. Image Classification deals with the classification of images. Images are grouped on the basis of their characteristics in respective groups [2].

Feature extraction is extraction of features from images. Features are nothing but attributes of an image that define or depict characteristics of the image in certain way. These features or attributes are attained from the image using various tools and techniques such as

---

*Corresponding Authors: pdas.nvs@gmail.com; dineshdobhal@gmail.com and manikamanwal17@gmail.com

DOI: 10.1201/9781003333500-1

1

Weka, Mazda etc. [3]. An image contains around four hundred features. But all features are not significant when we perform categorization. Only significant features, which contribute in grouping, should be taken into account. On the basis of significance in contributing for categorization features can be grouped into three types: strongly relevant, weakly relevant and irrelevant [4,5]. If a feature is significantly affecting the performance of a classifier then it is known as strongly relevant feature. If a feature is not highly affecting the performance of the classifier then they can be termed as weakly relevant features. If a feature is not affecting the efficiency then we can say that it is irrelevant feature. Selecting the significant features that are contributing in grouping is known as feature selection. Feature selection is the process of selecting features for classification [6–10]. This is achieved using various evolutionary approaches such as genetic algorithm, particle swarm optimization etc. [11]. An evolutionary approach evaluates the features and on the basis of fitness function the features are selected for classification. Optimization techniques are of three types: filter, embedded and wrapper. Filter approaches are good in speed. Wrapper approaches are slow but possess great efficiency in large computations. While embedded feature consist of mixed properties taken from both the approaches [12–14]. Classification strategies include machine learning and deep learning techniques namely neural network, CNN, SVM, multi-SVM etc. [15–20].

There are many applications of classification starting from medical, scientific, engineering, marketing, industry to digital spaces etc. Medical applications include use in detection of various diseases such as heart disease, breast cancer, hepatitis, brain tumor etc. Nowadays, machine learning approaches are prominent and are widely spreading in all the walks of life for betterment [21–24]. This problem is identified on the basis of complexity present in feature selection and classification methods. We are trying to propose a hybrid model for resolving the problem and providing satisfactory results. Hybrid model consist of evolutionary approach and classification techniques for yielding results [25,26].

2   LITERATURE REVIEWED

Various papers have been surveyed of classification, image classification, dimensionality reduction, evolutionary approaches, and feature space reduction using evolutionary approaches. Xu et al. [1] proposed an approach to save the spatial structures of clinical images. The information is then used to predict the categorization of infectious keratitis. Proposed methodology in this paper yields 80% efficiency. Wang, Y. and Wang, Z. [2] have discussed various approaches of image classification and prepared a detailed summary. Various networks such as ResNeXt, Cifar10 are observed. Long et al. [16] stated a multi-chaining approach for encouraging the segmentation performance. Disease recognition is achieved through graphs and natural language processing used together. On the basis of quadratic training of feature vectors a new optimization model is given. The segmentation of diseased area is investigated and automation process of knee joint illness is studied to help doctors. Region of Interest (ROI) are used in the earlier phase. A combined approach using Resnet and quadratic training is stated. Classification performance is enhanced by 11% in contention to Resnet model.

Manikandan, G. and Abirami, S. [27] gave a full coverage on the methods used for selecting a feature with latest advanced technologies. Filter, wrapper and embedded approaches are studied and covered in detail. Taxonomy of various techniques of dimensionality reduction and fuzzy logic are given in detail. Significance of selecting a feature is discussed and their use in different areas such as intrusion detection systems, audio, video and text analytics. Xue et al. [28] gave a detailed survey on latest and advanced evolutionary methods of feature selection. Contribution of various evolutionary computational approaches is studied and identified. They have also identified challenges, issues and significant areas of research for future. Lee et al. [29] has given a detailed survey on DNA motif prediction. Genetic algorithm is used for DNA motif recognition in this work. On the basis of

GA-based motif recognition the methods can be grouped into two categories: search by position and search for consensus. On the basis of their view, interpretations, fitness etc. comparisons are drawn for various algorithms. Pro and cons of different approaches are discussed and optimal algorithms are recommended for use. This survey work is beneficial for other researchers working in DNA motif estimation.

Sharma, V. and Juglan, K.C. [30] proposed a CAD approach for the recognition of liver ultrasound images. Seven approaches are used for extracting texture features. Mutual information is used for finding most differentiating features. The stated methodology provides 95.55% accuracy. Proposed method yield satisfactory results and displays that this approach can be used for the recognition of fatty and normal liver ultrasound images with higher accuracy. Toğaçar et al. [31] stated a method to improve deep learning techniques for White Blood Cells (WBC) recognition using image processing approaches. Two methods are used in conjunction: Maximal Information Coefficient and Ridge feature selection methods. These methods are used to attain most significant features. AlexNet, GoogLeNet, and ResNet-50 are used for extracting features. Quadratic discriminant analysis is used as a classifier. Results for the recognition of WBC are 97.95% using proposed approach. The practical results achieved displays that use of CNN models with other approaches such as feature selection improves the classification performance for WBC estimation.

Bharti, P. and Mittal, D. [25] proposed a hybrid feature selection method. This method chooses prime features out of various available features and categorizes in 4 groups: normal, chronic, cirrhosis, and heptocellular carcinomas. This hybrid method is attained using filter approach and wrapper approach. ReliefF and sequential forward selection (SFS) approaches are working in conjunction. ReliefF used for ranking the features. SFS selects the best ranked features to form a optimized feature set. This makes classification performance better and efficient. Acquired features are judged and checked along with given HFS approach. Efficiency of various methods such as given approach, ReliefF, SFS, sequential backward selection is compared. Hybrid model yields 95.2% accuracy with k-nearest neighbor.

Wang et al. [32] have explored that traditional machine learning methods poses better results for small datasets and deep learning techniques display efficient results when the dataset is large. SVM and CNN are compared and results are observed. SVM gives accuracy −0.88 and CNN provides efficiency −0.98 for a large mnist dataset. For a small COREL 1000 dataset, SVM provides accuracy −0.86 and CNN gives efficiency −0.83.

3   MATERIALS AND METHODOLOGY

The dataset of "Heart failure clinical records dataset" is attained from UCI repository [33]. In this dataset, we have 299 medical patients who have heart issue and fiasco. The dataset is gathered in the analysis phase. There are 13 features that are measured for each patient. There are 13 attributes in total. 12 attributes are for input and 1 is target attribute. The details are given here for the above mentioned dataset [34,35]:

Table 1:

Table 1.   Dataset description.

Types of Dataset:	Multivariate, Univariate	Occurrences:	299	Extent:	Life
Type of Attributes:	Integer, Real	Attribute's count:	13	Bestowed Date:	05-02-2020
Tasks:	Classification, Regression, Clustering	Values are missing?	Not Applicable		

3

Table 2.   Attribute information.

	Input Features		Output Features
Time	Smoking	Serum Sodium	Deceased or not?
Serum creatinine	Gender	Platelets	
Ejection fraction	Diabetes	Creatinine phosphokinase (CPK)	
High blood pressure	Anaemia	Age	

Attribute Information: Thirteen (13) clinical features are given below written in a sequence and in Table 2 [33]:

1. Time (Days),
2. Smoking (Yes / No),
3. Serum Sodium (Level in mEq/L),
4. Serum creatinine (mEq/L),
5. Gender (Male/Female/Bi),
6. Platelets (Kp/mL),
7. Ejection fraction (%),
8. Diabetes (Yes / No),
9. Creatinine phosphokinase (CPK) (Level in mcg/L),
10. High blood pressure (Yes / No),
11. Anaemia (Yes / No),
12. Age (in Years),
13. Deceased (Yes / No).

A hybrid approach is used to obtain results here. This hybrid approach is based on feature subset selection and classification. Feature subset selection is performed using genetic algorithm and classification is done by neural network.
Steps:

(i)   UCI repository is used for getting the dataset. Heart failure clinical records dataset is used here.
(ii) A Hybrid model is used containing evolutionary approaches and neural network:
   a) Evolutionary approach for feature reduction is used to attaining a reduced feature set. Fitness function is used for evaluating the quality of features to select the features which distinguishes most for categorizing the groups.
   b) Neural network is used for categorization the groups (nprtool).

(iii) Results are achieved on the dataset of Heart failure clinical records and are convincing for grouping the classes.

Feature reduction and classification using hybrid model is given below in Figure 1:

4   RESULTS AND DISCUSSION

Above mentioned methodology is used to find out the results for Heart failure clinical records dataset having statistics of 299 peoples who had heart issue and alarming condition using 13 features. There are twelve input features and one output features.

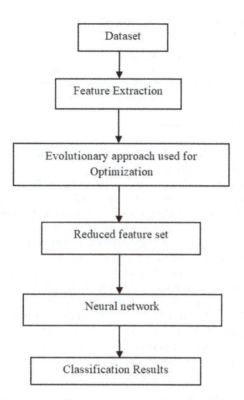

Figure 1.   Feature reduction and classification using hybrid model.

We have taken three scenarios:

(i)   Classification results obtained without feature optimization
(ii)   Classification results obtained, when a subset of four features is selected: (10, 12, 6, 2),
(iii)   Classification results obtained, when a subset of three features is selected: (12, 2, 10).

In the first case we have attained 79.4%, 64.4%, 66.7 and 75.3 for training, test, validation and final classification results respectively (Table 3). In the second case we have obtained 80.0%, 80.0% and 83.3% for test, validation and final classifications respectively (Table 4). In the third case, we have attained 82.2%, 86.7% and 82.9% for test, validation and final classifications respectively (Table 5).

Classification results are obtained on the used data set and three cases are considered. In the first case we are taking all the features into consideration, in the second case a feature subset of four features is selected and in the final case we have attained classification results for a feature set of three features. The classification results obtained are given in tables Tables 3–5 respectively.

Table 3.   Classification results on original feature set using hybrid approach.

Data	Total Chosen attributes	Selected Features	Results (%)			
			Training	Test	Validate	Final
Heart failure clinical records	12	ALL	79.4	64.4	66.7	75.3

Table 4.   Selected feature subset using hybrid approach.

Data	Total Chosen attributes	Selected Features	Results (%)			
			Training	Test	Validate	Final
Heart failure clinical records	4	(10, 12, 6, 2)	84.7	80.0	80.0	83.3

Results attained on Heart failure clinical records dataset by the method are in Table 4:

In Table 4, classification results were attained for a feature set of four features. Four features that are selected for attaining results are (10, 12, 6, 2). In the next section, classification results are attained in form of confusion matrix, roc plot, validation performance, gradient, error histogram and fitness function. Confusion matrix is measure of classification problem for two or more classes. 4 combinations of actual and predicted value are given in confusion matrix as true positive, true negative, false positive and false negative. Confusion matrix is helpful in calculating: recall, precision, accuracy and F-measure. True positive refers to the values which are predicted positive and that's true. True negative refers to the values which are predicted negative and that's true. False positive refers to the values which are predicted positive but they are not. False negative values are predicted negative but they are positive actually. ROC (Receiver operating characteristic) curve show the classification results at all thresholds of classifications. ROC plots two curves: true positive rate and false positive rate. ROC plot, Confusion matrix, gradient, validation performance, fitness function, and error histogram are given in Figures 2a–2f:

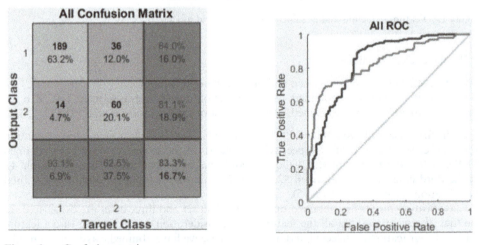

Figure 2a.   Confusion matrix.   Figure 2b.   ROC plot.

ROC plot, Confusion matrix, gradient, validation performance, fitness function, and error histogram are given in Figures 3a–3f:

The aim of performance evaluation is to learn how the system evaluates the new data. Training of the model creates the model that should precisely predict for unknown values. Validation check refers to the concept that neural network is not memorizing the values for training data. It has been trained and can be used later for new data. Error histogram is the histogram of errors between actual value and predicted value. It is done using feed forward network training. These error values denote the difference of actual and predicted values.

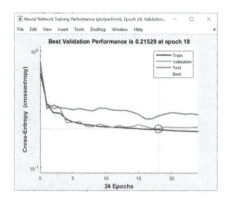

Figure 2c.   Validation performance.

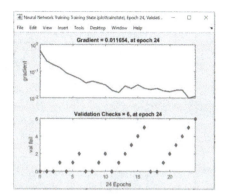

Figure 2d.   Gradient.

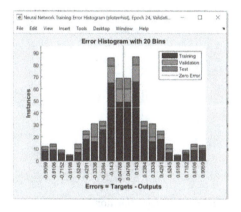

Figure 2e.   Error Histogram.

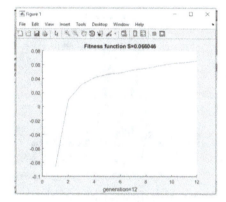

Figure 2f.   Fitness function.

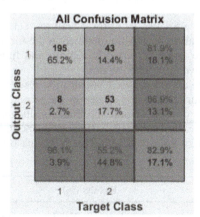

Figure 3a.   Confusion matrix.

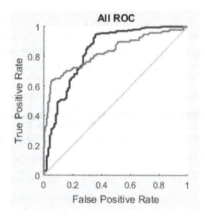

Figure 3b.   ROC plot.

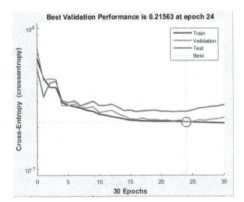

Figure 3c. Validation performance.

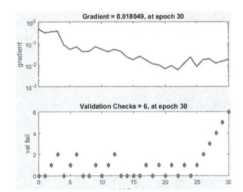

Figure 3d. Gradient.

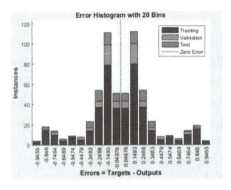

Figure 3e. Error Histogram.

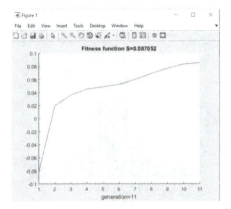

Figure 3f. Fitness function.

Fitness function takes a value as input and yields output. The performance of the function is evaluated by its fitness. It must measure the solution for its fitness and how fit solutions can be obtained.

In Table 5, classification results were attained for a feature set of three features. Selected features for attaining results are (12, 2, 10). Classification results are obtained in the form of confusion matrix, roc plot, validation performance, gradient, error histogram and fitness function. The detailed description of all these are given in earlier sections.

In the first scenario, all the features are taken into consideration. Classification results are attained without feature optimization (Table 3). Four features are chosen in the second scenario: (10, 12, 6, 2) and 80.0%, 80.0% and 83.3% for test, validation and final classifications respectively (Table 4). Three features are selected in the third case: (12, 2, 10) and 82.2%, 86.7% and 82.9% for test, validation and final classifications respectively (Table 5). In the below

Table 5. Selected feature subset using hybrid approach.

Data	Total Chosen attributes	Selected Features	Results (%)			
			Training	Test	Validate	Final
Heart failure clinical records	3	(12, 2, 10)	82.3	82.2	86.7	82.9

8

table there are four columns: feature-count, training, test, validate and final. In the feature-count column, total number of features for each case is given. Training column gives the classification results for learning phase. Similarly, test and validate cases gives results for respective cases. In the last column final classification results are given. Classification results attained on Heart failure clinical records dataset by the method are – Table 6:

Table 6.   Comparison of classification results.

Feature's Count	Training	Test	Validate	Final
12	79.4	64.4	66.7	**75.3**
3	82.3	82.2	86.7	**82.9**
4	**84.7**	**80.0**	**80.0**	83.3

5   CONCLUSION AND FUTURE SCOPE

UCI repository is used to get the dataset of Heart failure clinical records. A hybrid methodology is proposed to be used on the dataset. An evolutionary approach and a classification strategy are combined to achieve this hybrid model. Two instances are considered here. We have taken a feature set of best features of feature count – three (12, 2, 10) and another instance of feature count – four (10, 12, 6, 2). In instance of feature count – three, we are achieving 82.2%, 86.7% and 82.9% classification results for test, validation and final respectively. In another case, 80.0%, 80.0% and 83.3% are attained for test, validation and final respectively. Satisfactory results are achieved but these results can be significantly improved by checking with other feature subset selection and classification techniques.

REFERENCES

[1]   Xu, Y., Kong, M., Xie, W., Duan, R., Fang, Z., Lin, Y., Zhu, Q., Tang, S., Wu, F. and Yao, Y.F. Deep Sequential Feature Learning in Clinical Image Classification of Infectious Keratitis. *Engineering*, 7(7), 1002–1010, 2021.

[2]   Wang, Y. and Wang, Z. A Survey of Recent Work on Fine-grained Image Classification Techniques. *Journal of Visual Communication and Image Representation*, 59, 210–214, 2019.

[3]   Cen, F., Zhao, X., Li, W. and Wang, G. Deep Feature Augmentation for Occluded Image Classification. *Pattern Recognition*, 111, 107737, 2021.

[4]   Liu, H. and Zhao, Z. *Manipulating Data and Dimension Reduction Methods: Feature Selection*, 2009.

[5]   Nguyen, H.B., Xue, B., Liu, I. and Zhang, M., July. Filter Based Backward Elimination in Wrapper Based PSO for Feature Selection in Classification. *In 2014 IEEE Congress on Evolutionary Computation (CEC)*. IEEE, 3111–3118, 2014.

[6]   Liu, Y., Xie, H., Chen, Y., Tan, K., Wang, L. and Xie, W. Neighborhood Mutual Information and its Application on Hyperspectral Band Selection for Classification. *Chemometrics and Intelligent Laboratory Systems*, 157, 140–151, 2016.

[7]   Xia, H., Zhuang, J. and Yu, D. Multi-objective Unsupervised Feature Selection Algorithm Utilizing Redundancy Measure and Negative Epsilon-dominance for Fault Diagnosis. *Neurocomputing*, 146, 113–124, 2014.

[8]   Vats, S., Singh, S., Kala, G., Tarar, R., and Dhawan, S., "iDoc-X: An Artificial Intelligence Model for Tuberculosis Diagnosis and Localization," *J. Discret. Math. Sci. Cryptogr.*, 24, 5, 1257–1272, 2021.

[9]   Vats, S. and Sagar, B.B., "Performance Evaluation of K-means Clustering on Hadoop Infrastructure," *J. Discret. Math. Sci. Cryptogr.*, 22, 8, 2019, doi: 10.1080/09720529.2019.1692444.

[10]   Vats, S., Sagar, B.B., Singh, K., Ahmadian, A., and Pansera, B.A., "Performance Evaluation of an Independent Time Optimized Infrastructure for Big Data Analytics That Maintains Symmetry," *Symmetry (Basel)*., 12, 8, 2020, doi: 10.3390/SYM12081274.

[11]   Long, Z., Zhang, X., Li, C., Niu, J., Wu, X. and Li, Z., 2020. Segmentation and Classification of Knee Joint Ultrasonic Image via Deep Learning. *Applied Soft Computing*, 97, 106765.

[12] Bhatia, M., Sharma, V., Singh, P., and Masud, M., "Multi-level P2P Traffic Classification Using Heuristic and Statistical-based Techniques: A Hybrid Approach," *Symmetry (Basel).*, 12, 12, 2117, 2020.

[13] Vats, S. and Sagar, B.B., "An Independent Time Optimized Hybrid Infrastructure for Big Data Analytics," *Mod. Phys. Lett. B*, 34, 28, 2050311, Oct. 2020, doi: 10.1142/S021798492050311X.

[14] Agarwal, R., Singh, S., and Vats, S., "Implementation of an Improved Algorithm for Frequent Itemset Mining Using Hadoop," *in 2016 International Conference on Computing, Communication and Automation (ICCCA)*, 2016, 13–18. doi: 10.1109/CCAA.2016.7813719.

[15] Harkat, H., Ruano, A., Ruano, M.G. and Bennani, S.D. Classifier Design by a Multi-objective Genetic Algorithm Approach for GPR Automatic Target Detection. *IFAC-PapersOnLine*, 51(10), 187–192, 2018.

[16] Parvathy, V.S., Pothiraj, S. and Sampson, J. Optimal Deep Neural Network Model Based Multimodality Fused Medical Image Classification. *Physical Communication*, 41, 101119, 2020.

[17] Agarwal, R., Singh, S., and Vats, S., *Review of Parallel Apriori Algorithm on Mapreduce Framework For Performance Enhancement*, vol. 654. 2018. doi:10.1007/978-981-10-6620-7_38.

[18] Bhati, J.P., Tomar, D., and Vats, S., "Examining Big Data Management Techniques For Cloud-based IoT Systems," in Examining Cloud Computing Technologies Through the Internet of Things, *IGI Global*, 2018, 164–191.

[19] Sharma, V. *et al.*, "OGAS: Omni-directional Glider Assisted Scheme for Autonomous Deployment of Sensor Nodes in Open Area Wireless Sensor Network," *ISA Trans.*, Aug. 2022, doi: 10.1016/j.isatra.2022.08.001.

[20] Sharma, V., Patel, R.B., Bhadauria, H.S., and Prasad, D., "NADS: Neighbor Assisted Deployment Scheme for Optimal Placement of Sensor Nodes to Achieve Blanket Coverage in Wireless Sensor Network," *Wirel. Pers. Commun.*, vol. 90, no. 4, pp. 1903–1933, 2016.

[21] Sharma, V., Patel, R.B., Bhadauria, H.S., and Prasad, D., "Policy for Planned Placement of Sensor Nodes in Large Scale Wireless Sensor Network," *KSII Trans. Internet Inf. Syst.*, vol. 10, no. 7, pp. 3213–3230, 2016.

[22] Sharma, V., Patel, R.B., Bhadauria, H.S., and Prasad, D., "Deployment Schemes in Wireless Sensor Network to Achieve Blanket Coverage in Large-scale Open Area: A review," *Egypt. Informatics J.*, vol. 17, no. 1, pp. 45–56, 2016.

[23] Vikrant, S., Patel, R.B., Bhadauria, H.S., and Prasad, D., "Glider Assisted Schemes to Deploy Sensor Nodes in Wireless Sensor Networks," *Rob. Auton. Syst.*, vol. 100, pp. 1–13, 2018.

[24] Vats, S. and Sagar, B.B., "Data Lake: A Plausible Big Data Science for Business Intelligence," 2019.

[25] Bharti, P. and Mittal, D., Hybrid Feature Selection-based Feature Fusion for Liver Disease Classification On Ultrasound Images. In *Advances in Computational Techniques for Biomedical Image Analysis*. Academic Press, 145–164, 2020.

[26] Jansi Rani, M. and Devaraj, D. Two-stage Hybrid Gene Selection Using Mutual Information and Genetic Algorithm for Cancer Data Classification. *Journal of Medical Systems*, 43(8), 1–11, 2019.

[27] Manikandan, G. and Abirami, S. Feature Selection is Important: State-of-the-art Methods and Application Domains of Feature Selection on High-dimensional data. In *Applications in Ubiquitous Computing*. Springer, Cham, 177–196, 2021.

[28] Xue, B., Zhang, M., Browne, W.N. and Yao, X. A Survey on Evolutionary Computation Approaches to Feature Selection. *IEEE Transactions on Evolutionary Computation*, 20(4), 606–626, 2015.

[29] Lee, N.K., Li, X. and Wang, D. A Comprehensive Survey on Genetic Algorithms for DNA Motif Prediction. *Information Sciences*, 466, 25–43, 2018.

[30] Sharma, V. and Juglan, K.C. Automated Classification of Fatty and Normal Liver Ultrasound Images Based on Mutual Information Feature Selection. *Irbm*, 39(5), 313–323, 2018.

[31] Toğaçar, M., Ergen, B. and Cömert, Z. Classification of White Blood Cells Using Deep Features Obtained From Convolutional Neural Network Models Based on the Combination of Feature Selection Methods. *Applied Soft Computing*, 97, 106810, 2020.

[32] Wang, P., Fan, E. and Wang, P. Comparative Analysis of Image Classification Algorithms based on Traditional Machine Learning and Deep Learning. *Pattern Recognition Letters*, 141, 61–67, 2021.

[33] UCI Repository: https://archive.ics.uci.edu/ml/datasets/Heart+failure+clinical+records

[34] Chicco, D. and Jurman, G. Machine Learning Can Predict Survival of Patients with Heart Failure From Serum Creatinine And Ejection Fraction Alone. *BMC Medical Informatics and Decision Making*, 20(1), 1–16, 2020.

[35] Ahmad, T., Munir, A., Bhatti, S.H., Aftab, M. and Raza, M.A. Survival Analysis of Heart Failure Patients: A case study. *PloS one*, 12(7), e0181001, 2017.

Automation and Computation – Vats et al. (Eds)
© 2023 the Author(s), ISBN 978-1-032-36723-1

Mathematical modelling of tuberculosis and COVID-19 with saturated incidence rate

Vijai Shanker Verma, Harshita Kaushik & Archana Singh Bhadauria
Department of Mathematics and Statistics, Deen Dayal Upadhyaya Gorakhpur University, Gorakhpur, U.P., India

ABSTRACT: Tuberculosis (TB) and COVID-19 are in the list of diseases with high concern for globally public health and with a negative impact on socio economic status. In our research paper, we have proposed an epidemiological compartment model to study the dynamics of two concomitant diseases. A compartment model has been developed as an expanded version of the traditional SIS model with a saturated incidence rate. The equilibrium points are also obtained after characterizing the non-negativity and invariant region of the model. After deriving the fundamental reproduction number R_0, the model's stability is examined. It has been found that the endemic equilibrium is only stable when R_0 is more than one and the disease-free equilibrium is stable whenever R_0 is less than one. When $R_0 > 1$, the Routh-Hurwitz criterion is employed to demonstrate the endemic equilibrium's local stability. Numerical simulation illustrates the theoretical findings and to study the transmission dynamics of both the concomitant diseases during the first and second waves respectively in context of India.

1 INTRODUCTION

To study the dynamics of the spread and control of infectious illnesses like tuberculosis, COVID-19, and many others like malaria, influenza, chickenpox, and the common cold, various epidemiological models have been presented. Investigators working in the field of epidemiology have incorporated various parameters to study several infectious diseases. Saturated incidence rate is one of the important parameters included in the list of parameters affecting the various infectious diseases. Mengistu and Witboii (Mengistu *et al.* 2020) have given and analysed a model on TB with saturated incidence rate $\beta SI/(1 + bI)$, where βI is used to measure the infection rate when the disease is occupying a place in a complete vulnerable class (population), and $1/(1 + bI)$ is the reticence effect from the behavioural change of susceptible persons. Additionally, the saturated incidence rate is far more plausible than the conventional bilinear incidence rate (βSI) that Alexander and Moghadas had previously proposed. The explanation is that it takes into account the behavioural changes that take place, the cumulative effect of the infected individuals, and avoids the unboundedness of the contact rate by choosing appropriate parameters (Zhang *et al.* 2014). We have included the saturated incidence rate to observe the analytical results because the incidence rate plays a significant part in the scientific investigation of the many proposed epidemic models (Holmdahl & Buckee 2020).

DOI: 10.1201/9781003333500-2

Modelling of infectious diseases is done to observe the effects of various transmission parameters, which are considered to predict the future trend of the disease and further helps in evaluating the strategies that will help in controlling the pandemic/endemic and the results will provide logical and factual information that will be useful for policy-makers in decision-making (Zeb *et al.* 2020). Tuberculosis (TB) and corona virus disease (SARS-CoV-2) are also known as COVID-19 both the diseases are severe and come under the class of infectious diseases that mainly attack the human respiratory system and especially damage the lungs. Patients infected with both TB and COVID-19 possesses almost common symptoms and signs such as fever, coughing and respiration with difficulty. The transmission mode of TB infection is through tiny droplets when TB active patients either cough, exhale, spit or shout. However, the transmission mode of the deadly COVID-19 infection is mainly via aerosol particles (tiny) of exhaled respiratory molecules when COVID-19 active patients either cough, sneeze, shout or sing.

The relationship between TB and COVID-19 is a high concern for public health, policy-makers, government authorities and the co-infection of these two highly lethal diseases exists when a person acquired both the disease simultaneously. It has been observed in the clinical evidences that the patients suffering with active TB are highly prone to acquire the diseases, and the progression signs for the COVID-19 infection is quite severe as a result the control strategies and the management of TB affects in various number of ways.

Data generation is an important task in the beginning of big data. The methods used by researchers to produce data are observational study and randomized experiment. (Vats *et al.* 2019) Data generation is nowadays most demanding study which helps in mathematical modelling with more accurate results. (Vats *et al.* 2020)

In view of the earlier epidemiological models, we have proposed a deterministic model for the concomitant diseases Tuberculosis (TB) and corona virus (COVID-19) with saturated incidence rate.

2 MATHEMATICAL MODEL

Here, we have proposed a deterministic compartmental model for concomitant diseases TB and COVID-19 by assuming a homogeneous mixing of individuals within the population under consideration. To formulate the model, we have divided the total population concerning their disease status into four mutually exclusive epidemiological states. Here, we denote the population susceptible to infectious disease TB and COVID-19 by $x_s(t)$, population infected with TB only by $y_{TB}(t)$, COVID-19 infected population by $z_c(t)$ and the population infected with both TB plus COVID-19 diseases by $w_{TC}(t)$. Let the total population at any time $'t'$ be $N_p(t)$ in the region under deliberation, which is sum of all the four sub-populations.

Thus, we have:

$$N_p(t) = x_s(t) + y_{TB}(t) + z_c(t) + w_{TC}(t)$$

It is assumed that both the diseases TB and COVID-19 spread via direct contact between susceptible and infected individuals. The individuals recovered from the infection are also assumed to re-enter the susceptible class. Thus, our model is based on the classical SIS model.

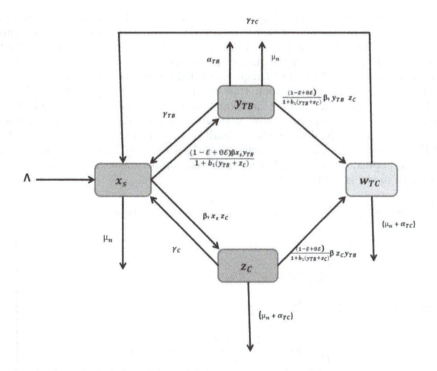

Figure 1.   A schematic depiction of deterministic compartmental model.

Table 1.   The list of description of variables in our suggested model.

Parameter	Description	Source
$\Lambda = 65937.74\ day^{-1}$	Recruitment rate for susceptible	https://www.macrotrends.net "Macrotrends"
$\alpha_{TB}= 0.004\ day^{-1}$	Death rate among those infected with TB	Nikshay Reports "https://nikshay.in"
$\alpha_c= 0.274\ day^{-1}$	Death rate among those infected with COVID-19	WHO(COVID-19) dashboard "https://covid19.who.int
$\alpha_{TC}= 0.272\ day^{-1}$	Death rate for people with TB and COVID-19 combined	Assumed
$\beta = 1.345399\times10^{-11}$	Transmission coefficient of TB infection from TB population	Assumed
$\beta_1 = 2.675 \times10^{-10}$	Transmission coefficient of COVID-19 infection from COVID-19 population	Assumed
$\gamma_{TB} = 0.0166\ day^{-1}$	Rate of recovery of TB population	https://www.medicinenet.com
$\gamma_C = 0.0714\ day^{-1}$	Rate of recovery of COVID-19 population	https://www.mohfw.gov.in
$\gamma_{Tc} = 0.0222\ day^{-1}$	Rate of population recovery from TB and COVID-19 infections	Assumed
$\mu_n = 0.00002\ day^{-1}$	Natural mortality rate of the population	https://knoema.com"Indian Death rate, 1950-2020-knoema.com"
$\epsilon = 0.715$	Vaccination coverage rate	(Zhang et al., 2014)
$\theta = 0.5$	Loss of protection for vaccination	(Zhang et al., 2014)
$b_1 = 0.0004$	Saturation rate	(Zhang et al., 2014)

Now, suppose that a class of susceptible population enters the system at a steady rate Λ. Susceptible moves to the infected class $y_{TB}(t)$ of TB at the transmission rate β and infected class $z_C(t)$ of COVID-19, at the transmission rate β_1. When TB infected individuals acquire COVID-19, they enter the class of individuals who have both TB and COVID-19 infections, also when COVID-19 acquire infected population TB, they enter to $w_{TC}(t)$. In addition, TB infected population recovers at the rate of γ_{TB}, COVID-19 infected population recovers at the rate of γ_C and population infected with both TB and COVID-19 after recovery re-enters the susceptible population at the recovery rate of γ_{TC}. Further, α_{TB}, α_C, and α_{TC} are the disease related death rate of TB, COVID-19 and both TB plus COVID-19 infected population respectively and μ_n is considered as the natural mortality rate of the population in each of our considered compartment. The mathematical formulation of model is as given below:

$$
\left.
\begin{aligned}
\frac{dx_s}{dt} &= \Lambda + \gamma_{TB}y_{TB} + \gamma_C z_C + \gamma_{TC}w_{TC} - \frac{1-\epsilon+\theta\epsilon}{1+b_1(y_{TB}+z_C)}\beta x_S y_{TB} \\
&\quad -\beta_1 x_S z_C - \mu_n x_S, \\
\frac{dy_{TB}}{dt} &= \frac{1-\epsilon+\theta\epsilon}{1+b_1(y_{TB}+z_C)}\beta_1 y_{TB} z_C - (\mu_n + \alpha_{TB} + \gamma_{TB})y_{TB}, \\
\frac{dz_C}{dt} &= \beta_1 x_S z_C - \frac{1-\epsilon+\theta\epsilon}{b_1(y_{TB}+z_C)}\beta z_C y_{TB} - (\mu_n + \alpha_C + \gamma_C)z_C, \\
\frac{dw_{TC}}{dt} &= \frac{1-\epsilon+\theta\epsilon}{b_1(y_{TB}+z_C)}\beta z_C y_{TB} + \frac{1-\epsilon+\theta\epsilon}{b_1(y_{TB}+z_C)}\beta_1 y_{TB} z_C - (\mu_n + \alpha_{TC} + \gamma_{TC})w_{TC},
\end{aligned}
\right\} \quad (1)
$$

with following initial conditions:

$$
x_s(0) = x_{s_0} > 0, \ y(0) = y_{TB_0} \geq 0, \ z_c(0) = z_{C_0} \geq 0, \ w_{TC}(0) = w_{TC_0} \geq 0. \quad (2)
$$

3 POSITIVITY AND BOUNDEDNESS OF THE MODEL

A physiologically relevant and practicable region will be taken into consideration while studying and conducting the TB plus COVID-19 model, which is described by system of equations (1), since the aforementioned model analyzes the human population, it is assumed that all of the variables and parameters are non-negative for all $t \geq 0$. We study the model in positively invariant set given below:

$$
\Omega = \left\{ x_S(t), \ y_{TB}(t), \ z_C(t), \ w_{TC}(t) \in R^4_+ \,\middle|\, 0 \leq x_S(t) + y_{TB}(t) + z_C(t) + w_{TC}(t) \leq (\Lambda)/\mu_n \right\}
$$

Now, we prove the following theorem:

Theorem 1: The system of equations (1) with initial conditions in R^4_+ has the set $\Omega \subset R^4_+$ and is positively invariant, or non-negative.

Proof: The total population for the given system of equation (1) is given by

$$
N_p(t) = x_S(t) + y_{TB}(t) + z_C(t) + w_{TC}(t)
$$
$$
\therefore \frac{dN_p(t)}{dt} = \frac{dx_S(t)}{dt} + \frac{dy_{TB}(t)}{dt} + \frac{dz_C(t)}{dt} + \frac{dw_{TC}(t)}{dt} \quad (3)
$$

Now, we substitute the values from system (1), we get

$$
\frac{dN_p(t)}{dt} \leq \Lambda - \mu_n N_p(t) - \alpha_{TB}y_{TB} - \alpha_C z_C(t) - \alpha_{TC}w_{TC} \leq \Lambda - \mu_n N_p(t)
$$

Integrating both sides of it and taking the result as $t \to \infty$, we get

$$0 \le N_p(t) \le \frac{\Lambda}{\mu_n} \qquad (4)$$

Therefore, the feasible solution for the system (1) is provided by

$$\Omega = \{x_S(t), y_{TB}(t), z_C(t), w_{TC}(t) \in R_+{}^4 | 0$$

$$\le x_S(t) + y_{TB}(t) + z_C(t) + w_{TC}(t) \le \frac{\Lambda}{\mu_n}\}.$$

Thus, our model is well-defined biologically and mathematically in the set Ω.

4 EXISTENCE OF EQUILIBRIUM POINTS

There are two equilibrium points, first disease-free equilibrium point $\left(\frac{\Lambda}{\mu_n}, 0, 0, 0\right)$ and other endemic equilibrium point $(x_S{}^*, y_{TB}{}^*, z_C{}^*, w_{TC}{}^*)$ for the system of equations (1).
The endemic equilibrium point must satisfy the following equations:

$$\Lambda + \gamma_{TB} y_{TB}{}^* + \gamma_C z_C{}^* + \gamma_{TC} w_{TC}{}^* - \frac{1 - \epsilon + \theta\epsilon}{1 + b_1(y_{TB} + z_c)} \beta x_S{}^* y_{TB}{}^* - \beta_1 x_S{}^* z_C{}^* - \mu_n x_S{}^* = 0,$$

$$(5)$$

$$\frac{1 - \epsilon + \theta\epsilon}{1 + b_1(y_{TB} + z_c)} \beta x_S{}^* - \frac{1 - \epsilon + \theta\epsilon}{1 + b_1(y_{TB} + z_c)} \beta_1 z_C{}^* - (\gamma_{TB} + \mu_n + \alpha_{TB}) = 0, \qquad (6)$$

$$\beta_1 x_S{}^* - \frac{1 - \epsilon + \theta\epsilon}{1 + b_1(y_{TB} + z_c)} \beta y_{TB}{}^* - (\mu_n + \alpha_C + \gamma_C) = 0, \qquad (7)$$

$$\frac{1 - \epsilon + \theta\epsilon}{1 + b_1(y_{TB} + z_c)} \beta z_C{}^* y_{TB}{}^* + \frac{1 - \epsilon + \theta\epsilon}{1 + b_1(y_{TB} + z_c)} \beta_1 y_{TB}{}^* z_C{}^* - (\mu_n + \alpha_{TC} + \gamma_{TC}) = 0, \qquad (8)$$

Now, from equation (6), we have

$$z_C{}^* = \frac{\beta x_S{}^* - (\gamma_{TB} + \mu_n + \alpha_{TB})}{\beta_2 \left\{\frac{1 - \epsilon + \theta\epsilon}{1 + b_1(y_{TB} + z_c)}\right\}} = f_1(x_S{}^*) \qquad (9)$$

Again, from equation (7), we have

$$y_{TB}{}^* = \frac{\beta_1 x_S{}^* - (\mu_n + \alpha_C + \gamma_C)}{\beta \left\{\frac{1 - \epsilon + \theta\epsilon}{1 + b_1(y_{TB} + z_c)}\right\}} = f_2(x_S{}^*) \qquad (10)$$

and from equation (8), we have

$$w_{TC}{}^* = \frac{(\beta + \beta_1) f_1(x_S{}^*) f_2(x_S{}^*)}{(\mu_n + \alpha_{TC} + \gamma_{TC}) \frac{1 - \epsilon + \theta\epsilon}{1 + b_1(y_{TB} + z_c)}}. \qquad (11)$$

15

Using (9), (10) and (11) in equation (5), we have

$$A = (\gamma_{TB} + \mu_n + \alpha_{TB}), B = (\mu_n + \alpha_C + \gamma_C)$$

$$g(x) = \Lambda + \gamma_{TB}\left\{\left[\frac{\beta_1 x_S^* - (B)](1+b_1(y_{TB}+z_c))}{\beta(1-\varepsilon+\theta\varepsilon)}\right]\right\} -$$

$$+ \gamma_C\left\{\left[\frac{\beta x_S^* - (A)](1+b_1(y_{TB}+z_c))}{\beta_1(1-\varepsilon+\theta\varepsilon)}\right]\right\} \tag{12}$$

$$+ \gamma_{TC}\frac{[(\beta+\beta_1)\left\{\left\{\beta x_S^* - \frac{\frac{[\beta x_S^* - A]}{\beta_1(1-\varepsilon+\theta\varepsilon)}[\beta_1 x_S^* - B]}{\beta(1-\varepsilon+\theta\varepsilon)}\right\}(1 + b_1(y_{TB}+z_c))}{(\mu_n+\alpha_{TC}+\gamma_{TC})(1-\varepsilon+\theta\varepsilon)}$$

$$\beta_1 x_S^{*2} + (B)x_S^* - \frac{[\beta x_S^{*2} - Ax_S^*](1+b_1(y_{TB}+z_c))}{(1-\varepsilon+\theta\varepsilon)}$$

$$\therefore g(0) = \Lambda + \gamma_{TB}\left\{\left[\frac{-(\mu_n + \alpha_C + \gamma_C)]1+b_1(y_{TB}+z_c)}{\beta(1-\varepsilon+\theta\varepsilon)}\right]\right\}$$

$$+ \gamma_C\left\{\left[\frac{-(\gamma_{TB} + \mu_n + \alpha_{TB})]1+b_1(y_{TB}+z_c)}{\beta_1(1-\varepsilon+\theta\varepsilon)}\right]\right\}$$

$$+ \gamma_{TC}\frac{[(\beta+\beta_1)\left[\left\{\left[\frac{-(\gamma_{TB}+\mu_n+\alpha_{TB})]1+b_1(y_{TB}+z_c)}{\beta_1(1-\varepsilon+\theta\varepsilon)}\right]\right\}\left\{\left[\frac{-(\mu_n+\alpha_C+\gamma_C)]1+b_1(y_{TB}+z_c)}{\beta(1-\varepsilon+\theta\varepsilon)}\right]\right\}\right]1 + b_1(y_{TB}+z_c)}{(\mu_n+\alpha_{TC}+\gamma_{TC})(1-\varepsilon+\theta\varepsilon)} \tag{13}$$

$$g\left(\frac{\Lambda}{\mu}\right) = \Lambda + \gamma_{TB}\left\{\left[\frac{\beta_1\frac{\Lambda}{\mu} - (\mu_n + \alpha_C + \gamma_C)]1+b_1(y_{TB}+z_c)}{\beta(1-\varepsilon+\theta\varepsilon)}\right]\right\}$$

$$+ \gamma_C\left\{\left[\frac{\beta\frac{\Lambda}{\mu} - (\gamma_{TB} + \mu_n + \alpha_{TB})]1+b_1(y_{TB}+z_c)}{\beta_1(1-\varepsilon+\theta\varepsilon)}\right]\right\}$$

$$+ \gamma_{TC}[(\beta+\beta_1)\frac{\left\{\left[\frac{\beta\frac{\Lambda}{\mu} - (\gamma_{TB}+\mu_n+\alpha_{TB})]}{\beta_1(1-\varepsilon+\theta\varepsilon)}\right]\right\}\left\{\left[\frac{\beta_1\frac{\Lambda}{\mu} - (\mu_n+\alpha_C+\gamma_C)]}{\beta(1-\varepsilon+\theta\varepsilon)}\right]\right\}1 + b_1(y_{TB}+z_c)}{(\mu_n+\alpha_{TC}+\gamma_{TC})(1-\varepsilon+\theta\varepsilon)}$$

$$- \beta_1\left(\frac{\Lambda}{\mu}\right)^2 + (\mu_n + \alpha_C + \gamma_C)\left(\frac{\Lambda}{\mu}\right) - \frac{\left[\beta\frac{\Lambda}{\mu} - (\gamma_{TB} + \mu_n + \alpha_{TB})\frac{\Lambda}{\mu}\right]1 + b_1(y_{TB}+z_c)}{(1-\varepsilon+\theta\varepsilon)} - \Lambda \tag{14}$$

Using equation (12), we have

$$g'(x_S) = \frac{\gamma_{TB}\beta_1}{\beta(1-\varepsilon+\theta\varepsilon)}1 + b_1(y_{TB}+z_c) + \frac{\gamma_C\beta}{\beta_1(1-\varepsilon+\theta\varepsilon)}1 + b_1(y_{TB}+z_c)$$

$$+ \frac{\gamma_{TC}(\beta+\beta_1)\{1+b_1(y_{TB}+z_c)\}^3}{\beta_1\beta(1-\varepsilon+\theta\varepsilon)^3(\mu_n+\alpha_{TC}+\gamma_{TC})}[\beta(\beta_1 x - (\mu_n+\alpha_C+\gamma_C) \tag{15}$$

$$+ \{\beta x - (\gamma_{TB}+\mu_n+\alpha_{TB})\}\beta_1]$$

$$- 2\beta_1 x + (\mu_n + \alpha_C + \gamma_C) - \left\{\frac{1+b_1(y_{TB}+z_c)}{(1-\varepsilon+\theta\varepsilon)}\right\}(2\beta x + (\gamma_{TB}+\mu_n+\alpha_{TB})) - \mu_n.$$

16

Clearly, $g(\Lambda/\mu_n) > 0$. Thus, there exists unique value x_S^* of x_S if $g(0) < 0$ and $'x_S > 0; \forall 0 < x_S < (\Lambda/\mu_n)$. Then, the corresponding y_{TB}^*, z_C^* and w_{TC}^* can be obtained from equations (12), (13) and (14) respectively.

5 DISEASE-FREE EQUILIBRIUM (DFE) AND BASIC REPRODUCTION NUMBER

The system of equation (1) has a DFE $P_0^* = (x_0^*, 0, 0, 0)$, where $x_0^* = \Lambda/\mu_n$. R_0 is traditionally defined as the average number of secondary infections that occur when a person who has the disease during the infected time period comes into contact with a population class that is wholly susceptible (Zhang $et~al.$ 2014). In an epidemiological analysis, R_0 is crucial for resolving the global dynamics of disease. It foretells and demonstrates whether the sickness will spread and persist or eventually disappear from the population. Using next-generation matrix method, we can obtain R_0 (Routh 1877).

Let $= \{y_{TB}, z_C, w_{TC}\}^T$, then the system (1) dictates that

$$dX/dt = F - K$$

Let us define

$$F'_1 = \left[\frac{\partial(R_1)_i}{\partial x_j}\right], and K'_2 = \left[\frac{\partial(R_2)i}{\partial x_j}\right]; for i,j = 1, 2, 3 \text{ at } P_0^*$$

$$F'_1 = \begin{bmatrix} \dfrac{(1-\epsilon+\theta\epsilon)\beta x_S - \{-1+b_1 z_C\}}{\{1+b_1(y_{TB}+z_c)\}^2}\beta\Lambda/\mu_n & -\dfrac{(1-\epsilon+\theta\epsilon)\beta y_{TB}\Lambda/\mu_n * b_1}{\{1+b_1(y_{TB}+z_c)\}^2} & 0 \\ 0 & \beta_1\Lambda/\mu_n & 0 \\ 0 & 0 & 0 \end{bmatrix}$$

$$and~K'_2 = \begin{bmatrix} (\gamma_{TB}+\mu_n+\alpha_{TB}) & 0 & 0 \\ 0 & (\mu_n+\alpha_C+\gamma_C) & 0 \\ 0 & 0 & (\mu_n+\alpha_{TC}+\gamma_{TC}) \end{bmatrix}$$

The non-negative nature of F'_1, the non-singularity of K'_2 as M-matrix, and therefore non-negative nature of $K'_2{}^{-1}$. Thus, we have

$$F'_1 K'_2{}^{-1} = \begin{bmatrix} \dfrac{(1-\epsilon+\theta\epsilon)-\{-1+b_1 z_C\}}{(\gamma_{TB}+\mu_n+\alpha_{TB})\{1+b_1(y_{TB}+z_c)\}^2}\beta\dfrac{\Lambda}{\mu_n} & -\dfrac{(1-\epsilon+\theta\epsilon)*b_1}{\{1+b_1(y_{TB}+z_c)\}^2(\gamma_{TB}+\mu_n+\alpha_{TB})}\beta y_{TB}\dfrac{\Lambda}{\mu_n} & 0 \\ 0 & \beta_1\dfrac{\Lambda}{(\mu_n+\alpha_C+\gamma_C)} & 0 \\ 0 & 0 & 0 \end{bmatrix}$$

Here, we note that $F'_1 K'_2{}^{-1}$ is also non-negative. Now, we find

$$R_0 = \beta_1\frac{\Lambda}{(\mu_n+\alpha_C+\gamma_C)}$$

As a result, when COVID-19 infection coexists with TB, it contributes the system's basic reproduction rate.

6 STABILITY ANALYSIS

We perform stability analysis of disease free and endemic equilibrium points as follows:

6.1 Local stability of disease-free equilibrium point

The variational matrix $V(E_0)$ of the system of equations (1) is computed in order to analyse the locally stability of an equilibrium point, and it is shown as follows:

$$V(E_0) = \begin{bmatrix} -\mu_n & \gamma_{TB} - (1-\epsilon+\theta\epsilon)\beta\dfrac{\Lambda}{\mu_n} & \gamma_C - \beta_1\dfrac{\Lambda}{\mu_n} & \gamma_C \\ 0 & (1-\epsilon+\theta\epsilon)\beta\dfrac{\Lambda}{\mu_n} - (\gamma_{TB}+\mu_n+\alpha_{TB}) & 0 & 0 \\ 0 & 0 & \beta_1\dfrac{\Lambda}{\mu_n} - (\mu_n+\alpha_C+\gamma_C) & 0 \\ 0 & 0 & 0 & -(\mu_n+\alpha_{TC}+\gamma_{TC}) \end{bmatrix}$$

Now, the eigen values of variational matrix corresponding to DFE are:

$$\lambda_1 = -\mu_n,$$

$$\lambda_2 = (1-\epsilon+\theta\epsilon)\beta\left(\frac{\Lambda}{\mu_n}\right) - (\gamma_{TB}+\mu_n+\alpha_{TB}), = -(\gamma_{TB}+\mu_n+\alpha_{TB})(1-R_1),$$

$$\lambda_3 = \beta_1\left(\frac{\Lambda}{\mu_n}\right) - (\mu_n+\alpha_C+\gamma_C), = -(\mu_n+\alpha_C+\gamma_C)(1-R_2),$$

$$\lambda_4 = -(\mu_n+\alpha_{TC}+\gamma_{TC}).$$

Clearly, the two eigen values λ_1 and λ_4 of variational matrix of DFE point are found to be negative and the remaining two eigen values λ_2 and λ_3 are having negative real parts if $R_1 < 1$ and $R_2 < 1$ respectively. Hence, by Routh-Hurwitz Criteria the DFE point is locally asymptotically stable [13], if $R_1 < 1$, $R_2 < 1$ else unstable if $R_1 > 1$, $R_2 > 1$.

6.2 Local stability of endemic equilibrium point

The effects of slight perturbations in the equilibrium state of the system can be better understood through the local stability analysis of equilibrium points. To study the stability of endemic equilibrium point, we linearize the system about the arbitrary equilibrium points $E^*(x_S, y_{TB}, z_C, w_{TC})$ and obtain the corresponding Variational matrix as given by

$$\begin{bmatrix} A_{11} & B_{12} & C_{13} & \gamma_c \\ D_{21} & E_{22} & F_{23} & 0 \\ G_{31} & H_{32} & I_{33} & 0 \\ 0 & J_{42} & K_{43} & L_{44} \end{bmatrix}$$

where $A_{11} = -\dfrac{(1-\epsilon+\theta\epsilon)}{1+b_1(y_{TB}+z_c)}\beta y_{TB} - \beta_1 z_C - \mu_n$, $B_{12} = \gamma_{TB} - \dfrac{(1-\epsilon+\theta\epsilon)\beta x_S\{-1+b_1 z_C\}}{\{1+b_1(y_{TB}+z_c)\}^2}$,

$$C_{13} = \gamma_C - \beta_1 x_S,$$

$$D_{21} = \frac{(1-\epsilon+\theta\epsilon)}{1+b_1(y_{TB}+z_c)}\beta y_{TB},$$

$$E_{22} = \frac{(1-\epsilon+\theta\epsilon)\beta x_S\{-1+b_1 z_C\}}{\{1+b_1(y_{TB}+z_c)\}^2} - \frac{(1-\epsilon+\theta\epsilon)\beta_1 z_C\{-1+b_1 z_C\}}{\{1+b_1(y_{TB}+z_c)\}^2} - (\gamma_{TB}+\mu_n+\alpha_{TB}),$$

$$F_{23} = \frac{(1-\epsilon+\theta\epsilon)\beta y_{TB}\{-1+b_1 y_{TB}\}}{\{1+b_1(y_{TB}+z_c)\}^2}, \quad G_{31} = \beta_1 z_C, \quad H_{32} = \frac{(1-\epsilon+\theta\epsilon)\beta z_C\{-1+b_1 z_C\}}{\{1+b_1(y_{TB}+z_c)\}^2},$$

$$I_{33} = \beta_1 x_S - \frac{(1 - \epsilon + \theta\epsilon)\beta y_{TB}\{-1 + b_1 y\}}{\{1 + b_1(y_{TB} + z_c)\}^2} - (\mu_n + a_C + \gamma_C),$$

$$J_{42} = \frac{(1 - \epsilon + \theta\epsilon)\beta z_C\{-1 + b_1 z_C\}}{\{1 + b_1(y_{TB} + z_c)\}^2} + \frac{(1 - \epsilon + \theta\epsilon)\beta_1 z_C\{-1 + b_1 z_C\}}{\{1 + b_1(y_{TB} + z_c)\}^2},$$

$$K_{43} = \frac{(1 - \epsilon + \theta\epsilon)\beta y_{TB}\{-1 + b_1 y_{TB}\}}{\{1 + b_1(y_{TB} + z_c)\}^2} + \frac{(1 - \epsilon + \theta\epsilon)\beta_1 y_{TB}\{-1 + b_1 y_{TB}\}}{\{1 + b_1(y_{TB} + z_c)\}^2},$$

$$L_{44} = -(\mu_n + a_{TC} + \gamma_{TC})$$

Then, the characteristic equation of the variational matrix is given by:

$$\lambda^4 + M_1\lambda^3 + M_2\lambda^2 + M_3\lambda + M_4$$

where

$$M_1 = \left[\left\{ (\mu_n + a_{TC} + \gamma_{TC}) - \frac{(1 - \epsilon + \theta\epsilon)}{1 + b_1(y_{TB} + z_c)}\beta y_{TB} - \beta_1 z_C - \mu_n \right.\right.$$

$$-\gamma_{TB} - \frac{(1 - \epsilon + \theta\epsilon)\beta x_S\{-1 + b_1 z_C\}}{\{1 + b_1(y_{TB} + z_c)\}^2} - \beta_1 x_S - \frac{(1 - \epsilon + \theta\epsilon)\beta y_{TB}\{-1 + b_1 y\}}{\{1 + b_1(y_{TB} + z_c)\}^2}$$

$$\left.\left. -(\mu_n + a_C + \gamma_C) \right\} \right],$$

$$M_2 = \left[\frac{1 - \epsilon + \theta\epsilon}{\{1 + b_1(y_{TB} + z_c)\}^2}\{-\beta x_s(-1 + b_1 z_c)(\mu_n + a_{TC} + \gamma_{TC}) + \beta_1 z_C(-1 + b_1 z_c)(\mu_n + a_{TC} + \gamma_{TC}) \right.$$

$$+ \beta y_{TB}(-1 + b_1 y_{TB})(\mu_n + a_{TC} + \gamma_{TC}) + \beta\beta_1 x_s{}^2(-1 + b_1 z_c) - \beta_1^2 x_s z_c(-1 + b_1 z_c)$$

$$+ \beta y_{TB}(-1 + b_1 y_{TB})(\mu_n + a_{TB} + \gamma_{TB}) - \beta x_s(-1 + b_1 z_c)(\mu_n + a_c + \gamma_c)$$

$$+ \beta_1 z_c(-1 + b_1 z_c)(\mu_n + a_C + \gamma_C)\} + \frac{1 - \epsilon + \theta\epsilon}{\{1 + b_1(y_{TB} + z_c)\}^3}\{-\beta^2 x_s y_{TB}(-1 + b_1 z_c)$$

$$- \beta_1 z_c - \mu_n + \beta\beta_1 \gamma_{TB} z_c(-1 + b_1 z_c) - \beta_1 z_C - \mu_n - \beta^2 x_s y_{TB}(-1 + b_1 z_c)$$

$$+ \beta^2 y_{TB}^2(-1 + b_1 y_{TB} - \beta_1 z_C - \mu_n) + \frac{(1 - \epsilon + \theta\epsilon)}{1 + b_1(y_{TB} + z_C)}\{\beta y_{TB} + \beta_1 z_c$$

$$+ \mu_n(\mu_n + a_{TC} + \gamma_{TC}) + \beta y_{TB} - \beta_1 z_c - \mu_n(\mu_n + a_{TB} + \gamma_{TB}) - \beta\beta_1 x_s y_{TB} - \beta_1 z_c$$

$$- \mu + \beta y_{TB} - \beta_1 z_c - \mu_n(\mu_n + a_c + \gamma_c) + \beta y_{TB}\gamma_{TB}\} + (\mu_n + a_{TB} + \gamma_{TB})$$

$$\times (\mu_n + a_{TC} + \gamma_{TC}) - \gamma_c(\mu_n + a_{TC} + \gamma_{TC}) + \beta_1 x_s(\mu_n + a_{TC} + \gamma_{TC})$$

$$-\beta_1 x_s(\mu_n + a_{TC} + \gamma_{TC})(\mu_n + a_c + \gamma_c) + (\mu_n + a_{TC} + \gamma_{TC})(\mu_n + a_c + \gamma_c)$$

$$\left. -\beta_1 x_s(\mu_n + a_{TB} + \gamma_{TB}) + (\mu_n + a_{TB} + \gamma_{TB})(\mu_n + a_c + \gamma_c)\} \right],$$

$$M_3 = \left[\frac{1-\epsilon+\theta\epsilon}{\{1+b_1(y_{TB}+z_c)\}^2} \{\beta_1\beta\gamma_{TB}z_c(-1+b_1y_{TB})\gamma_c + \beta_1^2\gamma_{TB}z_c(-1+b_1z_c)(\mu_n+\alpha_{TC}+\gamma_{TC}) \right.$$

$$-\beta_1^2z_cx_s(-1+b_1z_c)(\mu_n+\alpha_{TC}+\gamma_{TC}) + 2\beta y_{TB}(-1+b_1y_{TB})(\mu_n+\alpha_{TB}+\gamma_{TB})$$

$$\times(\mu_n+\alpha_{TC}+\gamma_{TC}) - \beta x_s(-1+b_1z_c)(\mu_n+\alpha_c+\gamma_c)(\mu_n+\alpha_{TC}+\gamma_{TC})$$

$$+(-1+b_1z_c)(\mu_n+\alpha_{TC}+\gamma_{TC})(\mu_n+\alpha_c+\gamma_c) - \beta\beta_1x_s^2(-1+b_1z_c)(\mu_n+\alpha_{TC}+\gamma_{TC})$$

$$-\beta_1^2x_sz_c(-1+b_1z_c)(\mu_n+\alpha_{TC}+\gamma_{TC}) + (1-\epsilon+\theta\epsilon)\beta\beta_1x_s^2y_{TB}(-1+b_1z_c)$$

$$-(1-\epsilon+\theta\epsilon)\beta^2x_sy_{TB}(-1+b_1z_c)(\mu_n+\alpha_c+\gamma_c) + \beta\beta_1y_{TB}z_c(-1+b_1y_{TB})$$

$$-(1-\epsilon+\theta\epsilon)\beta^2y_{TB}z_c\gamma_c(-1+b_1z_c) + \beta\beta_1x_sz_c\gamma_c(-1+b_1z_c) - \beta\beta_1x_s^2(-1+b_1z_c)$$

$$-\beta_1^2z_c^2\gamma_c(-1+b_1z_c) + \beta_1^3x_sz_c^2(-1+b_1z_c)\} + \frac{(1-\epsilon+\theta\epsilon)^2}{\{1+b_1(y_{TB}+z_c)\}^3}$$

$$\times\{\beta^2y_{TB}z_c(-1+b_1z_c) + \beta\beta_1y_{TB}z_c(-1+b_1z_c)\gamma_c + 2\beta_1\beta^2x_s^2y_{TB}(-1+b_1z_c)$$

$$-\beta_1z_c - \mu_n - 2\beta^2y_{TB}^2(-1+b_1y_{TB}) - \beta_1z_c - \mu_n(\mu_n+\alpha_{TB}+\gamma_{TB})$$

$$+\beta^2y_{TB}z_c(-1+b_1y_{TB}) - \beta_1z_c - \mu_n(\mu_n+\alpha_{TC}+\gamma_{TC}) - \beta^2x_sy_{TB}(-1+b_1y_{TB})$$

$$\times(\mu_n+\alpha_{TC}+\gamma_{TC}) + \beta^2y_{TB}^2\gamma_c(-1+b_1y_{TB}) - \beta_1\beta^2y_{TB}z_cx_s(-1+b_1y_{TB})(-1+b_1z_c)$$

$$+\beta_1\beta^2y_{TB}z_cx_s(-1+b_1z_c)\} + \frac{(1-\epsilon+\theta\epsilon)}{\{1+b_1(y_{TB}+z_c)\}^4}\{-(1-\epsilon+\theta\epsilon)\beta^2y_{TB}x_s(-1+b_1y_{TB})$$

$$\times(-1+b_1z_c)(\mu_n+\alpha_{TC}+\gamma_{TC}) + (1-\epsilon+\theta\epsilon)\beta\beta_1y_{TB}z_c(-1+b_1z_c)(-1+b_1y_{TB})$$

$$\times(\mu_n+\alpha_{TC}+\gamma_{TC}) - (1-\epsilon+\theta\epsilon)\beta^2y_{TB}x_s(-1+b_1z_c)(\mu_n+\alpha_{TC}+\gamma_{TC}) - \beta\beta_1y_{TB}z_c$$

$$\times(-1+b_1z_c)(-1+b_1y_{TB})(\mu_n+\alpha_{TC}+\gamma_{TC}) - (1-\epsilon+\theta\epsilon)^2\beta^3y_{TB}^2x_s(-1+b_1y_{TB})$$

$$\times(-1+b_1z_c)\} - 2\frac{(1-\epsilon+\theta\epsilon)^2}{\{1+b_1(y_{TB}+z_c)\}^5}\{-(1-\epsilon+\theta\epsilon)\beta^3y_{TB}^2x_s(-1+b_1y_{TB})$$

$$\times(-1+b_1z_c) - \beta_1z_c - \mu_n + \beta^3y_{TB}^2z_c(-1+b_1z_c)(-1+b_1y_{TB}) - \beta_1z_c - \mu_n\}$$

$$+\frac{(1-\epsilon+\theta\epsilon)}{1+b_1(y_{TB}+z_c)}\{(\beta y_{TB} - \beta_1^2z_cx_s - \mu_n) - \beta\beta_1y_{TB}x_s - \beta_1z_c - \mu_n(\mu_n+\alpha_{TC}+\gamma_{TC})$$

$$+(\beta y_{TB} - \beta_1z_c - \mu_n)(\mu_n+\alpha_{TC}+\gamma_{TC})(\mu_n+\alpha_c+\gamma_c) + (\mu_n+\alpha_{TC}+\gamma_{TC})\gamma_{TB}$$

$$-\beta\beta_1y_{TB}x_s\gamma_{TB} + \beta y_{TB}\gamma_{TB}(\mu_n+\alpha_c+\gamma_c)\} - 2(\mu_n+\alpha_{TB}+\gamma_{TB})\beta_1x_s(\mu_n+\alpha_{TC}+\gamma_{TC})$$

$$-\gamma_{TB}\beta_1z_c(\mu_n+\alpha_{TC}+\gamma_{TC}) + \beta_1^2x_sz_c(\mu_n+\alpha_{TC}+\gamma_{TC}) - (\mu_n+\alpha_{TB}+\gamma_{TB})\beta_1z_c\gamma_c$$

$$\left. +(\mu_n+\alpha_{TB}+\gamma_{TB})\beta_1^2x_sz_c\}\right]$$

$$M_4 = \left[\frac{1-\epsilon+\theta\epsilon}{\{1+b_1(y_{TB}+z_c)\}^2} \left\{ -\beta\beta_1 x_s z_c \gamma_c (\mu_n + \alpha_{TC} + \gamma_{TC}) - \beta_1^2 \beta z_c^2 y_{TB} \right. \right.$$

$$\times (-1+b_1 z_c)(-1+b_1 y_{TB})\gamma_c + \beta_1^2 z_c^2 \gamma_c (\mu_n + \alpha_{TC} + \gamma_{TC})$$

$$-\beta\beta_1 y_{TB} z_c (-1+b_1 y_{TB})\gamma_c \, \mu_n + \alpha_{TB} + \gamma_{TB}\beta_1^2 y_{TB} z_c (-1+b_1 y_{TB})\gamma_c$$

$$- \beta_1 \beta z_c^2 \gamma_c^2 (-1+b_1 z_c) - \beta_1^2 z_c^2 \gamma_c^2 (-1+b_1 z_c) + \beta\beta_1^2 x_s z_c^2 \gamma_c (-1+b_1 z_c)$$

$$+ \beta_1^3 x_s z_c^2 (-1+b_1 z_c)\gamma_c - \beta\beta_1 y_{TB} z_c \gamma_c \gamma_{TB}(\mu_n + \alpha_{TC} + \gamma_{TC})(-1+b_1 y_{TB}) + 1-\epsilon$$

$$+ \theta\epsilon\beta_1\beta^2 y_{TB} x_s z_c (-1+b_1 z_c)(-1+b_1 y_{TB})(\mu_n + \alpha_{TC} + \gamma_{TC}) - \beta\beta_1 x_s z_c \gamma_c (\mu_n + \alpha_{TC} + \gamma_{TC})$$

$$\times (-1+b_1 z_c) + \beta_1^2 z_c^2 \gamma_c (\mu_n + \alpha_{TC} + \gamma_{TC})(-1+b_1 z_c) + \beta_1^2 \beta^2 x_s^2 z_c (-1+b_1 z_c)$$

$$\left. \times (\mu_n + \alpha_{TC} + \gamma_{TC}) - \beta_1^3 z_c^2 x_s (-1+b_1 z_c)(\mu_n + \alpha_{TC} + \gamma_{TC}) \right\}$$

$$+ \frac{(1-\epsilon+\theta\epsilon)^2}{\{1+b_1(y_{TB}+z_c)\}^3} \left\{ \beta_1 \beta^2 y_{TB} z_c x_s (-1+b_1 z_c)\gamma_c + \beta\beta_1^2 x_s y_{TB} z_c (-1+b_1 z_c)\gamma_c \right.$$

$$- \beta^2 y_{TB} z_c (-1+b_1 z_c)(\mu_n + a_c + \gamma_c) - \beta\beta_1 y_{TB} z_c \gamma_c (-1+b_1 z_c)(\mu_n + a_c + \gamma_c)$$

$$+ \beta^2 \beta_1 x_s^2 y_{TB}(-1+b_1 z_c)\beta_1 z_c - \mu_n(\mu_n + \alpha_{TC} + \gamma_{TC})(\mu_n + \alpha_{TB} + \gamma_{TB})$$

$$- \beta^2 y_{TB} x_s (-1+b_1 z_c)(\mu_n + \alpha_{TC} + \gamma_{TC})(\mu_n + a_c + \gamma_c) - \beta_1 z_c - \mu_n$$

$$+ \beta\beta_1 y_{TB} z_c (-1+b_1 z_c)(\mu_n + \alpha_{TC} + \gamma_{TC})(\mu_n + a_c + \gamma_c) - \beta_1 z_c - \mu_n$$

$$+ \beta^2 y_{TB}^2 (-1+b_1 y_{TB})(\mu_n + \alpha_{TC} + \gamma_{TC}) + \gamma_{TB}\beta^2 y_{TB}^2 (-1+b_1 y_{TB})(\mu_n + \alpha_{TC} + \gamma_{TC})$$

$$- \beta^2 y_{TB} x_s (-1+b_1 z_c)(\mu_n + \alpha_{TC} + \gamma_{TC})(\mu_n + a_c + \gamma_c) + \beta_1 \beta^2 y_{TB} x_s z_c (-1+b_1 z_c)$$

$$\left. \times (\mu_n + \alpha_{TC} + \gamma_{TC}) \right\} - \frac{(1-\epsilon+\theta\epsilon)}{\{1+b_1(y_{TB}+z_c)\}^4} \left\{ (1-\epsilon+\theta\epsilon)\beta_1\beta^2 y_{TB} x_s z_c (-1+b_1 z_c) \right.$$

$$\left. \times (-1+b_1 y_{TB})\gamma_c + \beta_1^3 z_c^2 y_{TB}(-1+b_1 z_c)(-1+b_1 y_{TB})\gamma_c \right\} + \frac{(1-\epsilon+\theta\epsilon)^2}{\{1+b_1(y_{TB}+z_c)\}^5}$$

$$\times \left\{ (1-\epsilon+\theta\epsilon)\beta^2 y_{TB}^2 z_c (-1+b_1 z_c)(-1+b_1 y_{TB}) - (1-\epsilon+\theta\epsilon)\beta^3 y_{TB}^2 z_c \right.$$

$$\times (-1+b_1 z_c)(-1+b_1 y_{TB}) - (1-\epsilon+\theta\epsilon)\beta^3 y_{TB}^2 z_c (-1+b_1 z_c)(-1+b_1 y_{TB})\gamma_c$$

$$- (1-\epsilon+\theta\epsilon)\beta^2 \beta_1 y_{TB}^2 z_c (-1+b_1 z_c)(-1+b_1 y_{TB})\gamma_c - \beta^3 y_{TB} x_s (-1+b_1 z_c)(-1+b_1 y_{TB})$$

$$\times (\mu_n + \alpha_{TC} + \gamma_{TC}) - \beta_1 z_c - \mu_n - \beta^2 \beta_1 y_{TB}^2 z_c (-1+b_1 z_c)(-1+b_1 y_{TB})$$

$$\left. \times (\mu_n + \alpha_{TC} + \gamma_{TC}) - \beta_1 z_c - \mu_n - \beta^3 y_{TB}^2 x_s (-1+b_1 z_c)(-1+b_1 y_{TB})(\mu_n + \alpha_{TC} + \gamma_{TC}) \right\}$$

$$+ \frac{(1-\epsilon+\theta\epsilon)}{1+b_1(y_{TB}+z_c)} \left\{ -\beta\beta_1 x_s y_{TB} - \beta_1 z_c - \mu_n(\mu_n + \alpha_{TC} + \gamma_{TC})(\mu_n + \alpha_{TB} + \gamma_{TB}) \right.$$

$$+ \beta y_{TB} - \beta_1 z_c - \mu_n(\mu_n + \alpha_{TC} + \gamma_{TC})(\mu_n + \alpha_{TB} + \gamma_{TB})(\mu_n + a_c + \gamma_c)$$

$$\times \ \{-\beta\beta_1 x_s y_{TB} \gamma_{TB}(\mu_n + a_c + \gamma_c)(\mu_n + a_{TC} + \gamma_{TC}) + 2(\mu_n + a_c + \gamma_c)$$

$$\times \ (\mu_n + a_{TC} + \gamma_{TC})\beta_1 z_c \gamma_c - (\mu_n + a_{TC} + \gamma_{TC})(\mu_n + a_{TB} + \gamma_{TB})\beta_1 x_s\}\Bigg]$$

By Routh Hurwitz criterion endemic equilibrium point is locally asymptotically stable if $M_1 > 0$, $M_2 > 0$, $M_4 > 0$, and $M_1 M_2 M_3 - M_3^2 - M_1^2 M_4 > 0$.

7 NUMERICAL SIMULATION OF THE MODEL

In this section, we have observed the quantitative behaviour of the transmission dynamics implementing saturated incidence rate of concomitant diseases TB and COVID-19 in India. We justify by implementing saturated incidence rate that the estimated result is much closer to the reality. We have also justified the analytical findings of impact of COVID-19 on TB infected population and vice-versa using **MATLAB** software. Table 1 gives the parameter values to be used to perform numerical simulation of the model. Most of the parameter values have been taken from health sites of Indian Government, some of them are assumed as well as considered from cited articles. Numerical simulation is done for the first wave of COVID-19 in India from June 1, 2020 to September 2, 2020 and for the second wave from March 1, 2021 to June 1, 2021 by the same parameter values except transmission coefficient rates β and β_1. R_0, the basic reproduction number, is calculated and determined to be 0.31.

We have plotted the figures for both the waves of pandemic in India and compared the dynamics of both. In Figure 2, For the first and second waves, respectively, we have plotted the variation of the TB-infected population with time in days for the estimated value of TB patients and the actual data of TB patients. We observe that incorporation of saturated incidence rate in our mathematical model brings our result much closer to the reality.

In Figure 3, We have presented the changes in the TB-infected population over time for various COVID-19 infection transmission rates β_1 for the first and second waves. We observe that as β_1 increases, TB-infected population decreases during both the waves. But numbers of TB infected population reported during second wave were slightly higher than that during the first wave. From the figure, we observe that number of TB infected population reported in this situation decline. This decrement in number of infected TB population may be due to diagnosing TB as COVID-19 and treating TB infected population as COVID-19 infected population.

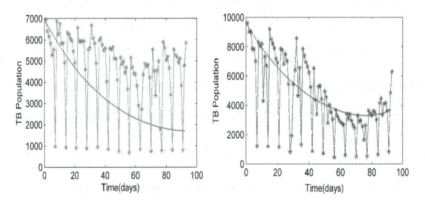

Figure 2.   Simulation of TB population for actual and estimated data for (a) first wave and (b) second wave.

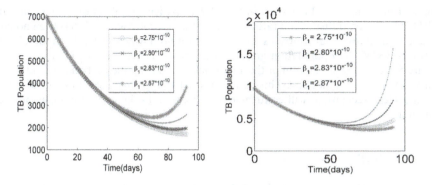

Figure 3. Simulation of TB population for different values of β_1 during (a) first wave of COVID-19 and (b) second wave of COVID-19.

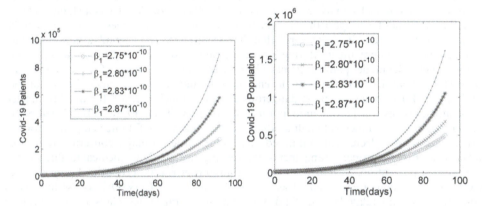

Figure 4. Simulation of COVID-19 population with time during (a) first wave of COVID-19 and (b) second wave of COVID-19.

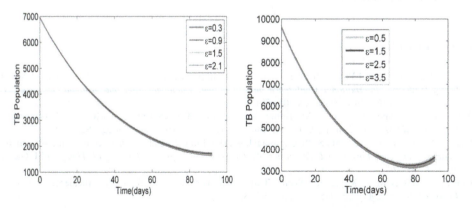

Figure 5. Simulation of TB population for different value of ϵ during (a) first wave of COVID-19 and (b) second wave of COVID-19.

Figure 4 illustrates the fluctuation in COVID-19 patient numbers over time for various COVID-19 infection transmission rates for the first and second waves. It has been noted that the graph's qualitative characteristics are the same for both waves. It has been found that for both waves, COVID-19 population grows as transmission coefficient β_1 increases.

Figure 5 depicts the simulation of the TB population during the first and second COVID-19 waves for various values of ϵ. This figure shows that applying a saturated incidence rate, which includes vaccination, will contribute to a decrease in the overall number of infected population.

8 CONCLUSION

A non-linear deterministic compartmental model for the co-infection of concomitant diseases TB and COVID-19 with saturated incidence rate is proposed to determine the dynamics of transmission of these diseases. For the proposed model, two equilibrium points i.e., the (DFE) and the EE point are computed. Additionally, we computed the basic reproduction number R_0, which will enable us to examine and monitor the dynamical behaviour of the suggested model. The DFE point is stable if $R_0<1$ is otherwise unstable, according to the model's analytical examination. Furthermore, $R_0<1$ has a stable endemic equilibrium point. To support the analytical conclusions and outcomes, numerical simulations are performed and it is discovered that they accord well. Additionally, the simulation's findings show that applying a saturated incidence rate makes our results considerably more realistic. Once the model parameters have been estimated on the basis of the available data on WHO and Nikshay portal, the model enables us to find out the decrement in the statistics of TB. COVID-19 being the most dangerous disease is much harmful for the TB infected population, because of the similar symptoms of TB and COVID-19, the TB infected population face problem in medical facilities due to which number of TB infected population are expected to rise but decrement in notified cases is the matter of major concern and must be taken seriously with proper arrangements of screening and supplying medical facilities to the TB infected population, so that we do not get an abrupt rise in TB notified patients once COVID-19 situations are normed. Disease related death rate of COVID-19 is also high when compared to disease related death of TB. Vaccination coverage rate ϵ plays important role in controlling TB and eradicate it from India. By considering the saturated incidence rate we observed that infected population obtained theoretically is closer to the notified infected population and thereby consideration of saturated incidence rate makes our model more realistic than any work done previously. No one has considered saturated incidence rate to the best of our knowledge.

REFERENCES

Agarwal R., Singh S., and Vats S. "Implementation of an Improved Algorithm for Frequent Itemset Mining Using Hadoop," *International Conference on Computing, Communication and Automation (ICCCA)*, pp. 13–18, 2016. doi: 10.1109/CCAA.2016.7813719.

Alexander M.E., and Moghadas S.M. Bifurcation Analysis of SIRS Epidemic Model with Generalized Incidence. *SIAM Journal on Applied Mathematics*, 65(5):1794–1816, 2005. [Google Scholar].

Bhati J.P., Tomar D., and Vats S. "Examining Big Data Management Techniques for Cloud-based IoT Systems," in *Examining Cloud Computing Technologies Through the Internet of Things*, IGI Global, pp. 164–191, 2018.

Esteva L., Matias M.A. Model for Vector Transmitted Diseases with Saturation Incidence. *Journal of Biological Systems*;9(4):235–245, 2001. [Google Scholar]

Gupta N., Ish P., Gupta A., Malhotra N., Caminero J.A., Singla R., Kumar R., Yadav S.R., Dev N., Agrawal S. and Kohli S. "A Profile of a Retrospective Cohort of 22 Patients with COVID-19 and Active/ treated Tuberculosis", *European Respiratory Journal*, 56(5), 2020.

Holmdahl I. and Buckee C. "Wrong but useful what COVID-19 Epidemiologic Models can and Cannot Tell us, *New England Journal of Medicine*, 383(4):303–305, 2020.

https://knoema.com/atlas/India/Birth-rate [google scholar]

https://mohfw.gov.in/ [google scholar]

https://nikshay.in.siteindices.com/ [google scholar]

https://www.macrotrends.net/countries/IND/india/birth-rate [google scholar]

https://www.medicinenet.com/ [google scholar]

https://covid19.who.int/ [google scholar]

Kuddus, Md Abdul, McBryde E.S., Adekunle A.I., White L.J., and Meehan M.T. "Mathematical analysis of a Two-strain Tuberculosis Model in Bangladesh." *Scientific Reports* 12(1): 1–13, 202.

McQuaid C.F., Vassall A., Cohen T., Fiekert K., and White R.G. "The Impact of COVID-19 on TB: A Review of the Data," *The International Journal of Tuberculosis and Lung Disease*, 25(6) 436–446. 2021.

Mengistu, Ashena & Kelemu, and Peter J. Witbooi. "Mathematical Analysis of TB Model With Vaccination and Saturated Incidence Rate." In *Abstract and Applied Analysis*. Hindawi, 2020.

Routh E.J. *"A Treatise on the Stability of a Given State of Motion: Particularly Steady Motion"*, Being the Essay to which the Adams Prize was Adjudged, in the University of Cambridge. Macmillan and Company, 1877.

Vats S. and Sagar B.B. "An Independent Time Optimized Hybrid Infrastructure for Big Data Analytics," *Mod. Phys. Lett. B*, vol. 34, no. 28, p. 2050311, 2019, Oct. 2020, doi: 10.1142/S021798492050311X.

Vats S. and Sagar B.B., *"Data lake: A Plausible Big Data Science for Business Intelligence,"* WHO. Global tuberculosis report. WHO/CDS/TB/2019.15, Geneva.

Zeb, Anwar, Ebraheem Alzahrani, Vedat Suat Erturk, and Gul Zaman. "Mathematical Model for Coronavirus Disease (COVID-19) Containing Isolation Class." *BioMed research international*. 2019.

Zhang, Jinhong *et al.* "Analysis of an SEIR Epidemic Model with Saturated Incidence and Saturated Treatment Function." *The Scientific World Journal*. p. 910421, 2014.

Automation and Computation – Vats et al. (Eds)
© 2023 the Author(s), ISBN 978-1-032-36723-1

Feature generation techniques for suicidal ideation detection from social media texts

K. Soumya*
School of Computer Science and Engineering, Lovely Professional University – Punjab, and Assistant Professor in VBIT – Hyderabad, India
ORCID ID: 0000-0001-7364-0795

Vijay Kumar Garg*
School of Computer Science and Engineering, Lovely Professional University – Punjab, India
ORCID ID: 0000-0001-5926-7162

ABSTRACT: Detection of suicide ideation (SI) from social media texts is challenging as it necessitates analyzing the content and context of the text utterances. The best set of features must be learnt from the content and context information which have higher correlation to SI. This work experiments with various content and context features and tests the correlation of these features to SI using clustering analysis. Natural language processing techniques (NLP) based methods were proposed to extract the features in terms of key phrases, key words, sentence incongruity, sentiment polarity and emotion tags. Through experimental analysis with twitter and reddit datasets, the features with higher correlation to SI were identified.

1   INTRODUCTION

The strong desire towards death, death thoughts and conceiving death plans is called as Suicidal ideation (SI). Worldwide there are about 8 lakhs people committing suicide every year [1, 2]. The suicide death rate is more at adolescence stage. Earlier, due to inexpressiveness of the subject, the SI becomes unnoticed and was difficult to detect. With dawn of social media revolution and its popularity increasing number of people express themselves on social media. People especially adults' use social media more often to express their life events and feelings. Increasing studies on mining of social media contents have found a higher correlation between SI and subject contents. More studies [3–9] have tested the effectiveness of social media contents in predicting SI and mining social media to predict SI have found to non-intrusive and easy compared to clinical observation or questionnaire based analysis. The studies have found a strong correlation between social media contents and SI. Mining social media posts from users and inferring SI from the emotion, thoughts and intentions expressed by the user looks promising for earlier detection of SI.

The social media texts have many content and context information. Once the most significant content and context information with higher correlation to SI is identified, machine learning classifiers can be designed to predict SI from social media texts. The features can be in multiple dimensions of key words, key phrases, sentiment and emotions. This work proposes a host of techniques based on NLP to extract features in multiple dimensions of key words, key phrases, sentiment and emotions. The features are represented in form of vector notation. The correlation between the feature vectors to the SI is analyzed with clustering analysis. The result is a feature vector embedding which can be used with machine learning techniques for SI detection. Following are the contributions of the works.

---

*Corresponding Authors: kothapallisowmya@gmail.com and vijay.garg@lpu.co.in

DOI: 10.1201/9781003333500-3

(i) Features in multiple dimensions of key words, key phrases, sentiments and emotions are extracted from social media texts applying NLP. Methods were proposed to extract sentiment information and sentence incongruity features from text.

(ii) Clustering based correlation analysis between the combination of features and the SI is done. Multiple features in different combinations are analyzed in terms clustering metrics of cohesion, separation and silhouette coefficient to detect the optimal feature combination with higher correlation to SI.

(iii) Identification of relevant and removal of redundant features and representing the most significant factors in features with higher correlation to SI.

The rest of the paper is organized as follows. The features and feature extraction techniques proposed for SI detection are explored in Section II. The proposed feature extraction technique to extract multi-dimensional content and context features is detailed in Section III. The correlation analysis and selection of most significant features are detailed in Section IV. Section V presents the novelties of proposed work. Section VI presents the conclusion and scope of future work.

2 RELATED WORK

Ji *et al.* [10] considered context features for SI detection. They applied statistical, syntactic, linguistic, work embedding and topic modeling for feature extraction. But the authors did not analyze the correlation of features to SI. Coppersmith *et al.* [11] extracted n gram character features from the document and used for SI detection. But considering only presence of n characters without considering context and semantic relationship introduced higher false positives. Mulholland *et al.* [12] extracted N grams, vocabulary, syntax and semantic features from the text and used it for classification of SI. There was no correlation analysis conducted against features to the SI. Huang *et al.* [13] used the psychological lexicon dictionary to extract linguistic features from the text. Lexicons used in this approach are specific to Chinese language. Liakata *et al.* [14] extracted n-gram, negative expressions, self-directed emotions from the text and correlated to different emotions of anger, sorrow, hopefulness, happiness, peacefulness, fear, pride, abuse and forgiveness. The emotions were in turn correlated to the SI. Pestian *et al.* [15] extracted bag of words and semantic features from text and used it for classification of SI. The features were not associated any weights and authors did not correlated to SI. Braithwaite *et al.* [16] extracted text properties, word categories and language markers from texts. Each of the feature values are normalized in range of 0 to 1 and used for SI ideation detection. No correlation analysis was conducted between the features and the SI. Nobles *et al.* [17] extracted term frequency- inverse document frequency (TF-IDF) and psycholinguistic features from the texts and used it for SI detection. Various dimensions of features were not correlated to SI. Coppersmith *et al.* [18] constructed Glove embeddings from texts and applied attention vector weights to the embeddings. The resulting feature is then used for SI detection using the LSTM classifier. Sawhney *et al.* [19] extracted three features of character n grams, TF-IDF and Bag of words. But there are many redundancy in the features and no redundancy removal was done. Tadesse *et al.* [20] extracted word embedding feature from the text and used it for SI detection. No filtering based on context or correlation analysis of word embedding to SI was implemented in this approach. Ji *et al.* [21] extracted sentimental and life event related topical indicators from text and used it to classify SI. But the work did not discriminate sarcasm from the sentimental indicators. Matero *et al.* [22] extracted word embeddings from text and used it for SI detection. The word embedding model proposed in this work did not applied any weights or importance to the words. Chen *et al.* [23] extracted sentiment, TF-IDF and posting behavior as input to classify SI. But the approach lacked correlation analysis to SI. Wicentowski *et al.* [24] extracted lexical and syntactic features from the text and used to classify the emotions. Emotions were correlated to SI. The approach lacked correlation analysis of change of emotions to SI. Schoene *et al.* [25]

extracted sentiment, linguistics and word frequency from text and classified it to SI. But the approach lacked sarcastic sentiment and weightage factors to words. Wang *et al.* [32] found that context features to be highly relevant in detecting SI compared to other features. But the work did not consider removal of sarcastic comments from the context features. Choudhury *et al.* [33] used statistical distribution of words to classify mental illness. Statistical distribution of words related to mental illness when combined with textual features was found to provide better accuracy. Guan *et al.* [34] used linguistic features and profile based information to detect SI. Social media profile information and textual content features extracted from text messages are used with random forest classifier to detect SI. But the accuracy is less than 70% in this method.

From the survey, it could be seen that existing solutions have considered many features like word embedding, TF-IDF, sentiment, key phrases and emotions etc. But most approaches lacked correlation analysis to SI. Sentiment indicators did not consider sarcastic opinions. Key phrases did not consider sentence incongruity. Without considering these problems, the features lacked higher discriminating ability resulting in lower accuracy and higher false positives for classifiers.

3   FETAURE GENERATION FOR SI

This work extracts features in two dimensions of content and context. The taxonomy of the features used in this work is given in Figure 1.

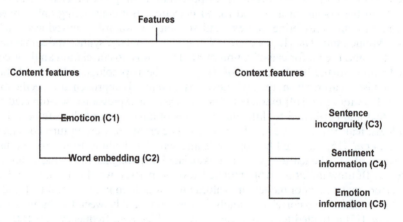

Figure 1.   Feature taxonomy.

3.1   *Content features*

These features are extracted from distribution of word and special symbols like emoticons. Word distribution and semantic relation between words is learnt using Glove embedding [26]. Glove is a powerful word embedding algorithm. In this unsupervised technique, the word vector representation for text corpus is learnt. It is done by reducing the dimension of the co-occurrence matrix. The vector for the word is constructed in such a way that similar words cluster together and different words repel against each other. Compared to other word embedding models like word2vec, glove embedding better captures local and global statistics.

Emoticons are the special symbols which carries information about the positive, negative and sarcastic expressions. The frequency of positive, negative and sarcastic symbols in the document is counted and emoticon feature vector is constructed.

3.2 Context features

Sentiment information is the measure of sentiment expressed in texts. . It has two components intrinsic and extrinsic. Intrinsic refers to the person's internal sentimental state. Extrinsic refers to event topic which acts as risk indicator. Intrinsic sentiment information is extracted for text using domain specific sentiment lexicon approach proposed in [27]. The output of intrinsic sentiment information is a work vector with polarity score (+1/-1) for each of domain specific word in the text. Extrinsic sentiment information is extracted using Latent Dirichlet Allocation (LDA) topic modeling approach proposed in [28]. The output of topic modeling is a vector of scores with each element representing the topic score.

Sentence incongruity is the concept of polarity contrast between the positive candidate term and negative phrase or negative candidate term with positive phrase. The order of occurrence is not important. Camp [39] detailed the incongruity patterns in English language sentences and the summary is presented in Table 1.

Table 1.   Sentence incongruity rules.

Candidate term (Verb positive/negative)	Positive/Negative patterns
Verb	Verb followed by Verb
Verb present participle	Verb followed by Adverb
Verb Gerund	Adverb followed by Verb
Verb past participle	Verb followed by proposition
Verb past form	Verb followed by adjective
Verb present participle third person singular	Verb followed by noun

The sentences are POS tagged and count of number of patterns as defined in Table 1 are found and given as sentence incongruity (*si*) feature.

Neural machine translation is used in this work to translate the sentence to named entity with emotion tag and emotion intensity value. Named entity is the span of sentence marked with one or more words which convey emotional content in the sentence. Emotion tag is one of following emotions of: happiness, sadness, anger, disgust, surprise and fear. Emotion intensity value is one of : low, medium and high. Some of the example of translation is given in Table 2.

Table 2.   Emotion tagging.

Text	Emotion tagging
I have to look at life in her perspective, and it would break anyone's heart	Break anyone's heart <**sadness, high**>
I hate it when certain people always seem to be better at me in everything they do	Hate it <**disgust, low**>
I felt bored and wanted to leave at intermission, but my wife was really enjoying it, so we stayed	Felt bored <**sadness, low**> enjoying it <**happiness, medium**>

Neural machine translation is trained to take the text as input and provide the named entity, its corresponding emotion tag and emotion intensity as the output.

For six emotions of: fear, anger, joy, sadness, disgust, and surprise, the intensity is measured and a feature vector representing the average intensity of each of the six emotions is calculated as six emotion feature.

From the features, the best set of features with higher correlation to SI is found using process given in Figure 2. Clustering analysis is conducted against five different features by

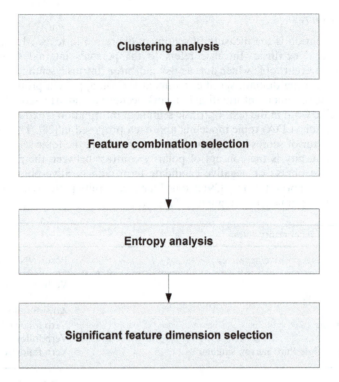

Figure 2.   Processing of features.

measuring average cohesion, separation and Silhouette coefficient. The combination with lowest value of cohesion, highest value of separation and silhouette coefficient is selected as optimal feature combination. Once the optimal feature combination is found, the best set of feature vlaues are found using entropy analysis.

The correlation between each of the feature values to the SI (positive/negative) is measured using symmetric uncertainty. The symmetric uncertainty (SU) between the input variable a and output variable b is calculated as

$$SU(a,b) = \frac{2 \times MI(a,b)}{H(a) + H(b)} \tag{1}$$

Where $MI(a,b)$ is the mutual information between the variable a and b. $H(a)$ is the entropy for the variable a.

Mutual information between variable a and b is calculated as

$$MI(a,b) = \sum_a \sum_b PDF(a,b) \log\log \frac{PDF(a,b)}{p(a) \times p(b)} \tag{2}$$

$PDF(a)$ is the probability density function for the variable a and $PDF(a,b)$ is the joint probability density function.

$H(a)$ is calculated in terms of shanon's entropy as

$$H(a) = -\int PDF(a) \log\log (PDF(a)) dx \tag{3}$$

SU is calculated for each of the 40 input variables against the Big five scores. The average value of SU is calculated as

$$\alpha = \frac{\sum_{\forall a} SU(a,b)}{N} \tag{4}$$

Where N is the total number of input features

The feature value a is selected as relevant feature when its $SU(a,b) \geq \alpha$.

The feature values selected in each feature is joined as a feature vector and this feature vector is used for SI classification.

4 RESULTS

The performance of proposed feature extraction algorithm is tested against UMD Reddit Suicidality Dataset [30] and Reddit suicide Dataset [31]. Both are the most used dataset for testing the effectiveness of suicide classification algorithms.

The feature (C1 to C5) are extracted from labeled dataset and clustering is done using K-means clustering algorithm with K as 2. From the resulting clusters, following metrics are calculated.

1. Average Cohesion
2. Average Separation
3. Silhouette coefficient

Cohesion is measurement of clustering compactness. The points in the cluster must have lower distance. It is measured as below

$$cohesion = \sum_i \sum_{x \in C_i} (x - m_i)^2 \tag{5}$$

Where m_i is the centroid of the cluster and x is the point in the cluster.

Separation is the measurement of how clusters are separated from each other. It is calculated as

$$separation = \sum_i |C_i|(m - m_i)^2 \tag{6}$$

Where $|C_i|$ is the size of the cluster i, and m is the centroid of whole feature set. For effective clustering, the separation distance must be higher.

Silhouette coefficient (SC) is a metric for measuring the effectiveness of the cluster. It is calculated as

$$SC = \left\{ 1 - \frac{a}{b}, \quad if \quad a < b \quad \frac{b}{a} - 1 \quad if \quad a \geq b \right. \tag{7}$$

For a individual point, a is average distance of i to the points in its cluster and b is minimum of average distance of i to points in another cluster. SC value ranges from 0 to 1 and when the value is towards 1, the data points are effectively clustered.

The results of clustering analysis for UMD Reddit Suicidality Dataset are given in Table 2. The lowest value of cohesion, highest value of separation and silhouette coefficient is achieved for combination of C1+C2+C3+C5. The results of clustering analysis for Reddit Suicidality Dataset are given in Table 3. The lowest value of cohesion, highest value of separation and silhouette coefficient is achieved for the combination of C1+C2+C3+C5.

The feature dimension before and after entropy analysis on the feature combination C1+C2+C3+C5 is given in Figure 3.

Table 3. Clustering analysis results for UMD Reddit dataset.

Feature combination	Average cohesion	Average separation	Silhouette coefficient
C1	725	421	0.52
C2	721	427	0.53
C3	719	430	0.56
C4	715	435	0.59
C5	700	450	0.58
C1+C2	694	455	0.59
C1+C3	695	460	0.61
C1+C4	696	466	0.62
C1+C5	696	465	0.62
C2+C3	683	464	0.63
C2+C4	682	465	0.64
C2+C5	680	466	0.65
C3+C4	673	464	0.65
C3+C5	668	450	0.66
C4+C5	662	452	0.67
C1+C2+C3	655	460	0.68
C2+C3+C4	653	462	0.69
C3+C4+C5	650	465	0.71
C1+C2+C3+C4	635	466	0.72
C1+C2+C4+C5	636	470	0.83
C1+ C2+C3+C5	599	490	0.91
C1+C2+C3+C4+C5	600	482	0.84

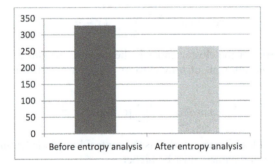

Figure 3.

The feature dimension reduced by 19.51% due to entropy analysis. This reduction is due to calculation of entropy of features and removing the features with entropy less than average entropy.

The accuracy and false positive of SI detection was measured with the reduced set of features using Random forest, SVM and Decision tree classifier. The results are given in Table 5 and 6. In the both datasets, 80% is used for training and 20% for testing.

The proposed feature able to achieve more than 93% SI classification accuracy even with traditional machine learning classifier and the false positives is less than 3 %. The higher accuracy is due to use of integration of both content and context features in the proposed solution and after integration too, the features with higher correlation to SI were detected using clustering analysis. Considering sentence congruity as a feature helped to remove the impact of sarcastic text. This has reduced the false positive in the proposed solution.

Table 4.   Clustering analysis results for Reddit suicide dataset.

Feature combination	Average cohesion	Average separation	Silhouette coefficient
C1	715	310	0.5
C2	710	315	0.523
C3	706	320	0.546
C4	702	325	0.569
C5	698	330	0.592
C1+C2	694	335	0.615
C1+C3	690	340	0.638
C1+C4	686	346	0.661
C1+C5	682	351	0.684
C2+C3	678	356	0.707
C2+C4	674	361	0.73
C2+C5	670	366	0.753
C3+C4	666	371	0.776
C3+C5	662	376	0.799
C4+C5	658	382	0.822
C1+C2+C3	654	387	0.845
C2+C3+C4	650	392	0.868
C3+C4+C5	646	397	0.891
C1+C2+C3+C4	642	402	0.89
C1+C2+C4+C5	631	407	0.86
C1+ C2+C3+C5	540	446	0.95
C1+C2+C3+C4+C5	630	407	0.92

The performance of the selected features in combination with random forest classifier is compared against LSTM-CNN algorithm proposed by Tadesse et al. [20] and attention relation networks solution proposed by Ji et al. [21]. The performance is tested against Reddit dataset. The results are given in Table 7.

Table 5.   Classification results with selected features in UMD Reddit dataset.

	Accuracy %	False positive %
Random Forest	94	3
SVM	93	2
Decision tree	93	2

Table 6.   Classification results with selected features in Reddit dataset.

	Accuracy %	False positive %
Random Forest	94.76	2.8
SVM	94	1.9
Decision tree	94	1.8

Table 7.   Results of performance.

Solutions	Precision	Recall	Accuracy	F1-score
Proposed	95.93	95.21	94.76	95.54
Tadesse et al.	94.8	90.5	91.7	92.6
Ji et al.	83.81	83.85	83.85	83.77

The proposed solution (selected features + random forest classifier) has atleast 3% higher accuracy compared to existing works. Consideration for features in both category of content and context with high discriminative ability to SI has increased the accuracy in proposed solution. But Tadesse *et al.* used only content features and Ji *et al.* used only context features for SI detection.

5 NOVELTY

This work presented a comprehensive evaluation of features with discriminative ability to detect SI. The work combined content based features extracted using Glove embedding with context features and evaluated the combinations in terms of their discriminative ability. Method to extract sentence incongruity and sentiment information features were proposed in this work. Clustering analysis based feature selection algorithm is proposed in this work. There has no existing works on analysis of discriminative ability of feature combination to SI in any of previous works.

6 CONCLUSION

An effective feature extraction technique is proposed for SI detection. The feature extraction technique considered both content and context features. The best feature combination with higher correlation to SI is found using clustering analysis. From the feature combination, the best feature values are found using symmetric uncertainty based entropy analysis. In the current work, the features considered were sentence based and this work did not explored the dependencies between sentences in long range over the entire text document. This context information can be extracted with deep learning based classifiers. Extracting sentence dependency information using deep learning techniques and exploring the relevance of features considered in this work with deep learning based features is in scope of future work.

REFERENCES

[1] Morese R. and Longobardi C. Suicidal Ideation in Adolescence: A Perspective View on the Role of the Ventromedial Prefrontal cortex. *Front Psychol.* (2020) 11:713.

[2] World Health Organization. World Health Statistics 2021: *Monitoring Health for the SDGs, Sustainable Development Goals. Industry and Higher Education.* Geneva, Switzerland: World Health Organization (2021).

[3] Klonsky E.D., May A.M. and Saffer B.Y. Suicide, Suicide Attempts, and Suicidal Ideation. *Annu Rev Clin Psychol.* (2016) 12:307–30.

[4] Kumar M., Dredze M. and Coppersmith G., De Choudhury M. Detecting Changes in Suicide Content Manifested in Social Media Following Celebrity Suicides. In: *The 26th ACM Conference.* ACM Press (2015).

[5] Carlyle K.E, Guidry J.P.D, Williams K., Tabaac A. and Perrin P.B. Suicide Conversations on InstagramTM: Contagion or Caring? *J Commun Healthc.* (2018)

[6] Jashinsky J., Burton S.H, Hanson C.L., West J., Giraud-Carrier C and Barnes M.D., *et al.* Tracking Suicide Risk Factors Through Twitter in the US. *Crisis.* (2014) 35:51–9.

[7] Wang J., Plöderl M., Häusermann M. and Weiss M.G. Understanding Suicide Attempts Among Gay Men from their Self-perceived Causes. *J Nerv Ment Dis.* (2015) 203:499–506.

[8] Jashinsky J., Burton S.H., Hanson C.L., West J., Giraud-Carrier C. and Barnes M.D., *et al.* Tracking Suicide Risk Factors Through Twitter in the US. *Crisis.* (2014) 35:51–9.

[9] Meshi D., Tamir D.I. and Heekeren H.R. The Emerging Neuroscience of Social Media. *Trends Cogn Sci.* (2015) 19:771–82.

[10] Ji S., Yu C.P., Fung S.F, Pan S. and Long G., "Supervised Learning for Suicidal Ideation Detection in Online User Content," *Complexity* (2018).

[11] Coppersmith G., Leary R., Whyne E. and Wood T., "Quantifying Suicidal Ideation Via Language Usage on Social Media," *In Joint Statistics Meetings Proceedings, Statistical Computing Section, JSM* (2015).

[12]   Mulholland M. and Quinn J., "Suicidal Tendencies: The Automatic Classification of Suicidal and Non-suicidal Lyricists Using Nlp." In *IJCNLP* (2013), pp. 680–684.

[13]   Huang X., Zhang L., Chiu D., Liu T., Li X. and Zhu T., "Detecting Suicidal Ideation in Chinese Microblogs with Psychological Lexicons," In *IEEE 11th International Conference on Ubiquitous Intelligence and Computing and Autonomic and Trusted Computing, and IEEE 14th International Conference on Scalable Computing and Communications and Its Associated Workshops (UTC-ATC-ScalCom). IEEE* (2014), 844–849.

[14]   Liakata M., Kim J.H., Saha S., Hastings J. and Rebholzschuhmann D., "Three Hybrid Classifiers for the Detection of Emotions in Suicide Notes," *Biomedical Informatics Insights*, vol. 2012, no. (Suppl. 1) (2012), 175–184.

[15]   Pestian J., Nasrallah H., Matykiewicz P., Bennett A. and Leenaars A., "Suicide Note Classification Using Natural Language Processing: A Content Analysis," *Biomedical Informatics Insights*, vol. 2010, no. 3 (2010), 19

[16]   Braithwaite S.R., Giraud-Carrier C., West J., Barnes M.D. and Hanson C.L., "Validating Machine Learning Algorithms for Twitter Data Against Established Measures of Suicidality," *JMIR Mental Health*, vol. 3, no. 2 (2016), e21

[17]   Nobles A.L., Glenn J.J., Kowsari K., Teachman B.A. and Barnes L.E., "Identification of Imminent Suicide Risk Among Young Adults Using Text Messages," In *Proceedings of the 2018 CHI Conference on Human Factors in Computing Systems* (2018), 1–11.

[18]   Coppersmith G., Leary R., Crutchley P., and Fine A., "Natural Language Processing of Social Media as Screening for Suicide Risk," *Biomedical Informatics Insights*, 2018.

[19]   Sawhney R., Manchanda P., Mathur P., Shah R. and Singh R., "Exploring and Learning Suicidal Ideation Connotations on Social Media with Deep Learning," In *Proceedings of the 9th Workshop on Computational Approaches to Subjectivity, Sentiment and Social Media Analysis* (2018), 167–175

[20]   Tadesse M.M., Lin H., Xu B. and Yang L., "Detection of Suicide Ideation in Social Media Forums Using Deep Learning," *Algorithms*, vol. 13, no. 1 (2020), 7

[21]   Ji S., Li X., Huang Z. and Cambria E., "Suicidal Ideation and Mental Disorder Detection with Attentive Relation Networks," arXiv preprint arXiv:2004.07601, 2020

[22]   Matero M., Idnani A., Son Y., Giorgi S., Vu H., Zamani M., Limbachiya P., Guntuku S. C. and Schwartz H.A., "Suicide Risk Assessment with Multi-level Dual-context Language and Bert," In *Proceedings of the Sixth Workshop on Computational Linguistics and Clinical Psychology* (2019), 39–44

[23]   Chen L., Aldayel A., Bogoychev N. and Gong T., "Similar minds post alike: Assessment of Suicide Risk using a Hybrid Model," In *Proceedings of the Sixth Workshop on Computational Linguistics and Clinical Psychology* (2019), 152–157

[24]   Wicentowski R. and Sydes M.R., "Emotion Detection in Suicide Notes Using Maximum Entropy Classification," *Biomedical Informatics Insights*, vol. 5 (2012), BII–S8972

[25]   Schoene A.M. and Dethlefs N., "Automatic Identification of Suicide Notes from Linguistic and Sentiment Features," In *Proceedings of the 10th SIGHUM Workshop on Language Technology for Cultural Heritage, Social Sciences, and Humanities* (2016), 128–133

[26]   George A., Ganesh H.B., Kumar M.A. and Soman K. Significance of Global Vectors Representation in Protein Sequences Analysis. *Computer Aided Intervention and Diagnostics in Clinical and Medical Images*: Springer (2019), 261–269

[27]   Hamilton W.L., Clark K., Leskovec J. and Jurafsky D. Inducing Domain-specific Sentiment Lexicons From Unlabeled Corpora. In *EMNLP*, vol. 2016 (2016), 595.

[28]   Ji S., Li X. and Huang Z. *et al.* Suicidal Ideation and Mental Disorder Detection with Attentive Relation Networks. *Neural Comput & Applic* (2021).

[29]   Camp E., *Sarcasm, Pretense, and the Semantics/Pragmatics Distinction.*. Noûs, vol. 46, no. 4 (2012), 587–634.

[30]   Ji S., Li X., Huang Z. *et al.* Suicidal Ideation and Mental Disorder Detection with Attentive Relation Networks. *Neural Comput & Applic* (2021).

[31]   http://users.umiacs.umd.edu/~resnik/umd_reddit_suicidality_dataset.html

[32]   Wang Y., Wan S. and Paris C., "The Role of Features and Context on Suicide Ideation Detection," In *Proceedings of the Australasian Language Technology Association Workshop* 2016 (2016), 94–102.

[33]   De Choudhury M., Kiciman E., Dredze M., Coppersmith G. and Kumar M., "Discovering Shifts to Suicidal Ideation from Mental Health Content in Social Media," In *Proceedings of the 2016 CHI Conference on Human Factors in Computing Systems. ACM* (2016), 2098–2110.

Automation and Computation – Vats et al. (Eds)
© 2023 the Author(s), ISBN 978-1-032-36723-1

A mathematical computation of LEACH and PEGASIS on the basis of energy consumed and overall time period in WSN

Ayushi Jain
Graphic Era Hill University/CSE, Dehradun, India

Kartik Bhatia
Pwc, India

ABSTRACT: The Wireless sensor networks (WSNs) are made up of spatially dispersed autonomous sensors that monitor environmental or physical elements like temperature, sound, and pressure. These sensors' data is collectively transmitted to a destination via network infrastructure. Because sensor nodes are energy-constrained devices, energy-efficient routing techniques have become more important. Regardless of how far wireless sensor networks (WSN) have progressed, efficient energy utilization is still required to extend the lifetime of the network. The network lifetime depreciates in real-time applications due to sensor node battery limits. Different routing techniques exist, and we have examined and interpreted the results of two hierarchical algorithms, Low Energy Adaptive Clustering Hierarchy (LEACH) and Power Efficient Gathering in Sensor Information Systems (PEGASIS) in this paper. The comparison is based on the amount of energy used each transmission and the WSN's overall lifetime, with simulations performed in MATLAB.

Keywords: Wireless Sensor Network, LEACH, PEGASIS

1 INTRODUCTION

Wireless Sensor Networks WSNs are by far the most advanced domains in terms of research work. They are collection of large mini wireless sensor nodes (SNs) which communicate with one another and collect relevant data, process it, and send it to a base station, which aids in health monitoring, environmental monitoring, border surveillance, and other applications. A sensor node, sometimes known as a mote, is a small device that collects, processes, and transmits sensory data to other linked nodes in the environment [1]. The processor unit, transceiver unit, sensing unit, and power unit are the four main components of the sensor node. The sensing unit uses sensors to collect data from the environment, and before sending the data to the processing unit, it uses the ADC to transform analogue signals received into digital signals. The processing unit is responsible for completing tasks, processing data, and is also connected to a tiny storage unit. A transceiver combines the functions of both the transmitter and the receiver into a single device. Connecting the node to the rest of the network is the transceiver's job. One of a sensor node's most important parts is the power unit. Though we use external power generators to power our sensors, the power units are the most concerned, and their energy is consumed at a faster rate. The collection and transmission process consumes energy, and the whole arrangement will function as long as the process has any remaining energy [2,3]. Because WSNs are frequently deployed in hostile areas with little or no maintenance, routing becomes crucial in both defense and civil applications to best utilize their power.

DOI: 10.1201/9781003333500-4

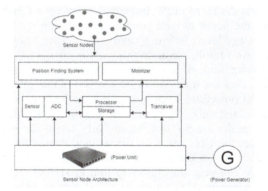

Figure 1. A typical sensor node.

The algorithms utilized determine the energy efficiency of the entire system; the better the algorithms used, the higher the efficiency and the longer the WSN's lifetime. Energy efficiency is a major challenge in wireless sensor networks, however Hierarchical routing strategies can help overcome it. As implied by their name, hierarchical routing protocols are built on the expansion of clusters or groups that are led by a cluster head [18]. The cluster head is often the sensor node in the cluster with the most energy. LEACH, PEGASIS, TEEN, APTEEN, VGA, and SOP are examples of some such algorithms which use hierarchical clustering techniques to nominate cluster heads who collect data from the environment and transfer data from the cluster to the base stations [17].

The findings of the experiments carried out on two such hierarchical procedures are presented in this paper. The paper's second section is a literature review on hierarchical routing protocols. The explanation of the two most prominent hierarchical protocols is covered in Section 3. LEACH and PEGASIS are two hierarchical routing protocols that are used in the experiments in Section 4. Finally, Section 5 represents the study's conclusion.

2 LITERATURE REVIEW/RELATED WORK

When numerous nodes sense, gather, and transmit data simultaneously in a network, the transmission may cause data collision, congestion, and a rapid decline in the energy of the network's sensors. Hierarchical routing techniques are thus used to limit data transmission between the nodes and the base station in order to lower the network's transmission overhead. Each node can deliver information straight to the cluster head and sink according to the self-organized routing mechanism LEACH [4].

A single-hop routing system called LEACH (Low Energy Adaptive Clustering Hierarchy) [4] protocol is self-organized and contains two phases: the setup phase and the steady phase. The network is organized into clusters in the first phase, which also searches for transmission. Data aggression, data compression, and data transmission are all done in the second phase. After a predetermined length of time, nodes start a new work round, choose a new cluster leader, and continue to circulate. This algorithm's flaw is that the residual energy isn't taken into account [17]. As a result, nodes with low beginning energy risk dying too soon or developing energy hole issues. Additionally, each time a cluster is formed, a large amount of energy is squandered. Additionally, nodes far from the base station may possibly experience an earlier demise since the cluster heads convey the gathered information directly to the base station.

The LEACH-C (LEACH centralized) method, which uses the centralized clustering algorithm in the first phase and the steady state phase of LEACH [17] in the second, is a better approach that was discussed in [5]. Instead of choosing a Cluster head on their own during the first phase, the nodes provide position information to the sink node. The sink considers the nodes' energy levels before adopting the cluster head's decision and broadcasting it to the nodes. As a result, the node with greater energy has a higher probability of being selected as the cluster head. However, because the nodes must send the data in a single hop to the base station, they use up a significant amount of energy.

Although the LEACH procedure saves energy, this method's primary flaw is the method of selecting cluster head randomly. The worst-case situation is that the data gathering may be impacted if the CH nodes are not distributed fairly across the nodes. A novel method called HEED [6] chooses the CHs based on the residual energy level and communication cost, which avoids the random selection of CHs. The intra-cluster communication is taken into consideration while choosing the cluster to join in HEED (Hybrid Energy Efficient Distributed Clustering Algorithm) [17]. Better load balancing and uniform CH distribution are provided by HEED. However, calculating the cost of intra-cluster communication requires information of the entire network. It does not randomly choose CH nodes, as LEACH does. Only very energetic sensors can develop into CH nodes. Overhead costs may result from the multiple iterations required for cluster creation in HEED. CHs close to the BS may potentially pass away sooner.

A different approach to cluster construction is called EECS (Energy Efficient Clustering Scheme), and it is based on the transmission distance between the CH and the normal node as well as the CH's distance from the sink [7]. To save energy during long-distance data transmission to the sink, a weighted function is applied to lower the size of the clusters that are the farthest away from the sink node. It is more likely that sensor nodes with greater remaining energy will be chosen as the CH.A predetermined number of nodes are first chosen as candidate nodes for CH. Then, in a specified amount of time, each candidate node tells its neighbors of their election. The candidate node will resign its candidacy if it discovers a stronger node—one with more energy left over—rather than becoming a cluster-head. Individuals are currently being invited to join each freshly established cluster by its cluster head. The energy consumption of regular nodes, the sink, and the CH are balanced at a particular point of the cluster formation process identified by the EECS. However, gathering data worldwide entails an additional burden on all sensors and necessitates a more thorough understanding of the gap between the CH and the sink. Additionally, CHs may perish as a result of the clusters around the BS becoming overloaded. Because they transmit a lot more traffic than remote nodes, CHs that are closer to the BS tend to perish sooner as a result of the previous algorithm's constraint.

A strategy to balance the energy usage among clusters was put out by EEUC (Energy Efficient Unequal Clustering) [8]. The goal was to keep cluster sizes close to the sink node significantly less than those of clusters farther away. More energy was conserved by doing this for both intra- and inter-cluster connections. The global information that each node in this scheme needs to know includes its coordinates and proximity to the sink node. It eliminates the issue of hot spots, lengthens the lifespan of the network, and distributes the load evenly among the nodes. The additional global data aggregation, however, raises overhead costs for all sensors and decreases network performance, particularly for multi-hop networks.

A heterogeneous network node is taken into account by EEHC (Energy Efficient Heterogeneous Clustered) [9] in terms of energy. The CH is selected in this scenario on the basis of the weighted election probabilities of each node and in accordance with the remaining energy. EEHC does not take in consideration the beginning energy of the nodes. Weighted probability is used to determine the requirements for becoming CH in each round. By taking into account the heterogeneous nodes and the leftover energy, the EEHC increases the network longevity. However, single hop interaction with the BS results in rapid energy depletion over time for the CHs.

The Distributed Fault Tolerant Clustering Algorithm (DFCA) [10] is a different approach that selects the CH based on how much energy the cluster's nodes still have. Sensor node with the most residual energy is selected as the CH. The procedure begins with boot-strapping, during which the node and gateways are given special identifiers. The hello message was produced by gateways and contains the gateway id, remaining energy, and distance. The two phases of the DFCA algorithm are the setup phase and the steady phase. Cluster formation process occurs during setup phase. Hello messages are sent by the gateway together with other data items. In steady phase networks, operations are further separated into rounds. Each round, the gateways collect information from member nodes and send it to the BS. The cluster nodes relocate to the neighboring CH for continued data flow if one CH fails for any reason [18]. This addresses the issue of the flaw. The strain on the cluster head and its energy consumption rise when cluster members who failed the CH shift to the neighboring CH. The network's lifespan is impacted by this.

As per [11], it claims that a scale-free topology evolution mechanism (SFTEM) and a regular hexagonal-based cluster scheme (RHCS) increase network survivability while maintaining energy balance. This method is more effective at handling faults and consuming less energy. For clustering sensor nodes, RHCS employs a regular hexagonal topology that provides at least 1-coverage fault-tolerance [16]. SFTEM connects clusters and builds a robust WSN that can endure a variety of flaws, including random failure and energy failure, by utilising the synergy between a dependable clustering strategy and topology development. The nodes of the network are represented using a Markov model in this method.

3 ROUTING ALGORITHMS

3.1 *LEACH (Low Energy Adaptive Clustering Hierarchy)*

According to its description, LEACH is a self-organizing and adaptive clustering protocol in which nodes create clusters using a distributed approach and independently decide on their own without any help from a central authority. There are two phases to this protocol [11]:

- Set-up phase
- Steady state phase

The network is organized into clusters in the first phase, which also advertises cluster heads and schedules transmission. For the following P rounds, where P is the required number of cluster heads, nodes that have already served as cluster heads are ineligible to do so. The probability of each node becoming a CH again is therefore 1/P [15]. Each node that isn't a cluster head chooses the closest cluster head, and at the conclusion of each round, that node joins that cluster. The cluster head is chosen using the formulas below:

$$T(n) = \begin{cases} \dfrac{p}{1 - p\left(r\left|\dfrac{1}{p}\right|\right)} & \text{if } n \in G \\ 0 & \text{otherwise} \end{cases}$$

Data aggregation, compression, and transmission to the sink all take place in the second phase. Before transferring an aggregated or fused packet directly to the base station or sink, the CHs compress data from nodes in the pertinent cluster [12]. Additionally, LEACH employs a TDMA/CDMA system to minimize collisions between and within clusters. The network returns to the setup phase and initiates a new round of CH elections after a pre-determined amount of time.

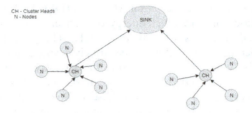

Figure 2.   Data transfer between the cluster head and the sink from the sensor nodes.

3.2 *PEGASIS (Power Efficient Gathering in Sensor Information System)*

In PEGASIS, each node takes turns acting as the leader for transmission to the BS and receives from and transmits to its immediate neighbors. The network's sensor nodes will get an equal share of the energy load using this method. The nodes are first distributed randomly across the field [18]. Then, the SNs themselves using a greedy method starting at some node, chain the nodes together or by using a third party. As an alternative, the BS may build this chain and spread it to all of the SNs. There are two stages [13]:

- Chain Formation: Using a greedy approach, the sensor nodes will chain together such that they can only connect with their nearest neighbours and relay information to the leader of all nodes. The farthest node from the sink is where it starts.

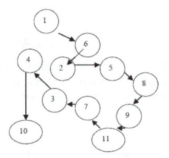

Figure 3.   Chain formation in PEGASIS.

- Data Gathering and Data Transmission: Each node on the chain makes a request for data from one of its neighbours during this phase, merges all of its own data with the requested data, and then sends the combined data to the other chain neighbour. The obtained data must be transmitted to the BS by the leader node [14]. Each communication round's leader will be chosen at random. In one round, the leader starts a straightforward control token passing procedure to start the transfer as from the chain's end.

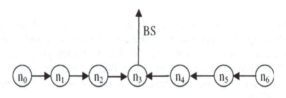

Figure 4.   Approach to pass token.

40

The algorithm outputs the graph below as when 100 nodes are deployed in a 100 x 200M deployment region.

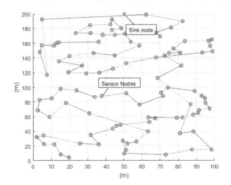

Figure 5.   PEGASIS WSN deployment.

4   IMPLEMENTATION OF ALGORITHMS

4.1   *Energy model*

Energy is needed to send and receive data every time. Energy dissipation is the energies dissipated in transmitting and receiving. The energy is calculated on the basis of following formulae.

$$Etx = Eelec * k + Eamp * k * d^2$$

$$Erx = (Eelec + EDA) * k$$

Where,

Etx: Transmitting dissipation (Joules/bit)
Erx: Receiving dissipation (Joules/bit)
Eamp: Power amplifier dissipation (Joules/bit/m^2)
EDA: Aggregation dissipation (Joules/bit)

4.2   *Simulation*

MATLAB is used to run our simulation. A network region measuring approximately 100m by 200m has 100 nodes dispersed evenly over it. The sensor nodes are regarded as working. The table below contains the simulation parameters.

4.3   *Observations*

As a result, using the above settings to simulate the network using the LEACH method, we obtain the graph shown below. The graph demonstrates that the LEACH algorithm can reach a maximum of 2932 iterations. At around 720 rounds, the first node died. Following the first sensor node failure, the remaining nodes slowly start to fail as well, and after about 2932 rounds, none are left alive. Energy is used by cluster heads for data transmission, reception, data compilation, and amplification.

41

Table 1. Simulation values.

Name	Value
Area of deployment	100M x 200m
Total nodes	100
Basic Energy	2J
Energy needed to operate circuitry	50*10^(-9)J
Energy used during transmission of data packet (Joules)	50*10^(-9)J
Energy used to receive the data packet (Joules)	50*10^(-9)J
Energy used in Amplifier (Joules)	100*10^(-12)J
Energy in Data Aggregation (Joules)	5*10^(-9)J
Size of Data(bits)	4000

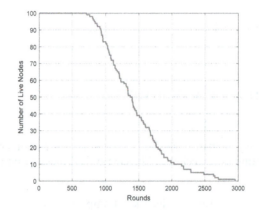

Figure 6. The number of nodes that remain alive per round in LEACH.

The following graph displays how much energy is used for each round. Nearly 0.1J is the value of least energy consumption. The maximum amount of energy used up until the first node died, which were roughly 720 rounds, was 0.69J. This algorithm uses a lot of energy as a result.

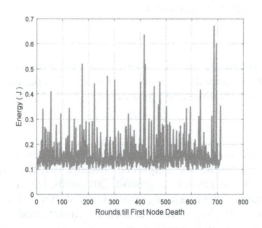

Figure 7. Amount of Energy used in each round in LEACH.

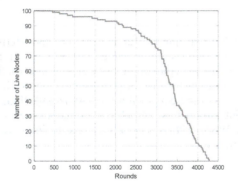

Figure 8.   Number of nodes that remain alive per round in PEGASIS.

Now, working on the network with the PEGASIS method and the identical settings as those given above, we obtain the graph displayed below. The graph demonstrates that the PEGASIS algorithm can reach a maximum of 4279 iterations. At around 470 rounds, the first node died. As soon as the first node fails, the other SNs also follow, and after 4279 rounds, none are still alive.

The following graph displays how much energy is used for each round. Nearly 0.06J is the value of the least amount of energy consumed. The maximum amount of energy used up until the first node died, which were roughly 470 rounds, was 0.075J.

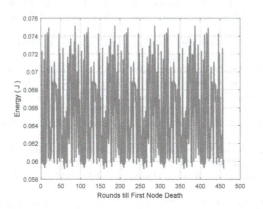

Figure 9.   Energy used in each round in PEGASIS.

4.4   *Results*

According to the observations, even though the first node fails in PEGASIS after 470 rounds, the algorithm continues to run for about 4200 rounds, while in LEACH, the first node fails after 720 rounds but the algorithm only lasts for about 2900 rounds. We can therefore see clearly from this observation that PEGASIS is more faults tolerant. The network becomes more dependable because to PEGASIS, which also lengthens the network's lifespan. Additionally, the hypothesis that PEGASIS uses less energy than the LEACH algorithm comes from comparing the energy use of the two methods. Therefore, PEGASIS is an algorithm that is both energy-saving and fault-tolerant.

5 CONCLUSION

It is particularly difficult to sustain longer due to the characteristics of WSNs and the environment in which sensor nodes are frequently put. But even in tough environments, we may extend network lifetime by deploying nodes using various routing algorithms. In wireless sensor networks, routing techniques help us save energy while also ensuring stability. The comparison of the two of these algorithms, LEACH and PEGASIS, reveals that, given the identical deployment parameters, PEGASIS produces superior outcomes to LEACH. With the expansion of WSN, efforts have been made to improve the algorithms' energy efficiency and fault tolerance.

REFERENCES

[1] Bian Q., Zhang Y., and Zhao Y., "Research on Clustering Routing Algorithms in Wireless Sensor Networks," 2010 *Int. Conf. Intell. Comput. Technol. Autom.* ICICTA 2010, vol. 2, pp. 1110–1113, 2010.

[2] Khediri S. E. L., Nasri N., Wei A., and Kachouri A., "A New Approach for Clustering in Wireless Sensors Networks Based on LEACH," *Procedia Comput. Sci.*, vol. 32, pp. 1180–1185, 2014.

[3] Baranidharan B. and Shanthi B., "A Survey on Energy Efficient Protocols for Wireless Sensor Networks," *Int. J. Comput. Appl.*, vol. 11, no. 10, pp. 35–40, 2010.

[4] Pandey S., "Energy Efficient Clustering Techniques for Wireless Sensor Networks-A Review," *Int. J. Sci. Res. Manag.*, vol. 5, no. 06, pp. 5454–5462, 2017.

[5] Boyinbode, O., Le, H. and Takizawa, M. (2011) 'A Survey on Clustering Algorithms for Wireless Sensor Networks', *Int. J. Space-Based and Situated Computing*, Vol. 1, Nos. 2/3, pp.130–136.

[6] Nooshin Nokhanji and Zurina Mohd Hanapi, "A Survey on Clustering Based Routing Protocols in Wireless Sensor Networks," *Journal of Applied Sciences*, vol.18, pp. 2011–2022, 2014.

[7] M. Kaur and P. Garg, "Improved Distributed Fault Tolerant Clustering Algorithm for Fault Tolerance in WSN," *Proc. – 2016 Int. Conf. Micro-Electronics Telecommun. Eng.* ICMETE 2016, pp. 197–201, 2017.

[8] Hu S. and Li G., "Fault-Tolerant Clustering Topology Evolution Mechanism of Wireless Sensor Networks," *IEEE Access*, vol. PP, no. c, p. 1, 2018.

[9] S. Misra and R. Kumar, "A Literature Survey on Various Clustering Approaches in Wireless Sensor Network," *2nd Int. Conf. Commun. Control Intell. Syst.* CCIS 2016, pp. 18–22, 2017.

[10] Priya Vyas and Manoj Chouhan, "Survey on Clustering Techniques in Wireless Sensor Network," *Int. J. Comput. Sci. ...*, vol. 5, no. July, pp. 238–243, 2014.

[11] Rajesh Chaudhary and Dr. Sonia Vatta, *"A Tutorial of Routing Protocols in Wireless Sensor Networks,"* vol. 3, no. 6, pp. 971–979, 2014.

[12] Gherbi C., Aliouat Z., and Benmohammed M., *A Survey on Clustering Routing Protocols in Wireless Sensor Networks*, vol. 37, no. 1. 2017.

[13] Chakravarty K., *"An Energy Balanced Algorithm of PEGASIS Protocol in WSN,"* no. 7, pp. 53–57, 2016.

[14] Lindsey S. and Raghavendra C. S., "PEGASIS: Power-efficient gathering in sensor information systems," *IEEE Aerosp. Conf. Proc.*, vol. 3, pp. 1125–1130, 2002.

[15] Wodajo, Hebron Hailu, *et al.* "Performance Evaluation of LEACH, PEGASIS and TEEN Routing Protocols in WSNs based on Network Life time." *International Journal of Advanced Networking an Applications"*, vol. 3, pp.4955–4961, 2021.

[16] Raut, Jaya, and Akhilesh Upadhyay. "Design of an Augmented Clustering Model using LEACH & PEGASIS for Low Complexity & High Network Lifetime." *2022 3rd International Conference for Emerging Technology (INCET)*, pp. 1–9, IEEE, 2022.

[17] Kumar, Deepak, and Thakur Y. S. "Analysis of Various Energy Efficiency Algorithms Used in Wireless Sensor Networks Leach, Sep, Teen, Deec, Pegasis And Fuzzy Logic," vol. 3, 2021.

[18] Bensaid R. and Boujemaa H., "A Combined Cluster-Chain based Routing Protocol for Lifetime Improvement in WSN," *2022 International Wireless Communications and Mobile Computing (IWCMC)*, 2022, pp. 542–547, doi: 10.1109/IWCMC55113.2022.9824293.

Automation and Computation – Vats et al. (Eds)
© *2023 the Author(s), ISBN 978-1-032-36723-1*

Anomaly detection in wireless sensor network using Generative Adversarial Network (GAN)

Vikas & B.B. Sagar
Computer Science & Engineering, Birla Institute of Technology, Mesra, Off Campus,
Noida, India

Manisha Manjul
Computer Science & Engineering, Delhi Skills & Entrepreneurship University, Delhi, India

ABSTRACT: Anomaly detection is a crucial concept across numerous research areas. Over the years, multiple efforts have been adopted to the arduous challenge of appropriately detecting and labelling anything unseen as unusual. Recently, this task has been solved with outstanding results utilizing the adversarial training method and generative adversarial networks (GANs). This article reviews the various GAN-based anomaly detection methods and discusses their advantages and disadvantages.

1 INTRODUCTION

Data patterns that deviate from a well-established definition of typical behaviors are known as anomalies. Wireless Sensor Networks (WSNs) [1–3] security is a difficult and important task. Anomaly detection is a critical barrier to assuring the security of WSNs. WSNs [4,5] are vulnerable to a variety of threats that could devastate the node and produce false results. The detection of such erroneous data is crucial to reduce false alarms. Machine learning [6] techniques are increasingly being used to identify suspicious data. The bulk of machine anomaly detection methods now in use require the node to retain all training data because they function in fixed environments. Since Ian Goodfellow's initial GAN proposal *et al.* in 2014 [7], many academics have come up with various creative concepts for applying GAN to find abnormalities in wireless sensor networks. The main benefit of adversarial nets is that no Markov chains are ever needed to compute gradients; just back-propagation will do. Additionally, no inference is required when learning, and the model may be simply expanded to include a variety of interactions and parameters.

This GAN property shows that, despite the fact that their application has only recently been investigated, they can be employed effectively for anomaly identification. The tasks involved in anomaly identification using GANs [8] include modelling normal behavior using an adversarial training process and identifying anomalies by assessing an anomaly score. To the best of our knowledge, the BiGAN architecture has been proposed in the Adversarial Feature Learning concept [9], which is the foundation for all GAN-based techniques to anomaly detection. The GAN framework initially learns a generator that maps samples from any latent distribution (noise prior) to data as well as a discriminator that aims to differentiate between real and generated samples.

DOI: 10.1201/9781003333500-5

2 GENERATIVE ADVERSARIAL NETWORK (GAN)

The first GAN framework includes the min-max game problem. The discriminator and the generator are constantly competing with one another in the min-max problem. The primary goal is to raise the likelihood that created images are phoney and actual photos are recognized as real by the discriminator, which returns a value D(x) indicating the possibility that x is an actual image. Cross-entropy is used to determine the loss. The goal is listed below:

$$\max_{D} V(D) = \mathbb{E}_{x \sim p_{\text{data}}(x)}[\log D(x)] + \mathbb{E}_{z \sim p_z(z)}[\log(1 - D(G(z)))]$$

recognize real images better   recognize generated images better

The objective function of the generator side wants the model to generate images with the highest value discriminator feasible in order to deceive the discriminator.

$$\min_{G} V(G) = \mathbb{E}_{z \sim p_z(z)}[\log(1 - D(G(z)))]$$

Optimize G that can fool the discriminator the most.

It is reasonable to think of the GAN as a minimax game in which G attempts to keep V as low as possible and D seeks to raise it as much as possible.

$$\min_{G} \max_{D} V(D, G) = \mathbb{E}_{x \sim p_{\text{data}}(x)}[\log D(x)] + \mathbb{E}_{z \sim p_z(z)}[\log(1 - D(G(z)))].$$

GAN implementation has taken many various forms, and the field of GAN research is now quite active. Following is a list of some of the significant ones that are now in use:

a) VANILLA GAN [7]: This GAN type is the most basic. The Generator and Discriminator in this scenario are straightforward multi-layer perceptron's. To try and optimise the mathematical challenge, vanilla GAN employs straightforward stochastic gradient descent.

b) Conditional GAN (CGAN) [10]: CGAN is a deep learning technique that employs a set of conditional parameters. In CGAN, the Generator is given an extra parameter, "y," to produce the necessary data. In order for the discriminator to aid distinguish between the real data and the phoney generated data, labels are additionally added to the input.

c) Deep Convolutional GAN (DCGAN) [11]: DCGAN is among the most well-liked and effective GAN implementations. ConvNets are used instead of multi-layer perceptron's in its construction. Convolutional stride actually replaces max pooling in the ConvNets implementation. The layers are also not entirely connected.

d) Laplacian Pyramid GAN (LAPGAN) [12]: The Laplacian pyramid is a linear invertible image representation made up of a group of band-pass images separated by an octave and a low-frequency residual. This method makes use of several Generator and Discriminator networks as well as various Laplacian Pyramid levels. The major reason this method is employed is that it results in photographs of extremely high quality. The image is initially downscaled at each pyramidal layer before being upscaled at each layer once again in a backward pass. Up until it reaches its original size, the image adds noise from the Conditional GAN at each of these layers.

e) Super Quality GAN (SRGAN) [13]: As the name implies, SRGAN is a method for creating a GAN that uses both to create higher quality photos, a deep neural network, and an adversarial network is used. This particular sort of GAN is very helpful in enhancing details in native low-resolution photos while minimising mistakes.

3   GANS FOR ANOMALY DETECTION

Security is a major issue in WSNs, as it affects the confidentiality, authenticity and reliability of sensor data transmission can be ensured. A discriminator plus a generator makes up a GAN. To deceive potential attackers, the generator produces phoney data with characteristics similar to the real data (in this situation, the WSN does not need to produce fictitious data to dramatically lower power usage.). The discriminator is configured accordingly to discriminate between true data and identify abnormalities for further processing. Although GAN is used in various research field and many researchers have come-up with many ideas to use GAN in their respective domain. To detect an anomaly using GAN are as follows:

a) The authors [14] suggest using deep generative adversarial networks for anomaly identification. By simultaneously training a generative model and a discriminator, it is possible to enable the detection of anomalies in previously unnoticed data utilising unsupervised learning algorithm on holistic data. Results demonstrate that our method can identify a variety of recognised anomalies, including retinal uid and HRF, which have never been observed during training. As a result, it is anticipated that the model will be able to identify new anomalies. Results show strong susceptibility and the capacity to group irregularities, despite the limitations of false positives do not take into account unique anomalies, it uses quantitative analysis using a selection of anomaly classifications. Scale-based anomaly detection enables data mining for applicants for markers, subject to later authentication.

b) The use of GAN has been investigated by authors to concurrently construct a deep learning network to simulate the classification of many sensor streaming data in a CPS under standard operation circumstances and another to identify abnormalities resulting from cyberattacks launched in an unsupervised manner against the CPS. Authors have suggested a unique GAN-based Anomaly Detection (GANAD) approach to use immediately challenging CPS dataset from a Secure Water Treatment Testbed as a researchers demonstrated that the suggested GAN-AD [8] was able to beat existing unsupervised detection methods using a discriminator and generator trained on multivariate time series to detect anomalies. techniques (SWaT), which was used to evaluate our methodology

c) The authors have looked into the usage of GAN for time-series data produced by CPSs multivariate anomaly detection. To train LSTM-RNNs on multivariate time-series data, we developed a unique MAD-GAN [15] (Multivariate Anomaly Detection with GAN) architecture. To discover anomalies, authors created a novel Discrimination and Reconstruction Anomaly Score using the discriminator and the generator (DR-Score). The authors used the Water Distribution System and the Secure Water Treatment Testbed (SWaT) (WADI) complex cyber-attack CPS datasets to test MAD-GAN, which outperformed other unsupervised detection techniques, entail a GAN-based solution.

d) Efficient GAN-based Anomaly Detection [16] Recent GAN models can be used to achieve state-of-the-art performance for anomaly detection on high-dimensional, complex datasets while being effective at test time. The usage of a GAN that concurrently learns an encoder serves as proof of this point because it does away with the necessity for a costly technique to retrieve the latent representation for a particular input.

Table 1 shows only supervised learning was supported by GANs like Vanilla GAN and Conditional GAN. Later versions added support for semi-supervised and unsupervised learning. Additionally, original adversarial frameworks used multilayer perceptron's; later, experiments included convolutional networks, autoencoders, and deep neural networks [17]. Additionally, for the majority of the models, optimization based on stochastic gradient descent was utilized to train both the generator and discriminator networks. Any adversarial network's main goal will always be to win a 2-player minimax game. Some models also have secondary goals like learning representations and features through associated semantic tasks, and leveraging the subsequently learned features for unsupervised classification or recognition. Laplacian pyramids and recurrent networks are used by the generator to produce visuals sequentially were also introduced by models like LAPGAN.

Table 1. Compares multiple GAN variants using a variety of metrics.

GAN Criteria	VANILLA GAN	CGAN	DCGAN	LAPGAN	INFOGAN	BIGAN
Type of Leaning	Supervised	Supervised	Unsupervised	Unsupervised	Supervised	Supervised & Unsupervised
Architecture	Multilayer	Multilayer	Convolution networks	Laplacian Pyramid	Multilayer	Neural Network
Objective	Reduce function value for G and increase D	Reduce function value for G and increase D	Classification representation for G and D	Generation of Images	Increase mutual infomlation and gain understanding of disentangled representation	Learn semantic function features, then apply them in unsupervised situations
Metrics	Log-Likelihood	Log-Likelihood	preciseness and error rate	Log-Likelihood and assessing persons	Learning from infomlation factors and depictions	preciseness

Additionally, older models used log-likelihood estimates to measure model performance, which was dropped in later iterations since they were poor estimates. Instead, the performance of a model was assessed using accuracy and error rates. Now days digital data [18–24] is being generated days by days we need to make sure to identify the fake an original data, GAN can help in that.

4 CONCLUSION

Based on their technique, design, and performance, the many Generative Adversarial Networks versions are compared in this research in a much-needed manner. It is clear that compared to the initial version, later adversarial networks are more reliable and have a wide range of applications. These networks can also be helpful in a variety of ways for picture classification, recognition, capture, and production.

REFERENCES

[1] Sharma V. *et al.*, "OGAS: Omni-directional Glider Assisted Scheme for Autonomous Deployment of Sensor Nodes in Open Area Wireless Sensor Network," *ISA Trans.*, Aug. 2022, doi: 10.1016/j.isatra.2022.08.001.

[2] Sharma V., Patel R.B., Bhadauria H.S., and Prasad D., "NADS: Neighbor Assisted Deployment Scheme for Optimal Placement of Sensor Nodes to Achieve Blanket Coverage in Wireless Sensor Network," *Wirel. Pers. Commun.*, vol. 90, no. 4, pp. 1903–1933, 2016.

[3]   Sharma V., Patel R.B., Bhadauria H.S., and Prasad D., "Policy for Planned Placement of Sensor Nodes in Large Scale Wireless Sensor Network," *KSII Trans. Internet Inf. Syst.*, vol. 10, no. 7, pp. 3213–3230, 2016.

[4]   Sharma V., Patel R.B., Bhadauria H.S., and Prasad D., "Deployment Schemes in Wireless Sensor Network to Achieve Blanket Coverage in Large-scale Open Area: A Review," *Egypt. Informatics J.*, vol. 17, no. 1, pp. 45–56, 2016.

[5]   Sharma V., Patel R.B., Bhadauria H.S., and Prasad D., "Glider Assisted Schemes to Deploy Sensor Nodes in Wireless Sensor Networks," *Rob. Auton. Syst.*, vol. 100, pp. 1–13, 2018.

[6]   Vats S., Sagar B.B., Singh K., Ahmadian A., and Pansera B.A., "Performance Evaluation of an Independent Time Optimized Infrastructure for Big Data Analytics That Maintains Symmetry," *Symmetry (Basel)*, vol. 12, no. 8, 2020, doi: 10.3390/SYM12081274.

[7]   Goodfellow I., Pouget-Abadie J., Mirza M., Xu B., Warde-Farley D., Ozair S., Courville A., and Bengio Y. *Generative Adversarial Nets.* pp. 2672–2680, 2014.

[8]   Schlegl T., Seebӧck P., Waldstein S.M., Schmidt-Erfurth U., and Langs G. *"Unsupervised Anomaly Detection with Generative Adversarial Networks to Guide Marker Discovery."* abs/1703.05921, 2017.

[9]   Donahue J., Krähenbühl P., and Darrell T. *"Adversarial Feature Learning."* abs/1605.09782, 2016.

[10]   Mirza M., and Osindero S. *"Conditional Generative Adversarial Nets."* arXiv preprint arXiv:1411. 1784, 2014.

[11]   Radford A., Metz L., and Chintala S. "Unsupervised Representation Learning with Deep Convolutional Generative Adversarial Networks", *4th International Conference on Learning Representations, ICLR*, 2016.

[12]   Denton E., Chintala S., Szlam A. and Fergus R. "Deep Generative Image Models using a Laplacian Pyramid of Adversarial Networks", *Advances in Neural Information Processing Systems*, 2015.

[13]   Maqsood M.H., Mumtaz R., Haq I.U., Shafi U., Zaidi S.M.H., and Hafeez M. "Super Resolution Generative Adversarial Network (SRGANs) for Wheat Stripe Rust Classification." *Sensors* 2021, 21, 7903. https://doi.org/10.3390/s21237903

[14]   Li D., Chen D., Goh J., and Ng S.-k. "Anomaly Detection with Generative Adversarial Networks for Multivariate Time Series", *7th International Workshop on Big Data, Streams and Heterogeneous Source Mining: Algorithms, Systems, Programming Models and Applications on the ACM Knowledge Discovery and Data Mining Conference*, August 2018.

[15]   Li D., Chen D., Jin B., Shi L., Goh J., and Ng S.-K., "MAD-GAN: Multivariate Anomaly Detection for Time Series Data with Generative Adversarial Networks" *Artificial Neural Networks and Machine Learning – ICANN 2019: Text and Time Series: 28th International Conference on Artificial Neural Networks*, Munich, Germany, September 17–19, 2019, Proceedings, Part IV Sep 2019. pp. 703–716, https://doi.org/10.1007/978-3-030-30490-4_56

[16]   Zenati H., Foo C.S., Lecouat B., Manek G., and Chandrasekhar V.R., "Efficient GAN-Based Anomaly Detection", *IEEE ICDM*, 2018, https://arxiv.org/abs/1802.06222

[17]   Vats S., Singh S., Kala G., Tarar R., and Dhawan S., "iDoc-X: An Artificial Intelligence Model for Tuberculosis Diagnosis and Localization," *J. Discret. Math. Sci. Cryptogr.*, vol. 24, no. 5, pp. 1257–1272, 2021.

[18]   Vats S. and Sagar B.B., "An Independent Time Optimized Hybrid Infrastructure for Big Data Analytics," *Mod. Phys. Lett. B*, vol. 34, no. 28, p. 2050311, Oct. 2020, doi: 10.1142/S021798492050311X.

[19]   Vats S. and Sagar B.B., "Performance Evaluation of K-means Clustering on Hadoop Infrastructure," *J. Discret. Math. Sci. Cryptogr.*, vol. 22, no. 8, 2019, doi: 10.1080/09720529.2019.1692444.

[20]   Agarwal R., Singh S., and Vats S., "Implementation of an Improved Algorithm for Frequent Itemset Mining Using Hadoop," *in 2016 International Conference on Computing, Communication and Automation* (ICCCA), 2016, pp. 13–18. doi: 10.1109/CCAA.2016.7813719.

[21]   Agarwal R., Singh S., and Vats S., *Review of Parallel Apriori Algorithm on Mapreduce Framework for Performance Enhancement*, vol. 654. 2018. doi: 10.1007/978-981-10-6620-7_38.

[22]   Bhati J.P., Tomar D., and Vats S., "Examining Big Data Management Techniques for Cloud-based IoT Systems," in *Examining Cloud Computing Technologies Through the Internet of Things*, IGI Global, 2018, pp. 164–191.

[23]   Bhatia M., Sharma V., Singh P., and Masud M., "Multi-level P2P Traffic Classification Using Heuristic and Statistical-based Techniques: A Hybrid Approach," *Symmetry (Basel)*, vol. 12, no. 12, p. 2117, 2020.

[24]   Vats S. and Sagar B.B., *"Data Lake: A Plausible Big Data Science for Business Intelligence,"* 2019.

Automation and Computation – Vats et al. (Eds)
© *2023 the Author(s), ISBN 978-1-032-36723-1*

Novel node deployment strategies for wireless sensor networks

Anvesha Katti

The NorthCap University, Gurugram, Haryana, India

ABSTRACT: Wireless Sensor Network (WSN) is a self configuring network and monitors conditions such as temperature and pressure in the physical environment and helps the sensors transfer the data through the sensor nodes to a sink. Deployment is one of the major challenges in WSN that has a demanding effect on coverage and connectivity of WSN. Sensor network has many characteristics amongst which the coverage is the most important. Coverage is defined to be a marker for effective monitoring of a network by sensors. We present deployment schemes which increase coverage in WSN. We also find the count of sensor nodes used in each scheme.

1 INTRODUCTION

We can define Wireless Sensor Networks (WSN) as a network containing sensors which are self governed and keep track of various conditions in the environment, like temperature, sound, pressure, and transfers the data through the network to a sink. Size and cost of the sensors are by far the most challenging areas wherein most of the research is being done [1].

Coverage indicates monitoring of a set of sensors, and it is one of the primary research areas in wireless sensor networks. Coverage can be distributed into following categories- Area Coverage and Point Coverage. While considering Area Coverage, we monitor a given area using sensors. On the other hand, considering point coverage entails monitoring sensors in a particular region. Usually, a disk or sphere models the range of sensing for a particular sensor. As is evident, the disk model encompasses 2D space while the sphere model points towards a 3D space. We shall assume that the sensor is located in the center. How the sensors are distributed plays a major role in increasing the efficiency of WSN and maximizing coverage [2].

This paper presents diverse sensor node deployment schemes such as, firstly, the square and secondly, the triangular sensor node deployment. For each of the aforementioned deployment schemes, we investigate, the coverage and total of sensor nodes required.

2 LITERATURE REVIEW

Deployment of sensors is one of the most researched areas which aims at coverage maximization, and connectivity improvement [3].

Ant Colony optimization has been used by Liu and He [4] for obtaining the maximum coverage in grid based WSN. Monte Carlo method has been used along with genetic algorithms for extracting the maximum amount of coverage by Yourim and Yong-Hyuk [5]. Abdelhamid *et al.* [6] has maximized the coverage by yet another novel technique using Artificial Potential Fields-based algorithm (APF). Ref. [9] makes use of geometry and graph theory for finding out coverage. It mainly deals with mobile sensor nodes and self-localization.

Ref [10] presents coverage maximization technique using radar and sonar perspective. In Ref [11,12] coverage maximization is done using ILP. This approach cannot be used for

    DOI: 10.1201/9781003333500-6

large problem instances. Further, Ref. [17] introduces corona and non corona based node deployment strategies. [18] introduces regular deployment in 3D terrains. The authors in Ref. [19] introduce a new approach to solve for deterministic and grid-based deployment. Ref. [20] uses the Glowworm swarm optimization approach (**LBR-GSO**) for effective node deployment.

3 THEORETICAL MODELLING FOR SPECIFIC SENSOR NODE DEPLOYMENTS

Coverage finds place in one of the extensive areas of research in WSN. It details how efficiently a network field is managed by sensor nodes. There is a finite sensing range for every sensor in WSN. The union of sensing range is known as sensing coverage area in a network. [14]. There are many applications of sensor networks such as monitoring of climate and natural surroundings to monitoring of wildlife habitat and surveillance in military and battlefields, etc. Deployment strategy forms the basis for maximum coverage. In a sensor network, coverage can also be interpreted to be a marker for the quality of service.(QoS). It relates to how well each point is covered by sensors in the sensing field. Deployment of sensors is either regular or random. Regular deployment consists of uniform sensor nodes across the field. When the sensors are dropped randomly it is known as random deployment. This is mainly done in hostile environments. If uniform coverage is required then regular deployment of nodes is used. We shall study the coverage for regular placement of sensor nodes. In addition, we shall consider immovable sensor nodes, i.e., the sensor nodes are not movable once deployed. Regular deployment is better when compared to random deployment, in terms of coverage and connectivity. Regular deployment of nodes helps maximise the coverage area due to the uniform placement of nodes. The technically correct definition of coverage prediction can be said to be as follows: it is the area enclosed by the node divided by the total area of deployment [15]. Deployment of nodes is used so that the coverage is maximized [3].

We have studied the most suitable distribution scheme of sensors with the total of sensor nodes required in each case.

3.1 *Triangular sensor node deployment*

In triangular node deployment a square shaped field has been considered with the edges. The whole square shaped area has been divided into various equilateral triangles. Figure 1 illustrates these concepts clearly. The midpoint of each side of the triangle has sensor placed on it as shown in Figure 1. Range of sensing for any node(r) can take value between 0 and Rmax. Here Rmax can be defined to be maximum radius of sensing and r = 0 is the minimum. We shall assume a similar sensing range for each sensor which can be represented as a circle radius. Coverage can be defined to be sensing area divided by the total area [15]. The coverage for triangular sensor deployment can be written as:

$$C_f = \frac{\pi r^2}{2\sqrt{3}R_{max}^{\,2}} \tag{1}$$

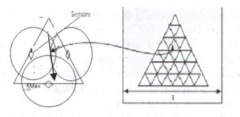

Figure 1.   Triangle node deployment.

3.2 Square sensor node deployment

In square node deployment a square shaped field has been considered with the edge s. The whole square shaped area has been further sub divided into various smaller squares. Figure 2 illustrates these concepts clearly. The midpoint of each side of the square has sensor placed on it as shown in Figure 2. Range of sensing for any node(r) can take value between 0 and Rmax. Here Rmax can be defined to be maximum radius of sensing and r = 0 is the minimum. We shall assume a similar sensing range for each sensor which can be represented as a circle radius.

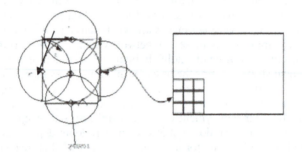

Figure 2.   Square node deployment.

Coverage can be defined to be sensing area divided by the total area [15]. The coverage for square sensor deployment can be written as:

$$C_f = \frac{3\pi r^2}{4Rmax^2} \tag{2}$$

3.3 Sensor nodes estimation

By putting r = Rmax in (1) and (2), the maximal coverage for triangle is $3\pi/(6\sqrt{3})$, and the maximal coverage prediction for square is $3\pi/4$ [15].

We can proximate the number of sensor nodes needed using Triangle distribution method as

$$N_{s(Triangle)} = 3a^2/2 \tag{3}$$

Number of sensor nodes needed using Square distribution as

$$N_{s(Square)} = 3a^2/4 \tag{4}$$

4   RESULTS AND DISCUSSION

This section presents the simulation results done on Matlab. The graphs depict the coverage with different different deployments on Coverage Prediction and the Number of Sensors required for coversge. In the simulation, the entire sensing field is considered to be square with area A = 500*500 m². The maximum sensing radius Rmax is estimated to be 20 m. The sensing field is distributed into small equilateral triangular, square sub-regions of side 20m.

In Figure 3, the coverage is higher in case of Triangle Distribution than Square distribution. From the figure, we can infer the coverage for each of the distributions for various values of the sensing radius. The figure clearly indicates that the coverage for the triangular

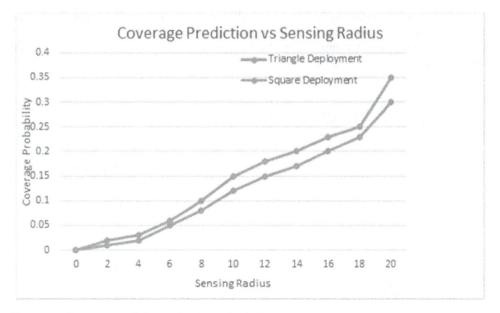

Figure 3. Coverage prediction vs Sensing radius(m).

distribution is more when compared to the square distribution. It is thus construed that the triangular deployment is most advantageous and results in better coverage when compared with the square deployment.

Figure 4 shows the variation of coverage with normalized sensing radius(defined as r/ Rmax) for diverse types of node distribution. Triangular distribution results in maximum coverage prediction and the square distribution results in comparatively lesser coverage.

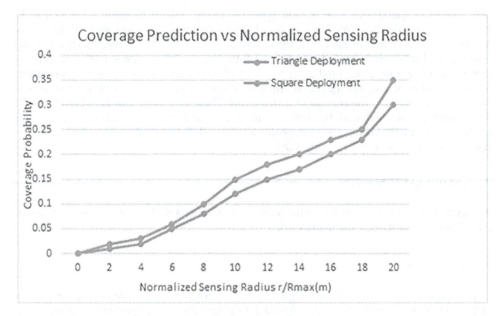

Figure 4. Coverage prediction vs Normalised sensing radius.

Figure 5 shows the relationship between number of sensors and sensing radius for the two distribution schemes. We observe that the Square distribution corresponds to a higher number of sensors when compared to the triangular distribution for covering a certain sensing range.

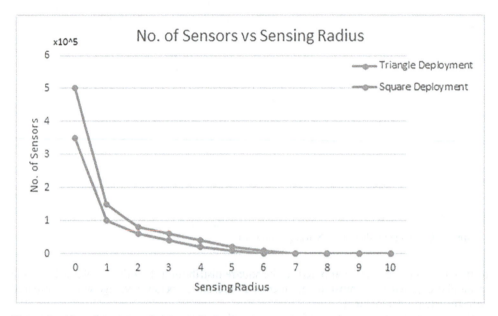

Figure 5.   No. of sensors vs Sensing radius.

5   CONCLUSION AND FUTURE SCOPE

We have investigated several important types of sensor node distribution schemes for the regular node distribution. We have explored the triangular and square node distributions. We have also compared in each case, the coverage and the number of sensors required. We find that higher coverage prediction has been found in case of the triangular node distribution while a lower coverage prediction has been obtained for the square node distribution. Square distribution uses the maximum number of sensors while the least number of sensors has been considered in the triangular distribution. There is a scope for studying many different adaptations which shall be attempted in future. Future work concerns deeper analysis of these distribution schemes, new proposals to design different methods, for deployment of sensors which can increase coverage further.

ACKNOWLEDGEMENTS

The author would like to acknowledge fruitful discussions with Dr. Aavishkar Katti which have helped in improving the manuscript.

REFERENCES

[1]   "Wireless sensor network," Wikipedia. [Online]: https://en.wikipedia.org/wiki/Wireless_sensor_network.
[2]   Li D. and Liu H. Sensor Coverage in Wireless Sensor Networks," Wirel. Networks Res. Technol. Appl., pp. 3–31, 2009.

[3] Abdollahzadeh S. and Navimipour N.J. Deployment Strategies in the Wireless Sensor Network: A Comprehensive Review. *Comput. Commun.*, vol. 91, pp. 1–16, 2016.

[4] Liu X. and He D. Ant Colony Optimization with Greedy Migration Mechanism for Node Deployment in Wireless Sensor Networks. *J. Netw. Comput. Appl.*, vol. 39, pp. 310–318, 2014.

[5] Yoon Y. and Kim Y.-H. An Efficient Genetic Algorithm for Maximum Coverage Deployment in Wireless Sensor Networks. *IEEE Trans. Cybern.*, vol. 43, no. 5, pp. 1473–1483, 2013.

[6] Senouci M.R., Mellouk A., and Assnoune K. Localized Movement-assisted Sensor Deployment Algorithm for Holedetection and Healing. *IEEE Trans. parallel Distrib. Syst.*, vol. 25, no. 5, pp. 1267–1277, 2014.

[7] Musman S.A., Lehner P.E., and Elsaesser C. Sensor Planning for Elusive Targets, *Math. Comput. Model.*, vol. 25, no. 3, pp. 103–115, 1997.

[8] Howard A., Matarić M.J., and Sukhatme G.S. *Mobile Sensor Network Deployment using Potential Fields: A Distributed, Scalable Solution to the Area Coverage Problem in Distributed Autonomous Robotic Systems 5*, Tokyo: Springer Japan, pp. 299–308, 2002.

[9] Meguerdichian S., Koushanfar F., Potkonjak M., and Srivastava M.B., *Coverage Problems in Wireless Ad-hoc Sensor Networks.*

[10] Priyantha N.B., Chakraborty A., and Balakrishnan H. The Cricket Location-support System. In *Proceedings of the 6th annual international conference on Mobile computing and networking - MobiCom' 00*, pp. 32–43, 2000.

[11] Chakrabarty K., Iyengar S.S., Hairong Qi, and Eungchun Cho. Coding Theory Framework for Target Location in Distributed Sensor Networks. In *Proceedings International Conference on Information Technology: Coding and Computing*, pp. 130–134.

[12] Chakrabarty K., Iyengar S.S., Hairong Qi, and Eungchun Cho. Grid Coverage for Surveillance and Target Location in Distributed Sensor Networks, *IEEE Trans. Comput.*, vol. 51, no. 12, pp. 1448–1453, 2002.

[13] Dhillon S.S., Chakrabarty K., and Iyengar S. S. Sensor Placement for Grid Coverage Under Imprecise Detections. In *Proceedings of the Fifth International Conference on Information Fusion. FUSION 2002. (IEEE Cat.No.02EX5997)*, vol. 2, pp. 1581–1587, 2002.

[14] Wang Y., Zhang Y., Liu J., and Bhandari R. Coverage, Connectivity, and Deployment in Wireless Sensor Networks. In *Recent Development in Wireless Sensor and Ad-hoc Networks*, Springer, pp. 25–44, 2015.

[15] Katti A. and Lobiyal D.K. Sensor Node Deployment and Coverage Prediction for Underwater Sensor Networks in Computing for Sustainable Global Development (INDIACom), *2016 3rd International Conference on*, pp. 3018–3022, 2016.

[16] Kumar S. and Lobiyal D.K. Sensing Coverage Prediction for Wireless Sensor Networks in Shadowed and Multipath Environment, *Sci. World J.*, vol. 2013, 2013

[17] Rahman, A.U., *et al.* Corona Based Deployment Strategies in Wireless Sensor Network: A Survey." *Journal of Network and Computer Applications*, vol. 64, pp. 176–193, 2016.

[18] Kumar, P., and Reddy S.R.N., Wireless Sensor Networks: A Review of Motes, Wireless Technologies, Routing Algorithms and Static Deployment Strategies for Agriculture Applications." *CSI Transactions on ICT*, vol. 8, no. 3, pp. 331–345, 2020.

[19] Liu, X. A Deployment Strategy for Multiple Types of Requirements in Wireless Sensor Networks." *IEEE Transactions on Cybernetics*, vol. 45, no. 10, pp. 2364–2376, 2015.

[20] Sampathkumar, A., Mulerikkal J., and Sivaram M. Glowworm Swarm Optimization for Effectual Load Balancing and Routing Strategies in Wireless Sensor Networks. *Wireless Networks*, vol. 26, no. 6, pp. 4227–4238, 2020.

Automation and Computation – Vats et al. (Eds)
© 2023 the Author(s), ISBN 978-1-032-36723-1

Fog-based IoT-enabled system security: A survey

Sanjay Kumar Sonker & Bharat Bhushan Sagar
CSE, Birla Institute of Technology Mesra, Off-Campus Noida, India

ABSTRACT: Internet of things is an emerging innovative technology that in order to increase the efficiency, potential, and smooth operation of devices regardless of physical barriers, connects them to internet. Presently, IoT technology is being applied to homes appliances, monitor our health easily and accurately and in the field of power energy distribution using with Smart grid system more easily and effectively. Areas that present the predominant application of IoT technology are homes and offices appliances, transportation, healthcare, telecommunications, agriculture, the rapid expansion of IoT devices presents some security related new challenges. Heightening uses of IoT devices leads to huge accumulation of data on clouds increasing its loads. Most importantly, this decline in capacity of IoT devices poses some grievous security threats. The goal of this paper is to present a detailed review of the security-related challenges solutions in Fog-based IoT.

Keywords: Internet of Things (IoT), Fog /Edge computing, Machine learning

1 INTRODUCTION

Internet of Things is a recent technological phenomenon which has led to wireless and remote access to and control of devices through internet. [1] Vision of IoT-"things or objects" becoming more intelligent, efficient, smart and behaving active. In this era of communication IoT devices are fast gaining popularity, having a substantial impact on daily life, and helping key decision-makers in the transportation, healthcare, and power grid sectors [4]. By 2025, it is predicted that there will be 25.4 billion Internet of Things (IoT) connected devices deployed globally, a significant increase from the 8.74 billion units anticipated in 2020 [5]. Today, the Internet of Things – a large network of connected objects gathering and analyzing data and autonomously completing activities. Thanks to the development of data communication technologies and data analytic [5]. Figure 1: show the many levels of blocks and layers that make up an IoT-enabled system.

1. Perception Layer	Sensor and Actuators, Mechanisms
2. Connectivity Layer	Ethernet, Wi-Fi, NFC, Bluetooth,ZigBee,LPN, Cellular Network, Messaging protocols
3. Fog/ Edge Computing Layer	Data Accumulation, Data Abstraction
4. Processing layer	Accumulating, sorting, processing
5. Application Layer	QoS, Data management
6. BusinessLayer	Deliberate plans based on accurate data
7. Security Layer	Fog/edge based IDS system, privacy preservation

Figure 1.   Buildings blocks and layers of IoT architecture.

DOI: 10.1201/9781003333500-7

Perception Layer: Another name for perception layer is the "Sensor" in IoT. Utilizing sensors and actuators, this layer's goal is to gather environmental data [6].

Connectivity Layer: The Connectivity Layer is responsible for data routing and transmission to various IoT devices over the internet. such as Wi-Fi, LTE, Bluetooth, Zigbee, 5G, etc. [7].

Fog/Edge Computing Layer: The Fog (Edge) computing layer is designed to store data quickly close to its sources.

Processing Layer: Once the data is collected through the fog computing layer, it is accumulating, storing and processing information [8].

Application Layer: The Application layer is responsible to provide data management such as cloud computing, Quality of Services (QoS), machine-to-machine (M2M) etc. [8].

Business Layer: The business layer assists businesses in fine-tuning their company strategies and developing more deliberate plans based on accurate data.

Security Layer: Security layer are need due to the massively data generated by the IoT infrastructure [9]. When IoT devices generate massive amounts of data, it is difficult to detect data breaches, malicious code injection, and hacking.

Adversaries will have greater opportunities to access connected devices and use them in major attacks such as network attacks, application attacks, and IoT device attacks as the number of linked devices increases. In this paper, we will focus on security challenges and solutions and recent attacks such as DoS (Denial of Service) and DDoS (Distributed Denial of Service) in IoT enabled systems. The organization of the paper is as follows: Section II discussed the related work, Issues of the Internet of Things in 21st Century in Section III. In Section IV, the classification of the IoT architecture is discussed. Security challenges and solutions are presented in Section V, with discussion on the classification of distributed denial of service attacks on fog-based IoT systems in Section VI. Finally, Section VII concludes the paper.

2 RELATED WORK

The security risks associated with data transfer through IoT devices and communication protocols need the development of an efficient IDS based on machine learning and deep learning from artificial intelligence. A. Samy *et al.* [10] proposed a framework for fog and cloud layer using with six DL model with five datasets, achieved binary classification accuracy 99.85% and multiclass classification 99.65% accuracy. M.A. Lawal *et al.* [11] proposed a framework for anomaly mitigation using a BoT-IoT dataset its accuracy on DT,k-NN 99.96%, 99.97% for binary classification and multi class classification accuracy; 99.96%, 93.00% Malik *et al.* [12] proposed DBM Classifier method based on IDS (Intrusion Detection System). Upon applying it to TON-IOT data set its accuracy was calculated to be 84%. Atul *et al.* [13] Proposed J48 and MLP model. Upon comparing it with MB-MLG methods it produces higher accuracy. Roopak *et al.* [14] Proposed an IDS system for IoT environment based on Machine Learning; CNN, LSTM, SVM, Random Forest method. His IDS system achieved 99.09% accuracy. De Souza *et al.* [15] developed an IDS system using Deep Neural Network and KNN that operates in Fog Layer. Upon applying it to NSL-KDD and CICIDS2017. Its accuracy was calculated to be 99.7%. Almiani *et al.* [16] proposed a Fog based IDS modal using Muti-layer Deep Recursive Neural Network method. Its detection speed being 66 micro second. Hwang *et al.* [17]

analyzed Packet Header and Traffic Flow with LSTM approach in order to detect suspicious activities. Hasan *et al.* [18] analyzed the Traffic Flow and Data on IoT network using different machine learning approach; SVM, Decision Tree, Random Forest on public data set. Random performed better comparatively. Khatun *et al.* [19] analyzed ANN methods for detecting malicious node in IoT environments with live wi-fi data and proposed an ANN method. It detects malicious nodes with 77% accuracy. Efstathopoulos *et al.* [20] One Class-SVM, Isolation Forests, Angle-Base Outlier Detection (ABOD), Stochastic Outlier Selection (SOS), Principal Component Analysis (PCA), and deep fully connected autoencoders were used to analyze Anomaly Based Detection on the Smart Grid. His suggestion was to use SVM, an ontology-based fuzzy analyzer. Compared to previous models, this model's detection accuracy has increased by 29%. Prabavathy *et al.* [21] They presented a detection method based on neighborhood fog nodes using the NSL-KDD dataset's online sequential extreme learning machine. Its accuracy was estimated at 97.36%. Zhang *et al.* [22] proposed a G-DBN for intrusion detection in IoT enabled network. He compared it with other methods; FCANN, TANN, BPNN. As a result of comparison, G-DBN was found to be high in detection rate in DoS attacks (99.45%), R2L (97.78%), Probe (99.37%), U2R (98.68%). Durairaj *et al.* [23] proposed a Enhanced DBN Based model for detecting attacks; DoS, False Data injection in smart microgrid. EDBN was found to be high detection accuracy (92%), false alarm rate less the (1%). The following Pie-chart provides the overview of the research papers belonging to various DDoS attacks detection techniques.

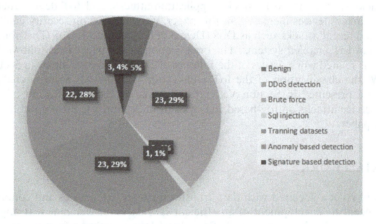

Figure 2.   Show the number of papers related to DDoS attack.

3   ISSUES OF INTERNET OF THINGS IN 21ST CENTURY

The Cloud and host computing models of today cannot adequately address many new issues brought on by the developing IoT. Here, we go over a few of these essential issues [25].

Connectivity: IoT development of applications such as smart grid, smart city, healthcare required a real time transmission of data, but is become a challenge because of poor connectivity or latency [26]. due to poor/unstable network connection that drops the device out of the system in the middle of processing. Especially in large-scale IoT applications [27].

Table 1.   Overview of related work.

Article	Year	Model	Dataset	Classification	Performance
[23]	2022	DBN	TON_IoT	Multiclass	Acc=92
[24]	2022	KNN,K-MEANS	NSL-KDD	Multiclass	Acc=99.6,99.7
[12]	2022	DBN	TON_IoT_Wwather	Multiclass	Acc=84–86
[13]	2021	J48 and MLP	KDD CUPP 99,real time	Multiclass	Acc=85
[14]	2020	CNN,LSTM,	CISIDS2017	Multiclass	Acc=99.06
[10]	2020	Bi-LSTM,GRU,	Five datasets:	Binary	Acc=99.8
		CNN,CNN_LSTM	UNSW-NB15,	Multiclass	Acc=99.6
			CICIDS-2017,RPLNIDS-		
			2017,N BaIoT-2018,		
			NSL-KDD		
[15]	2020	DNN,KNN	NSL-KDD,	Multiclass	Acc=99.7
			CICIDS-2017		
[16]	2020	RNN	NSL-KDD	Multiclass	Acc=92.18
[11]	2020	DL,KNN	BoT-IoT	Binary	Acc=99.96
				Multiclass	Acc=93.00
[17]	2019	CNN,LSTM	ISCX2012,USTC	Multiclass	Acc=99.97,99.88
			TFC2016Mirai-RGU,		Acc=99.98,99.36
			Mirai CCU		

Security and Privacy: Security is on everyone's mind today; lack of the security increases the risk of leaking the confidential information and data. In IoT enabled Networks many devices are in a loop. If one device in a loop gets attacked, rest of the devices in the networks can easily be commanded and controlled by the attacker. Presently, IoT standards are not competent enough to deal with emerging security threats. Here are few factors that affect IoT security system.to name few such factors, insecure authentication and authorization, weakness of transport encryption, absences of secure code practices in IoT technology. Security and privacy are major challenge as each layer of IoT enable system is prone to cyber-attack and malware.

Power Requirements: All the IoT devices require more power to work continuously but most of them are battery operated. Due to the above challenge, we are moving to renewable resources of energy such as solar energy, wind energy and tidal energy etc.

Heterogeneity: A heterogeneous network is a network connecting embedded computers and other devices where the standards and protocols have significant differences [28].

Big Data: In IoT based applications required a large number devices are connected and communicate which, each other, large data is accumulated out of which only a little bit is useful, managing this big data is big challenge [29]. Data lake has the same quality of Big Data in big data environment, data analytic has to allow to breaking down volume of data from various IoT based networks [30, 31].

IoT Network Bandwidth Requirement: The generation of data is accelerating due to the vast and quickly expanding number of connected devices [32]. The amount of network bandwidth needed to send all the data to the cloud will be prohibitively high. Because of rules and worries about data privacy, Fog computing must be added to cloud computing [25].

4 CLASSIFICATION OF INTERNET OF THINGS ARCHITECTURE

Industries and crucial applications are using the expanding classification of IoT architecture for monitoring and controlling [33]. IoT and cloud computing are two separate technologies with numerous real-time applications such as smart power grid, smart home automation, smart city, e-healthcare monitoring system etc. IoT has provides large number of devices and embedded system to users. Cloud computing offers an efficient way to process, store, and manage resources that can be accessed from any location and is widely used by many applications [34]. In IoT generate a stream to massive data which increase the load on the networks and cloud integration have ushered in a new era of potential and challenges for more efficient data processing, storage, management, and security. In addition, industry has always been concerned about the challenges that IoT networks face when integrating with the cloud Integration of IoT and the cloud cannot address all of the problems that IoT networks face. Therefore, the extension of cloud is known as fog computing. Fog computing provides storage, processing, controlling, and networking, much like the cloud. Edge-cloud architecture is divided into two parts: sensor to cloud and fog enable-cloud framework [35]. The sensor layer and cloud layer make up the older structure, whereas the sensor layer, fog layer, and cloud layer make up the new framework. The purpose of integration of layer to perform high computational powers [36]. The sensor, fog and cloud communicate with each other via wired or wireless means. The description of under-laying architecture of sensor to fog and fog to cloud framework is shown in Figure 3.

A fog-enabled IoT application for a smart grid, for example, collects data from various sensors and offloads some computations to the fog layer. This could also include delay-sensitive data analysis and processing at the fog layer. Furthermore, rest of the data can be sent to cloud those are not delay sensitive [37].

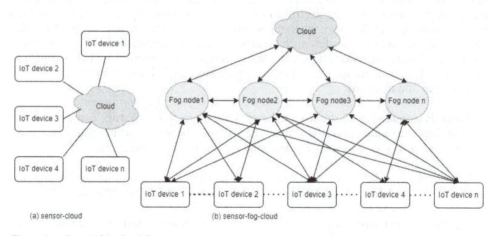

Figure 3. Sensor-fog-cloud framework.

5 SECURITY CHALLENGES AND SOLUTIONS

Sensor-Fog-Cloud frameworks are more secure than the sensor-cloud frameworks as it has less reliance on internet, only processing data storage is required not extra storage and more efficient for the information exchange between the cloud and the end devices such as sensor. It uses various interlinked networks, for example wireless network and conventional network turning them into easy target for attacker [33]. There are three major categories of fog security challenges. (a) Network Security; (b) Data Security; and (c) IoT Devices Security.

Network Security: Fog computing performs computation near to the generation locations to enable real-time data transmission in IoT systems. Identification /Verification; The process of utilizing real-time services, such as fog nodes, involves a number of different entities and services, all of which must be trusted in order for data to be communicated. This presents security concerns for IoT security. Intrusion detection; If an adequate intrusion detection system is not established, policy breaches on fog nodes won't be found. Attackers can not only damage the fog base architecture but also take command and control of the fog nodes. according to Zhang *et al.* [38], access control is seen as a flexible instrument to secure and manage requirements such as latency, efficacy, generality, aggregation, privacy, protection, and trust management.

Data Security: For the security point of view, requiring encrypted data transmission from one fog node to another, many features are compromised due to the not properly designed encryption algorithms. These include data searching, sharing, and aggregation [39].

IoT Devices Security: Unfortunately, a lot of IoT devices are weak, so remote software update functions must be developed to manage security updates. It is necessary for fog nodes to create a reliable mechanism for updating the software on edge layer devices [40]. In this section, we also discuss the solution provided by Fog on a network level, such as an intrusion detection system. Traditional security software cannot be installed in IoT systems due to a lack of computational and storage resources. Detecting attacks in a fog-based IoT environment is very difficult due to heterogeneous connectivity of devices, and as the growth of IoT devices leads to an increase in their vulnerability to attacks, in fog-based IoT contexts, standard intrusion detection systems (IDS) are proven to be largely useless. Signature-based, anomaly-based, and hybrid-based detection of attacks on IoT-enabled systems are the three main methods. Anomaly-based intrusion systems employ traffic patterns to identify attacks, whereas signature-based intrusion detection systems rely on knowledge of known threats. This is the difference between the two types of systems. IoT networks generate traffic that can be either benign or malicious [24]. One drawback of the signature-based approach is the length of time needed to update the signature database. It gets computationally expensive to compare input with the database as it grows in size. Because of this incapacity of the signature-based approach, identifying zero-day or previously unknown threats goes beyond the purview of the technique. The main justification for this is that this method defends against known attacks. An anomaly-based approach is used when an unusual traffic pattern is found because it examines normal traffic flow and either raises an alert or restricts traffic. The use of anomaly-based detection algorithms has the benefit of being skilled at spotting novel threats and zero-day vulnerabilities, but it can also result in a large number of false positives.

6   DDOS ATTACKS CLASSIFICATION ON FOG-BASED IOT

Numerous security risks have been posed to the fog-based IoT. They do, however, become vulnerable targets due to their heterogeneous nature, strong independence, capacity to connect to and exercise control through the internet, and resource limitations. In a Figure 4, shows the classification of DDoS attacks.

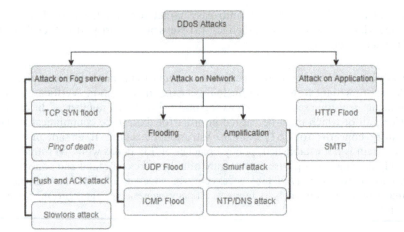

Figure 4.   A classification of DDoS attacks.

In DDoS attack, an Attacker can take a command and control over the fog integrated network, bots through which attacker can perform a various malicious activity or launch a large-scale attack such as spam emails [41].

DDoS attacks are classified into three major categories namely Attack on Fog server, Attacks on Network, and Attack on Application. And each of these major categories has its own subcategories. The above chart of DDoS attacks and its types are described in detail below [42].

TCP Flooding Attacks: During a TCP flooding attack, the attacker repeatedly sends SYN packets to multiple ports using fake IP addresses. The source never sends the anticipated ACK message when the destination sends a SYN-ACK in response to a SYN request, leaving the target waiting until its connection limit is reached [43].

Ping of Death Attacks: When an attacker assaults computers or services with huge data packets and causes them to crash or disrupt this is known as a DoS attack [44].

Push and ACK Attacks: A hacker can disable a server from responding to legitimate traffic by bombarding it with bogus PUSH and ACK requests. This technique is referred to as a "PUSH" or "ACK" [45]. Slowloris attacks In the slowloris attack, the attacker accesses the targeted web server. slowloris establishes many connections and keeps them open for as long as necessary [46].

Flooding Attacks: Another type of large-scale attack is packet flooding. Because fog-nodes connected to each other have no control over the which packets they have to received, these attacks can occur. Because fog nodes have internet connectivity, they are vulnerable to flooding attacks, and IoT traffic can travel through multiple hops before reaching a gateway or an IDS [47]. Attackers can overwhelm the target with spurious packets using a false source IP address, causing the target to crash-down. UDP flooding attacks: Attackers can transmit UDP data-grams through networks connected to fog in UDP flooding attacks, which also increase network traffic. ICMP flooding attacks: In the ICMP flooding attack, attacker continuous ping the target from the different sources without wait for response, monopolized the link or network resources [48].

Amplification Attacks: Any attack that can use an amplification factor to multiply its power is referred to as an amplification attack. Amplification attacks are" asymmetric," which means that an attacker needs only a little number of low-level resources to make a much larger amount or higher level of target resources malfunction or fail [49].

Smurf Attack: In this attack, the victim's IP address is changed using ICMP packet spoofing[50]. Network time protocol (NTP) amplification Attackers use bandwidth from a depend-able NTP source to overwhelm victims during an attack on the network time protocol (NTP). DNS Amplification Attack, attackers can send domain name system (DNS) requests to the servers using a forged client IP address [50].

Attack on Application: Attackers frequently look at the application layer first while looking for application vulnerabilities. HTTP flood: The attacker continuously sends many HTTP GET request messages to the server. The server satisfies these requests and waits for an acknowledgement because the attacker never replies.

7 CONCLUSION

The related works helps to identifying the security concerns in fog-enabled IoT system. Detecting attacks is a major challenging task in fog-enabled IoT system. It is vulnerable to the threats that can harm the fog nodes and attacker can easily take a command and control over the fog nodes. The subject of IoT applications is quite exciting and full of intrigue. However, very little empirical research has been done regarding the QoS and security requirements for time-sensitive IoT systems.

REFERENCES

[1]   Evans D., "*The Internet of Things how the Next Evolution of the Internet is Changing Everything,*" pp. 1–11, 2011.

[2]   Sharma V., Patel R., Bhadauria H.S. and Prasad D., "Policy for Planned Placement of Sensor Nodes in Large Scale Wireless Sensor Network," *KSII Transactions on Internet and Information Systems (TIIS)*, vol. 10, no. 7, pp. 3213–3230, 2016.

[3]   Vats S., Singh S., Kala G., Tarar R. and Dhawan S., "idoc-x: An Artificial Intelligence Model for Tuberculosis Diagnosis and Localization," *Journal of Discrete Mathematical Sciences and Cryptography*, vol. 24, no. 5, pp. 1257–1272, 2021.

[4]   Sharma V., Patel R., Bhadauria H. and Prasad D., "Deployment Schemes in Wire-less Sensor Network to Achieve Blanket Coverage in Large-scale Open Area: A Review," *Egyptian Informatics Journal*, vol. 17, no. 1, pp. 45–56, 2016.

[5]   Vailshery L.S., "Number of Internet of Things (IoT) Connected Devices Worldwide from 2019 to 2030," pp. 1–2, 2022.

[6]   Mahmoud R., Yousuf T., Aloul F. and Zualkernan I., "Internet of things (iot) Security: Current Status, Challenges and Prospective Measures," *In 2015 10th International Conference for Internet Technology and Secured Transactions (ICITST)*, pp. 336–341, IEEE, 2015.

[7]   Leo M., Battisti F., Carli M. and Neri A., "A Federated Architecture Approach for Internet of Things Security," *In 2014 Euro Med Telco Conference (EMTC)*, pp. 1–5, IEEE, 2014.

[8]   Anand S. and Sharma A., "Assessment of Security Threats on Iot based Applications," *Materials Today: Proceedings*, 2020.

[9]   Vats S. and Sagar B., "Performance Evaluation of k-means Clustering on hadoop Infrastructure," *Journal of Discrete Mathematical Sciences and Cryptography*, vol. 22, no. 8, pp. 1349–1363, 2019.

[10]   Samy A., Yu H. and Zhang H., "Fog-based Attack Detection Framework for Internet of Things using Deep Learning," *IEEE Access*, vol. 8, pp. 74571–74585, 2020.

[11]   Lawal M.A., Shaikh R.A. and Hassan S.R., "An Anomaly Mitigation Framework for iot using fog Computing," *Electronics*, vol. 9, no. 10, p. 1565, 2020.

[12] Malik R., Singh Y., Sheikh Z.A., Anand P., Singh P.K. and Workneh T.C., "An Improved Deep belief Network Ids on Iot-based network for Traffic Systems," *Journal of Advanced Transportation*, vol. 2022, 2022.

[13] Atul D.J., Kamalraj R., Ramesh G., Sankaran K. S., Sharma S. and Khasim S., "A Machine Learning based Iot for Providing an Intrusion Detection System for Security," *Microprocessors and Microsystems*, vol. 82, p. 103741, 2021.

[14] Roopak M., Tian G.Y., and Chambers J., "An Intrusion Detection System Against DDoS Attacks in Iot Networks," *In 2020 10th Annual Computing and Communication Workshop and Conference (CCWC)*, pp. 0562–0567, IEEE, 2020.

[15] de Souza C. A., Westphall C. B., Machado R. B., Sobral J.B.M. and dos San- tos Vieira G., "Hybrid Approach to Intrusion Detection in Fog-based iot Environments," *Computer Networks*, vol. 180, p. 107417, 2020.

[16] Almiani M., AbuGhazleh A., Al-Rahayfeh A., Atiewi S. and Razaque A., "Deep recurrent neural network for iot Intrusion Detection System," *Simulation Modelling Practice and Theory*, vol. 101, p. 102031, 2020.

[17] Hwang R.-H., Peng M.-C., Nguyen V.-L. and Chang Y.-L., "An lstm-based deep Learning Approach for Classifying Malicious Traffic at the Packet Level," *Applied Sciences*, vol. 9, no. 16, p. 3414, 2019.

[18] Hasan M., Islam M. M., Zarif M.I.I. and Hashem M., "Attack and Anomaly Detection in iot Sensors in Iot Sites using Machine Learning Approaches," *Internet of Things*, vol. 7, p. 100059, 2019.

[19] Khatun M.A., Chowdhury N. and Uddin M.N., "Malicious Nodes Detection based on Artificial Neural Network in Iot Environments," *In 2019 22nd International Conference on Computer and Information Technology (ICCIT)*, pp. 1–6, IEEE, 2019.

[20] Efstathopoulos G., Grammatikis P.R., Sarigiannidis P., Argyriou V., Sarigiannidis A., Stamatakis K., Angelopoulos M.K. and Athanasopoulos, "Operational data based Intrusion Detection System for Smart Grid S.K.," *In 2019 IEEE 24th International Workshop on Computer Aided Modeling and Design of Communication Links and Networks (CAMAD)*, pp. 1–6, IEEE, 2019.

[21] Prabavathy S., Sundarakantham K., and Shalinie S.M., "Design of Cognitive Fog Computing for Intrusion Detection in Internet of Things," *Journal of Communications and Networks*, vol. 20, no. 3, pp. 291–298, 2018.

[22] Zhang Y., Li P. and Wang X., "Intrusion Detection for Iot based on Improved Genetic Algorithm and Deep Belief Network," *IEEE Access*, vol. 7, pp. 31711–31722, 2019.

[23] Durairaj D., Venkatasamy T.K., Mehbodniya A., Umar S. and Alam T., "Intrusion Detection and Mitigation of Attacks in Microgrid using Enhanced Deep Belief Network," *Energy Sources, Part A: Recovery, Utilization, and Environmental Effects*, pp. 1–23, 2022.

[24] Roy S., Li J. and Bai Y., "A Two-layer Fog-cloud Intrusion Detection Model for Iot Networks," *Internet of Things*, p. 100557, 2022.

[25] Vikrant S., Patel R., Bhadauria H.S. and Prasad D., "Glider Assisted Schemes to Deploy Sensor Nodes in Wireless Sensor Networks," *Robotics and Autonomous Systems*, vol. 100, pp. 1–13, 2018.

[26] Bhatia M., Sharma V., Singh P. and Masud M., "Multi-level p2p Traffic Classification using Heuristic and Statistical-based Techniques: A Hybrid Approach," *Symmetry*, vol. 12, no. 12, p. 2117, 2020.

[27] Sharma V., Patel R., Bhadauria H.S., and Prasad D., "Nads: Neighbor Assisted deployment Scheme for Optimal Placement of Sensor Nodes to Achieve Blanket Coverage in Wireless Sensor Network," *Wireless Personal Communications*, vol. 90, no. 4, pp. 1903–1933, 2016.

[28] Vats S., Sagar B.B., Singh K., Ahmadian A., and Pansera B.A., "Performance Evaluation of an Independent Time Optimized Infrastructure for Big Data Analytics that Maintains Symmetry," *Symmetry*, vol. 12, no. 8, p. 1274, 2020.

[29] Vats S. and Sagar B., "An Independent Time Optimized Hybrid Infrastructure for Big Data Analytics," *Modern Physics Letters B*, vol. 34, no. 28, p. 2050311, 2020.

[30] Vats S. and Sagar B., "Data Lake: *A Plausible Big Data Science for Business Intelligence,*" In *Communication and Computing Systems*, pp. 442–448, CRC Press, 2019.

[31] Agarwal R., Singh S. and Vats S., "Implementation of an Improved Algorithm for Frequent Itemset Mining Using Hadoop," *In 2016 International Conference on Computing, Communication and Automation (ICCCA)*, pp. 13–18, IEEE, 2016.

[32] Mearian L., "*Internet of Things Data to Top 1.6 Zettabytes by 2022,*" 2016.

[33] Tariq N., Asim M., Al-Obeidat F., Zubair Farooqi M., Baker T., Hammoudeh M., and Ghafir I., "The Security of Big Data in Fog-enabled iot Applications Including Blockchain: A Survey," *Sensors*, vol. 19, no. 8, p. 1788, 2019.

[34] Bhati J.P., Tomar D. and Vats S., "Examining Big Data Management Techniques for Cloud-based Iot Systems," In *Examining Cloud Computing Technologies Through the Internet of Things*, pp. 164–191, IGI Global, 2018.

[35] Chang V., Golightly L., Modesti P., Xu Q.A., Doan L.M.T., Hall K., Bodd S. and Kobusinska A., "A Survey on Intrusion Detection Systems for Fog and Cloud Computing," *Future Internet*, vol. 14, no. 3, p. 89, 2022.

[36] Hassija V., Chamola V., Saxena V., Jain D., Goyal P. and Sikdar B., "A survey on Iot Security: Application Areas, Security Threats, and Solution Architectures," *IEEE Access*, vol. 7, pp. 82721–82743, 2019.

[37] Sharma V., Vats S., Arora D., Singh K., Prabuwono A.S., Alzaidi M.S. and Ahmadian A., "Ogas: Omni-directional Glider Assisted Scheme for Autonomous Deployment of Sensor Nodes in Open Area Wireless Sensor Network," *ISA Transactions*, 2022.

[38] Zhang P., Chen Z., Liu J.K., Liang K. and Liu H., "An Efficient Access Control Scheme with Outsourcing Capability and Attribute Update for Fog Computing," *Future Generation Computer Systems*, vol. 78, pp. 753–762, 2018.

[39] Haider N. and Azad C., "Data Security and Privacy in fog Computing Applications," *Cloud and Fog Computing Platforms for Internet of Things*, pp. 57–66, 2022.

[40] Sicari S., Rizzardi A. and Coen-Porisini A., "Insights Into Security and Privacy Towards fog Computing Evolution," *Computers & Security*, p. 102822, 2022.

[41] Kumar V. and Kumar R., "An Adaptive Approach for Detection of Blackhole Attack in Mobile ad Hoc Network," *Procedia Computer Science*, vol. 48, pp. 472–479, 2015.

[42] Kogo S. and Kanai A., "Detection of Malicious Communication using DNS Traffic Small Features," In 2019 *IEEE 8th Global Conference on Consumer Electronics (GCCE)*, pp. 125–126, IEEE, 2019.

[43] Haris S., Ahmad R., Ghani M. and Waleed G.M., "Tcp Syn Flood Detection based on Payload Analysis," in 2010 *IEEE Student Conference on Research and Development (SCOReD)*, pp. 149–153, IEEE, 2010.

[44] Yusof M.A.M., Ali F.H.M. and Darus M.Y., "Detection and Defense Algorithms of Different Types of DDoS Attacks using Machine Learning," In *International Conference on Computational Science and Technology*, pp. 370–379, *Springer*, 2017.

[45] Ubale T. and Jain A.K., "Taxonomy of DDoS Attacks in Software-defined Networking Environment," In *International Conference on Futuristic Trends in Network and Communication Technologies*, pp. 278–291, *Springer*, 2018.

[46] Sharma S., Manuja M. and Kishore K., "Vulnerabilities, Attacks and their Mitigation: An Implementation on Internet of Things (Iot)," *Int. J. Innov. Technol. Explor. Eng*, vol. 8, no. 10, pp. 146–150, 2019.

[47] Sonar K. and Upadhyay H., "A Survey: DDoS Attack on Internet of Things," *International Journal of Engineering Research and Development*, vol. 10, no. 11, pp. 58–63, 2014.

[48] Bijalwan A., Wazid M., Pilli E.S. and Joshi R.C., "Forensics of Random-UDP Flooding Attacks," *Journal of Networks*, vol. 10, no. 5, p. 287, 2015.

[49] Zhou L., Guo H. and Deng G., "A Fog Computing based Approach to DDoS Mitigation in Iiot Systems," *Computers & Security*, vol. 85, pp. 51–62, 2019.

[50] Ubale T and Jain A.K., "Survey on DDoS Attack Techniques and Solutions in Software-Defined Network," in *Handbook of Computer Networks and Cyber Security*, pp. 389–419, *Springer*, 2020.

Automation and Computation – Vats et al. (Eds)
© 2023 the Author(s), ISBN 978-1-032-36723-1

Experimental assessment of flicker in commercially available white-LED indoor luminaires

Vishwanath Gupta

Research Scholar, Electrical Engineering Department, Jadavpur University, Kolkata, India

Shubham Kumar Gupta

Programmer Analyst Trainee, CTS Pvt. Ltd., Bengaluru, India

Biswanath Roy

Professor, Electrical Engineering Department, Jadavpur University, Kolkata, India

ABSTRACT: The proposed work deals with experimental assessment of flicker phenomenon, a type of Temporal Light Artefact (TLA) in commercially available white-LED indoor luminaires. Flicker assessment is done as 'percentage flicker (%F)', for six commercially available w-LED luminaires of different makes (out of six, two are dimmable), procured from market. Flicker is measured as the variation in illuminance value, at a setpoint below the test w-LED luminaires at rated supply for all six luminaires as well as at different dimming levels for the two dimmable luminaires. The experimental results showed that %F at rated conditions varies from 8% to 36%for the six test samples, whereas the recommended maximum allowable upper limit is 30%. From the experimental results in case of dimming, it was observed that %F increases with increase in dimming levels for the two test luminaires.

1 INTRODUCTION

LEDs has emerged as one of the most used artificial light sources for indoor, outdoor and road lighting applications due to their cost effectiveness, longer lifetime, design flexibility and energy efficiency. However, LED lights with high flicker (a type of TLA) are detrimental for end-users because they can cause eye strain, blurred vision, reduced visual task performance and neurological complications. In India, the flicker percentage of various commercially available LED lights varies from 40% to 80% which is much higher than the prescribed threshold limit of 30% prescribed by IEEE standards PAR 1789, California (Orient electric 2015). Hence, it is extremely important to do an extensive qualitative study on commercially available LEDs to correctly evaluate and assess the impact of using them for the overall well-being of the end-users. Flicker metrics like percentage flicker and flicker index are not supplied in the datasheet of LED Luminaires manufacturers and to correctly assess the quality of light by LEDs, the flicker phenomenon must be considered (Collin 2019; Kobav 2018; Kumar 2020; Sailesh 2017). The research work done by Sailesh & Sailesh (2017) presents a procedure to define a minimum performance criterion for flicker measurement using a mathematical formulation for calculating various indices in relation to the measurement of flicker. In another study done by Kobav & Colarič (2018), the measurements of luminous flux of different artificial light sources is proposed to get an insight on the

DOI: 10.1201/9781003333500-8

flicker or TLA present in those lamps and an experimental arrangement was developed to study the flicker of various lamps with the variation of the supply voltage signal. Collin *et al.* (2019) in their research work introduced an extensive labeling framework for comparing LED lamp performance based on light flicker and power factor that will help in selecting correct LED drivers to reduce flicker levels in LED lighting systems. In the research carried out by Kumar and Kumar (2020), comparison of voltage flicker indices of low power warm white and cool white LEDs with the latest Indian Standard IS 14700-3-3 (2003) is done.

However, in the above mentioned research articles, the flicker phenomenon was not studied when LED modules were dimmed. But, in current energy scenario, most of the LED modules are also operated in dimmed states to reduce energy consumption. So, studying the flicker phenomenon of luminaires under rated conditions as well as dimmed condition is important for qualitative analysis of luminaires. The present work deals with the study of flicker phenomenon in commercially available w-LED indoor luminaires at rated conditions as well as at dimmed conditions.

2   EXPERIMENTAL TESTING AND MEASUREMENT

In the proposed work, 6 different commercially available w-LED indoor luminaires given in Table 1 are considered for experimental measurement of flicker phenomenon. The present work quantifies flicker in terms of '%F' by measuring the fluctuations in the illuminance level at a set-point due to the LED luminaires at rated condition and also at different dimming levels. %F is calculated as given by Equation 1.

$$\%F = \frac{A - B}{A + B} \times 100 \qquad (1)$$

where, A is the peak light output value and B is the trough light output value.

The block diagram of the flicker measurement procedure followed in the proposed work is shown in Figure 1. A programmable power supply [GwINSTEK APS 1102] is used to power the LED lighting systems and an Arduino based dimming circuit developed by Gupta et al. (2020) is used to control the dimming levels of the LEDs. The light from the LED luminaire whose flicker measurement is to be done is made to fall on an ambient light sensor (TEMT6000X01) placed at a set point below the luminaire. TEMT6000 has a special ambient light detector whose spectral responses matches to a human eye and it converts the light falling on it to proportionate voltage. The variation of voltage, corresponding to light output, with time is then observed using a GwINSTEK make Digital Storage Oscilloscope [DSO GDS 2202-E]. The experimental setup installed for flicker assessment is shown in Figure 2.

Table 1.   Ratings of test w-LED luminaires.

w-LED luminaire	dimmable	Rated Power (W)	CCT (K)
LED 1	No	12	6500
LED 2	No	12	3000
LED 3	No	10	6000
LED 4	Yes	6	3000
LED 5	No	15	6500
LED 6	Yes	6	6500

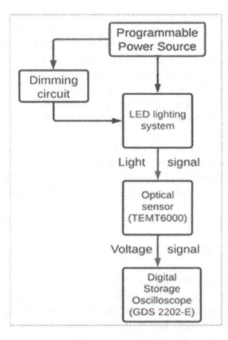

Figure 1.   Block diagram of the measurement procedure.

Figure 2.   Experimental setup.

3 OBTAINED RESULTS AND ANALYSIS

3.1 *Rated condition*

The flicker percentage of 6 different commercially available LED lighting systems at rated power was calculated from the corresponding voltage waveforms obtained from the DSO and is given in Figure 3. The calculated percentage flicker varies from approximately 8%–6%. For LED1, LED2 and LED6 it is found to be greater than 30% (maximum prescribed value) and for LED3, LED4 and LED5 the percentage flicker is less than 30%. LED4 has the minimum flicker percentage of 7.46%.

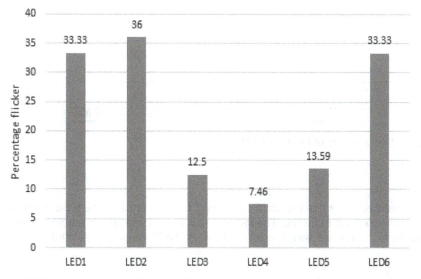

Figure 3.   Flicker percentage of different LED lighting systems under test.

For each test LED luminaire, two voltage waveforms are observed on the DSO. First voltage waveform is the one that is the output of the TEMT6000 light sensor corresponding to the light output from the LED luminaire. The second waveform is the voltage waveform corresponding to the output current from the driver of the LED luminaire under test. The obtained voltage waveforms corresponding to light output at 100Hz frequency for the LED lighting systemLED4 under test is shown in Figure 4. The waveform obtained at Channel 1 is the voltage waveform from TEMT6000 and that obtained at Channel 2 is the voltage waveform corresponding to the LED driver output current.

3.2 *Dimmed condition*

Out of the 6 LED lighting systems under test, the dimmable LED luminaires LED4 and LED6 were tested under dimmed conditions to obtain information about the variation of flicker percentage in case of dimming. PWM technique is utilized for dimming the LED luminaires. The frequency of PWM is 500Hz. The variation of flicker percentage of LED4 and LED6 with variation of light output (100% light output to 20% light output) is shown in Figure 5.

From Figure 5, it is observed that %F increases with increase in dimming levels (or decrease in light level). This is because with increase in dimming levels, the duty cycle of

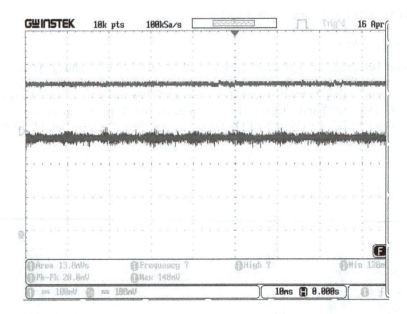

Figure 4.   Flicker profile of LED2 at 100Hz.

PWM is reduced and therefore the time interval between consecutive pulses increase which results in the increase in flicker. The increase in %F with increase in dimming levels is also evident from the voltage waveforms of LED4 at PWM frequency of 500Hz obtained for

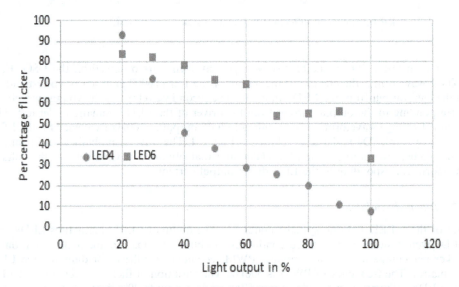

Figure 5.   Variation of flicker percentage of LED4 and LED6 with variation of light output.

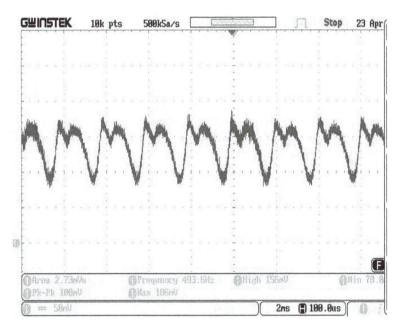

Figure 6. Flicker profile of LED4 at 70% light output.

70%, 50% and 30% of light output as shown in Figures 6–8. From Figures 6–8, it is observed that the Pk-Pk value of the voltage waveform is 108mV, 178mV and 214mV for 70%, 50% and 30% of light output respectively.

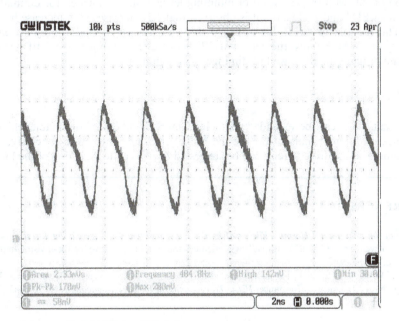

Figure 7. Flicker profile of LED4 at 50% light output.

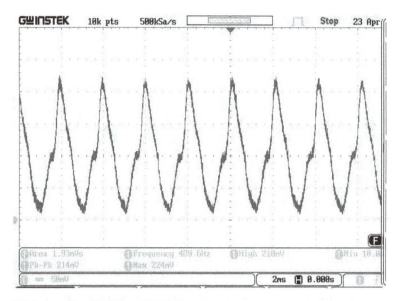

Figure 8. Flicker profile of LED4 at 30% light output.

4 CONCLUSIONS

In this proposed work, flicker measurement of six different commercially available LED lighting system, varying between 6W to 15W, is carried out in the laboratory. The percentage flicker (%F) at rated light output for the 3 LED lighting systems were below 30% which is the maximum allowable limit and more than 30% for the remaining 3 LED lighting system. The variation of %F with the variation of dimming levels is also presented for dimmable LED luminaires LED4 and LED6 and it is observed that %F increases with increase in dimming levels. This increase in %F is due to PWM dimming and can be reduced by increasing the frequency of PWM. Finally, the percentage flicker of LED lighting system LED4 is found to be minimum under rated as well as dimmed light conditions.

5 FUTURE SCOPE

The present work can be extended to include more numbers of dimmable indoor LED luminaires of different makes to make the % flicker comparison more conclusive. Further modifications in the driver circuit can be suggested to reduce the flickering of the luminaires.

REFERENCES

Be Wary of the Invisible but Harmful Flicker of LED Lights, Internal Webpage, Orient Electric, New Delhi, India, viewed on 28/05/2022. https://www.orientelectric.com/blog/be-wary-of-the-invisible-but-harmful-flicker-of-led-lights

Collin A., Djokic S., Drapela J., Langella R. and Testa A. 2019. Light Flicker and Power Factor Labels for Comparing LED Lamp Performance. *IEEE Transactions on Industry Applications* 55(6): 7062–7070.

Gupta V., Basak B., and Roy B. 2020. A Fault-Detecting and Motion-Sensing Wireless Light Controller for LED Lighting System. *In 2020, IEEE Calcutta Conference* (CALCON), Kolkata, India, 28–29 February, pp. 462–466.

Kobav, M.B. & Colarič, M. 2018. *Flicker Experimental Set up and Visual Perception of Flicker.* In 2018, Seventh Balkan Conference on Lighting (BalkanLight), Sofia, Bulgaria, 4–6 June, pp. 1–6.

Kumar K. J. & Kumar R.S. 2020. Voltage Flicker From Warm and Cool White LED Bulbs. In 2020, *IEEE PES Innovative Smart Grid Technologies Europe* (ISGT-Europe), Delpht, The Netherlands, 26–28 October, pp. 934–938.

Shailesh K.R. & Shailesh T. 2017. Review of Photometric Flicker Metrics and Measurement Methods for LED Lighting. *In 2017, 4th International Conference on Advanced Computing and Communication Systems (ICACCS)*, Coimbatore, India, 6–7 January, pp. 1–7.

TEMT6000X01, *Ambient Light Sensor, Vishay Semiconductors*, document number 81579, viewed on 28/05/2022. https://www.vishay.com/docs/81579/temt6000.pdf

Automation and Computation – Vats et al. (Eds)
© 2023 the Author(s), ISBN 978-1-032-36723-1

Cognitive intelligence-based model for health insurance premium estimation

Ankita Sharma*
Assistant Professor, Chandigarh University, Chandigarh, India

Bhuvan Sharma*
Scholar, Chandigarh University, Chandigarh, India

ABSTRACT: Health has become the major contribution for the growth of any country. Therefore, Artificial intelligence and Machine learning are the appropriate approaches to make human life easier by anticipating and diagnosing with prediction of the future. Still many researchers are investigating which approach is best for predicting medical insurance. This research uses multiple algorithms to identify which cognitive approach is better for predicting healthcare insurance premium estimation. The proposed research uses PCA for feature extraction and then trains the model by using Linear Regression, Support Vector Regression, Decision Tree, Random Forest Regression, and k-Nearest Neighbors and Artificial neural network are all used in the suggested study methodology. The dataset is acquired from the KAGGLE repository and different machine learning and neural network methods are used to show the analysis and to compare the model accuracy. The results show that the ANN is the best model which gives 96% accuracy for predicting health insurance.

Keywords: Health insurance, Prediction, Medicare, Analyze

1 INTRODUCTION

Predicting how much people will spend on healthcare in the future is a powerful tool for raising standards of care transparency. Numerous healthcare sectors generate the data related to patient diseases and diagnoses but very few tools provide the significance of patient healthcare insurance [1]. Medical insurance is a coverage that helps pay at least reduces the out-of-pocket costs associated with medical emergencies and other types of loss. Insurance and medical care costs may be affected by a number of variables [2]. Accurately estimating individual healthcare bills using prediction models is crucial for a range of stakeholders and health administrations [3]. The resources dedicated to care management are limited, and reliable cost estimates could help healthcare providers and insurers make future plans [2]. In addition, having an estimate of future costs might help patients choose insurance policies with reasonable deductibles and rates, factors contribute to the evolution of insurance plans [4].

Machine learning can improve the effectiveness of insurance policy language. In healthcare, one area where ML algorithms really shine is in the prediction of high-cost, high-need patient spending. [5]. There are three main categories of machine learning: supervised

---

*Corresponding Authors: ankita.ceh@gmail.com and bhuvansharma0908@gmail.com

DOI: 10.1201/9781003333500-9

machine learning, used for classification and regression when all data is labeled; unsupervised machine learning, used for cluster analysis when all data is unknown and evolutionary computation, used for deciding by learning from previous errors [16].

Here, we employ supervised ML models to assess and contrast the performance of several different types of regression analysis, including the support vector machine, k-nearest Neighbor, Decision Tree, Random Forest Regression, and Multiple Linear Regression. The following is a summary of the work's contributions:

1. Looking into the viability of using a computational intelligence method based on machine learning to forecast healthcare insurance premiums.
2. Evaluating how the most well-known machine learning algorithms for price prediction fare against one another

Below is the outline for the remaining portions of the paper. In Section 2, highlight the work done in this area. Section 3 discusses the working methodology followed in the implementation. Results from the experiments are presented in Section 4. In the last, Section 5 contains the conclusion of the entire paper.

2   RELATED WORK

The author identify that field have utilized a variety of machine learning (ML) algorithms to examine healthcare data and calculate premiums [7]. In this Studies [8–13] variety of ML techniques to analyze medical data. The author implemented the Health insurance premium forecasting using a neural network-based regression model. These authors used character-istics to make predictions about how much people would have to pay for health insurance. The author found the possible downside of employing predictive algorithms to determine insurance premiums was examined at the outset of one article. Would this lead to new forms of bias and insurance discrimination that threaten the concept of risk mutualization? In the second phase, the authors analyzed how the customer's perception of the business changed once she learned that the organization routinely collected and analyzed data about her actual behavior [14]. The author was using the XGB model for predicting Health insurance pre-miums [15] with flexible imputation of missing data. According to a study by the authors of [16, 17], compared the accuracy of using logistic regression (LR) and extreme learning machine (XGB) to forecast the existence of a limited number of accident claims, the former is a superior model due to its interpretability and good prediction. The author produced a paper in which they predicted insurance prices based on people's health records. The per-formance of several algorithms was evaluated and compared by means of analysis. The dataset was used for model training, and the resulting models were used for prediction. To ensure the model was accurate, it was put to the test by comparing the calculated result to the observed data [18].

Healthcare cost forecasting using ML models, including hierarchical Decision Trees, is discussed in [19,20]. Additionally, They said that the healthcare industry relies heavily on machine learning tools and methods, and that these systems are employed only for the purpose of determining how much medical insurance would cost. Similarly, the insurance company's underwriting procedure and the medical examinations required to profile the applicants' risks may be time-consuming and expensive [21].

Insurance companies, as stated in [22], spend their time gathering application infor-mation because of its complexity. In most cases, an applicant for health insurance will be required to provide results from a battery of medical exams or other supporting evidence to the insurance company. The insurance company thinks about the applicant based on their history with the company, and then accepts or rejects the application. When that's done, premiums may be calculated [23]. The author stated that the effectiveness of each of the ML algorithms we used here, looked into an insurance dataset from the Zindi Africa competition that was purported to be from Olusola Insurance Company in Lagos, Nigeria. These results showed that Zindi data showed insurance regulators, shareholders, management, financial experts, banks, accountants, insurers, and consumers were worried about the bankruptcy of insurance businesses. This concern resulted from a perceivable need to lessen management and auditing obligations while also protecting the general public from the effects of insurer insolvencies [24]. This article's author presents a plan for avoiding the bankruptcy of insurance companies. Previously, methods like logit analysis, multiple regression, recursive partitioning algorithm and another were used to tackle prediction problems. XGBoost was utilized to address the problem of claim prediction and assess its efficacy. Additionally, we evaluated how well XGBoost performed in comparison. Additionally, there are online learning methods and various ensemble learning strategies including AdaBoost, Stochastic GB, Random Forest, and Neural Network. According to our simulations, XGBoost outperforms other techniques in terms of normalized Gini [25]. The author was focused on the scammers take advantage of the growing market for this kind of insurance. Both the policyholder and the insurance provider might be guilty of insurance fraud. Exaggerated claims and out-of-date policies are two forms of client-side insurance fraud. Fraud is committed on the insurance vendor side, among other things, by enforcing rules from fictitious companies and failing to submit premiums. Several classification strategies are compared and contrasted in this work [26]. The author was showing their concern related to the healthcare industry and among their patients agree that doctors and nurses play an essential role in the healthcare system. In the field. On the other hand, healthcare expenses It's quite difficult to provide an accurate estimate since a great bulk of the funds come from people with rare disorders. The field of machine learning has advanced to the point where there is a wide variety of methods of forecasting. confidence in the precision of these methods. However, the confidence in the forecasted outcome is low [27].

Unfortunately, the training time required for machine learning models prevents them from immediately being put to use from active use in the present. Because of this, studies aim to create novel ensembles for calculating individual insurance premiums, with the goal of improving prediction accuracy [28]. The author identify the different ensemble models were developed using various methods, such as boosting, bagging, and assembling. The use in the estimation of health insurance costs challenges this research. The findings of the experiments provide evidence that the revised assembly method relies on The prediction accuracy of machine learning approaches for completing the stated task is greater than that of prior version, form [29].

3 FRAMEWORK

In this study, we have used machine learning methods to analyze the Health insurance information. A database of medical insurance costs is acquired from KAGGLE's [30] repository, and executed preparation of the data. After preparing the data, we choose the traits by means of engaging in feature engineering. The dataset is split in two: a training set and a testing set. Typically, the former contains around 70% of the data and the latter 30%. To forecast medical insurance costs, regression models are first trained using the training dataset. Figure 1 depicts our process for getting things done.

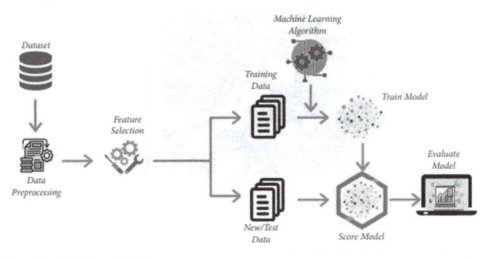

Figure 1. Working methodology.

3.1 *Dataset*

One, a collection of data individual healthcare expenditure datasets may be downloaded from the KAGGLE. Miri Choi submitted this seven-attribute dataset in 2018 [31] show in Table 1 provides an overview of the dataset.

Table 1. An explanation of the dataset.

Sr. No.	Feature name	Description
1	Age	Age is a major consideration in medical care.
2	Sex	Gender
3	BMI (Body mass index)	Knowing the human body: excessive weight disproportionately tall or short in comparison to height
4	Children	Dependents/Children Count
5	Smoker	Smoking state
6	Region	Area of residence
7	Charges	Medical expenses covered by health insurance

3.2 *Using a correlation matrix and feature engineering*

Feature engineering is the method used in machine learning to enhance the efficiency of ML algorithms via step-by-step procedure involving the application of know-how in a specific field to raw data. Attributes like smoking status, body mass index, and patient age play a crucial role in the medical insurance cost statistics. Additionally, we see the following: sex, children, and region have no bearing on the costs. It's possible that we'll decide not to include these three columns after generating the heat map graph to establish the connection between the dependent value and the independent attributes.

With the help of the heat map, it is simple to see which characteristics are most connected to one another or to the dependent variable of interest. The results are shown in Figure 2.

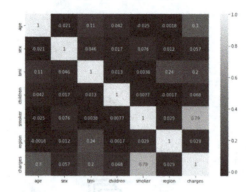

Figure 2.   Correlation matrix with the heat map.

3.2.1   *Feature selection and extraction*

Principal Component Analysis (PCA) is often used to extract features from data and decrease the dataset's dimensionality. With this method, high-dimensional data is projected onto a new subspace with the same number of dimensions as the original data set, or a smaller number of dimensions if the data set is very sparse [31].

From Figure 3, We can identify the correlation between the different features and prove that age, BMI and smoker play the important role for training the model and providing the correct data.

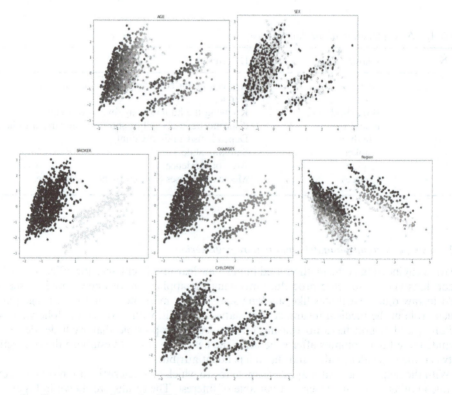

Figure 3.   Feature extraction using PCA method.

4   RESULTS AND DISCUSSION

Results from real-world applications of machine learning models are described in the section. This allows us to go forward with exploratory data analysis in which features are plotted against each other (in terms of costs) visualization.

4.1   *Age vs. charges*

Figure 4 shows that insurance premiums will rise as people become older. We can see that the insurance premium increases to $6,3770 after a person reaches the age of 54. The x-axis represents age and the y-axis represents fees.

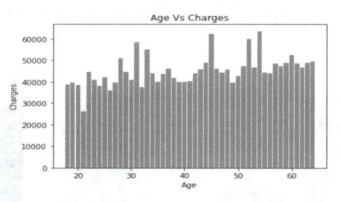

Figure 4.   Details of the insurance premium based on age.

4.2   *Region vs. charges*

In Figure 5 we see how the cost of insurance varies throughout the country. In comparison to other parts of the country, health insurance premiums in the South are higher. The x-axis indicates the location while the y-axis depicts the accumulated charges.

4.3   *Costs vs. body mass index*

Specifically, in Figure 6, females are represented by the zero value, while men are represented by the one value. On the x-axis are male and female body mass index numbers, and on the y-axis are the fees. When demonstrated in Figure 5, insurance premiums will change as body mass index values are adjusted.

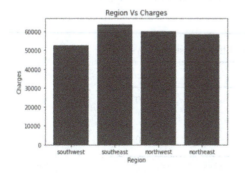
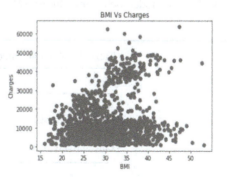

Figure 5.   Region vs insurance charges feature.   Figure 6.   Index of body mass to insurance.

4.4 Smoker count

Figure 7 displays the range of premiums a typical smoker might expect to pay for health insurance. Health insurance premiums for women are higher than those for men because women tend to live longer than men do and because women are less likely to smoke. Figure 7 shows that when men's smoking rates rise, their insurance premiums will fall but women's will rise. On the x-axis, smokers' values are plotted against associated fees.

4.5 Sex vs. charges

Figure 8 displays the sex categories on the x-axis and the associated medical insurance premiums on the y-axis, clearly demonstrating that males pay more than females for health coverage. The statistic shows that the average insurance premium for women is 12569, while for men it is roughly 13956.

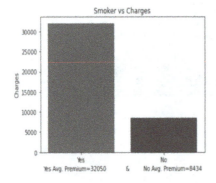

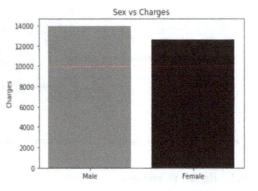

Figure 7.   Smoker vs charges feature.   Figure 8.   Sex vs insurance charges features.

4.6 Performance of ML algorithm

Root mean squared error (RMSE), training, and test scores, and cross-validations are shown for each method in Table 2. Figure 9 illustrates the precision and accuracy values of all machine learning (ML) algorithms for easier comprehension. These ML models' RMSE values can be compared to those of other ML models, and ANN offers a high RMSE value of 96.03%. We found that Support Vector Machine, Random Forest, Linear Regression, and Natural Algorithm performed better than the other Machine Learning (ML) techniques, with these models achieving almost 82%, 87%, and 74% accuracy, respectively.

Table 2.   Accuracy model for different insurance system.

Model	Accuracy
ANN	96.03
Random Forest	87
SVM	82
Linear Regression	74.08

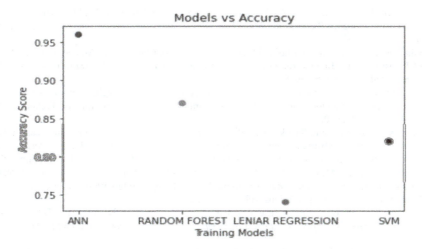

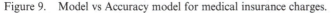

Figure 9.   Model vs Accuracy model for medical insurance charges.

5   CONCLUSION

The machine learning is well suited in the field of health insurance. Artificial intelligence and machine learning are capable to predicting a large amount of data of health insurance. The research and development are still needed to address the problem of precisely determining the costs related to health insurance. By utilizing a variety of ML methods, this study applies computational intelligence to the challenge of forecasting costs associated with health insurance. To train and test the support vector machine, random forest, linear regression, decision tree, k-Nearest neighbor, and natural algorithm, the medical insurance dataset was made available via the KAGGLE repository. Before regression analysis, this dataset underwent preprocessing, feature engineering, data splitting, regression, and evaluation procedures. The final result showed that ANN is providing 96.03% greater accuracy than other algorithms. There are numerous insurance prediction which has not been explored fully in this research.

REFERENCES

[1]   Sommers B.D., "Health Insurance Coverage: What Comes After the ACA?" *Health Affairs*, vol. 39, no. 3, pp. 502–508, 2020
[2]   Milovic B. and Milovic M., "Prediction and Decision Making in Health Care Using Data Mining," *Kuwait Chapter of the Arabian Journal of Business and Management Review*, vol. 1, no. 12, 2012.
[3]   Morid M.A., Kawamoto K., Ault T., Dorius J. and Abdelrahman S., "Supervised Learning Methods for Predicting Healthcare Costs: Systematic Literature Review And Empirical Evaluation," In *Proceedings of the AMIA Annual Symposium Proceedings, American Medical Informatics Association*, Washingdon, DC, USA, vol. 2017, November 2017.
[4]   Kumar M., Ghani R. and Mei Z.S., "Data Mining to Predict and Prevent Errors in Health Insurance Claims Processing," In *Proceedings of the 16th ACM SIGKDD International Conference on Knowledge Discovery and Data Mining*, Washington, DC, USA, pp. 65–74, July, 2010.
[5]   Yang C., Delcher C., Shenkman E. and Ranka S., "Machine Learning Approaches for Predicting High Cost High Need Patient Expenditures in Health Care," *BioMedical Engineering Online*, vol. 17, no. 1, pp. 131–220, 2018.
[6]   Iqbal M. and Yan Z., "Supervised Machine Learning Approaches: A Survey," *ICTACT Journal on Soft Computing*, vol. 5, no. 3, 2015, https://www.potentiaco.com/what-is-machine-learningdefinition-types-applications-and-examples/.

[7]   Panay B., Baloian N., Pino J.A., Peñafiel S., Sanson H., and Bersano N., "Predicting Health Care Costs using Evidence Regression," *Multidisciplinary Digital Publishing Institute Proceedings*, vol. 31, no. 1, p. 74, 2019.

[8]   Ghani M.U., Alam T.M. and Jaskani F.H., "Comparison of Classification Models for Early Prediction of Breast Cancer," In *Proceedings of the International Conference on Innovative Computing (ICIC)*, Lahore, Pakistan, November.2019.

[9]   Shaukat K., Iqbal F., Alam T. M. *et al.*, "+e impact of Artificial Intelligence and Robotics on the Future Employment Opportunities," Trends in Computer Science and Information Technology, vol. 5, no. 1, pp. 50–54, 2020.

[10]   Yang X., Khushi M. and Shaukat K., "Biomarker CA125 Feature Engineering and Class Imbalance Learning Improves Ovarian Cancer Prediction," *In Proceedings of the IEEE AsiaPacific Conf. on Computer Science and Data Engineering (CSDE)*, pp. 1–6, Gold Coast, Australia, December 2020.

[11]   Alam T.M., Khan M.M.A., Iqbal M.A., Abdul W. and Mushtaq M., "Cervical Cancer Prediction through Different Screening Methods using Data Mining," International Journal of Advanced Computer Science and Applications, vol. 10, no. 2, 2019.

[12]   Mohamed H., "Predict Health Insurance Cost by using Machine Learning and DNN Regression Models," *International Journal of Innovative Technology and Exploring Engineering (IJITEE)*, vol. 10, 2021.

[13]   Kaushik, K., Bhardwaj, A., Dwivedi, A.D. and Singh, R. Machine Learning-Based Regression Framework to Predict Health Insurance Premiums. *Int. J. Environ. Res. Public Health*,19, 7898, 2022

[14]   Cevolini, A. and Esposito, E. *From Pool to Profile: Social Consequences of Algorithmic Prediction in Insurance. Big Data Soc.*, 2020

[15]   Fauzan M.A. and Murfi H., "Accuracy of XGBoost for Insurance Claim Prediction," *International Journal of Advanced Software Computer Applications*, vol. 10, no. 2, 2018.

[16]   Van B.S., Flexible Imputation of Missing Data, *CRC Press*, Boca Raton, FL, USA, 2018.

[17]   Pesantez-Narvaez J., Guillen M. and Alcañiz M., "Predicting Motor Insurance Claims using Telematics Data-XGBoost Versus Logistic Regression," *Risks*, vol. 7, no. 2, 2019

[18]   Bhardwaj, N. and Anand, R. "Health Insurance Amount Prediction," *Int. J. Eng. Res.* 2020

[19]   Nithya B. and. Ilango V, "Predictive Analytics in Health Care using Machine Learning Tools and Techniques," In *Proceedings of the International Conference on Intelligent Computing and Control Systems (ICICCS)*, IEEE, Madurai, India, pp. 492–499, June 2017.

[20]   Tike A. and Tavarageri S., "A Medical Price Prediction System Using Hierarchical Decision trees," In *Proceedings of the IEEE International Conference on Big Data (Big Data)*, IEEE, Boston, MA, USA, pp. 3904–3913, December 2017.

[21]   Boodhun N. and Jayabalan M., "Risk Prediction in Life Insurance Industry Using Supervised Learning Algorithms," *Complex & Intelligent Systems*, vol. 4, no. 2, pp. 145–154, 2018.

[22]   Carson J.M., Ellis C.M., Hoyt R. E., and Ostaszewski K., "Sunk Costs and Screening: Two-part Tariffs in Life Insurance," *Journal of Risk & Insurance*, vol. 87, no. 3, pp. 689–718, 2020.

[23]   Wuppermann A.C., "Private Information in Life Insurance, Annuity, and Health Insurance Markets," *Scandinavian Journal of Economics*, vol. 119, no. 4, pp. 855–881, 2017.

[24]   Ejiyi, C.J., Qin, Z., Salako, A.A., Happy, M.N., Nneji, G.U., Ukwuoma, C.C., Chikwendu, I.A. and Gen, J. Comparative Analysis of Building Insurance Prediction Using Some Machine Learning Algorithms. *Int. J. Interact. Multimed. Artif. Intell.* 7, 75–85, 2022.

[25]   Rustam, Z. and Yaurita, F. Insolvency Prediction in Insurance Companies Using Support Vector Machines and Fuzzy Kernel C-Means. *J. Phys. Conf. Ser.*, 1028, 012118, 2018

[26]   Fauzan, M.A. and Murfi, H. The Accuracy of XGBoost for Insurance Claim Prediction. *Int. J. Adv. Soft Comput. Appl.*, 10, 159–171, 2018.

[27]   Rukhsar, L., Bangyal, W.H., Nisar, K. and Nisar, S. Prediction of Insurance Fraud Detection Using Machine Learning Algorithms. *Mehran Univ. Res. J. Eng. Technol.* 2022

[28]   Kumar Sharma, D. and Sharma, A. Prediction of Health Insurance Emergency Using Multiple Linear Regression Technique. *Eur. J. Mol. Clin. Med.* 2020

[29]   Choi M., "*Medical Cost Personal Datasets*," 2018, https://www.kaggle.com/mirichoi0218/insurance

[30]   *Principal Component Analysis*, Available on: https://vitalflux.com/feature-extraction-pca-python-example/

Automation and Computation – Vats et al. (Eds)
© 2023 the Author(s), ISBN 978-1-032-36723-1

An empirical study on the implications of Artificial Intelligence (AI) in hotels of Uttarakhand

Ashish Dhyani
Assistant Professor, Department of Hospitality, Graphic Era Deemed to be Dehradun, India

Yashveer Singh Rawat
Assistant Professor, Renaissance College of Hotel Management and Catering Technology, Ramnagar, Kumaun University, Uttarakhand, India

Resham Taluja
Assistant Professor, CSE, Graphic Era Hill University, Dehradun, India

Ishita Uniyal
Assistant Professor, EEE, Graphic Era Hill University, Dehradun, India

Shiv Ashish Dhondiyal
Assistant Professor, CSE, Graphic Era Deemed to be University, Dehradun, India

Sumeshwar Singh
Assistant Professor, CSE, Graphic Era Hill University, Dehradun, India

ABSTRACT: Although intelligent technologies are widely used in many facets of contemporary society, services need to adopt certain strategies. Basic interactions involve helping clients and those in charge of providing services to them. The paper's objective is to examine and assess theimplementation of Artificial Intelligence (AI) in the hotel industry in its advanced form. This research reveals that personalizedand customized requirements of the guests, chatbots and messaging apps, machine learning-powered business intelligence tools, augmentedand virtual reality are the key deployments. Additionally, we conducted a survey (n = 238 respondents) and asked participants regarding opinions and mindset concerning AI, particularly how they felt about its use in hotels. The paper tries to address how potential customers might encourage the hotel business to use particular AI implications. However, as the findings are not definitive, more study on this subject is still required. The development of models to: a) quantify the pros and cons of implementations of AI; b) identify and assess the aspects influencing customer acceptance of AI; and c) assess the overall perception of the guests regarding the AI applications in hotels may be the subject of future studies using both qualitative and quantitative methods.

Keywords: Artificial Intelligence, HM, Machine Learning, Deep Learning, Data Science.

1 INTRODUCTION

The service sector has assumed a bigger role in contemporary life over the previous few decades. The customer-focused, innovations aimed at the hospitality sector, as well as the various other industries where intelligent technologies have been effectively deployed, seem to be the most promising and well-received by customers. Businesses in the hotel sector can use existing on-site services and procedures and enhance visitor experiences with the aid of

AI. It's crucial to stay in touch with visitors and attend to their demands in order to maintain overall quality. Since AI makes it possible to give individualized and customized experiences, the idea of a "smart hotel" has itself gained significant attention from the academic and corporate worlds [1]. To our knowledge, however, very few researches have examined the penetration of the variousapplications of AI (AI) in the hotel sector. This study uses a qualitative technique, specifically a literature review and logic analysis, to investigate and assess the most recent studies in the field. In this regard, we made use of the information sources that were accessible. It is important to note that these suppliers, which cover academic literature from virtually every area, are well-known and well regarded by the scientific community.

2 LITERATURE REVIEW

No matter the sector, the implementation of automatic technologies and robotics in business operations has demonstrated to give significant competitive edge by incorporating AI into corporate planning and decision-making processes. Furthermore, there is strong evidence that AI will promote creativity by hastening the ability to offer of new services and goods. But now, the subject of what AI actually is comes up. [2]

Let's look back to 1998, when the identical query was posed by John McCarthy, quoted, "creating intelligent machines, particularly clever computer programmes, is the science and engineering of it. Although it is related to the related job of utilizing computers to comprehend human intelligence, AI need not be limited to observable processes. [3] The interested reader is directed to the writings of Leg and Hutter, Kok *et al.*, and Dobrev for a more thorough summary. Owoc *et al.*, Pondel *et al.*, and Hernes *et al.* studies's on applications of AI are also fascinating. AI (AI) has been encroaching commercial sectors recently, including hospitality. It is a technical aid that assists in processing and analyzingdata pertaining to customer satisfaction. Actually, the ability to improve organizational performance and expedite the achievement of strategic goals seems to support the AI application justification [4].

For instance, Ruel and Njoku argue that the character, structure, and circumstances of labour are likely to be greatly affected by AI and robotics. Even though AI is still in the research and deployment stages, there are already effective real-world applications in the hotel industry [5].

2.1 *In-person customer service*

Natural language processing algorithms and a short learning curve enable AI systems to "speak" in a language that is comparable to that of people. The basic responsibilities that may be carried out with no staff include making reservations, checking guests out, answering consumer concerns, addressing common issues, or even assisting with hospitality [6].

The Hilton group's Connie AI robot, which was used as the best example of this thus far, is one such case. Customers who interact with the robot can obtain tourist information from it. The ability to adapt to different people and learn from human speech is what is most astounding. In the end, this means that as more users interact with it, the better it will become [7].

2.2 *Chatbots and messaging*

AI chatbots have been employed on social media sites to allow users to ask questions and get answers almost quickly, seven days a week, twenty-four hours a day. It helps in such a way so as to save time and provide relevant information. For exampleSam, an intelligent travel chatbot assisting frequent travelers and corporate tourists [8].

2.3 Business Intelligence tools powered by Machine Learning

Business Intelligence (BI) is "a collection of approaches, processes, architectures, and technology that transform unstructured data into actionable knowledge and usable information." According to recent data, the hotel sector is seeing a surge in studies on BI and Big Data. Machine Learning (ML) techniques are frequently coupled with contemporary Business Intelligence (BI) platforms to find deep insights and hidden patterns in data. Thanks to improvements in computing power, the hospitality industry can now more easily create and use complicated data models. An organisation can use this technology to sort through data gathered from surveys, online evaluations, and other sources. The AI can then analyse this data to make judgments about the performance of the entire business [9].

2.4 Virtual reality and augmented reality

AI can also be helpful with combination with other technologies like virtual reality. A virtual tour through a 360-degree video provides virtual experience to the prospective customers. Not only the flight, the arrival, and some of the most memorable views are just a few of the travel-related experiences that users can practically recreate, customers can create their personalized and customized products and services as per their requirements through the applications of AI [10].

3 FINDINGS

AI is playing a bigger and bigger role in the management of the hotel sector since it can perform typically human tasks whenever it wants. Furthermore, it is claimed that the use of AI can produce advantages including reductions in operating costs, minimizing human mistakes, and effective, efficient, prompt and standardized delivery of products and services.

It is significant to note that both autonomous and on-site solutions as well as mobile apps that connect visitors with a hotel's services and technology are covered by the developments. A growing trend in hotels is the installation of speech-activated support staff and voice control technologies [11].

4 METHODOLOGY

There questionnaire consisted of 12 questions. The survey's objectives were to investigate guests' attitudes towards AI used in the hospitality sector as well as public understanding of and openness to them. Online surveying for the study took place during May and June 2022. The respondents ranged in age, resided in cities and towns of various sizes. Without the interviewer asking them any questions, each responder completed the questionnaire on their own.

An online survey created with Google Forms was used to gather the data. The interested users received the link through email. A total of 238 replies were collected, and they were then the subject of additional analysis using the well-known spreadsheet program.

Figure 1 shows that maximum respondents (n=78) are in the 19–25 age range. Over 56s and those under 55 were the groups with the least number of members.

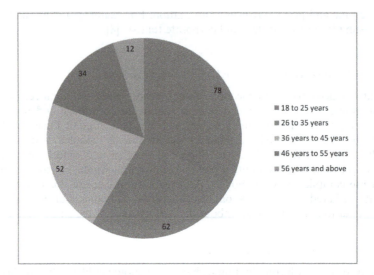

Figure 1.　The distribution of the respondents' age.

5   RESULTS AND DISCUSSION

5.1   *Knowledge regarding the implication of AI*

Figure 2 shows that more than half (n=122) of respondents think their knowledge of AI is sufficient, while roughly 36.91% think they possess sufficient knowledge about AI. The least amount of responders (n=28) believe they lack expertise. Since the majority of people are knowledgeable about AI, the results are straightforward. Given the range of age groups, a different conclusion may be anticipated if these participants had grown up with technology.

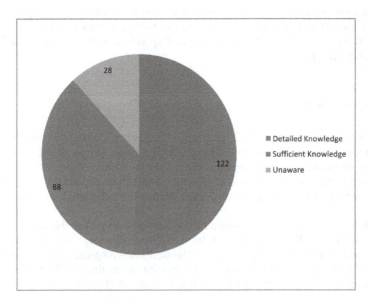

Figure 2.　Respondents knowledge about AI.

5.1.1 *Technological understanding of AI among respondents*

The three options that, in the respondents' judgement, best define the phrase "AI," were presented to the respondents. Maximum respondents (n=92) chose this option the most frequently: "technology assisting humans." Least respondent (n=10) followed by (n=18) cast a vote for the least preferred response, which is "technology express" and "technology with logical reasoning abilities" respectively. A "technology that can work without human indulgence" (n=54) and "technology that can replace humans" was opined by the majority of respondents.

"Self-reliant technology" was also suggested in a (n=22) answers. The topic is summarized in Figure 3.

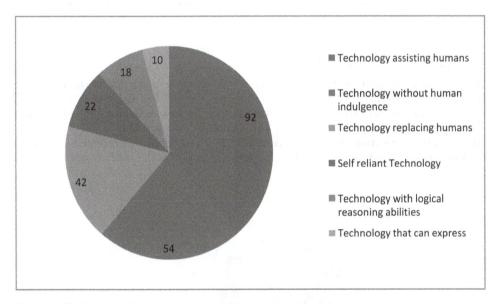

Figure 3.   Technological interpretations of AI in respondents' opinions.

5.2 *Implementation of AI in the hotel sector*

On a scale of 0-3 (where 0 = unnecessary, 1= Can't say, 2 = Useful 3 = Very much necessary), the responses were recorded to analyze the data pertaining to the guest stay in a hotel obtained through the Customer feedback form.

The findings demonstrate that Electronic Locking System and the advantages it provides is the factor that matters most to the respondents. Nearly 78% of the respondents in a combined form rated the presence of Electronic Locking Systemimportantand very much necessary followed by EPABX Electronic Private Automatic Branch Exchange, hotel's own telephone exchange with approx 70% responses. A significant portion of responders believe that KIOSK Self check-in and check-out terminals component is entirely important with a total of around 67% respondents opting for the same. The findings reveal that robotic assistance and meticulous technologies like scanning goes for disregard for very sophisticated technology components was also evident.

The following inquiry focused on the eye-based payment mechanism (PayEye [35]), and it was expanded to include three questions about the convenience, security, and decision-making process.

Table 1. Mindset of respondents regarding the variousapplied AI solutions in hotels.

| | Answers | | Total (100%) | |
Solutions	0	1	2	3
Electronic Locking System	9.2%	13%	38.2%	39.4%
EPABX Electronic Private Automatic Branch Exchange	11.7%	18.4%	32.3%	37.3%
KIOSK Self check-in and check-out terminals	13.4%	19.3%	31.5%	35.7%
Robotic assistance and automation in air conditioning, TV, Wi-Fi	18%	21.8%	28.5%	31.5%
CCTV, Safety and Security measures like scanners, X ray machines etc	21.4%	26%	23.5%	28.9%

5.3 Supporting hotels through AI solutions

Another query centered on the application of AI during a pandemic is the most contemporary and relevant question. 161 (67.64%) respondents think it is absolutely vital and 22 (9.24%)respondents marked it as unimportant to look for new technology solutions. More than 23% of respondents i.e. 55 said they were undecided on the matter.

It's important to remember that the AI is allegedly a crucial factor to combat COVID-19 pandemic which is undoubtedly the biggest public health crisis the modern world has ever experienced.

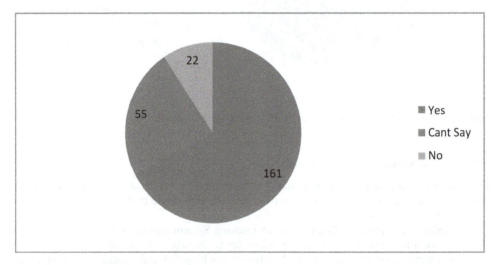

Figure 4. Mindset of respondents on necessity of using intelligent technology during pandemic.

6 CONCLUSIONS

It appears that there is a clear focus on assisting clients by providing a variety of services through the use of AI technologies. The application of technologies pertaining to AI in the hotel industry can also draw guests and benefit employees too. Present study was conducted during the unlock period of pandemic, so for the customers we still have now, it was undoubtedly out of the ordinary, but the use of applied technology improved.

Our findings are merely a few research components of into the implementation of AI in hotels. Second, while there are variety of AI solutions used in hotels, using breakthroughs in hospitality industry networks appears superior. Thirdly, despite the respondents being relatively young, their knowledge of AI and its applications is fairly restricted. According to a guest, automation of manual labour and leisure services are most desired, whereas translation, document preparation, and creative pursuits are not significant [12]. The self-service check-in/check-out process and assistance for guests checking in or out of the hotel are the two most crucial AI solutions in the hotel industry. However, according to our respondents, humanoid robot assistance is still an unappealing service.

With both implication of qualitative and quantitative methods, models can be developed further emphasizing on the advantages and disadvantages of AI and to ascertain the extent to which such technologies and its application is acceptable to the customers. The overall perspective of the customers regarding AI can be a addressed at a larger scale in order to identify the need for the aspects to be inculcated [13].

REFERENCES

[1] Buhalis D., Leung R. Smart Hospitality—interconnectivity and Interoperability Towards an Ecosystem. *International Journal of Hospitality Management* 71, 41–50, 2018.

[2] Cockburn I.M., Henderson R., Stern S. The Impact of AI on Innovation. *Technical Report. National Bureau of Economic Research*, 2018.

[3] Czerniachowska K., Hernes M. A Heuristic Approach to Shelf Space Allocation Decision Support Including Facings, Capping, and Nesting. *Symmetry* 13, 314, 2021.

[4] Dananjayan S., Raj G.M. AI During a Pandemic: The Covid-19 Example. *The International Journal of Health Planning and Management*, 2020.

[5] Franczyk B., Hernes M., Kozierkiewicz A., Kozina A., Pietranik M., Roemer I., Schieck M. Deep Learning for Grape Variety Recognition. *Procedia Computer Science* 176, 1211–1220, 2020.

[6] Gaafar H.A.A.S.M. AI in Egyptian Tourism Companies: Implementation and Perception. *Journal of Association of Arab Universities for Tourism and Hospitality* 18, 66–78, 2020.

[7] Gawlik-Kobylinska M., Maciejewski P. New Technologies in Education for Security and Safety, In: *Proceedings of the 2019 8th International Conference on Educational and Information Technology*, pp. 198–202, 2019.

[8] Kalinowski M., Baran J., Weichbroth P. The Adaptive Spatio-temporal Clustering Method in Classifying Direct Labor Costs for the Manufacturing Industry, In: *Proceedings of the 54th Hawaii International Conference on System Sciences*, pp. 236–243, 2021.

[9] Kamola M., Arabas P. Improving Time-series Demand Modeling in Hospitality Business by Analytics of Public Event Datasets. *IEEE Access* 8, 53666–53677, 2020.

[10] Neuhofer B., Buhalis D., Ladkin A. Smart Technologies for Personalized Experiences: A Case Study in the Hospitality Domain. *Electronic Markets* 25, 243–254, 2015.

[11] Batinić I. Organization of Business in Hotel Housekeeping. *Journal of Process Management – New Technologies, International* Vol. 3, No.1, 51–54, 2015.

[12] Marić D., Marinković V., Marić R. & Dimitrovski D. Analysis of Tangible and Intangible Hotel Service Quality Components. *Industrija*, Vol. 44, No.1, 7–25, 2016.

[13] Olsen Michael D. and Connolly Daniel J., Experience-based Travel: How Technology Is Changing the Hospitality Industry. *Cornell Hospitality Quarterly*, Vol. 41, No. 1, 30–40, 2000.

Automation and Computation – Vats et al. (Eds)
© 2023 the Author(s), ISBN 978-1-032-36723-1

The study of grid-connected induction generators applicable to wind power

Ishita Uniyal
Assistant Professor, Department of Electrical Engineering, Graphic Era Hill University, Dehradun, India

Shiv Ashish Dhondiyal
Assistant Professor, Department of Computer Science & Engineering, Graphic Era Deemed to be University, Dehradun, India

Manisha Aeri, Manika Manwal & Sonali Gupta
Assistant Professor, Department of Computer Science & Engineering, Graphic Era Deemed to be University, Dehradun, India

ABSTRACT: The use of induction generators in wind power application has become increasingly common, particularly in recent decades. A prime mover drives the rotor at a higher speed than synchronous speed in generator operation. On the basis of reactive power we can divided induction generator in two categories: first one is the standalone generator and second one is grid-connected induction generator. In the case of a grid-connected generator, magnetizing current is drawn from the grid, whereas in the case of a standalone induction generator, the magnetizing flux is produced by a capacitor bank connected to a machine. The base of induction generator is faraday's law of electromagnetic induction. If there is current flowing through the stator coil, then the stator will generate flux; as a result, the rotor will induce current. A grid-connected induction generator is the subject of this study, where frequency and voltage are determined by the grid. DFIG (dual-fed induction generator) wind turbines are one type of IG. They benefit from the capability to provide energy at a consistent voltage and frequency. In MATLAB-SIMULINK, we look at modern control approaches such as vector control and MFC (magnitude and frequency control) and model the performance of several systems.

Keywords: The wind power scenario in India, Grid-connected renewable energy generation, tariff, VC - vector control, MFC - magnitude and frequency control, DFIG - dual-fed induction generator.

1 INTRODUCTION

Wind power is the process of transforming wind power into a usable form of power, like generating electricity with wind turbines, providing mechanical energy with windmills, pumping water with wind pumps, or propelling ships with sails [1]. The total amount of economically extractable wind energy available surpasses current human energy consumption from all sources put together. Wind power is the world's rapidly-growing energy source. Instead of more environmentally friendly technologies like hydroelectric power, Coal is used to create the majority of electricity [3]. CO_2 and other hazardous pollutants from the burning of coal cause significant environmental damage. The energy sector is undergoing significant transformations.

Wind power has a variety of edges that make it a viable energy source for both big-scale and small-scale dispersed generating station. Wind energy provides a number of advantages:

• Wind power is an emission-free and never-ending source of energy since it emits no pollutants and does not decrease with time. In a single year, a one-megawatt (1 MW) wind

     DOI: 10.1201/9781003333500-11

turbine may displace around 1,500 tonnes of CO_2, 6.5 tonnes of SO_2, 3.2 tonnes of NO and 60 pounds of Hg. Technology that is modular and scalable—Wind applications come in a variety of shapes and sizes [4–6].

1.1 *Statistics*

India, like other countries, has taken a step ahead as a result of the rapidly expanding need for power and an emphasis on renewable energy. By the end of 2011, total installed wind capacity was slightly shy of 238 GW, according to the 2011 Global Wind Report. India has 16085MW of installed wind power producing capacity, accounting for 6.8% of global wind power capacity [6].

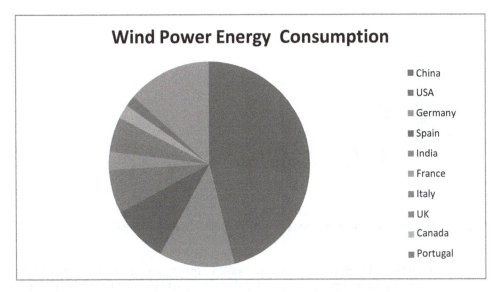

Figure 1.   Cumulative wind power statistics for the world.

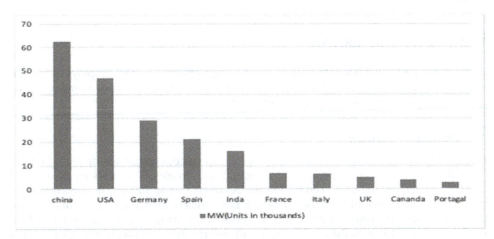

Figure 2.   Cumulative wind power statistics in terms of Megawatt (MW) units.

2   AN INDIAN PERSPECTIVE ON WIND ENERGY

India's Ministry of Non-Conventional Energy Sources (MNES) was established in the early 1980s to promote energy diversification and fulfill the countries rapidly expanding energy demands. Afterwards ministry was renamed as the Ministry of New and Renewable Energy (MNRE) in 2006. India overtook China as Asia's second largest wind power market in the first decade of the twenty-first century. Its total installed capacity is approaching 13 GW, and the market has expanded at a pace of more than 20% per year over the last three decades [7–9]. During the fiscal year 2010–11, more than 2,100 MW of wind capacity was installed. By December 2010, installed capacity had risen from 41.3 MW in 1992 to 13,065.78 MW. In 2019, India's wind energy capacity increased to around 37 thousand megawatts, up from over 35 thousand megawatts in 2018. During the year 2019, the total wind energy capacity of all countries on the planet was over 100 Giga watts.

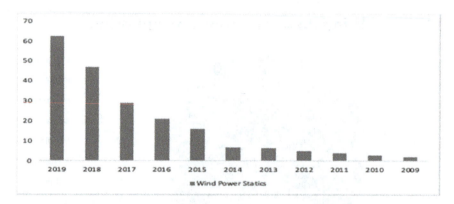

Figure 3.   Wind power statistics in India (capacity from 2009 to 2019 in megawatts).

Up till the end of 2017, global wind power installed capacity was at a high of 539581 MW. The top 10 ranked countries generate 455950 MW of wind energy. The remaining countries produce only 15% of worldwide wind power. As a result, China, the US, Germany, India, Spain, the UK, Canada, France, Italy, and Brazil are the leading countries in terms of wind installations (in 2017).

3   INDUCTION GENERATOR

An asynchronous generator often call as an induction generator is that employs the principles of induction motors to produce energy. Induction generators generate negative slip by spinning its rotor mechanically in generator mode.

3.1   *Principle of operation*

Electrical power is generated when the rotor of an induction generator or motor rotates faster than the synchronous frequency. A conventional 4-pole motor working on a 60 Hz grid has an operating speed of 1800 revolutions per minute while the same motor working on a 50 Hz grid has a speed of 1500 rpm. In typical motor functioning, the spinning of the stator flux is quick than the rotor rotating. This aims rotor currents to be triggered by stator flux, resulting in rotor flux with the opposite magnetic polarity as the stator [10,11]. As a result, a value equal to slip drags the rotor behind the stator flux.

3.2 Induction generator with grid connection

The utility grid provides the excitation for induction generators. The initiated energy is supplied to the supply system when the IG is run above synchronous speed. Generator with cage rotors has a low negative slip rate because they only feed through the stator. Wound rotor machines, also known as **DFIT** (Doubly Fed Induction Machines) [12,13], on the other hand can supply power to the bus via both the stator and the rotor.

3.3 Wind turbine generator with a fixed speed

Mechanical sub-circuits, such as pitch control time constants; influence shows the formation of grid-connected wind turbines in Figure 4.

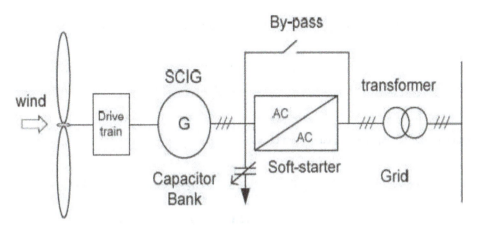

Figure 4.   Squirrel-cage induction generator setup for wind turbine at a fixed speed.

3.4 Doubly Fed Induction Generator (DFIG)

The Figure 5 below depicts a typical DFIG system: the converters of rotor side (Crotor), the grid side (Cgrid) and comprise AC/DC/AC. They are VSC that employ electrical compo-nents with forced commutation to generate AC voltage from a DC voltage source (IGBTS). The IG turns the turbine's captured wind energy into electricity, which is then sent to the grid via stator and rotor windings [14].

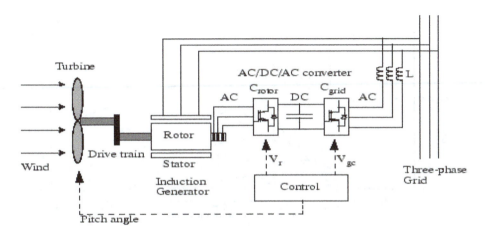

Figure 5.   A setup that includes a DFIG and a wind turbine.

93

3.5 Induction generator mathematical modeling

The system operates on the theory of electromagnetism, and power is transferred through transfer action. As a result, the machine can be compared to a transformer, except it is rotatory rather than stationary.

3.6 DFIG in a synchronously rotating frame model

Figures 6 and 7 show the comparable circuit design of an induction machine. In this picture, the model is portrayed as a 2-φ model; as previously mentioned, a 3-φ model may be depicted as a 2-φ model that functions according to specific rules [15–17]. To model DFIG in a synchronously rotating frame, the 2-φ ds-qs (stator) and dr-qr (rotor) circuit variables represented:

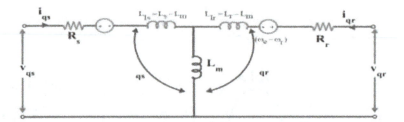

Figure 6. DFIG's q-axis (d-q equivalent circuit).

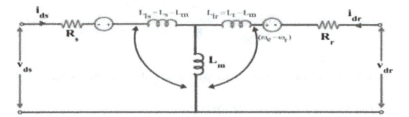

Figure 7. DFIG's d-axis (d-q equivalent circuit).

4 SIMULATIONS AND RESULTS

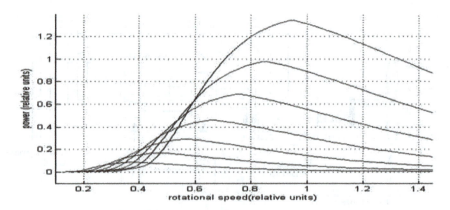

Figure 8. Power production of wind turbines vs. rotational speed and wind speed.

94

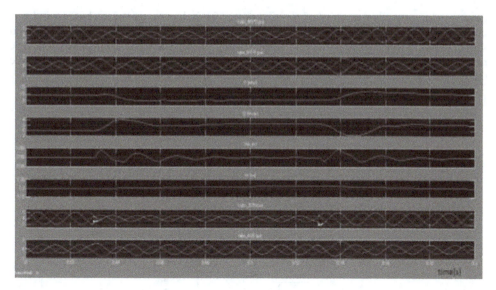

Figure 9.   Simulation results of DFIG model

In the above Figure 13 we simulated grid side and wind turbine side parameters and the corresponding results have been displayed.

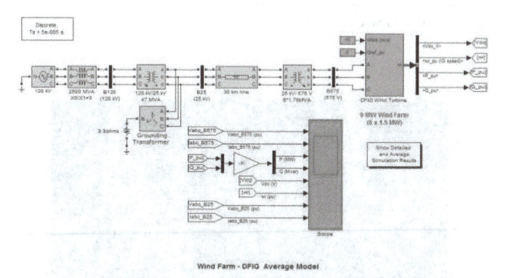

Wind Farm - DFIG Average Model

Figure 10.   Wtdfig evaluation in a 9mw wind farm integrated with a 25KV, 60 Hz system.

The above figure shows the plot for real power and reactive power generating 9 MW power. Figure 11 shows the real power increasing rapidly towards 1 per unit (6 × 1.5 MW). Coming to plot second in Figure 12 reactive power is settling down to almost zero. So we have to taking very less amount of reactive power for induction generator based wind turbine. Considering the results it can be said that DFIG proved to be more reliable and stable system when connected to grid side with the proper converter control systems. The rotor side converter usually provides active and reactive power control of the machine while the grid side converter keeps the voltage of the the DC-link constant.

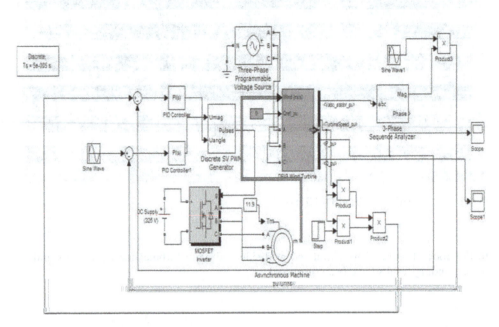

Figure 11.   Model for MFC in Simulink.

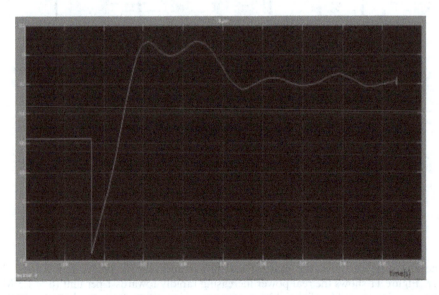

Figure 12.   Active power.

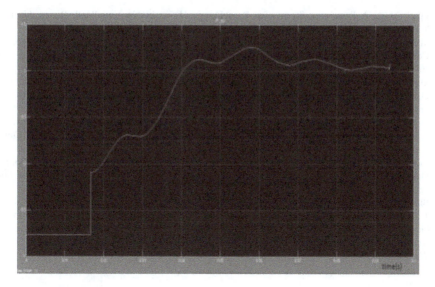

Figure 13.   Reactive power.

5   CONCLUSION

DFIGs are widely employed in wind systems because they can produce power at a constant voltage and frequency. In the MATLAB environment, the characteristics of DFIG are explored. The DFIG's control approaches have been investigated. A Simulink model was proposed to examine the regulation of magnitude and frequency. The MFC methodology calculates the magnitude and frequency of the rotor voltage, unlike older systems. This enhances system reliability while simplifying the control system architecture.

REFERENCES

[1]   Indian Wind Energy Outlook 2011, *Global Wind Energy Council*, April, 2011.
[2]   Shaheen S.A., Hasanien H.M., and Abd-El-Latif Badr M., Study on Doubly Fed Induction Generator Control, IEEE press, 2010.
[3]   Wang Z., Sun Y., Li G. and Ooi B.T., Magnitude and Frequency Control of Grid connected Doubly Fed Induction Generator based on Synchronised Model for Wind Power Generation, *IET Renewable Power Generation*, 2010.
[4]   Zhixin Miao, Yuvarajan S. and Glower J., A Comparison of Slip Control, FMA Control and Vector Control in DFIG Converter, *IECON 2008, 34th Annual Conference of IEEE*, 2008.
[5]   Satish Choudhury, Kanungo Barada Mohanty and Chitti Babu B., Performance Analysis of Doubly-Fed Induction Generator for Wind Energy Conversion System, *Proc. PSUUNS International Conference on Engineering and Technology (ICET-2011)*, Phuket, Thailand, pp. 532–536, May 2011.
[6]   Bhadra S.N., Kastha D. and Banerjee S., *Wind Electrical Systems*, Oxford University Press, New Delhi, 2009.
[7]   Lara O.A., Jenkins N., Ekanayake J., Cartwright P. and Hughes M., Wind Energy Generation: Modeling and Control, John Wiley and Sons, UK, 2009.
[8]   Miller N.W., Sanchez-Gasca J.J., Price W.W. and Delmerico R.W., "Dynamic Modelling of GE 1.5 and 3.6 mw Wind Turbine-generators for Stability Simulations," *GE Power Systems Energy Consulting, IEEE WTG Modelling Panel*, SessionJuly 2003.
[9]   Hansen A.D. and Hansen L.H., Wind Turbine Concept Market Penetration over 10 years (1995–2004), *Wind Energy*, pp. 81–97, 2007.

[10] Iqbal, M.T., A Feasibility Study of a Zero Energy Home in Newfoundland, Renewable Energy, Vol. 29, No. 2, pp. 277–289, 2004.

[11] Machado, R.Q., Marra. and E.G., Pomilio., Bi-directional Electronic Interface of Induction Generator Connected to a Single Phase Feeder, *International Journal of Renewable Energy Engineering*, August 2002.

[12] Miao, L., Wen, J., Xie, H., Yue, C. and Lee, W.-J., Coordinated Control Strategy of Wind Turbine Generator and Energy Storage Equipment for Frequency Support. *IEEE Trans. Ind. Appl.*, pp. 2732–2742, 2015.

[13] Badmasti B. and Bevrani H., On Contribution of DFIG Wind Turbines in the Secondary Frequency Control. In *Proceedings of the 1st Conference on New Research Achievements in Electrical and Computer Engineering*, Tehran, Iran, pp. 1–7, May 2016.

[14] Fleming, P.A., Aho, J., Buckspan, A., Ela, E., Zhang, Y., Gevorgian, V., Scholbrock, A., Pao, L. and Damiani, R., Effects of Power Reserve Control on Wind Turbine Structural Loading. *Wind Energy*, pp. 453–469, 2016.

[15] Blaabjerg, F. and Ma, K. Wind Energy Systems, *Proc. IEEE*, pp. 2116–2131, 2017.

[16] Torkaman, H. and Keyhani, A. A Review of Design Consideration for Doubly Fed Induction Generator based Wind Energy System, *Electr. Power Syst. Res.*, pp. 128–141, 2018.

[17] Xia, Y., Chen, Y., Song, Y. and Strunz, K. Multi-Scale Modeling and Simulation of DFIG-Based Wind Energy Conversion System, *IEEE Trans. Energy Convers*, pp. 560–572, 2020.

Automation and Computation – Vats et al. (Eds)
© 2023 the Author(s), ISBN 978-1-032-36723-1

An image encryption and decryption process using singular value decomposition

Vineet Bhatt
Department of Mathematics, Chandigarh University, Mohali, Punjab, India

Raj Kishor Bisht*
Department of Mathematics and Computing, Graphic Era Hill University, Dehradun, Uttarakhand, India

Rajesh Kumar Tripathi*
Department of Mathematics, Graphic Era Hill University, Dehradun, Uttarakhand, India

ABSTRACT: An image is a collection of numbers in the form of matrix that constitutes different light intensities in different areas of the image. A secure communication of images over network is very important for different purposes. There are many ways to hide image data or transmission of image information secretly. In the present paper, we propose a secure image transmission technique based on matrix decomposition technique from linear algebra. Singular value decomposition (SVD) method factorizes a matrix into three different matrices. We factorize a gray image stored in the form of a matrix into three matrices using SVD and define a new method for secure encryption and decryption of digital image that can be used in various ways. MATLAB programming is used as an exemplary basis for coding purpose to show the utility of the proposed method. The proposed method provides a new and effective method for secure transmission of digital image over a network; therefore, it may be quite useful for many real intelligence applications.

1 INTRODUCTION

A computer network is used as a media for communication of messages and data. Secure transmission of messages through computer networks is an important issue and as the technology is changing rapidly day by day, challenges of secure transmission are also increasing. Thus, this area is attracting researchers to work continuously towards secure transmission. Transmission of images through networks is essential for many purposes such as army operations, medical reports etc. Thus, confidentiality of the image data is an important issue. Image encryption is a technique for sending image over public networks without any concerns about leak of secret image. In case, if someone hacks the network traffic between two communicating parties, the receiver may not be able to get the original image, because the encrypted image would be far different from the original image. Thus, the need of secure transmission is ever growing. There are different techniques available in the literature and still the work is continuous. Several researchers have done works for secure transmission of images. Here we elaborate some of the works. Pareek *et al.* (2013) discussed a gray image decryption scheme based on diffusion and substitution. Mandal (2012) used logistic map to define image encryption. Pakshwar *et al.* (2013) described various available

_____
*Corresponding Authors: bishtrk@gmail.com and pvats2009@gmail.com

DOI: 10.1201/9781003333500-12

image encryption and decryption techniques. El-Samie (2013) in their work examined various image encryption algorithms for secure wireless communication using permutation and substitution based two different approaches. Martin (2005) discussed coding a color image using C-SPIHT algorithm for compression. Steganography is quite useful in secure data transmission. Kini *et al.* (2019) used steganography concepts to transmit an image through another image. Borra (2019) presented a hybrid method for secure transmission of radiological color images utilizing the compressive sensing theory. Wang *et al.* (2018) proposed a chaotic image encryption scheme utilizing Josephus traversing and mixed chaotic map. Teng *et al.* (2018) proposed image encryption by converting a color image from three to one bit-level image. Singh *et al.* (2018) reviewed the various work done in the field of chaotic image encryption. Shankar and Lakshmanaprabu (2018) proposed a secure transmission utilizing homomorphic encryption. Yadav & Singh, K. (2017) used Arnold transform and singular value decomposition for enhancing security to images. For purpose of colour image encryption and decryption, modified Chua's circuit is used by (Batuhan A. 2020). The author also represented the importance and efficiency of this method and also tested the encrypted coloured image obtained through their method by a variety of tests including the secret key size and sensitivity analysis, information entropy analysis, noise attack analysis, occlusion analysis etc. Author claimed that this method is a good, time saving and secure communication technique. (Batuhan A. *et al.* 2020) proposed an approach for secure transmission and protection of an image data information. (Pengfei Fang 2021) proposed a block encryption algorithm utilizing two approaches generative adversarial networks and DNA sequence coding.

To ensure the security of digital images during transmission and storage, (Zhen Li *et al.* 2020) proposed a chaos-based color image encryption method. They have shown that their method has a good security performance and a speed in comparison to other works. (Zakaria S. B. & Navi K. 2022) investigated image encryption algorithms based on chaotic systems and asserted that the approximate adder circuit performance in the image encryption was indicative of an improvement in system speed, security, and power consumption. For digital image encryption and decryption, (Zhang Y. *et al.* 2022) presented a thumbnail-preserving encryption algorithm that reduced the processing time and obtained excellent visual quality of encrypted photos. They further proved that their method has resistance over several popular assaults like differential attack and face identification. The author (Gao X. *et al.* 2022) described a multiple-image encryption technique based on chaotic, diffuse, and single-channel scrambling systems. Author (Xian, Y., & Wang, X. (2021)) presented a new encryption method using an irregular and infinitely iterative fractal sorting matrix (FSM) for improving encryption techniques which is used in many areas of image processing. (Zou C. *et al.* (2021)) proposed a technique based on matrix algebra to improve the image encryption process and confirmed that proposed algorithm can effectively encrypt image data. (Acharya *et al.* 2008) used a randomly generated self-invertible matrix as an encryption key for the image data encryption and also used this for video encryption. (Rasras *et al.* 2019) used color image decryption-encryption technique based on matrix manipulation.

The present work utilizes a simple and a different approach from previous existing works. In this work, the authors first decompose RGB (3-dimensional image) image matrix into a gray image matrix A(2 dimensional), then apply singular value decomposition on A to get three different matrices U, S and V such that $A = USV^T$. Here, we propose encryption on different cells of S and send the keys in a different file to the receiver. This provides a new approach of encrypting an image.

1.1 *Singular value decomposition*

SVD is a well know method of linear algebra (De Leeuw 2006; Henry & Hofrichter 1992;) that factorizes a square matrix into three different matrices. Let us consider a real matrix A

of order $m \times n$. SVD factorizes the matrix A into three matrices, $U = [u_{ij}]_{m \times m}$, $S = [\sigma_{ij}]_{m \times n}$ and $V = [v_{ij}]_{n \times n}$ such that $A = USV^T$, where S is a diagonal matrix with non-negative numbers as diagonal elements and U and V are orthogonal matrices. The columns of U and V are left singular and right singular vectors of A and the diagonal elements of S are the singular values of A. Singular values of a square or non-square matrix A can also be calculated using the square root of non-zero eigen values of $A^T A$ (Akritas & Malaschonok 2004; Bhatt & Petwal 2013; De Lathauwer *et al.* 2000).

For any matrix $A \in (a_{ij})_{m \times n}$, the SVD factorize matrix A in to USV^t, where $U = [u_1, u_2, \ldots \ldots u_r, u_{r+1}, \ldots u_m]$ is a $(m \times m)$ orthogonal matrix such that set of column vectors u_i for $i = 1, 2, 3 \ldots, m$ generate an orthonormal set and column vectors of matrix $V = [v_1, v_2, \ldots \ldots, v_n]$ form an orthonormal set of vectors. Matrix S is a singular matrix of order $(m \times n)$, in which the singular values (σ_i) satisfies the conditions $\sigma_1 \geq \sigma_2 \geq \ldots \sigma_r > 0$, and $\sigma_{r+1} = \sigma_{r+2} = \ldots = \sigma_N = 0$.

Let us take an example, for a matrix $A = 1\,3\,2\,4\,5\,6$, the SVD of A is

$$U = -0.2298 \quad 0.8835 \quad 0.4082 \quad -0.5247 \quad 0.2408 \quad -0.8165 \quad -0.8196 \quad -0.4019 \quad 0.4082,$$

$$S = 9.5255\,0\,0\,0.5143\,0\,0, \quad V = -0.6196 \quad -0.7849 \quad -0.7849 \quad 0.6196$$

This concept of decomposing a matrix into three matrix using SVD, provides an important method to decompose any information stored in the form of a matrix into different parts. An image can be defined as a two-dimension function $f(x, y)$ where x and y are spatial coordinates of a two-dimensional image, and the amplitude of f at any pair of (x, y) is grey level of the image at that point. For example, a grey level image can be represented as f_{ij} where $f_{ij} = f(x_i, y_j)$. An image is called a digital image when x, y and the amplitude value of f are finite, discrete quantities, A digital image is a large data matrix so if we want to hide an image data than we can factorize the image data into different matrices using any matrix decomposition technique.

In the next section, we define our method which is based on a well-known method SVD of linear algebra. All the previous methods are based on complicated mathematical and computational processes, but the proposed method is quite simple and effective. Our method is quite different then above because we used a secure private key for decryption of an image. The singular value matrix S contained many real numbers accordingly image size in principle diagonal and we can use one or more than one key for encryption of a digital image data. This method is based on matrix algebra using MATLAB coding hence it is suitable, secure and time saving technique.

2   THE PROPOSED METHOD

In this section we propose a new secure image transmission process using singular value decomposition of matrices. The procedure uses MATLAB for encryption and decryption of images. The proposed procedure is as follows: for encryption, first, we convert an RGB image into a grey image and the image data is stored in the form of a matrix, called image matrix A. Using SVD, we decompose the image matrix A into three matrices U, S and V such that $A = USV^T$. Since the matrix S is a unique diagonal matrix representing the singular values on diagonals, thus one or more diagonal values can be encrypted to secure the image transmission. After encryption of some diagonal values, we find an encrypted diagonal matrix S_{en}. The keys are stored in a separate file. Four different files F_1, F_2, F_3 and F_4 containing the information of U, S, V and the keys respectively are sent to the receiver separately. The files can be sent through any suitable format like excel, csv, word etc. For decryption, at the receiver's end, the matrices are retrieved from the files $F_1, F_2,$ and F_3. The matrix S_{en} is decrypted using the keys in file F_4 to get the matrix S and finally, the product of three matrices USV is obtained to get the grey image matrix. The above procedure can be summarized as follows:

2.1 The encryption procedure

1. Convert the RGB (3-dimensional image) image into a gray image matrix A(2 dimensional), where A is a matrix of order $m \times n$. (Appendix A)
2. Apply singular value decomposition to A and find $U = [u_{ij}]_{m \times m}$, $S = [\sigma_{ij}]_{m \times n}$ and $V = [v_{ij}]_{n \times n}$ such that $A = USV^T$. (Appendix A)
3. Choose a set of diagonal elements (key values) from S and apply an encryption scheme to the values while keeping other unchanged and form an encrypted diagonal matrix S_{en} (See Figures 1 and 3).
4. Save the information of U, S_{en}, V and the key values in four files F_1, F_2, F_3 and F_4. The format of the files may be different like excel, csv, word etc.
5. Send files F_1, F_2, F_3 and F_4 to receiver separately (See Figure 1).

2.2 The decryption procedure

1. Retrieve the matrices from files F_1, F_2 and F_3 and decrypt the matrix S_{en} into S using the keys from file F_4 (See Figure 2).
2. Calculate $USV^T = A'$ to get the decrypted image matrix (See Figure 2).
3. Apply reverse procedure to convert A' to grey image. (Appendix A)

The proposed process can be understood with the help of following two Figures. Figure 1 shows the process at sender's end and Figure 2 shows the process at receiver's end. We elaborate the process through an example.

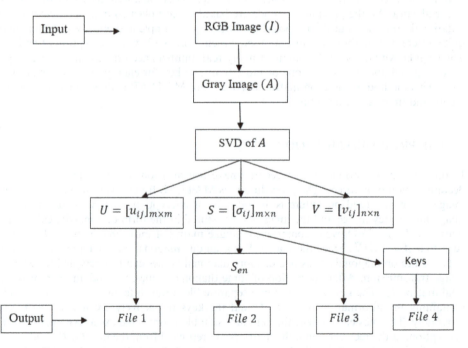

Figure 1. Encryption process at sender's end.

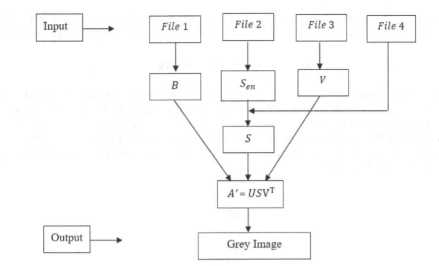

Figure 2. Decryption process at receiver's end.

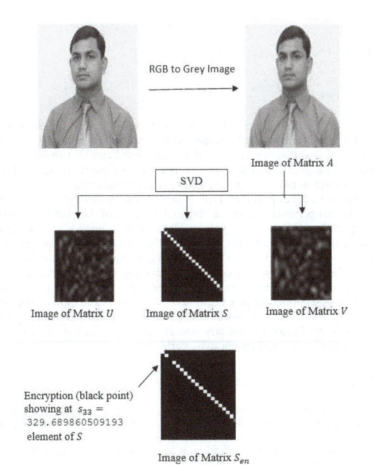

Image of Matrix U Image of Matrix S Image of Matrix V

Encryption (black point) showing at $s_{33} =$ 329.689860509193 element of S

Image of Matrix S_{en}

Figure 3. Encryption of an image.

Example 1: We apply the proposed procedure to an image. After applying singular value decomposition, we apply a simple encryption method for demonstration purpose defined as follows:

$$es_{ii} = \{0 \quad if \quad i = 3 \ s_{ii} \ otherwise. \tag{1}$$

Figure 3 shows the images after singular value decomposition of the given image and applying encryption. First the colour image is converted to a grey image, then we can see the three images corresponding to three matrices U, S and V. Then we apply encryption (1) to the image S. In the encrypted image of S_{en}, the encrypted diagonal elements are shown with black colour. If anyone decrypt image without key values then receiver cannot find clear and actual image (See Figure 4).

Figure 4.   Image without applying decryption on s_{en}.

3   CONCLUSION

In the present work, we have proposed a new mathematical method for secure transmission of image data with the help of singular value decomposition. The method is based on sending an image data in three different matrices with one encrypted matrix. From Figure 3, it can be observed that the image of U, S and V are totally different from the image of A, thus in case of any security breach in any message on cannot get the actual image from a single matrix. In the given example, we have applied encryption on only one element of the matrix S, however we can find different permutations of the diagonal elements of the matrix S to be encrypted, thus providing different ways of encryption. Here we have applied the simple encryption method, still the image given in Figure 4 is quite different form original image. Any advance encryption method will further distort the image. Thus, the proposed method is simple and effective. Since the proposed method decomposes an image data into three matrices, thus we can apply three different permutations also to send the image data to a receiver: three matrices can be sent separately or product of first two matrices and third matrices can be sent separately or first matrix and product of last two matrices can be sent separately. This technique may be applicable for military operations, secure transmission of image data, online data communication and corporative world etc.

REFERENCES

Acharya, B., Patra, S. K., & Panda, G. Image encryption by novel cryptosystem using matrix transformation. In 2008 *First International Conference on Emerging Trends in Engineering and Technology*. IEEE, 77–81, 2008, July.
Akritas, A. G., & Malaschonok, G. I., Applications of Singular-value Decomposition (SVD). *Mathematics and Computers in Simulation*, 67(1–2), 15–31, 2004.

Batuhan Arpacı Erol Kurt Kayhan Çelik and Bünyamin Ciylan, Colored Image Encryption and Decryption with a New Algorithm and a Hyperchaotic Electrical Circuit, *Journal of Electrical Engineering & Technology*, 15, 1413–1429, 2020.

Batuhan Arpacı Erol Kurt Kayhan Çelik, A New Algorithm for the Colored Image Encryption via the Modified Chua's Circuit, *Engineering Science and Technology, An International Journal*, 23(3), 595–604, 2020.

Borra, S., Thanki, R., Dey, N., & Borisagar, K., Secure Transmission and Integrity Verification of Color Radiological Images using Fast Discrete Curvelet Transform and Compressive Sensing. *Smart Health*, 12, 35–48, 2019.

De Lathauwer, L., De Moor, B., &Vandewalle, J., A Multilinear Singular Value Decomposition. *SIAM Journal on Matrix Analysis and Applications*, 21(4), 1253–1278, 2000.

De Leeuw, J., Principal Component Analysis of Binary Data by Iterated Singular Value Decomposition. *Computational Statistics & Data Analysis*, 50(1), 21–39, 2006.

El-Samie, F. E. A., Ahmed, H. E. H., Elashry, I. F., Shahieen, M. H., Faragallah, O. S., El-Rabaie, E. S. M., & Alshebeili, S. A., *Image Encryption: A Communication Perspective*. CRC Press, 2013.

Gao, X., Mou, J., Xiong, L., Sha, Y., Yan, H., & Cao, Y. A Fast and Efficient Multiple Images Encryption based on Single-channel Encryption and Chaotic System. *Nonlinear Dynamics*, 108(1), 613–636, 2022.

Henry, E. R., & Hofrichter, J., Singular Value Decomposition: Application to Analysis of Experimental Data. In *Methods in Enzymology*. Academic Press, 210, 129–192, 1992.

Kini, N. G., Kini, V. G. & Gautam, A Secured Steganography Algorithm for Hiding an Image in an Image. In *Integrated Intelligent Computing, Communication and Security*. Springer, Singapore, pp. 539–546, 2019.

Mandal, M. K., Banik, G. D., Chattopadhyay, D., & Nandi, D., An Image Encryption Process based on Chaotic Logistic Map. *IETE Technical Review*, 29(5), 395–404, 2012.

Martin, K., Lukac, R., & Plataniotis, K. N., Efficient Encryption of Wavelet-based Coded Color Images. *Pattern Recognition*, 38(7), 1111–1115, 2005.

Narendra K. Pareek, Vinod Patidar, Krishan K. Sud, Diffusion–substitution based Gray Image Encryption Scheme, *Digital Signal Processing*, 23(3), 894–901, May 2013.

Pakshwar, R., Trivedi, V. K., & Richhariya, V., A Survey on Different Image Encryption and Decryption Techniques. *International Journal of Computer Science and Information Technologies*, 4(1), 113–116, 2013.

Pengfei Fang, Han Liu, Chengmao Wu and Min Liu, A Secure Chaotic Block Image Encryption Algorithm Using Generative Adversarial Networks and DNA Sequence Coding, *Mathematical Problems in Engineering, Article ID* 6691547, 2021.

Rasras, R. J., Abuzalata, M., Alqadi, Z., Al-Azzeh, J., & Jaber, Q. Comparative Analysis of Color Image Encryption-Decryption Methods Based on Matrix Manipulation. *International Journal of Computer Science and Mobile Computing*, 8(3), 14–26, 2019.

Shankar, K., & Lakshmanaprabu, S. K., Optimal Key based Homomorphic Encryption for Color Image Security Aid of ant Lion Optimization Algorithm. *International Journal of Engineering & Technology*, 7(9), 22–27, 2018.

Singh, C., Pandey, B. K., Mandoria, D. R. H. L., & Kumar, A., A Review Paper on Chaotic Map Image Encryption Techniques. *International Research Journal of Engineering and Technology (IRJET) e-ISSN*, 2395–0056, 2018.

Singh, P., Yadav, A. K., & Singh, K., Phase Image Encryption in the Fractional Hartley Domain using Arnold Transform and Singular Value Decomposition. *Optics and Lasers in Engineering*, 91, 187–195, 2017.

Teng, L., Wang, X., & Meng, J., A Chaotic Color Image Encryption using Integrated Bit-level Permutation. *Multimedia Tools and Applications*, 77(6), 6883–6896, 2018.

Vineet Bhatt and K.C. Petwal, Survey of Image Compression and Face Recognition, *Inventi Rapid: Image & Video Processing*, Issue 2, 2013.

Wang, X., Zhu, X., & Zhang, Y., An Image Encryption Algorithm based on Josephus Traversing and Mixed Chaotic Map. *IEEE Access*, 6, 23733–23746, 2018.

Xian, Y., & Wang, X. (2021). Fractal Sorting Matrix and its Application on Chaotic Image Encryption. *Information Sciences*, 547, 1154–1169.

Zakaria, S. B., & Navi, K. (2022). Image Encryption and Decryption using Exclusive-OR based on Ternary Value Logic. *Computers and Electrical Engineering*, 101, 108021.

Zhang, Y., Zhao, R., Zhang, Y., Lan, R., & Chai, X. (2022). High-efficiency and Visual-Usability Image Encryption based on Thumbnail Preserving and Chaotic System. *Journal of King Saud University-Computer and Information Sciences*.

Zhen Li, Changgen Peng, Weijie Tan and Liangrong Li, A Novel Chaos-Based Color Image Encryption Scheme Uning Bit-Level *Permutation*, *Symmetry*, 12, 1497, 2020; doi:10.3390/sym12091497.

Zou, C., Wang, X., & Li, H. (2021). Image Encryption Algorithm with Matrix Semi-tensor Product. *Nonlinear Dynamics*, 105(1), 859–876.

APPENDIX A

Encryption Process:

```
Im1=imread('C:\Users\VINEET BHATT\Desktop\M1.JPG');
A1=imresize(Im1,[112,92]); % Resize image Im1.
s1 = size(A1);% find out the size of the image.
ss1 =size(s1);
ss1(:,2)= 3; % If the image is a color image in jpeg or jpg format it will covert to the
grey scale.
A1 = rgb2gray(A1);
[u1,s1,v1]=svd(double(A1)); % Find SVD of the images.
```

% s1 before encryption at a_{33}

```
s1=[4370.41594561188,0,0,0,0,0,0,0,0,0,0,0,0,0,0,0,0,0,0,0,0;
0,543.527644714021,0,0,0,0,0,0,0,0,0,0,0,0,0,0,0,0,0,0,0;
0,0,329.689860509193,0,0,0,0,0,0,0,0,0,0,0,0,0,0,0,0,0,0;
0,0,0,184.273067209306,0,0,0,0,0,0,0,0,0,0,0,0,0,0,0,0,0;
0,0,0,0,148.941187492170,0,0,0,0,0,0,0,0,0,0,0,0,0,0,0,0;
0,0,0,0,0,105.678647490898,0,0,0,0,0,0,0,0,0,0,0,0,0,0,0;
0,0,0,0,0,0,87.2074837596001,0,0,0,0,0,0,0,0,0,0,0,0,0,0;
0,0,0,0,0,0,0,81.0846084692255,0,0,0,0,0,0,0,0,0,0,0,0,0;
0,0,0,0,0,0,0,0,52.3304992634119,0,0,0,0,0,0,0,0,0,0,0,0;
0,0,0,0,0,0,0,0,0,42.9447432862839,0,0,0,0,0,0,0,0,0,0,0;
0,0,0,0,0,0,0,0,0,0,39.6826128421427,0,0,0,0,0,0,0,0,0,0;
0,0,0,0,0,0,0,0,0,0,0,38.9981697939725,0,0,0,0,0,0,0,0,0;
0,0,0,0,0,0,0,0,0,0,0,0,28.7666490870281,0,0,0,0,0,0,0,0;
0,0,0,0,0,0,0,0,0,0,0,0,0,21.4201489964419,0,0,0,0,0,0,0;
0,0,0,0,0,0,0,0,0,0,0,0,0,0,18.8459854748084,0,0,0,0,0,0;
0,0,0,0,0,0,0,0,0,0,0,0,0,0,0,9.74855860404886,0,0,0,0,0;
0,0,0,0,0,0,0,0,0,0,0,0,0,0,0,0,5.23168144463489,0,0,0,0;
0,0,0,0,0,0,0,0,0,0,0,0,0,0,0,0,0,3.20211269233394,0,0,0;
0,0,0,0,0,0,0,0,0,0,0,0,0,0,0,0,0,0,2.05261273687488,0;
0,0,0,0,0,0,0,0,0,0,0,0,0,0,0,0,0,0,0,0.443520503727879;0,0,0,0,0,0,0,0,0,0,0,0,0,0,0,0,0,0,0,0,0;
0,0,0,0,0,0,0,0,0,0,0,0,0,0,0,0,0,0,0,0,0;0,0,0,0,0,0,0,0,0,0,0,0,0,0,0,0,0,0,0,0,0;
0,0,0,0,0,0,0,0,0,0,0,0,0,0,0,0,0,0,0,0,0;0,0,0,0,0,0,0,0,0,0,0,0,0,0,0,0,0,0,0,0,0];
```

% s1 after encryption at a_{33}

```
s1=[4370.41594561188,0,0,0,0,0,0,0,0,0,0,0,0,0,0,0,0,0,0,0,0;
0,543.527644714021,0,0,0,0,0,0,0,0,0,0,0,0,0,0,0,0,0,0,0;0,0,0,0,0,0,0,0,0,0,0,0,0,0,0,0,0,0,0,0,0;
0,0,0,184.273067209306,0,0,0,0,0,0,0,0,0,0,0,0,0,0,0,0,0;
0,0,0,0,148.941187492170,0,0,0,0,0,0,0,0,0,0,0,0,0,0,0,0;
0,0,0,0,0,105.678647490898,0,0,0,0,0,0,0,0,0,0,0,0,0,0,0;
0,0,0,0,0,0,87.2074837596001,0,0,0,0,0,0,0,0,0,0,0,0,0,0;
0,0,0,0,0,0,0,81.0846084692255,0,0,0,0,0,0,0,0,0,0,0,0,0;
0,0,0,0,0,0,0,0,52.3304992634119,0,0,0,0,0,0,0,0,0,0,0,0;
0,0,0,0,0,0,0,0,0,42.9447432862839,0,0,0,0,0,0,0,0,0,0,0;
0,0,0,0,0,0,0,0,0,0,39.6826128421427,0,0,0,0,0,0,0,0,0,0;
0,0,0,0,0,0,0,0,0,0,0,38.9981697939725,0,0,0,0,0,0,0,0,0;
0,0,0,0,0,0,0,0,0,0,0,0,28.7666490870281,0,0,0,0,0,0,0,0;
```

0,0,0,0,0,0,0,0,0,0,0,0,0,21.4201489964419,0,0,0,0,0,0;
0,0,0,0,0,0,0,0,0,0,0,0,0,0,18.8459854748084,0,0,0,0,0;
0,0,0,0,0,0,0,0,0,0,0,0,0,0,0,9.74855860404886,0,0,0,0;
0,0,0,0,0,0,0,0,0,0,0,0,0,0,0,0,5.23168144463489,0,0,0;
0,0,0,0,0,0,0,0,0,0,0,0,0,0,0,0,0,3.20211269233394,0,0;
0,0,0,0,0,0,0,0,0,0,0,0,0,0,0,0,0,0,2.05261273687488,0;
0,0,0,0,0,0,0,0,0,0,0,0,0,0,0,0,0,0,0,0.443520503727879;0,0,0,0,0,0,0,0,0,0,0,0,0,0,0,0,0,0,0,0;
0,0,0,0,0,0,0,0,0,0,0,0,0,0,0,0,0,0,0,0;0,0,0,0,0,0,0,0,0,0,0,0,0,0,0,0,0,0,0,0;
0,0,0,0,0,0,0,0,0,0,0,0,0,0,0,0,0,0,0,0;0,0,0,0,0,0,0,0,0,0,0,0,0,0,0,0,0,0,0,0];

Decryption process
 u1 = [- - - -]%matrix of u1
v1 =[- - - -]%matrix of v1
s1=[- - - -]%matrix of s1 %decrypted matrix.
combinf2=uint8(u1*s1*v1');%

Automation and Computation – Vats et al. (Eds)
© 2023 the Author(s), ISBN 978-1-032-36723-1

A survey: Smart healthcare monitoring system using IOT and ML techniques

K. Puri & S. Kumar

Tula's Institute, Uttarakhand Technical University, Dehradun, Uttarakhand, India

ABSTRACT: The foundation of every human being is health. Human health has to be closely checked and must be treated with the proper medications. Proactively keeping an eye on one's health helps lower the risk of many illnesses. The rapid advancement of technology in recent decades has led to the market availability of several wearable gadgets and health monitoring equipment. These technologies enable the monitoring of a person's condition and the immediate delivery of advice. The Kaggle's or UCI dataset can be used for the machine learning algorithm for initial training and validation phase. During the testing phase, the person's temperature, blood pressure, and heart rate are measured utilizing an IoT setup. In this Survey paper we will study about Smart Healthcare Monitoring System, which can use best suitable classifiers such as Decision Tree, Logistic Regression, SVM, and K-NN based on their performances instead most other traditional classifiers.

1 INTRODUCTION

As the primary cause of an unhealthy lifestyle in humans, a number of variables, including improper sleep, diet, irregular exercises, intense work, excess stress levels etc [3]. These extensive Busy Schedules has held us back in making our lifestyle impacted with greater health hazards. The provision of healthcare is a necessity for life. The preservation and enhancement of health via illness prevention and diagnosis is known as health care. Diagnostic tools like CT, MRI, PET, SPECT, and others can be used with the purpose of locating any deep-skin rips or irregularities. Additionally, some aberrant disorders, Epilepsy and heart attacks, for example, can be recognised even before they appear. Modern healthcare systems are under pressure due to the population's constant growth and the unpredictability with which chronic illnesses are spreading throughout the populace, thus there is a huge need for everything from physicians and nurses to hospital beds. By enabling patient observation beyond the usual clinical settings (such as at home), such a remote patient monitoring arrangement increases availability of human services offices while decreasing costs [4]. Many recent studies have concentrated on the Internet of Things (IoT), which might reduce the pressure on healthcare systems. Both the infrequent monitoring of diabetic individuals and the monitoring of patients with certain diseases like Perkinson disease are mentioned in [5]. Using a Smart Healthcare Monitoring System, the doctor may keep an eye on the patient's cardiac impulses, which aids in accurate diagnosis. To offer dependable wireless data transfer, a number of wearable solutions have been suggested [8]. There are several benefits to using the internet of things in healthcare, but there are also some drawbacks. Concerns over data security and IoT device management are shared by IT and healthcare administrators.

DOI: 10.1201/9781003333500-13

2 IOT ARCHITECTURE FOR DISEASE DETECTION

With the use of sensor networks, this system offers a platform for patient monitoring and care. There are parts for the hardware and the software in the design. The Raspberry Pi board, sensors for temperature, blood pressure, and heart rate are all part of the hardware area. The steps in Figure 1. IOT System Architecture are as follows: gathering sensor data, storing it in the cloud, and then analysing it to look for anomalies in the patient's health.

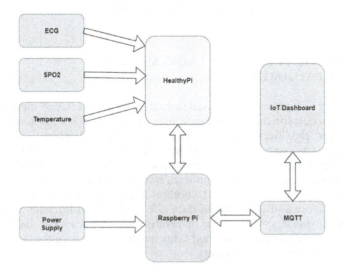

Figure 1.   IOT system architecture.

Usually, disorders arise from undetected occupation in the bodily components. The incidence of epileptic seizures in the brain might quicken pulse. The heart rate may be determined using heart rate sensors. It is possible to determine the heart's rate at any given moment. The sensor is connected to a Raspberry Pi board so that the output readings may be seen. The values can be seen visually by connecting an LCD display or a serial monitor. All data that has been collected forwarded to the cloud due to the size of the data volume. The data that sent to the cloud is examined locally. For any problematic situations, the heart rate typically tends to progressively rise.

3 MACHINE LEARNING

Modern healthcare sectors generate a vast amount of data as a result of the development of modern healthcare systems (For example, the diagnosis of an illness, the state of the patient, etc). To create a predictive analytic model, these data are employed. Data from multiple senses are analysed using the machine learning (ML) approach, which condenses the data into useful knowledge. The field of machine learning emerged from artificial intelligence (AI). We can create more intelligent and superior machines by utilising artificial intelligence. Machine learning is a strategy for passively learning from examples and experience. Rather than creation of code, the generic algorithm is fed with data, and construction of logic using data provided. Stocks Tradings, Web searches, filtering spams, placements of ads, and other processes all employ machine learning [1]. By evaluating these large pieces of data and making the job of data scientists simpler through an automated process, machine learning gets the same relevance and respect as big data and cloud computing. The majority of conventional way of storing data using machine learning algorithms are designed exclusively in memory [2]. Learning from this wealth of information is likely to lead to significant advancements in science and engineering as well as improvements in the quality of our lives, but it also provides us with a wealth of new experiences [7]. According to a McKinsey Global Institute research, the next significant wave of innovation will be driven by machine learning [6]. In most remote areas, residents would not have easy access to the medical facilities [10].

4 HEALTH STATE PREDICTION USING MACHINE LEARNING ALGORITHM

Machine learning techniques may be used to analyze most datasets. A methodical strategy for generating models for categorization from input data sets is known as a classification methodology (or classifier). Decision Tree, K Nearest Neighbor (K-NN), Logistic Regression, Support Vector Machine (SVM), are a few instances of Machine Learning Techniques. Each method offers a learning process to identify a model that closely matches the relationship between the input data's class label and attribute set. Model of a Learning Algorithm ought to be able to both accurately predict the record class labels it has never seen before and fit the input data effectively. Each learning algorithm's main goal is to create models with strong generalisation abilities. The dataset for Smart Healthcare Monitoring System is trained using machine learning techniques, and analysis is carried out depending on the training.

4.1 Support vector machine

Regression and classification analyses are carried out using this approach of supervised learning. It is a model that uses classification algorithms for two group classification problems. It does categorization utilising parallel lines between data analysis of vast amounts of data [9,11]. It divides a single line into flat, linear segments, sometimes known as hyperplanes. SVM models can be majorly used for text classifications using labelled training data for each category and also to predict the presence or absence of diabetes in a person based on few parameters. SVM has major benefit of faster processing and improved performance with fewer samples.

$$(w^*,b^*)\max\frac{2}{||w^T||}y_i\begin{cases} +1 & where \quad w^Tx_i + b >= 1 \\ -1 & where \quad w^Tx_i + b <= -1 \end{cases} \tag{1}$$

4.2 Logistic regression

Under the category of "Supervised Learning Technique," logistic regression is used to forecast categorical dependent variables using a predetermined set of independent variables. The result will be presented as a discrete or categorical value, such as True or False, 0 or 1, or Yes or No, but it will actually take the form of probabilistic values that fall between 0 and 1. In order to determine the ideal weight or convert anticipated values to probabilities, a logistic function or sigmoid function is used. Where **w*** is a Vector or weights for each features.

$$w^* = argmin_w \sum_{i=1}^{n} \log\left(1 + \exp(-y_i * w^T x_i)\right) \qquad (2)$$

4.3 K Nearest Neighbor (K-NN) classifier

Supervised learning using non-parametric methods approach in which the training set is used to classify the data into a certain category. By searching the whole training set for the k instances (neighbors) that are the most similar, and then summarizing the output variables for those k examples, a new instance (x) is predicted. This represents the mode class value in the categorization.

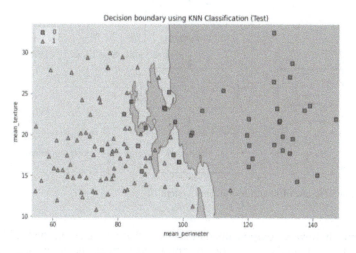

Figure 2.   Overfitting case when K is small.

$$P(y = a|, X = x) = \frac{1}{K} \sum_{i \in A} I\left(y^{(i)} = a\right) \qquad (3)$$

The most important challenge is choosing the K-value in the K-nearest neighbor algorithm. A Low K values suggest that noise will have a bigger influence on the result, which raises the possibility of overfitting as shown in Figure 2. When K is high, it becomes computationally costly and undermines the fundamental principle of KNN. Cross Validation is a tool we may utilize to improve the outcomes. The cross-validation method allows us to test the KNN algorithm with various K values. The model that provides excellent accuracy might be regarded as the best option.

5   METHODOLOGY

According to the suggested system, the microcontroller will gather and process the data from the sensor network. The suggested outcomes are kept in the cloud. The processed data may be accessed from the cloud and used for analysis. The analyzed data is once more stored in the cloud for access by the doctors. The results and the patient's condition are made public on the hospital website. The flow diagram of the whole arrangement is shown in Figure 3. The total system may be divided into three major sections. The three main areas of the plan are the emergency alert system, health monitoring system, and health condition prediction system. The data gathered and processed must be kept private because the technique deals with health-related concerns. The system is given credit for using encryption technologies to provide security and secrecy.

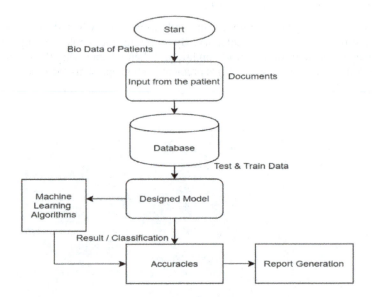

Figure 3.   Flow diagram of health monitoring system.

The system's hardware components that allow IoT are included in the health monitoring module, which uses a number of sensors to capture the patient's health information. Here, all of the sensors are connected to the Raspberry Pi through the GPIO ports or another central gateway server, if their output is analogue and the Raspberry Pi only works with digital signals, via the MCP3008 analog-to-digital converter.

One of the most promising components of the suggested system is health state prediction. The database is used in this module to store the patient's health information that is gathered from the sensory nodes. The Best classifier, which categorizes different health statuses in individuals, is used to test the database's data. The classifier makes an accurate categorization and seldom needs manual rechecking.

6   REVIEW SYSTEM WORKFLOW

It illustrates Data resources used, Sensor system analysis, cloud service integration for IoT devices, machine learning algorithm analysis, and automated Smart Healthcare Monitoring System. Kaggle's or UCI Diabetes Disease Dataset any can be used to analyze the Performance Measures of all the classifiers. The following characteristics have been given to assist us in predicting whether or not a person has diabetes. At the initial phase few parameters can be very crucial for proper diagnosis of a disease like temperature, Blood Pressure etc.

Table 1. Diabetics dataset.

Features	Average	Least	High	Critical
Pregnancies	3<=5	1<=3	6<=9	=>10Above
Body Mass Index (BMI)	18.50<=25	25<= 30	30<=35	=>35
Age	Woman 21–28	28–34	35–45	46–50
Blood Pressure	<80mm Hg	80mm Hg<=90	90mm Hg<=100	=>100mmHg
Insulin	Fasting < 25 ml U/L	30 min, 30–230ml U/L	1hr18–276 ml U/L, 2hr 16–166ml U/L	>3hr,<25ml U/L
Skin Thickness	25<=30	0<=25	30<40	40<=50
Glucose Levels	70mg/dL	110mg/dL<130	130mg/dL<=160	=>160 Above
Diabetes Predict Function	Float 0.001–0.324	Float 0.324–0.999	Float 1.0–1.99	Float=>2.00

The dataset belongs to various age group of pregnant women. The review study found that SVM had an overall accuracy of 80.51%, outperforming all other classifiers on several performance metrics including f1-score, precession, and recall, while K-nn classified with an accuracy of 71.42%, and Decision Tree accuracy was 70.22% [3].

7 CONCLUSION

Internet of things plays a vital role in today's modern Healthcare Sector which has generated greater impact's over our society. Remote based Health monitoring is a valuable approach which can be used for conventional routine examinations that can be further utilized for diagnosis of appropriate disease by a doctor at any distance and given point of time. Unnoticed human health variables can lead to major problems and even put a person's life in jeopardy. As a novel solution, IoT-based Smart Healthcare Parameter Monitoring auto-mation is addressed.

This research survey addressed a number of methods and algorithms for a smart health-care monitoring system utilizing some ML techniques. The system has the tendency to capture all the relevant data via. sensors like Body Temperature, Blood Pressure, ECG etc. that is sent over cloud using a gateway. The data is further processed and used for analysis, now the Best Algorithm can be chosen to identify a certain disease. By improving upon the present process and making it a more effective way to track patient health data, this system solves its drawbacks.

REFERENCES

[1] Fuqaha, Guizani, Mohammadi, Aledhari, Moussaayyash, Internet Of Things: A Survey on Enabling Technologies, Protocols and Applications. *IEEE Communications Surveys & Tutorials*, vol. 17, no. 4, *Fourthquarter*, 2015.

[2] Sandryhaila, Moura, Big Data Analysis with Signal Processing On Graphs: Representation and Processing of Massive Datasets with Irregular Structure. *IEEE Signal Proc Mag*, vol. 31, no. 5, pp. 80–90, 2014.

[3] Godi, Viswanadham, Muttipati, Samantray & Gadiraju "E-Healthcare Monitoring System Using Iot with Machine Learning Approaches" *Conference : 2020 International Conference On Computer Science, Engineering and Applications (ICCSEA)*

[4]   Pasluosta, Gassner, Winkler, Klucken, & Eskofier, "An Emerging Era In The Management Of Parkinson's Disease:Wearable Technologies and The Internet Of Things," *IEEE Journal of Biomedical and Health Informatics*, vol. 19, no. 6, pp. 1873–1881, 2015

[5]   Joyia, Liaqat, farooq, & rehman, Internet of Medical Things (Iomt): Applications, Benefits and Future Challenges in Healthcare Domain, *Journal of Communications*, Vol. 12, no.4, April 2017.

[6]   Janmenjoynayak, Bighnarajnaik, Behera, "A Comprehensive Survey on Support Vector Machine in Data Mining Tasks", *International Journal of Database Theory and Application*, Vol. 8. No. 1, 2015

[7]   Qiu, Wu, Ding, Xu, Feng, "A Survey of Machine Learning for Big Data Processing". *Eurasip Journal on Advances in Signal Processing*.

[8]   Kslavakis, Gbgiannakis, Mateos, "Modeling and Optimization for Big Data Analytics: (Statistical) Learning Tools for our Era of Data Deluge". *IEEE Signal Proc Mag*, vol. 31, no. 5, pp. 18–31, 2014.

[9]   Sangeetha, Venkatesan, Shitharth, 'Security Appraisal Conducted on Real Time Scada Dataset using Cyber Analytic Tools', *Solid State Technology*, vol. 63, no. 1, 2020, pp. 1479–1491.

[10] Riazulislam, Daehankwak, M, Kwak, K: *"The Internet of Things for Health Care: A Comprehensive Survey". in: IEEE Access*, 2015.

[11] Shitharth, Manimala, Bhavani, Nalluri, 'Real Time Analysis of Air Pollution Level in Metropolitan Cities by Adopting Cloud Computing based Pollution Control Monitoring System using Nano Sensors', *Solid State Technology*, Vol. 63, no. 2, pp. 1031–1045, 2020.

Automation and Computation – Vats et al. (Eds)
© 2023 the Author(s), ISBN 978-1-032-36723-1

Correlation feature selection based BFTree framework for diagnosing Chronic Obstructive Pulmonary Disease severity grade

Akansha Singh[1]*
University School of Information, Communication, and Technology, GGSIPU, Delhi, India

Nupur Prakash
Northcap University, Gurgaon, India

Anurag Jain
University School of Information, Communication, and Technology, GGSIPU, Delhi, India

ABSTRACT: Chronic Obstructive Pulmonary Disease (COPD) is a noxious disease, causing millions of deaths every year. Hence, the early prediction of COPD is crucial. In this regard, this study has attempted to provide a framework for early prediction of COPD severity Grade using Machine Learning (ML) models. The grades of COPD severity categorization have been declared by Guidelines for Obstructive Lung Disease (GOLD) as mild, moderate, severe, and very severe. Expected Maximization (EM) Imputation has been used for handling missing values. An optimal minimal set of features has been provided by Correlation-based feature selection (CFS). This study uses various ML classifiers: NB, RBFNetwork, SMO, Bagging, and BFTree. For model evaluation purposes several metrics have been used such as accuracy, kappa statistics, precision, recall, etc. The experiments with modified features and missingness treatment using EM Imputation and BFTree gives the best performance with accuracy, precision, and recall of 99%, 99.1%, and 99% approximately respectively in multi-classifying the severity grade of COPD patients.

1 INTRODUCTION

Chronic Obstructive Pulmonary Disease (COPD) is a Chronic Respiratory Disease (CRD) causing millions of deaths every year. It causes airway inflammation making them narrower which further makes it difficult for a person to breathe. It damages the air sacs. It is generally caused by smoking, air pollution, and alpha anti-trypsin deficiency. According to GOLD[1], 65 million people suffer from COPD every year and 3 million people die each year because of it. The factor that makes COPD a noxious disease is that it's a progressive disease that has no cure. The early detection of COPD is not an easy task because of its similar and overlapping symptoms with other CRDs. The main diagnosis for COPD is done through a Spirometry test. During the Spirometry test, two factors normally measured are: Forced Vital Capacity (FVC) and Forced Expiratory Volume (FEV1). FVC defines the amount of air that can be breathed in and out. FEV1 is the measure of the amount of air exhaled by a person in 1 second. If the ratio (FEV1/FVC) <=70%, then the person is said to have COPD. GOLD has further defined certain criteria for categorizing the severity of COPD based on

---

*Corresponding Author: akansha.trar03@gmail.com
[1]Global Initiative for Obstructive Diseases,Chronic Obstructive Pulmonary Disease (COPD), {"https://goldcopd.org/world-lung-day-2019-healthy-lungs-for-all/"}, accessed on 30/09/2022

the FEV1 value. A patient has mild COPD if FEV1 >=80% predicted, moderate if 50%< =FEV1<80%, severe if 30%<=FEV1<50%, and very severe if FEV1<=30% predicted.

It has been estimated that by 2050, the prevalence of COPD will reach 592 million people. Hence, the early detection of COPD is crucial. In recent years, Machine learning (Kor *et al.* 2022) has shown great potential in the healthcare domain especially in respiratory by diagnosing and detecting a particular disease. There have been various studies conducted for diagnosing COPD and the associated factors with it such as predicting mortality rate (Liao *et al.* 2021; Nam *et al.* 2022), hospitalization rate (Tadahiro G. *et al.* 2019), and exacerbations (Ali H. *et al.* 2021; Wang *et al.* and Peng *et al.* 2020), etc. A lot of researches were conducted to determine the severity of COPD (Choi *et al.* 2021). However, there was only one study which has determined COPD severity using the same dataset that has been used in this study (Dritsas *et al.* 2022). The limitation of the previous study is that the dataset consisted of missing values that were left untreated which also further affected the results. Hence, the main aim of this study is to propose a framework for diagnosing the COPD severity grade as per GOLD standard efficiently and accurately. It also aims to solve the data missingness problem and to investigate a minimal set of features that can help in diagnosing COPD accurately and efficiently.

The paper is further divided into the following sections: Section 2 describes the previous works done in diagnosing COPD using Machine learning methods. Section 3 describes the proposed methodology in detail focusing on pre-processing, feature selection, and classification models. It also describes the dataset used. It addresses the imbalance dataset and missingness problem of the dataset. Section 4 explains the results obtained and discusses them in detail. Section 5 concludes the paper.

2 RELATED WORK

This section discusses the various state-of-art works done in diagnosing COPD using machine learning models. The authors utilized different kinds of data for predicting COPD such as clinical indicators, laboratory features, and respiratory sounds. Based on such data, the authors have predicted the mortality rate, hospitalization rate, exacerbation rate, etc.

In (Moll *et al.* 2020) the authors created a machine learning mortality prediction (MLMP) model which is a Cox regression model for predicting all-cause mortality in COPD. They included 2,632 participants from the COPDGene study and 1268 from ECLIPSE. Random survival forest (RSF) was utilized for feature selection. The model achieved a c-index of 0.74. In (Dritsas *et al.* 2022) the authors predicted COPD severity in elderly patients using MLN techniques. They embedded these models in an AI framework of the GATEKEEPER system for the risk assessment of the elderly population. The dataset collected from Kaggle consisted of 101 patients and 23 features. The authors utilized Lr and SVM classifiers with 3 different kernels. The results showed that the SVM classifier achieved the most prominent outcomes. The authors (Spathis *et al.* 2019) have compared various machine learning classifiers such as NB, LR, NN, SVM, KNN, DT, and RF for diagnosing asthma and COPD. They utilized a dataset provided by a pulmonologist doctor consisting of 132 entries. It consisted of 22 attributes including two classes: asthma and COPD. The result showed that Random Forest outperformed the other models by achieving a precision of 97.7% for COPD and 80.3% for asthma. The most important features for COPD diagnosis were: smoking, forced expiratory volume I, age, and forced vital capacity.

In (Ali H *et al.* 2021), the authors utilized different supervised classifiers such as RF, SVM, DT, GB, KN, XG, and LDA for diagnosing COPD and also differentiating the early stage of COPD (ESCP) and advanced stage of COPD (ASCP). For feature selection, the authors used the Recursive Feature Elimination, cross-validated (RFECV) method. The features selected using the doctor's recommendation were labeled as FRDR. Among all the classifiers, SVM performed the best. SVM along with RFECV provided an accuracy of 96%

and FRDR provided an accuracy of 90%. Similarly, the authors in (Zarrin *et al.* 2020) utilized different ML classifiers such as XGBoost, SVM, Gradient Naïve Bayes (GNB), LR, and ANN for classifying COPD and healthy patients based on saliva samples. The results showed that XGBoost outperformed the other classifiers and achieved the highest accuracy of 91.25%. The authors (Haider *et al.* 2019) also utilized different ML classifiers such as SVM, KNN, LR DT, and Discriminant Analysis (DA) to classify COPD and healthy patients using respiratory sounds. 39 lung audio features were utilized. The result showed that SVM and LR classifiers obtained an accuracy of 100%. In (Ali H. *et al.* 2021) the authors proposed a voting ensemble classifier to identify the severity of COPD patients. The authors utilized 5 ML classifiers such as SVM, gradient boosting machine (GBM), XGBoost, and KNN. The data consisted of 54 features and after feature selection, 24 features were finally selected. The outputs from different classes were combined using a soft voting ensemble. This method achieved an accuracy of 91.0849%.

3   PROPOSED METHODOLOGY

This section provides a brief description of the dataset used and further explains the proposed methodology in detail. Figure 1. shows the basic framework of the proposed work.

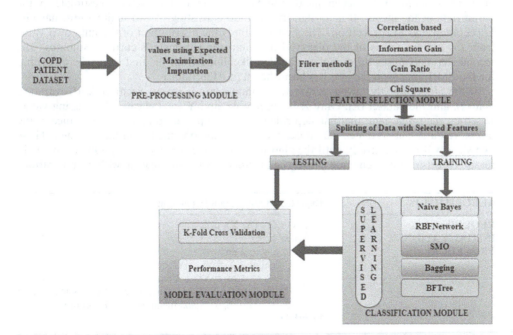

Figure 1.   Proposed framework representing the complete procedure of multi-classification of COPD severity.

The framework starts with the pre-processing of the incoming raw data. The dataset used in this study suffers from a missingness problem. As it is known that the better the data quality, the better will be the performance. Hence, it is of utmost importance to treat the data before feeding it to the classifiers. For tackling the missingness problem, EM (Expectation Maximization) Imputation has been used. The pre-processed data further went through the feature selection phase wherein different feature selection techniques have been compared. The selected number of features is then used for the classification of instances into different categories. The feature selection and classification process have been described in further sections.

3.1 *Dataset description*

To detect COPD severity, this study has utilized COPD patient dataset. It is collected from the Kaggle dataset repository and is publicly available[2]. To manage COPD, a group of patients was invited to participate in a rehabilitation program. Based on the measurements taken, each of these patients was classified into different COPD severity categories. There are four severity levels defined, hence making it a multi-classification study. The dataset consists of 101 instances and 23 attributes. The attributes are: ID, Age, PackHistory, MWT1, MWT2, MWT1Best, FEV1, FEV1PRED, FVC, FVCPRED, CAT, HAD, SGRQ, AGEquartile, COPD, gender, smoking, Diabetes, muscular, hypertension, AtialFib, IHD, COPD SEVERITY. The attribute COPD SEVERITY defines the class. The dataset suffers from a missingness problem.

3.2 *Pre-processing*

As mentioned in the above section, the dataset used in this study suffers from a missingness problem. The dataset has 4 missing values. To tackle the missingness problem, the Expectation Maximization (EM) Imputation technique has been used. It is a multiple imputation method. It is a technique that finds the maximum log-likelihood estimates of the parameters of the model. It is an iterative method consisting of two major steps, namely, Expectation and Maximization. The first step finds the distribution of the unknown value from the known values of the observed data and the parameter's current estimates. The second step maximizes the complete-data log-likelihood from the first step. The algorithm for the EM Imputation method has been shown in Algorithm 1.

In the algorithm, the given parameter estimates are μ (mean vectors), Σ (Covariance matrix), and the dataset (Z). Only X is observable, and Yi_{miss} indicates the missing value. Based on the given data, the first step calculates the expectation and the covariance of the unknown data. The next step maximizes the expectation obtained from the first step. These two steps will run iteratively until the algorithm converges or $E_{new} = E_{old}$. One of the main reasons for using this technique is that the likelihood is non-decreasing with each iteration.

ALGORITHM 1:	**Expected Maximization Imputation**
REQUIRE:	Dataset (Z)= (X, Y), μ, Σ
INITIALIZATION:	For each case i
PROCEDURE:	While $E_{new} = E_{old}$ **do**
	$\quad$ Calculate $E_{old} = Yi_{miss}\|Yi_{obs}$, μ, Σ
	$\quad\quad$ Cov $= Yi_{miss}\ Yi_{obs}$, μ, Σ
	$\quad$ /*Calculate the EM and covariance of the missing value*/
	$\quad$ E_{new}= Max E_{old} /*maximizing the old Expectation*/
	Endwhile

3.3 *Feature selection*

After pre-processing, the next step is to select an optimal minimum set of features. In the state-of-art studies, only a few studies have used feature selection (FS). It is an important step especially in dealing with healthcare data. Because in healthcare, the data is usually the

---

[2]COPD Patient Dataset, https://www.kaggle.com/datasets/prakharrathi25/copd-student-dataset, accessed on 1/10/2022.

number of tests that have been performed to diagnose a particular disease. However, not all these tests are necessary for diagnosing a particular disease.

Hence, to determine the set of a minimum number of tests and factors that can also help in diagnosing a disease accurately in an efficient way, feature selection is crucial. In this study, filter methods have been used for feature selection. These methods use measurement techniques such as correlation, loss factor, and squared values for finding a good subset of features from the original set of features. Here, we have utilized four different filter methods for comparison purposes. These methods include Correlation-based feature selection (CFS), Information Gain (IG), Chi-square, and Gain Ratio (GR). The number of features selected by all these methods has been listed in Table 1.

Table 1. The number of features selected by different feature selection algorithms.

Feature Selection Techniques	No. of Features Selected	Selected Features
Correlation-based	16	ID, PackHistory, MWT1, MWT2, MWT1Best, FEV1, FEV1PRED, FVC, FVCPRED, CAT, HAD, COPD, Diabetes, Muscular, IHD
Information	21	COPD, FEV1PRED, FEV1, FVC, FVCPRED, MWT1, MWT2, MWT1Best, ID, CAT, HAD, SGRQ, PackHistory, Age, Hypertension, Muscular, Diabetes, AtrialFib, AGEquartiles, gender, IHD
Gain Ratio	21	COPD, FEV1PRED, FEV1, MWT1, FVCPRED, FVC, MWT2, MWT1Best, CAT, ID, Hypertension, SGRQ, Age, HAD, Diabetes, Muscular, IHD, PackHistory, AGEquartiles, AtrialFib, gender
Chi-Square	21	COPD, FEV1PRED, FEV1, FVC, MWT2, MWT1, FVCPRED, ID, MWT1Best, HAD, CAT, SGRQ, Age, PackHistory, Hypertension, AtrialFib, Diabetes, Muscular, Gender, AGEquatiles, IHD

All these features are listed in the order of their rank as defined by the respective feature selection technique. Out of all these methods, CFS resulted in an optimal minimal set of features. CFS filter finds the relevance between the features and classes and also between features (Singh A. & Jain. A. 2021). It assigns a rank to each feature based on that relevance. The aim is to find features that have high class-feature correlation and low feature-feature correlation.

3.4 *Supervised classifiers*

After performing pre-processing and FS, the dataset is divided into training and testing sets using K-fold cross-validation where K value is set to 10. The training data is fed into different supervised classifiers. The dataset is a labeled dataset hence supervised classifiers have been used for classification. All these classifiers have been discussed briefly and the hyperparameters that were set for all these classifiers have also been defined below:

- **Naïve Bayes**: This classification algorithm works on the Bayes theorem which uses conditional probability. Bayes Theorem provides the conditional probability of an event X when event Y has already occurred. In the case of machine learning, given a class variable Z, and feature set (f1, ...fn). The conditional probability of class variable Z can be defined as shown in equation 1:

$$P(Z|, f1, \ldots \ldots, fn) = \frac{P(Z)P(f1, \ldots \ldots fn|, Z)}{P(f1, \ldots \ldots fn)} \quad (1)$$

The hyperparameters used for this technique are batchSize=100, and super-vised_discretion ="false".

- **RBFNetwork-** It is also a feed-forward neural network consisting of mainly three layers: input layer, hidden layer, and output layer. It is different from the standard neural net-work in that it has only a single hidden layer whose computation is different from the standard hidden layer. In the hidden layer, each neuron has a prototype vector which is a vector form of the training set. Each neuron computes the similarity between the input vector and its prototype vector using the formula shown in equation (2) where X is the input vector, u is the prototype vector, σ is the bandwidth of the neuron, and $\emptyset$ is the output of the neuron.

$$\emptyset_i = e\left(-\frac{||X - u_i||^2}{2\sigma_i^2}\right) \tag{2}$$

The final layer calculates the final predicted output which is the summation of the weighted outputs from the hidden layer as shown in equation (3) where w is the weight connection, and y is the predicted output.

$$y = \sum_i^n w_i \emptyset_i \tag{3}$$

Here, the Gaussian radial basis function has been used as the activation function. K-means clustering has been utilized to provide a radial basis. The hyperparameters for this technique are seed=1, batchSize=100, maximum no. of iterations for LR = −1, no. of clusters=2, and standard deviation for cluster = 0.1.

- **Sequential Minimal Optimization (SMO)-**It is a technique for training Support Vector Machines (SVMs). It is mostly used because of its ability to solve large quadratic pro-gramming (QP) optimization problems into smaller QP problems that further save a lot of time. It can also be used for training larger datasets. At every step, SMO chooses two Lagrange multipliers and optimizes them, and then heuristically chose which multiplier to optimize. The objective functions for two endpoints of the diagonal line are shown in equation (4) and (5) where L and H are two endpoints of the diagonal, x is the input vector, vectors with subscript 1 and 2 refers to Lagrange multiplier 1 and 2 respectively. K is the kernel function and φ is the objective function.

$$\varphi_L = L_1 f_1 + L f_2 + \frac{1}{2} L_1^2 K(x_1, x_1) + \frac{1}{2} L^2 K(x_2, x_2) + sL L_1 K(x_1, x_2) \tag{4}$$

$$\varphi_H = H_1 f_1 + H f_2 + \frac{1}{2} H_1^2 K(x_1, x_1) + \frac{1}{2} H^2 K(x_2, x_2) + sH H_1 K(x_1, x_2) \tag{5}$$

The hyperparameters for this technique includes: seed=1, batchSize=100, complexity parameter (c)=1.0, calibrator= logistic, kernel=polykernel, numfolds = −1.

- **Bagging-**It is a type of ensemble classifier that combines various weak learners to build a strong learner. The idea is to train all the weak learners using the same learning algorithm. However, the training data that is fed to these weak learners are different. In the end, the bagged prediction is obtained either through hard voting or soft voting. The formula for bagging is shown in equation (6) where f_{bag} indicates the bagged prediction and $f_{1\ldots\ldots b}$ indicates different individual learners.

$$f_{bag} = f_1(X) + f_2(X) + \ldots + f_b(X) \tag{6}$$

Each classifier is given a different sample of the training data so that it can provide different outputs during testing. The hyperparameters for this technique include bagSizePercent=100, batchSize=100, seed=1, and number of iterations= 10.

- **BFTree**-It is a best-first decision tree. It can be used for both classification and regression. In the best first tree, the best node i.e., the node whose split leads to the maximum reduction of impurity is expanded first. It uses binary split for both nominal and numeric attributes. To measure the node impurity, it uses two commonly used splitting criteria as Information Gain (IG) and Gini Index (GI). The information is based on the distribution of the classes. IG is calculated using entropy. If pi indicates the probability of each class and there are n classes then entropy, IG, and GI can be defined as shown in equations (7), (8), and (9):

$$entropy(p_1, p_2 \ldots p_n) = -p_1 \log p_1 - p_2 \log p_2 - \ldots p_n \log p_n \tag{7}$$

$$info([p_1, p_2, \ldots p_n]) = entropy \left(\frac{p_1}{\sum_{k=1}^{n} p_k}, \frac{p_2}{\sum_{k=1}^{n} p_k}, \ldots \frac{p_n}{\sum_{k=1}^{n} p_k} \right) \tag{8}$$

$$gini(p_1, p_2, \ldots p_n) = \sum_{j \neq i} p_i p_j \tag{9}$$

The hyperparameters for this technique include seed=1, batchSize=100, splitting_criterion=Gini_index, pruning=post_pruning, numfoldspruning=5, and the minimum number of instances at terminal nodes=2.

4 RESULTS AND DISCUSSION

This paper aims to predict the severity of COPD in patients with great accuracy and efficiency. We have been able to get such results after performing pre-processing and FS on the dataset. The results are shown in Table 2. The step pre-processing provided us with data with great quality without any missingness problems. The performance metrics of all the classifiers were obtained corresponding to different feature sets obtained through different feature selection techniques. However, among all those results, the CFS technique helped in achieving the best results with a lesser number of attributes as compared to the other feature selection techniques.

Table 2. Comparison of ML classifiers after pre-processing and feature selection in terms of different performance measures.

Classifiers	Accuracy	Precision	Recall	F-measure	Kappa statistics
Naïve Bayes	96.0396	96.2	96.0	96.0	94.22
RBFNetwork	99.0093	99.5	99.5	99.5	99.35
SMO	96.0396	96.6	96.0	95.5	94.19
Bagging	99.0099	99.1	99.0	99.0	98.57
BFTree	99.0099	99.1	99.0	99.0	98.57

As can be seen from the table above, all the classifiers have been compared in terms of different measures such as accuracy, precision, recall, F-measure, and also kappa statistic.

121

The reason for using kappa statistics is that in the case of multi-classification, kappa statistics gives the best measure of comparison of the classifiers. The best performance has been achieved by Bagging and BFTree, both with accuracy and kappa statistics of 99.0099% and 98.57% respectively. However, the training time taken by BFTree (0.01s) is less as compared to Bagging (0 s). In this study, the comparison of final results with and without pre-processing and FS have also been compared. The comparison is shown in Figure 2.

ACCURACY COMPARISON OF CLASSIFIERS WITH AND WITHOUT FS AND PRE-PROCESSING

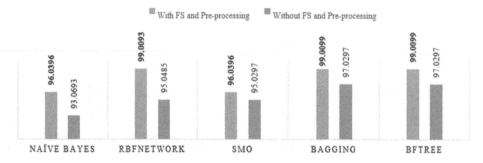

Figure 2.   Accuracy comparison of various classifiers with pre-processing and FS, and without pre-processing and FS.

It can be seen from Figure 2, that pre-processing and FS have greatly helped in achieving great accuracy. In the results obtained without pre-processing and FS, BFTree has achieved the highest performance. The ability of BFTree that always chose the best node for splitting has helped in achieving the highest performance. Hence, the BFTree classifier along with the CFS technique has provided the best results for multi-classifying COPD severity grade. In the literature, only one study has been done on the same dataset and compared with the proposed work as shown in Table 3.

Table 3.   Comparison of the proposed work with the previous work done on the same dataset.

Paper id	Technique	Accuracy	Limitation/Remark
Dritsas *et al.*, 2022	SVM	83%	Untreated missing values
Proposed Work	BFTree with Correlation-based feature selection	99%	Missing values treated using EM Imputation

This study has some limitations which are as follows: Firstly, the dataset is very small. Secondly, the dataset is not balanced. The number of instances is unevenly distributed among all four classes which also affected the results. Because of the imbalance issue, the results are also somewhat biased, hence, causing the problem of accuracy paradox. In the future, this study will aim to use oversampling techniques such as the Synthetic Minority Oversampling Technique (SMOTE), Generative Adversarial Networks (GANs), and data augmentative techniques which will not only provide unbiased results but also increases the number of instances.

5 CONCLUSION

In this study, multi-classification of COPD severity grade has been done using Machine learning models. Because of the availability of labeled datasets, various supervised classifiers have been utilized. The data collected from Kaggle was not of good quality due to the presence of missing values. To solve this problem, this study has utilized Expectation Maximization Imputation. For feature selection, 4 different feature selection techniques have been compared to provide an optimal minimum set of features. Among all the techniques, CFS provided the minimal set which is further used in the classification process. All the supervised classifiers were compared in terms of various performance metrics. The results showed that Bagging and BFTree outperformed the other classifiers in terms of all performance metrics. However, the training time of BFTree was less as compared to Bagging. Hence, it can be concluded that for the multi-classification of COPD severity grade, CFS +BFTree has given the best performance.

REFERENCES

Ali H., et al. "Forecast the Exacerbation in Patients of Chronic Obstructive Pulmonary Disease with Clinical Indicators Using Machine Learning Techniques." *Diagnostics* 11(5): 829, 2021.

Ali H et al. "Detection of Different stages of COPD Patients Using Machine Learning Techniques." In *2021 23rd International Conference on Advanced Communication Technology (ICACT)*, 368–372, 2021.

Choi et al."Prediction of COPD Severity based on Clinical Data using Machine Learning." In *2021 IEEE International Conference on Bioinformatics and Biomedicine (BIBM)*, 1646–1648, 2021.

Dritsas et al. "COPD Severity Prediction in Elderly with ML Techniques." In *Proceedings of the 15th International Conference on PErvasive Technologies Related to Assistive Environments*, 185–189, 2022.

Haider et al. "Respiratory Sound based Classification of Chronic Obstructive Pulmonary Disease: A Risk Stratification Approach in a Machine Learning Paradigm." *Journal of medical systems* 43(8): 1–13, 2019.

Kor, Chew-Teng, et al. "Explainable Machine Learning Model for Predicting First-Time Acute Exacerbation in Patients with Chronic Obstructive Pulmonary Disease." *Journal of personalized medicine* 12(2): 228, 2022.

Liao, et al. "Machine Learning Approaches for Predicting Acute Respiratory Failure, Ventilator Dependence, and Mortality in Chronic Obstructive Pulmonary Disease." *Diagnostics* 11(12): 2396, 2021.

Moll et al.. "Machine Learning and Prediction of All-cause Mortality in COPD." *Chest* 158(3): 952–964, 2020.

Nam, Ju Gang, et al. "Deep Learning Prediction of Survival in Patients with Chronic Obstructive Pulmonary Disease Using Chest Radiographs." *Radiology*: 212071, 2022.

Peng et al. "A Machine-learning Approach to Forecast Aggravation Risk in Patients with Acute Exacerbation of Chronic Obstructive Pulmonary Disease with Clinical Indicators." *Scientific Reports*, 1: 1–9, 2020.

Spathis et al. "Diagnosing Asthma and Chronic Obstructive Pulmonary Disease with Machine Learning. "*Health Informatics Journal* 25(3): 811–827, 2019.

Singh A. and Jain. A. "Hybrid Bio-inspired Model for Fraud Detection with Correlation-based Feature Selection. "*Journal of Discrete Mathematical Sciences and Cryptography*, 5: 1365–1374, 2021.

Tadahiro, G., et al. "Machine Learning-based Prediction Models for 30-day Readmission After Hospitalization for Chronic Obstructive Pulmonary Disease." *COPD: Journal of Chronic Obstructive Pulmonary Disease*, 5–6: 338–343, 2019.

Wang et al. "Comparison of Machine Learning Algorithms for the Identification of Acute Exacerbations in Chronic Obstructive Pulmonary Disease." *Computer Methods and Programs in Biomedicine*: 105267, 2020.

Zarrin et al. "In-vitro Classification of Saliva Samples of COPD Patients and Healthy Controls Using Machine Learning Tools." *IEEE Access* 8: 168053–168060, 2020.

Automation and Computation – Vats et al. (Eds)
© *2023 the Author(s), ISBN 978-1-032-36723-1*

Analysis of traditional businesses in the Himalayan state using unsupervised learning

Raj Kishor Bisht
Department of Mathematics and Computing, Graphic Era Hill University, Dehradun, India

Ila Pant Bisht
Department of Economics and Statistics, Government of Uttarakhand, Dehradun, India

ABSTRACT: Traditional businesses are the identity of any place that not only help people of that place to get recognition but also to attain livelihood for their families. Traditional businesses need to be encouraged in order to solve socio-economic problems, like unemployment, migration of local people, maintain the heritage of unique places etc. In this paper an attempt has been made to analyze the condition of traditional businesses of the people of Uttarakhand state of India. Keeping the importance of the traditional businesses in view, five districts have been selected where the products of traditional works are in demand and these traditional businesses are the main source of livelihood for these people. The data collected by the Directorate of Economics and Statistics, Uttarakhand is used for analysis. First, we put the present scenario of traditional businesses through descriptive statistics and then apply the unsupervised clustering methods to find the various clusters by considering income, the status of the business, age, etc. The cluster analysis helps to find the businesses that are performing well and generating employment and further to identify the traditional businesses that are not performing well. These findings are quite useful for policymakers to take steps for preservation of traditional businesses and employment generation.

1 INTRODUCTION

Traditional businesses are the works that are carried out by the people of certain geographical regions as per the availability of resources and requirements of traditional products. These traditional businesses also play an important role in providing livelihood to the people of these places. Since traditional businesses need manpower and the products are costly, thus growth of these traditional businesses is not up to the expectations, hence people are forced to leave their works and adopt other means of livelihood. This situation compelled people to leave their places and move to big cities for better employment opportunities. For those people, who are still doing the same work, too many challenges are there to sustain their work and maintain their livelihood. In the present era where the population is increasing and providing job opportunities is a challenge for governments, it is very important to keep alive the traditional businesses of any place. This will not only keep alive the heritage of that place but also help generate employment opportunities and maintain socio-economic balance in the society.

Uttarakhand is a hill state situated in the northern part of India in the foothills of the Himalaya. This state shares international boundaries with China in the north and Nepal in the east. The state has thirteen districts and two commissionaires 'Kumaun' and 'Garhwal'. As per the census of 2011, the state has a population of 10.1 million. The state has a total of 53483 Square kilometers of area, out of which 86% part is hilly region and 65% is covered by forests. The state is blessed with natural beauty, mountains, lakes, forests, and peace. It is

DOI: 10.1201/9781003333500-15

also known as 'Dev Bhoomi', a land of God. The state is worldwide famous for its culture, sacrament and traditions and it also reflects in its art and craft. A sample of wood carving can be seen in the temples and main doors of the houses in the villages. As per the sixth economic census, the state has 9398 handicrafts and handlooms businesses, out of which 2710 are run by females. The total number of families involved in handicraft and handloom business is 11096 and 18438 people are working in them out of which 4001 are females. The state is the center of traditional businesses since ancient times. 'Ringal', a particular kind of bamboo is found in adequate amounts in some high-altitude areas of Garhwal region and it is used to construct different items like baskets, chairs, mats etc. Knitting works like preparing woolen shawls, caps, mufflers, scarfs, carpets, making carpets, etc. are the traditional businesses of many regions of Uttarakhand state. Constructing copper vessels is a very renowned work of the Almora district of the Kumaun region. Wood carving is also the traditional work in some of the places. 'Aipan', a particular kind of drawing used during festivals, marriage ceremony is a growing traditional work. For many people, these works are the main source of income. Though at present, many people are doing the traditional businesses, but frequent migration is affecting rural economy and due to which there is a decline in the number of workers in the traditional businesses. Thus, there is a need to analyze the present status of work in different aspects so that policymakers can work accordingly to improve the traditional businesses. Machine learning approaches play a vital role in discovering meaningful information from data. Unsupervised learning methods like, clustering is quite useful for finding patterns in data.

Machine learning techniques are playing quite important role in almost every field of human life and providing solutions of different real-life problems. Clustering is one of the most important unsupervised machine learning techniques. Clustering is a natural phenomenon; we find similar small groups in large groups naturally. In clustering, we try to find meaningful underlying patterns in a data and it is quite important for many real-life problems like, prediction of diseases (Kondeti *et al.* 2019); decision making in organizations to improve the creativity of employees (Shrestha *et al.* 2021); air pollution studies to find the pollutant pathway (Govender & Sivakumar 2020); human resource management to identify efficient workers (Estrada-Cedeno *et al.* 2019); human resource management (Zhao 2020); decision making in management (Sun *et al.* 2017) etc. Clustering can be also be used as a feature extraction method for classification (Piernik & Morzy 2021).

There are various clustering algorithms available in the literature. K-mean and Hierarchy algorithms are two popular algorithms. K-mode algorithm is used for categorical data. The advancement of K means clustering is the cross-entropy clustering (Tabor & Spurek 2014). Wu summarized big data, clustering algorithms (Wu 2019). Jain presented an overview and summary of different clustering algorithm (Jain 2010). Finding patterns in a data is a challenging task. There are many issues that needs to be addressed. Different research works are going on for further improvement in clustering algorithms. An efficient clustering method that can handle mixed variables and plenty of null values is proposed in (Maione *et al.* 2019). Ahmad and Lipika proposed a clustering algorithm based on k-mean algorithm is proposed that can perform well for mixed numeric and categorical features (Ahmad & Lipika 2007). Clustering is a challenge when the data contains a large amount of noise, outliers and other redundant features. A new clustering algorithm incorporating feature selection and K-means clustering, flexible subspace clustering, is proposed in (Long *et al.* 2021). Li used principal component analysis to improve accuracy and speed of multivariate time series clustering (Li 2019). Finding clustering in a text document is a challenging task as it contains high dimensions. A new approach based on similarity function for text classification is proposed in (Kotte *et al.* 2020). Behzadi defined a new clustering algorithm for mixed data type (Sahar *et al.* 2020).

For clustering analysis, it is important that the data should have clusters. The study of the ability of having clusters is performed in (Adolfsson *et al.* 2019). To understand the data, in order to find the clusters, is necessary as one should be sure that the data contains clusters. Thus, after studying the data, a suitable clustering can be applied. The study of characteristics of clustering is performed in (Hennig 2015). Statistical significance of clustering is studied in (Liu *et al.* 2008). It is also important to know whether the selected clusters really represent underlying important structure or not. Statistical significance in hierarchical clustering is tested using Monte Carlo based approach in (Kimes *et al.* 2017).

Finding optimal number of clusters is another important aspect for getting valuable information. Different mathematical methods are available for this purpose. Still researchers are engaged in finding the best method for optimal number of clusters. For large amount of data, a new procedure for approximating the number of clusters is defined in (Estiri *et al.* 2018). A new algorithm, I-nice is defined in (Masud *et al.* 2018). An improved fuzzy C-means clustering is proposed in (Wang *et al.* 2020).

Classification comes under supervised machine learning. There are several methods in use for classification (Dunham 2006; Dwivedi 2018). Here we discuss some of the classification methods which we use for our study purpose. K-Nearest Neighbors (KNN) algorithm based on the similarity between a data set point and available representatives of different classes. It uses Euclidean distance between data points. Naïve Bayes classification is based on Bayes' theorem. Decision tree classification is based on finding subsets of a dataset and simultaneously creating decision tree. Support Vector Machine (SVM) based on defining a hyperplane to separate two classes. Random Forest classification is based on a number of decision trees. Classification techniques are quite useful in different domains, particularly in health care, these techniques are playing an important role in predicting the chances of a disease (Naushad *et al.* 2018; Tougui *et al.* 2020).

2 METHODOLOGY

Directorate of Economics and Statistics, Uttarakhand conducted a survey to know the status of traditional businesses in Uttarakhand state. In their survey, districts were selected on the basis of traditional works in the districts whose products were in high demand for last some years and these businesses were the main source of income for the persons running these businesses and at present, these businesses are in running state and main source of income for the concerned persons. Table 1 shows the selected districts with details of traditional work.

Table 1. Selected districts for the study and status of their traditional businesses.

Sr. No.	District	Number of workers in traditional businesses as per 6th economic census		Nature of Traditional work
		Male	Female	
1	Pithoragarh	1266	720	Knitting woolen cloths and 'Ringal' work
2	Chamoli	470	737	Knitting woolen cloths and 'Ringal' work
3	Almora	1117	119	Construction of Copper Vessels, 'Aipan' work
4	Uttarkashi	607	247	Knitting woolen cloths
5	Haridwar	2317	900	Knitting woolen blankets and bed sheets Antique Items

After the selection of districts, it was proposed to select 5 clusters representing the villages or towns or areas where maximum family run the business from each district. But due to the non-availability of 5 clusters, the number of clusters was different in each district. The selected clusters from different districts were as follows: Pithoragarh-7, Almora-3, Chamoli-3, Haridwar-6 and Uttarakshi-2. From these clusters, a minimum of 15 families were selected randomly. A total of 321 entities have been recorded, each entity represents the main worker of a family. Though the data contains quite detailed information, yet we consider a few important variables only. Table 2 shows the variables under study and their description. There are different types of traditional businesses performed in different districts of Uttarakhand. Table 3 shows the name, description and working areas of different traditional businesses.

Table 2.   Variables under study and their description.

Sr. No.	Variable Name	Description	Data type
1	District	Name of the district	Categorical
2	Age	Age of the main worker in the family	Integer
3	Sex	Sex of the main worker in the family	Categorical
4	HHS	Household size	Integer
5	DW	Description of the traditional work	Categorical
6	DE	Demand of products as per their expectation	Categorical
7	SB	Present status of the business	Categorical
8	TE	Total employment generated by the traditional business	Integer
9	MD	Manufacturing days in a year (working days)	Integer
10	TOE	Total expenditure per year in running the traditional business	Integer
11	YIPM	Yearly income per member from the traditional work.	Integer

Table 3.   Description of traditional businesses.

Sr. No.	Name of traditional work	Description
1	Copper work	It includes formation of copper vessels.
2	'Aipan' work	It is a traditional name used for a special drawing, painting in front of doors, walls, courtyard
3	'Ringal' work	A type of 'Chimnobambusa falcata' used to form mats, baskets, flower pots etc.
4	Knitting work	It includes making of different woolen items like cap, hand gloves, sweaters, mufflers, scarps, blankets, carpets etc.
5	Antique items work	Creation of antiques items

3   DESCRIPTIVE STATISTICS

In this section, we present the descriptive statistics of the traditional businesses and related information. The data of 321 main workers were collected, out of which 198 were female and 123 were male. Age-wise distribution of the main worker is given in Figure 1. From Figure 1,

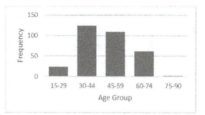

Figure 1.   Age-wise distribution of main workers.

we observe that majority of the workers are from 30-44 age group. Figure 2 shows the percentage of workers in different traditional businesses. From Figure 2, we observe that the maximum number of workers are in Knitting work indicating that Knitting work is quite popular traditional business.

Figure 3 shows the sex-wise distribution of workers in different traditional businesses. From Figure 3, we observe that majority of female workers are involved in Knitting and 'Aipan' works. Figure 4 shows the number of workers in different traditional businesses in different districts. From Figure 4 we observe that Knitting work is done in majority of the district, while some districts have unique traditional works; Copper work in Almora, Antique items work in Haridwar, 'Ringal' work in Chamoli and Pithoragarh. Figure 5 shows the status of different traditional businesses in percentage. From Figure 5, we observe that 'Ringal' work, Knitting work, and 'Aipan' works are increasing while copper work and antique item works are decreasing.

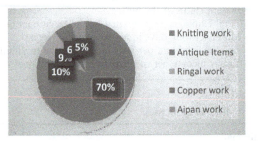

Figure 2.   Percentage of workers in different traditional businesses.

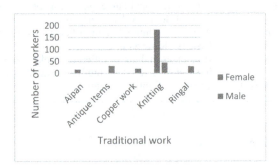

Figure 3.   Sex-wise distribution of workers in different traditional businesses.

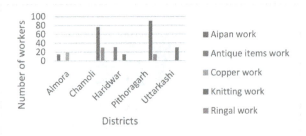

Figure 4.   District-wise distribution of workers in different traditional businesses.

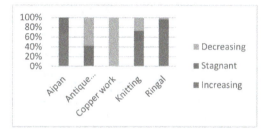

Figure 5.   Status of different traditional businesses.

Figure 6 shows the average manufacturing days of different traditional businesses. From Figure 6 we observe that Antique work is the work that is carried out almost throughout the year. Knitting, Copper work and 'Ringal' work also provide works for nearly completely year, while 'Aipan' work is the seasonal work. Figure 7 describes the average yearly income per member from the traditional businesses for different types of works.

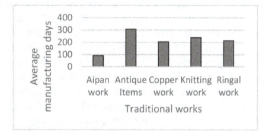

Figure 6.   Average manufacturing days of different traditional businesses.

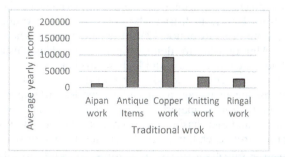

Figure 7.   Average yearly income per member in different traditional businesses

4   CLUSTERING PATTERNS AND ANALYSIS

We choose four prominent numerical variables 'TE', 'MD', 'TOE' and 'YIPM' for performing K-means clustering. First, we perform three-dimensional clustering by considering three different variables. To find the optimal number of clustering, we apply the elbow method and, in all cases, we find the number of optimal clusters is 3. Figure 8 shows the clustering pattern based on employment generation, manufacturing days and yearly income

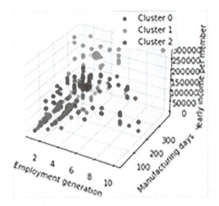

Figure 8.   Clusters in employment generation, manufacturing days and yearly income.

per member. We find three different clusters. Following is the detailed analysis of different clusters.

Cluster 0 represents 74% female and 26% male workers who have an average working days of 203 in a year having an average income of Rs. 17726 per year and an average investment of Rs. 36466. For the majority of the families, however, they are not getting the demand as per their expectations, yet their traditional business is increasing with an average of 2 persons of employment generation from each family. In this cluster, 'Knitting' work is the major contributor followed by 'Ringal' work and then 'Aipan' work. In this cluster majority of workers are from Chamoli district (49%) and Pithoragarh district (32%). A minor contribution is from Almora and Uttarakashi (7%) and Haridwar (4%). This cluster includes persons with fewer working days and low annual income.

Cluster 1 represents 13% female and 87% male workers who have an average working days of 310 in a year having an average income of around Rs. 226325 per year and an average investment of Rs. 434533. For the majority of the families, they are not getting the demand as per their expectations and their business is also decreasing with an average of 3 persons of employment generation from each family. In this cluster, a high majority of male workers (79%) are from Haridwar district who are involved in 'Antique work'. From the remaining workers, a minor contribution 12% of female workers who are doing 'Knitting' work and 8% from Almora who are doing 'Aipan' work. This cluster included persons having work for almost a complete year and highest-earning.

Cluster 2 represents 56% female and 44% male workers who have an average working days of 287 in a year having an average income of around Rs. 80114 per year and an average investment of Rs. 250801. For the majority of the families, they are not getting the demand as per their expectations and their business is also decreasing with an average of 3 persons of employment generation from each family. In this cluster, the majority of workers (39%) are from Pithoragarh district who are involved in 'Knitting' work followed by 'Ringal' work. An almost equal number of workers are from Almora, Haridwar and Uttarakashi performing 'Copper' work, 'Antique items' work and 'Knitting' work respectively. This cluster includes persons having moderate working days and annual income.

This information is quite useful for policymakers to promote these traditional businesses in the form of promotion of these works so that demand for these items can be increased and hence, income and employment generation can also be increased. Figure 9 shows the clustering pattern based on total expenditure, manufacturing days and yearly income per member. We find three different clusters. Following is the detailed analysis of different clusters.

Cluster 0 includes the workers having low expenditure and low income. In this cluster, there are 72% female and 28% male workers who have an average-working days of 224 in a

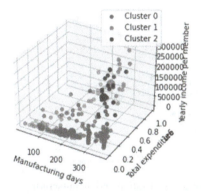

Figure 9.   Clusters in expenditure, manufacturing days and yearly income.

year having an average income Rs. 30992 per year and an average investment of Rs. 20179. For the majority of the families, however, they are not getting the demand as per their expectations, yet their traditional business is increasing with an average of two persons employment generation from each family. This cluster includes majority of workers in 'Knitting' work from Chamoli (39%) and Pithoragarh (38%). This cluster represents the persons with average yearly working days and very low income. The main works under this cluster are 'Knitting work' , 'Ringal work', 'Aipan work' and 'Copper work'.

Cluster 1 represents the persons in traditional businesses having high expenditure and moderate yearly income. Cluster 1 represents 5% female and 95% male workers who have an average-working days of 255 in a year having an average income of Rs. 133542 per year and an average investment of Rs. 709881. For the majority of the families, they are not getting the demand as per their expectations and their traditional business is decreasing with an average of 6 persons employment generation from each family. In this cluster majority of workers are involved in 'Antique items work' (52%) followed by 'Knitting work' (48%) and these workers are from 'Haridwar' district only. One another interesting fact of this cluster is that the average expenditure and average income of 'Antique items work' is higher than the average expenditure and average income of 'Knitting work' and further, the majority of the workers in 'Antique items work' admitted that the work is stagnant while 'knitting work' is decreasing.

Cluster 2 represents the workers having moderate expenditure but little high yearly income than cluster 1. In this cluster, there are 13% female and 87% male workers who have an average-working days of 293 in a year having an average income of Rs. 150677 per year and an average investment of Rs. 354781. This cluster includes the majority of workers from Haridwar district involved in 'Antique items' work (58%) and 'Knitting' work (13 %) followed by Almora district involved in 'Copper' work (16%) and then Pithoragarh district involved in 'Knitting' work. For the majority of the families, they are not getting the demand as per their expectations and their traditional business is decreasing with an average of 4 persons employment generation from each family. This cluster is the best cluster with regards to expenditure, manufacturing days, income and employment generation. In this cluster, the majority of workers are involved in 'Antique items work' followed by 'Knitting work' and 'Copper work'.

Clusters based on 'TE', 'MD', 'TOE' and 'TE', 'TOE' and 'YIPM' are same as 'TOE'-'MD'-'YIPM'. Further, we use K-modes (kprototypes) clustering for getting clustering patterns based on categorical and numerical variables. We find same clusters as in previous cases. We also try to find the clusters based on two variables. In almost all cases, we find almost same clusters except in the case when 'TE' and 'MD' are considered. Figure 10 shows the clustering pattern based on employment generation and manufacturing days. We find three different clusters. Following is the detailed analysis of different clusters.

131

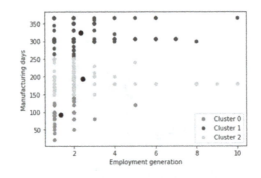

Figure 10.   Clusters in employment generation and manufacturing days.

Cluster 0 represents 67% female and 33% male workers who have average-working days of 92 in a year having an average income Rs. 13309 per year and an average investment of Rs. 13200. For the majority of families, however, they are not getting the demand as per their expectations, yet their traditional business is increasing with an average of one person employment generation from each family. This cluster includes the majority of the workers from Chamoli district involved in 'Knitting' and 'Ringal' work (nearly 22% in each case) followed by Almora district involved in 'Aipan' work (26%) and 'Copper' work (7%) and then the workers involved in 'Knitting' work from Uttarakashi (20%) and Pithoragarh(4%). This cluster represents the persons with very few yearly working days and very low income and only one employment generation.

Cluster 1 represents 61% female and 39% male workers who have an average-working days of 323 in a year having average income Rs. 83044 per year and an average investment of Rs. 157070. For the majority of families, they are not getting the demand as per their expectations and their traditional business is decreasing with an average of 2 persons employment generation from each family. This cluster includes a majority of workers from Pithoragarh district involved in 'Knitting' work (49%) and 'Ringal' work (11%). This cluster is the best cluster with regards to manufacturing days and employment generation.

Cluster 2 represents 61% female and 39% male workers who have average-working days of 193 in a year having an average income of Rs. 27302 per year and an average investment of Rs. 127024. For the majority of families, however, they are not getting the demand as per their expectations, yet their traditional business is increasing with an average of 2 persons employment generation from each family. In this cluster majority of workers are from Chamoli district involved in 'Knitting' work (46%) and 'Ringal' work(14%) followed by Pithoragarh district involved in 'Knitting' work (15%) and then Haridwar district involved in 'Knitting' work (12%). This cluster is the average cluster with regards to manufacturing days and employment generation.

5   DISCUSSION AND CONCLUSION

In this section, we discuss the outcomes of the analysis. Descriptive analysis reveals that the majority of traditional workers are involved in 'Knitting' work and this work can be considered as the work adopted by most people of the state. 'Ringal' work is the second most popular work in the state. 'Copper' work and 'Aipan' work are the traditional work of Almora district only and 'Antique item' work is the traditional work of Haridwar district only. 'Knitting' and 'Aipan' works have female workers in the majority while in other works male workers are in majority. 'Copper' work is almost decreasing while for the majority of cases 'Aipan' and 'Ringal' and 'Knitting' works are increasing. 'Antique' items work is either decreasing or stagnant. 'Antique items' work is providing employment for almost complete year. 'Copper', Knitting, and 'Ringal' works are also providing employment for more than half a year but 'Aipan' is the seasonal work only. 'Antique' work is the best traditional work with respect to income generation followed by 'Copper' work.

Clustering analysis reveals different patterns in the data of traditional work. A few families involved in 'Antique' work from Haridwar, 'Copper' work from Almora and 'Knitting' work from Pithoragarh are performing very well in terms of employment generation, expenditure, manufacturing days and income. There is a need to study their business model so that other workers can get help to improve their business. A few families in 'Antique items' work have more expenditure and less income in comparison to the workers in the same work having same income within less expenditure. There is a need to investigate the reasons for the same so that the business work of these people can be improved. Traditional works 'Knitting' and 'Ringal' in Chamoli district, 'Knitting' work in Uttarkashi are not performing well. The majority of workers in 'Knitting' work from Pithoragarh are performing well in terms of employment generation and manufacturing days, however, they are lacking in income generation. Chamoli and Pithoragarh districts have the maximum number of workers with a low number of yearly working days and low income. 'Antique' work is the traditional work which is giving maximum employment generation and income. A good number of workers in Chamoli and Pithoragarh districts are involved in 'Knitting' work. Though they have work for more than half the year, yet their income is low.

In the present work, the data of traditional works in Uttarakhand state has been analyzed. The present analysis is quite useful for the identification of the traditional works those who are performing well, average or need improvements in different aspects. In future, various factors affecting the performances of traditional works may be analyzed so that policy can be implemented accordingly.

ACKNOWLEDGEMENT

The authors are thankful to Director, Economics and Statistics, Dehradun for providing permission to use the primary data of the survey for analysis and research publication.

REFERENCES

Adolfsson A., Ackerman M., & Brownstein N. C. To Cluster, or not to Cluster: An Analysis of Clusterability Methods. *Pattern Recognition*, 88, 13–26, 2019. doi:10.1016/j.patcog.2018.10.026

Ahmad A., & Lipika D. *A K-mean Clustering Algorithm for Mixed Numeric and Categorical Data. Data and Knowledge Engineering*, 63(2), 503–527, 2007. doi:10.1016/j.datak.2007.03.016

Dunham M. *Data Mining Introductory and Advance Topics.* Pearson Education India, 2006.

Dwivedi A. K. Performance Evaluation of Different Machine Learning Techniques for Prediction of Heart Disease. *Neural Computing and Applications*, 29(10), 685–693, 2018. doi:10.1007/s00521-016-2604-1

Estiri H., Abounia Omran B., & Murphy S. N. Kluster: An Efficient Scalable Procedure for Approximating the Number of Clusters in Unsupervised Learning. *Big Data Research*, 13, 38–51, 2018. doi:10.1016/j.bdr.2018.05.003

Estrada-Cedeno P., Layedra F., Castillo-Lopez G., & Vaca C. The Good, the Bad and the Ugly: Workers Profiling through Clustering Analysis. *2019 6th International Conference on eDemocracy and eGovernment, ICEDEG*, Quito, Ecuador: IEEE, 101–106, 2019. doi:10.1109/ICEDEG.2019.8734453

Govender P., & Sivakumar V. Application of K-means and Hierarchical Clustering Techniques for Analysis of Air Pollution: A Review (1980–2019). *Atmospheric Pollution Research*, 11, 40–56, 2020. doi:10.1016/j.apr.2019.09.009

Hennig C. What are the True Clusters?*Pattern Recognition Letters*, 64(C), 53–62, 2015.

Jain A. K. Data Clustering: 50 Years Beyond K-means. *Pattern Recognition Letters*, 31(8), 651–666, 2010. doi:10.1016/j.patrec.2009.09.011

Kimes P. K., Liu Y., Neil Hayes D., & Marron J. S. Statistical Significance for Hierarchical Clustering. *Biometrics*, 73(3), 811–821, 2017. doi:10.1111/biom.12647

Kondeti, P. K., Ravi, K., Mutheneni, S. R., Kadiri, M. R., Kumaraswamy, S., Vadlamani, R., & Upadhyayula, S. M. Applications of Machine Learning Techniques to Predict Filariasis using Socio-economic Factors. *Epidemiology and Infection*, 147, e260, 2019. doi:10.1017/S0950268819001481

Kotte V. K., Rajavelu S., & Rajsingh E. B. A Similarity Function for Feature Pattern Clustering and High Dimensional Text Document Classification. *Foundations of Science*, 25(4), 1070–1094, 2020. doi:10.1007/s10699-019-09592-w

Li H. Multivariate Time Series Clustering based on Common Principal Component Analysis. *Neurocomputing*, 349, 239–247, 2019. doi:10.1016/j.neucom.2019.03.060

Liu Y., Hayes D. N., Nobel A., & Marron J. S. Statistical Significance of Clustering for High-Dimension, Low–Sample Size Data. *Journal of the American Statistical Association*, 103(483), 1281–1293, 2008.

Long Z. Z., Xu G., Du J., Zhu H., Yan T., & Yu Y. F. Flexible Subspace Clustering: A Joint Feature Selection and K-Means Clustering Framework. *Big Data Research*, 23(100170), 0–0, 2021. doi:10.1016/j.bdr.2020.100170

Maione C., Nelson D. R., & Barbosa R. M. Research on Social Data by Means of Cluster Analysis. *Applied Computing and Informatics*, 15(2), 153–162, 2019. doi:10.1016/j.aci.2018.02.003

Masud M. A., Huang J. Z., Wei C., Wang J., Khan I., & Zhong M. I-nice: A new Approach for Identifying the Number of Clusters and Initial Cluster Centres. *Information Sciences*, 466, 129–151, 2018. doi:10.1016/j.ins.2018.07.034

Naushad S., Hussain T., Indumathi B., Samreen K., Alrokayan S., & Kutala V. Machine Learning Algorithm-based Risk Prediction. *Molecular Biology Reports*, 45, 901–910, 2018.

Piernik, M., & Morzy, T. (2021). A Study on using Data Clustering for Feature Extraction to Improve the Quality of Classification. *Knowledge and Information Systems*, 63, 1771–1805. doi:10.1007/s10115-021-01572-6

Sahar B., Nikola S. M., & Claudia P. Clustering of Mixed-type Data Considering Concept Hierarchies: Problem Specification and Algorithm. *International Journal of Data Science and Analytics* volume, 10, 233–248, 2020. doi:10.1007/s41060-020-00216-2

Shrestha Y. R., Krishna V., & von Krogh G. Augmenting Organizational Decision-Making with Deep Learning Algorithms: Principles, Promises, and Challenges. *Journal of Business Research*, 123, 2021. doi:10.1016/j.jbusres.2020.09.068

Sun L., Chen G., Xiong H., & Guo C. Cluster Analysis in Data-Driven Management and Decisions. *Journal of Management Science and Engineering*, 2(4), 227–251, 2017. doi:10.3724/SP.J.1383.204011

Tabor J., & Spurek P. Cross-entropy Clustering. *Pattern Recognition*, 47(9), 3046–3059, 2014.

Tougui I., Jilbab A., & El Mhamdi J. Heart Disease Classification using Data Mining Tools and Machine Learning Techniques. *Health and Technology*, 10, 1137–1144, 2020.

Wang Z., Wang S., & Du H. Fuzzy C-means Clustering Algorithm for Automatically Determining the Number of Clusters. *Proceedings - 2020 16th International Conference on Computational Intelligence and Security*, CIS 2020, pp. 223–227. Guangxi, China: IEEE. doi:10.1109/CIS52066.2020.00055

Wu C. (2019). Research on Clustering Algorithm Based on Big Data Background. *Journal of Physics: Conference Series*. doi:10.1088/1742-6596/1237/2/022131

Zhao Y. (2020). Application of K-means Clustering Algorithm in Human Resource Data Informatization. *Proceedings of the 2020 International Conference on Cyberspace Innovation of Advanced Technologies*, pp. 12–16. doi:doi.org/10.1145/3444370.3444540

Automation and Computation – Vats et al. (Eds)
© 2023 the Author(s), ISBN 978-1-032-36723-1

Intrusion detection and plant health monitoring smart system

Rahul Chauhan & Soumili Tapadar
Graphic Era Hill University, Uttarakhand, India

ABSTRACT: Farmers are the backbone of any nation. But life is not a bed of roses for them as they withstand bitter cold and scorching heat to provide us the necessities of life. Farmer treat their crops as their own child as they nurture those lifeless seeds to bloom to life. But unfortunately, even at the odd hours of night they must look for any possible intrusion by wild animals or even humans with malicious intentions as well as make sure that their crops are healthy and disease-free. So, why not build a smart hybrid system that not only guards the field from any suspected intrusion 24*7 and analyses the crop health and growth regularly.

1 INTRODUCTION

In this high-tech era where everyone is as busy as a bee from dawn to dusk, gone are the days when people used to handle events manually. With the ever increasing in population the surge of demand of food grains is inevitable. Specially for an agriculture-based country such as India, to cope up with the exponentially increasing demand of the market many studies are being done. Any nation's economy relies heavily on agronomy, which is seen as the backbone of every society. Identification of plant health is crucial since, as an agricultural country, the nation's food supply depends on an adequate supply of healthy plants. Crop production loss is caused by pathogens, pests, weeds, and animals (Abisha 2021) . With this gradual increase in pressure on the farmers to meet the surplus production size many new innovations have been bought up. But farmers today face 2 major hurdles today.

- Protecting crops from contaminating deadly pathogens.
- Preventing unwanted intrusion at odd hours which may potentially damage the crops.

Unfortunately, still there is no proven strategy yet to prevent unwanted intrusion in farms at night by any potential threat for the crops such as any wild animals. Although there are so many features available for safety prevention yet every alternate day, we see headlines about how crops set across acres of land caught fire or got infected by any deadly disease. Moreover, most of the culprits run away unnoticed resulting in unsuccessful diagnosis of the root cause involved. Farmers are the backbone of any nation. But life is not always fair to them as they withstand bitter cold and scorching heat to provide us the necessities of life. They responsibly utilize natural resources using primitive and advanced level technologies for surplus production. Additionally, there are numerous valuable ways in which farmers contribute for the upliftment of nation. At the end of a tough day, a farmer deserves a good sleep, but unfortunately even at the odd hours of night they must look for any possible intrusion by wild animals or even humans with malicious intentions. Either they spend whole night guarding the field or wake up occasionally, for the sake of their dear crops. Farmer treat their crops as their own child as they nurture those lifeless seeds to bloom to life. Unfortunately, plants are still at a constant threat of pathogens. Some of them even prove to be fatal if not diagnosed and treated on time. And in recent years another challenge has emerged for the farmers, i.e., wild animals. Animals and rodents such as wild boars, elephants, monkeys, moths etc. cause irreversible damage to crops.

As per Hindustan times, Weeds lead to India losing an average of $11 billion each year in 10 major crops, shows data from 1,581 farm trials in 18 states. Below is a graph representing the annual damage data in India caused due to weeds.

In the US, feral boars have an annually negative economic effect of $2.5 billion on agribusiness, natural flora, vertebrate and invertebrate fauna, soil characteristics, and aquatic habitat. The University of Georgia's Warnell School of Forestry and Natural Resources disseminated a six-page wild pig census. to landowners and ranchers in January 2012. A total of 1200 questionnaires were distributed to randomly chosen respondents, and 471 of these were viable, resulting in a response rate of 39.25%.

Since the most recent survey in 2014, animal destruction has risen by a factor of twofold. In contrast to 88 thousand tons in 2014, 165 000 tons of cereal were damaged by animals in 2020.Animals such as birds, rodents, insects, and other mammals are the main causes of this mishap. Rodents and insects can destroy up to 20% or more of crops. It's important to note that losing a percentage of 25% or more can cause a loss of profitability. This often means that harvests are abandoned, in order to avoid the expense that's caused if you can't deal with the pests. At the end of the day, animals only look for an accessible food source, which is why they see a real buffet in these harvests.

In the lineage of finding a one stop solution to all these problems we have come up with a hybrid solution having a purpose of assisting farmers and landlords with both surveillance, security along with real time plant health and growth rate monitoring. It aids in preventing unwanted intrusion in farmland by creating an invisible fence around it. In case any abnormal motion is detected, an alarm is raised along with real time HD image capturing of intruder shared on user's portable device such as mobile phone etc. using high tech PIR sensors and Raspberry pi integrated camera. We have also integrated high accuracy AI based advance plant health recognition system specific to each agricultural species. While will help in tracking growth rate and development of crops along with comparison with the ideal rate by AI with data saved on cloud as per fed by machine learning. In case any anomality is found, troubleshooting any potential damage and notifying the owner on their

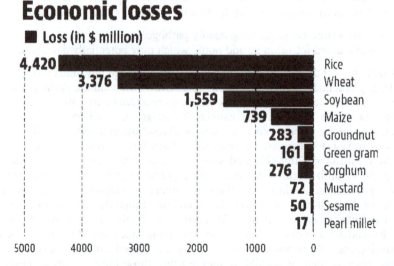

Figure 1.   Factual data representing economic loss annually for different agricultural species in millions.

local device along with an instantaneous image clicked by the Raspberry pie integrated camera of the potential intruder.

2 RELATED WORK

2.1 *AID: A Prototype for Agricultural Intrusion Detection Using Wireless Sensor Network*

The sensors and actuators in the AID technology demonstrator are utilized to track and detect trespassers who violate the agricultural land. After a field invader has been discovered, the detected input is sent either straight to the sink or through additional intermediary nodes. The farmer is further alerted by the sink via text messages on his or her smartphone. The gadget inside the farmer's home triggers an alert in addition to the text message. This alert and the messaging app signal an intruder's entry onto the farmer's cropland. Based on their functionality, the four tiers of the AID architecture involving the use of passive infrared (PIR) and ultrasonic sensors for detecting activities. The PIR sensor detects an intruder as they enter the field, activating the node. The distance between the invader and field border is then measured by the ultrasonic sensor. The micro-controller found in Layer 2 processes the data obtained from PIR and ultrasonic sensors. With the aid of ZigBee, information is sent to the sink through single- or multihop connection following the detection of an intrusion. Layer 3 is in charge of directing the information from the node to the sink. The GSM framework is able in Layer 4 to trigger an alert in the farmer's home while also sending an SMS to the farmer's cell phone. The problem with this model is that it only sends a text message whereas in our proposed model an image of the intruder will also be sent to the farmer. Also, this model is only for intrusion detection whereas our proposed mode is a hybrid model that serves the purpose of an intrusion detection system as well as a plant health monitoring system.

2.2 *Smart intrusion detection system for crop protection*

The system's methodology attempts to create a secure system for protecting farmlands against intruders such as wild animals and humans, sometimes without human participation. The GSM module is used as an additional security feature that sends an email to the owner if an intruder is discovered. Also, for the farmer to know better about the intrusion system contains a camera which continuously surveys the area and captures the images whenever the movement of any intruder is sensed by the sensors, and those images can be displayed on the monitor display. At night the flashlight allows the camera to capture an almost readable or understandable pictures.

Unfortunately, unlike to its claims this model is only for intrusion detection whereas our proposed mode is a hybrid model that serves the purpose of an intrusion detection system as well as a plant health monitoring system providing an overall solution.

2.3 *Plant health monitoring system using image processing*

This is an algorithm-based solution proposed based on image processing management and analysis. The algorithm of proposed system consists of various steps and stages the first one being training stage. Under this stage the first step is to select or upload various images and its labels. Secondly, they apply Pre-Processing using Histogram Equalization and Denoising on whole image datasets. Next, they apply Colour and Cluster Based Combine Segmentation approach. Fourthly they Extract Shape, Colour and Texture Features for all images. And the last step for training module is applying machine Learning Approach SVM, RF and making a database.

Next comes the Testing stage starting from selecting or upload an image and applying Pre-Processing using Histogram Equalization and Denoising. Thirdly, applying Colour and Cluster Based Combining Segmentation approach. Next step is extracting Shape, Colour and Texture Features followed by applying machine Learning classification Approach SVM, RF using database. Lastly, Classifying Disease type.

The drawback is this model only classifies the disease type and doesn't suggest a possible solution for that. Also, this model is only for plant health monitoring whereas our proposed mode is a hybrid model that serves the purpose of an intrusion detection system as well as a plant health monitoring system and provides an efficient one stop solution.

While researching on the internet regarding possible solutions to this problem was a Herculean problem as not much insight was available on the internet regarding the same. Even the ones we found had scattered ideas. We decided to create a one stop all in one solution for our honorable farmers so they do not to opt for various third-party tools as they are not that much tech friendly and handling multiple systems can be stressful. Alternatively, if we provide them a one stop solution from a single vendor, they will be free from any possible mishap. Surveillance cameras are highly utilized in several public utilities already such as airports, supermarkets, stadiums, and railway stations to deter crime or catch offenders after they've done it. By being more than generally deployed in retail outlets for video surveillance, raspberry pie-based surveillance moves this lineage ahead. Safety-related applications, such as those that monitor corridors, doors, entrance areas, and exits, such as emergency exits, are other crucial ones. Additionally, under normal working settings, our suggested technology guarantees good performance, and these devices are vulnerable to visual impairments (Paul & Ronald 2016).

In recent years, application-oriented studies focusing on video surveillance systems have attracted a lot of attention. The most current research tries to include artificial intelligence, image processing, and computer vision technologies into video surveillance systems. Even though numerous datasets, methodologies, and frameworks have been acquired and published, there aren't many articles that can give a complete overview of the status of video surveillance system research at the moment, specifically for farming and agricultural lands [3].

Additionally, we propose a one-step solution Automatic processing software inspects the photos upon receipt for abnormalities, assesses crop performance, and forecasts the number of days till harvest for each crop tray. The picture analysis technology automatically alerts the agronomic team to abnormalities or inferior behavior, which they then debate, share, or correct using available countermeasures. The invention, in one aspect, offers a technique for enhancing plant health because the methodology may also include exposing the plant to either biotic or abiotic stress before any possible damage to its health (Patil *et al.* 2021). Potential biotic disease examples include fungal illnesses like Soybean Rust, bacterial, viral, and insect infestations as well as weed infestations.

3   METHODOLOGY

A surveillance system, surveillance radar system, software application, and non-transitory computer-readable trackable form of communication, where the surveillance system is largely comprised of an electro - optic camera that utilises received radiance to capture images and has a first area of view, a radar sensor that emits and receives electromagnetic radiation and has a second field of view, and in which the primary field of view is movable (Shidik *et al.* 2019).

The suggested system's block diagram is shown here. The whole diagram can be divided into two parts the first part basically explains the working of an intrusion detection system and the second one explains the working of a plant health monitoring system. The intrusion detection part has a Passive Infrared Sensor (PIR) as the main component which detects any kind of intrusion in the field. In case any intrusion is detected then two processes take place simultaneously, the first is that an alarm goes off so that it may woo away the intruder and the second is that an image is captured. The captured image is sent to the user/farmer through cloud, it will help the user/farmer to remotely check whether it is a false alarm or not or the severity of the intrusion.

It frequently gets challenging for investigation authorities to trace the incidents. We are putting out a framework to track and investigate undesired events at strange times, even at poor resolution at nighttime and over vast distances (Shariff & Hussain 2017). The creation and implementation of a clever, all-encompassing system for big farmlands and cultivated fields that provides constant outdoor security is the main topic of this essay. The Raspberry Pi is the name of this outside security system (RPI). When the camera board is used, the Raspberry Pi, a credit card-sized computer with an attached camera board, may be converted into a visual surveillance system. This device may detect any change and provide notifications in the form of warnings to the client's dashboard, especially on a cell phone, advising them to pay more attention to what they are checking. When you tell the arrangement of your absence from the farmland, the sensors are activated. When a change is seen, it is possible to set the security system to record activity through the surveillance camera (Kumar. Natraj *et al.* 2017).

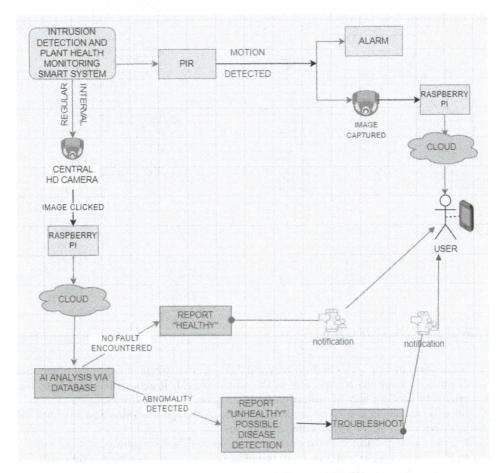

Figure 2. Block diagram of the proposed system.

The second part of the system is a plant health monitoring system. It is based on image capturing, processing and analysis. Pictures are taken on regular basis and then they are analyzed based on the algorithm. The training stage is the first in a series of phases and stages that make up the algorithm of the proposed system. The initial step in this stage is to choose or upload a variety of photographs together with their labels. Second, they pre-process entire image collections using histogram equalization and denoising. They then use a method called Color and Cluster Based Combine Segmentation. Fourth, all photos are extracted for shape, color, and texture features. The final step in the training module is to create a database and use the machine learning approaches SVM and RF.

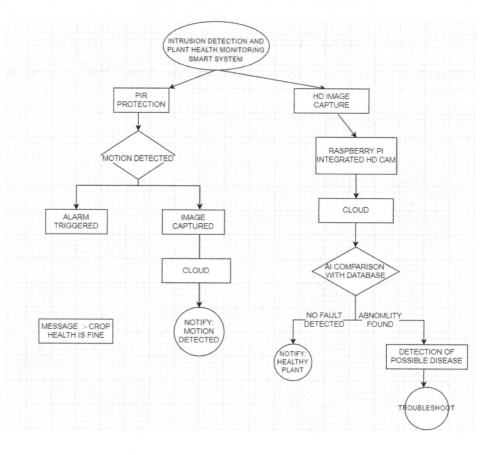

4 WORKING ALGORITHM

1. A landlord can install this high-tech hybrid field security system in their field of required dimensions along the borders and at the center of field. It acts as a virtual fence that provides a modern way of protection from intruders and plant diseases.
2. Initially the camera and alarm are in sleep mode. Although, the Infra-Red layer will still be active and look for any possible disruptions.
3. If any motion is detected in the given range of any of the four sensors.
4. In case the motion sensed is from an insignificant entity such as a bird or an insect, it is ignored to prevent fake alarming alert.

5. Alternatively, for large/significant entities the PIR sensor triggers buzzer to raise an alarm to frighten away wild animals (if any) along with motion sensor raspberry pi camera to click instantaneous image of intruder.
6. The captured image via raspberry pi is shared on cloud from where an intruder alert notification and an instantaneous image clicked is shared with the user so he can prevent any potential damage as per his own choice.
7. Along with this we have a central plant health monitoring system in this hybrid. It clicks an HD image from proximity of crops in a regular interval. This image is shared on cloud via raspberry pi.
8. The image data is compared with standard images respective to various growth stages present in database via machine learning and artificial intelligence.
9. In case any anomality is detected, the user is notified about the potential diseases along with troubleshooting the issue and suggesting solutions accordingly.
10. For no abnormality found, user is notified a message stating "Your plants are healthy!"

5 CONCLUSION

Every country is built on its farmers. However, they do not have it easy because they must endure extreme temperatures to give us the things we need to survive. Farmers nurture those dead seeds to burst into life, treating their harvests as if they were their own children. Unfortunately, they must check for any intrusions by wild creatures or even humans with nefarious intents at strange hours of the night, as well as ensure that their crops are healthy and free of illness. Why not create a smart hybrid system that not only constantly monitors the field for any suspected intrusions but also routinely evaluates the crop's health and growth?

The advancement of agricultural surveillance technology has kept pace with the Internet of Things' advancement. A monocular camera and Raspberry Pi-based visual navigation system for a monitored farming field are built. The greying factor, intermediate filtering, Holt threshold segmentation, and morphological processing are examined and used to separate the greying picture from the backdrop and green vegetation. A practical configural extraction algorithm and province of relevance selection technique are presented. Furthermore, it is proposed that navigation parameters be calculated to regulate the computer's forward and steering man oeuvres. The design offers a wide variety of agricultural production applications, high dependability, powerful anti-interference capabilities, and low cost (Mao *et al.* 2019).

So, the time has come to implement modern farming techniques. A significant agricultural revolution has been ushered in by artificial intelligence in farming. We have worked on both illness detection and a surveillance system based on raspberry pi in this project. Three methods to use the rapid and adaptable microcontroller board. The Raspberry Pi Pico offers you the capability to supervise a wide assortment of domestic, recreational, and commercial activities, including signs, IoT devices, light displays, and production procedures (raspberry-pi 2017). We have trained and modelled a Content-based filtering method on the dataset statistics for the crops recommendation system. In this we make recommendations for crop health considering several factors.

To establish a real-time smart outdoor monitoring system that increases security, this research study primarily integrates Raspberry Pi with the Internet of Things, the foundation based on artificial intelligence, and other technologies. People can utilize a remote-control technique to instantly evaluate the security of their own agriculture. The actualized smart system is built on a Raspberry Pi, which utilises the Internet of Things architecture and integrates several sensors and sensors for farm security to create a more robust green environment (Lee *et al.* 2020).

Although the proposed implementation is yet to be tested physically so the accuracy cannot be predicted yet statistically. In future, there are numerous improvement scope available to this noble cause to support the backbone of our nation.

REFERENCES

Abisha A. and Bharathi N., "Review on Plant Health and Stress with Various AI Techniques and Big Data," *2021 International Conference on System, Computation, Automation and Networking (ICSCAN)*, 2021, pp. 1–6, doi:10.1109/ICSCAN53069.2021.9526370.

Feng P.C.C., Brinker R. J., Patent no: US9155300B2,2016

Huang B., "Research on the Application of Internet of Things Monitoring System in Greenhouse Flowers," *2021 International Wireless Communications and Mobile Computing (IWCMC)*, pp. 1856–1859, 2021, doi:10.1109/IWCMC51323.2021.9498699.

Kumar K. N. K., Natraj H. and Jacob T. P., "Motion Activated Security Camera Using Raspberry Pi," *2017 International Conference on Communication and Signal Processing (ICCSP)*, pp. 1598–1601, 2017, doi:10.1109/ICCSP.2017.8286658.

Lee Y.-C. and Lee C.-M., "Real-Time Smart Home Surveillance System of Based on Raspberry Pi," *2020 IEEE Eurasia Conference on IOT, Communication and Engineering (ECICE)*, 2020, pp. 72–74, doi:10.1109/ECICE50847.2020.9301929.

Mao J., Guo Z., Geng H., Zhang B., Cao Z. and Niu W., "Design of Visual Navigation System of Farmland Tracked Robot Based on Raspberry Pie," *2019 14th IEEE Conference on Industrial Electronics and Applications (ICIEA)*, pp. 573–577, 2019, doi:10.1109/ICIEA.2019.8834077. https://www.raspberrypi.org/

Patil N., Kelkar S., Ranawat M. and Vijayalakshmi M., "Krushi Sahyog: Plant Disease Identification and Crop Recommendation using Artificial Intelligence," *2021 2nd International Conference for Emerging Technology (INCET)*, pp. 1–6, 2021, doi:10.1109/INCET51464.2021.9456114.

Shidik G.F., Noersasongko E., Nugraha A., Andono P.N., Jumanto J. and Kusuma E.J., "*A Systematic Review of Intelligence Video Surveillance: Trends, Techniques, Frameworks, and Datasets,*" *in IEEE Access*, 7, pp. 170457–170473, 2019, doi:10.1109/ACCESS.2019.2955387.

Shariff S.U., Hussain M. and Shariff M.F., "Smart Unusual Event Detection Using Low Resolution Camera for Enhanced Security," *2017 International Conference on Innovations in Information, Embedded and Communication Systems (ICIIECS)*, pp. 1–6, 2017, doi:10.1109/ICIIECS.2017.8275833.

Automation and Computation – Vats et al. (Eds)
© 2023 the Author(s), ISBN 978-1-032-36723-1

Mathematical model for chilled food with temperature dependent deterioration rate

Manoj Kumar*, Amit Kumar Mishra* & Sandeep Kumar*
Graphic Era Hill University Dehradun, Dehradun, India

Manoj Kumar Sharma*
Uttaranchal University Dehradun, Dehradun, India

ABSTRACT: In many cases the perishable items in which deterioration does not depend on temperature more significantly, there is no need to worry about it. But some items such as chilled food products are highly sensitive about the surrounding temperature. A slight increase in temperature may deteriorate the products up to the level which is not acceptable in the market. In the present study, the case of chilled pork is discussed. As the temperature increases slightly the pork starts to spoil, and it may affect the supply chain causing the loss to the vendors. In consequence, to develop a new model, we mixed the quality prediction model and the traditional deterioration model. In the proposed model, we used temperature dependent linear deterioration rate. To illustrate the model, a numerical example is also used. In this study, it is found that the present model is highly sensitive about the temperature. As the temperature increases, the profit decreases.

Keywords: Pork, deterioration, chilling, time, demand

1 INTRODUCTION

Inventory policies play an important role for building a nation. By following these policies, any supplier and vendor may be benefitted. If the policies are not appropriate, it may burden the supply chain. Most of the inventory items deteriorate in storage and customers show less interest to buy the items such as non – veg foodstuffs, drugs, and chemicals. Such items will decline or lost its value. If these items are sold, then vendors have to reduce the price of such items. Microorganisms are responsible for contamination and spoilage of foodstuff. As the surrounding temperature rises, a sudden rise in the microorganism count is noticed. Health may always be down if such food is consumed. The growth of microorganism depends on properties of food, such as temperature, *pH,* and storage condition. By considering all these factors, a mathematical model is formulated. Food and Agriculture Organization's reports of United Nations shows that one – third or more of all food is spoiled and wasted before use. Packaging, storage and transportation are responsible for spoilage of food. Sometimes in transportation a lot of time is wasted, such as in Delhi (India) a large amount of packed food is spoiled due to farmer's agitation at the boundary of Delhi state. To save food, a

*Corresponding Authors: docmanojkumar64@gmail.com; amitmishraddun@gmail.com; drsk79@gmail.com and manojsharmamath1984@gmail.com

single vendor system web – based online tool is used to manage all activities related to business in a good manner. The demand of chilled food is raised according to the lifestyle of human beings.

In past few decades, many innovative ideas were come into effect for controlling the inventory of perishable items used in real life. Perishable items such as stuff of pork sandwich is more sensitive about storage temperature. In the process of chilling foodstuffs are preserved at the temperature higher than freezing point and lower than 15°C. Duun [1] studied quality changes during storage of pork roast and concluded that super chilling is better than traditional chilling. Super chilling at −2°C improves shelf life compared to traditional chilling at +3.5°C. In 2008 Monatari [2] studied on a system of refrigerated goods of cold chain of uninterrupted-temperature controlled transport and storage between suppliers and consumers. Nunes, M.C.N. [3], concluded in his paper that environmental conditions affect fruit and vegetable quality during a typical consumer retail display.

In 2010 Parfitt [4] studied how to control the wastage of food within the supply chain and concluded that infrastructure of developing countries and growing techniques are responsible for food wastage. Koutsoumanis [5] worked on Probabilistic model for Listeria monocytogenes growth and on pasteurized milk. Lee [6] in 2011 worked on Korean rice cake for storage as chilled food. Tragethon [7] performed an experimental work for storage of meat and poultry products through vacuum cooling. Shakila [8] in 2012 works on Quality and safety of fish curry processed by sous vide cook chilled and hot filled technology process during refrigerated storage. Food Sci Technol Int. 18:261. [9] Jaczynski in 2012 wrote a Handbook of Frozen Food Processing and Packaging. Islam [10] in 2014 studied physicochemical properties of mushrooms, in which he studied about the effect of ultrasound-assisted immersion freezing on selected physicochemical properties. James [11] performed an experiment on small and large packets of frozen products stored in a large store room.

In 2017 Mudgil [12] studied to slow down the process of growth of bacteria and microorganisms refrigeration and freezing technologies are used to stop the spoilage of food for reducing the economic losses. Goedhals [13] in 2017 found that a Broken cold chain in products of fruits and vegetables 81% and 41.5% of product temperature rose 2°C for longer time. To know the characteristics of products during the distribution process, temperature should be maintained inside the cargo was studied by H. Rafik [14]. Yang [15] published an article based on chilled food and found that the model demonstrate the high temperature reduces the profit. An idea is taken from [16] and [17] about remaining ratio as used in equation (9). Win [18] established a model based on joint pricing and inventory control. Giusti [19] studied about the evaluation of microbial safety of meat and vegetables which can be used directly.

2 ASSUMPTIONS

(i) A single item vendor system is followed.
(ii) A case of non – veg pork sandwich is considered.
(iii) Temperature dependent linear deterioration rate is considered.
(iv) Deterioration rate of pork sandwich Θ is temperature dependent.
(v) Time is taken as infinite for planning horizon.
(vi) Demand is price dependent $D(p) = ap^{-\gamma}$, $a > 0 \ and \ \gamma > 0$.
(vii) No shortage policy is adopted.
(viii) Lead time is taken as zero.

3 NOTATIONS

These are the notations used in the formulation of this model.

(i)	p	: unit selling price
(ii)	c	: the purchasing cost per unit item
(iii)	$D(p) = ap^{-\gamma}$	: the unit purchase cost dependent demand rate, $a > 0$ and $\gamma > 0$
(iv)	a	: demand at $l = 0$
(v)	l	: time
(vi)	o	: the ordering cost
(vii)	A_0	: the initial bacteria count
(viii)	$\tilde{G}$	: the growth of bacteria count
(ix)	M	: the time of maximum growth
(x)	τ	: the storage temperature
(xi)	$\hbar$	: the inventory holding cost per unit per unit time
(xii)	L	: time interval of each cycle
(xiii)	$N(l)$	: the end bacterial count of food at time l, $0 \le l \le L$
(xiv)	$\Re$	: the remaining ratio value at the end of deterioration
(xv)	$I(l)$	: the inventory level at any time l, $0 \le l \le L$ with deterioration
(xvi)	$\tilde{I}(l)$	: the inventory level at any time l, $0 \le l \le L$ with no deterioration
(xvii)	$Z(L)$	: the stock loss caused by deterioration in $[0, l]$
(xviii)	$C(L, p)$	: the total cost per cycle
(xix)	$\overline{C}(L, p)$	: the total cost per unit time
(xx)	$P(L, p)$	: the total profit per unit time
(xxi)	$\Theta(\tau)$	: the temperature dependent deterioration rate $\Theta(\tau) = \theta\tau$
(xxii)	θ	: deterioration, $\theta > 0$

4 MATHEMATICAL FORMULATION

After reviewing the different inventory models and observing the deterioration of chilled food, the inventory level of chilled food at any time can be represented by the equation

$$\frac{dI(l)}{dl} + \Theta(\tau)I(l) = -D(p), \ 0 \le l \le L, \tag{1}$$

The boundary conditions are

$$I(L) = 0 \tag{2}$$

The solution of (1) can be expressed as

$$I(l) = \frac{ap^{-\gamma}}{\theta\tau}\left(e^{\theta\tau(L-l)} - 1\right), \ 0 \le l \le L \tag{3}$$

At $l = 0$

$$I(0) = \frac{ap^{-\gamma}}{\theta\tau}\left(e^{\theta\tau L} - 1\right) \tag{4}$$

$I(l)$ shows the inventory level at any time l ($l \ge 0$), $I(0)$ shows the initial inventory level and $I(L)$ shows end inventory level.

Suppose $Z(L)$ be the loss in stock due to deterioration in the interval $[0, l]$ with $I(L) = 0$. Then loss in stock is given by equation (5) and detail is shown in appendix B.

$$Z(L) = \frac{ap^{-\gamma}}{\theta\tau}\left(e^{\theta\tau L} - 1\right) - D(p)L \tag{5}$$

$Z(L)$ represents the difference in the inventory levels with and without deterioration at the end of cycle time. The quantity order ordered per cycle can obtained by the following function:

$$Q_L = Z(L) + D(p)L$$

$$Q_L = \frac{ap^{-\gamma}}{\theta\tau}\left(e^{\theta\tau L} - 1\right) \tag{6}$$

For $l = 0$, equations (4) and (6) represent the same equations.

Total cost per cycle

$$C(L,p) = O + cQ_L + \hbar \int_0^L I(l)\,dl$$

$$C(L,p) = O + \frac{cap^{-\gamma}}{\theta\tau}\left(e^{\theta\tau L} - 1\right) + \hbar \int_0^L \frac{ap^{-\gamma}}{\theta\tau}\left(e^{\theta\tau(L-l)} - 1\right)dl$$

$$C(L,p) = O + \frac{cap^{-\gamma}}{\theta\tau}\left(e^{\theta\tau L} - 1\right) + \frac{\hbar ap^{-\gamma}}{\theta\tau} \int_0^L \left(e^{\theta\tau(L-l)} - 1\right)dl$$

$$C(L,p) = O + \frac{cap^{-\gamma}}{\theta\tau}\left(e^{\theta\tau L} - 1\right) + \frac{\hbar ap^{-\gamma}}{\theta\tau} \left[\frac{e^{\theta\tau(L-l)}}{-\theta\tau} - l\right]_0^L$$

$$C(L,p) = O + \frac{cap^{-\gamma}}{\theta\tau}\left(e^{\theta\tau L} - 1\right) + \frac{\hbar ap^{-\gamma}}{\theta^2\tau^2}\left(e^{\theta\tau L} - \theta\tau L - 1\right) \tag{7}$$

The total cost per unit item is

$$\overline{C}(L,p) = \frac{C(L,p)}{L}$$

$$\overline{C}(L,p) = \frac{O}{L} + \frac{cap^{-\gamma}}{\theta\tau L}\left(e^{\theta\tau L} - 1\right) + \frac{\hbar ap^{-\gamma}}{\theta^2\tau^2 L}\left(e^{\theta\tau L} - \theta\tau L - 1\right) \tag{8}$$

For selling the product predictive quality model [16,17] is to be used for remaining value of non – veg pork sandwiches after deterioration.

$$N(l) = A_0 + \bar{G}e^{-e^{-\Theta(\tau)(l-M)}}$$

Remaining ratio after deterioration

$$\Re = \frac{7 - N(l)}{7}, \quad 0 < \Re < 1 \tag{9}$$

Now remaining value of each pork sandwich after deterioration

$$\Re p D(p) = pD(p)\left(\frac{7 - N(l)}{7}\right)$$

$$\Re p D(p) = ap^{1-\gamma}\left(\frac{7 - N(l)}{7}\right) \tag{10}$$

Let $P(L,p)$ be the total profit per unit item then profit function can be expressed as

$$P(L,p) = \Re D(p)p - \overline{C}(L,p)$$

$$= \Re D(p)p - \frac{O}{L} - \frac{cap^{-\gamma}}{\theta \tau L}\left(e^{\theta \tau L} - 1\right) - \frac{\hbar ap^{-\gamma}}{\theta^2 \tau^2 L}\left(e^{\theta \tau L} - \theta \tau L - 1\right) \qquad (11)$$

Substitute $D(p) = ap^{-\gamma}$.

$$P(L,p) = \Re ap^{1-\gamma} - \frac{O}{L} - \frac{cap^{-\gamma}}{\theta \tau L}\left(e^{\theta \tau L} - 1\right) - \frac{\hbar ap^{-\gamma}}{\theta^2 \tau^2 L}\left(e^{\theta \tau L} - \theta \tau L - 1\right).$$

$$P(L,p) = \left(1 - \frac{A_0}{7} - \frac{\bar{G}}{7e^{1/e^{\theta r(L-M)}}}\right)ap^{1-\gamma} - \frac{O}{L} - \frac{cap^{-\gamma}}{\theta \tau L}\left(e^{\theta \tau L} - 1\right) - \frac{\hbar ap^{-\gamma}}{\theta^2 \tau^2 L}\left(e^{\theta \tau L} - \theta \tau L - 1\right)$$

$$(12)$$

For maximization of profit function $P(L,p)$, a concave function is required for $L > 0$ and $p > 0$ as proved in Appendix C. Take partial derivative of profit function $P(L,p)$ with respect to p we have

$$\frac{\partial P(L,p)}{\partial p} = \Re(1 - \gamma)ap^{-\gamma} + \frac{ca\gamma p^{-\gamma-1}}{\theta \tau L}\left(e^{\theta \tau L} - 1\right) + \frac{\hbar a\gamma p^{-\gamma-1}}{\theta^2 \tau^2 L}\left(e^{\theta \tau L} - \theta \tau L - 1\right) \qquad (13)$$

For optimal price set $\frac{\partial P(L,p)}{\partial p} = 0$.

$$\Re(1 - \gamma)ap^{-\gamma} + \frac{ca\gamma p^{-\gamma-1}}{\theta \tau L}\left(e^{\theta \tau L} - 1\right) + \frac{\hbar a\gamma p^{-\gamma-1}}{\theta^2 \tau^2 L}\left(e^{\theta \tau L} - \theta \tau L - 1\right) = 0$$

Multiply (13) by p^γ throughout and substitute $\Re = \frac{7-N(l)}{7}$

$$1 - \frac{A_0 + \bar{G}e^{-e^{-\theta r(L-M)}}}{7}(1 - \gamma)a + \frac{ca\gamma p^{-1}}{\theta \tau L}\left(e^{\theta \tau L} - 1\right) + \frac{\hbar a\gamma p^{-1}}{\theta^2 \tau^2 L}\left(e^{\theta \tau L} - \theta \tau L - 1\right) = 0$$

$$\frac{A_0 + \bar{G}e^{-e^{-\theta r(L-M)}}}{7}(1 - \gamma)a - \frac{ca\gamma p^{-1}}{\theta \tau L}\left(e^{\theta \tau L} - 1\right) - \frac{\hbar a\gamma p^{-1}}{\theta^2 \tau^2 L}\left(e^{\theta \tau L} - \theta \tau L - 1\right) - 1 = 0 \qquad (14)$$

$$p^{-1}\left[\frac{ca\gamma}{\theta \tau L}\left(e^{\theta \tau L} - 1\right) + \frac{\hbar a\gamma}{\theta^2 \tau^2 L}\left(e^{\theta \tau L} - \theta \tau L - 1\right)\right] = -\Re(1 - \gamma)a$$

$$p = \frac{1}{\Re(\gamma - 1)}\left[\frac{c\gamma}{\theta \tau L}\left(e^{\theta \tau L} - 1\right) + \frac{\hbar \gamma}{\theta^2 \tau^2 L}\left(e^{\theta \tau L} - \theta \tau L - 1\right)\right] \qquad (15)$$

Substitute $\Re = \frac{7-N(l)}{7}$ and $N(l) = A_0 + \bar{G}e^{-e^{-\Theta(r)(l-M)}}$.

$$p = \frac{1}{\left(1 - \frac{N(l)}{7}\right)(\gamma - 1)}\left[\frac{c\gamma}{\theta \tau L}\left(e^{\theta \tau L} - 1\right) + \frac{\hbar \gamma}{\theta^2 \tau^2 L}\left(e^{\theta \tau L} - \theta \tau L - 1\right)\right]$$

$$p = \frac{1}{(\gamma - 1)\left(1 - \frac{A_0}{7} - \frac{\bar{G}}{7e^{1/e^{\theta r(L-M)}}}\right)}\left[\frac{c\gamma}{\theta \tau L}\left(e^{\theta \tau L} - 1\right) + \frac{\hbar \gamma}{\theta^2 \tau^2 L}\left(e^{\theta \tau L} - \theta \tau L - 1\right)\right]$$

Now differentiate (12) w.r.t. L partially.

$$\frac{\partial P(L,p)}{\partial L} = -\frac{\bar{G}\theta\tau ap^{1-\gamma}}{7e^{1/e^{\theta\tau(L-M)}}}e^{-\theta\tau(L-M)} + \frac{O}{L^2} - \frac{cap^{-\gamma}}{\theta\tau L^2}\left(L\theta\tau e^{\theta\tau L} - e^{\theta\tau L} + 1\right)$$

$$- \frac{\hbar ap^{-\gamma}}{\theta^2\tau^2 L^2}\left(L\theta\tau e^{\theta\tau L} - e^{\theta\tau L} + 1\right)$$

For optimal cycle time, set $\frac{\partial P(L,p)}{\partial L} = 0$.

$$\frac{\bar{G}\theta\tau ap^{1-\gamma}}{7e^{1/e^{\theta\tau(L-M)}}}e^{-\theta\tau(L-M)} - \frac{O}{L^2} + \frac{cap^{-\gamma}}{\theta\tau L^2}\left(L\theta\tau e^{\theta\tau L} - e^{\theta\tau L} + 1\right) + \frac{\hbar ap^{-\gamma}}{\theta^2\tau^2 L^2}\left(L\theta\tau e^{\theta\tau L} - e^{\theta\tau L} + 1\right) = 0 \quad (16)$$

5 NUMERICAL EXAMPLE

To illustrate the preceding model an example is presented. Following random data is used. Initially the demand is taken as $a = 1000$ units, index $\gamma = 0.9$, the inventory holding cost per order $\hbar = \$2$, the storage temperature $\tau = 6°C$, initial bacteria count $A_0 = 5$, the time of maximum growth $M = 0.05$ hours, the ordering cost $O = \$250$, the purchasing cost per unit item $c = 40$, and the growth of bacteria count $\bar{G} = 6.8$colony forming units then we get optimal value of $p = \$500.83$, $L = 554.00$ hours.

6 SENSITIVITY ANALYSIS

Parameter	% Changes in Parameter	Value of Parameter	P ($)	L (hours)	Total Profit ($)	% Change in Profit
τ	−10%	5.4	500.9221	554.0064	1794.7	−1.47672%
	−5%	5.7	500.8739	554.0058	1808.8	−0.70268%
	0%	6	500.8305	554.0052	1821.6	0.00000%
	5%	6.3	500.7912	554.0047	1833.2	0.63680%
	10%	6.6	500.7555	554.0043	1843.7	1.21322%
A_0	−10%	4.5	500.8305	554.0052	1821.6	0.00%
	−5%	4.75	500.8305	554.0052	1821.6	0.00%
	0%	5	500.8305	554.0052	1821.6	0.00%
	5%	5.25	500.8305	554.0052	1821.6	0.00%
	10%	5.5	500.8305	554.0052	1821.6	0.00%
M	−10%	18	500.8305	554.0052	1821.6	0.00%
	−5%	19	500.8305	554.0052	1821.6	0.00%
	0%	20	500.8305	554.0052	1821.6	0.00%
	5%	21	500.8305	554.0052	1821.6	0.00%
	10%	22	500.8305	554.0052	1821.6	0.00%
θ	−10%	0.045	500.9221	554.0064	1794.7	−1.48%
	−5%	0.0475	500.8739	554.0058	1808.8	−0.70%
	0%	0.05	500.8305	554.0052	1821.6	0.00%
	5%	0.0525	500.7912	554.0047	1833.2	0.64%
	10%	0.055	500.7555	554.0043	1843.7	1.21%
c	−10%	36	500.8305	554.0052	1821.6	0.00%
	−5%	38	500.8305	554.0052	1821.6	0.00%
	0%	40	500.8305	554.0052	1821.6	0.00%
	5%	42	500.8305	554.0052	1821.6	0.00%

(*continued*)

Continued

Parameter	% Changes in Parameter	Value of Parameter	P ($)	L (hours)	Total Profit ($)	% Change in Profit
	10%	44	500.8305	554.0052	1821.6	0.00%
O	−10%	225	500.8305	554.0052	1821.6	0.00%
	−5%	237.5	500.8305	554.0052	1821.6	0.00%
	0%	250	500.8305	554.0052	1821.6	0.00%
	5%	262.5	500.8305	554.0052	1821.6	0.00%
	10%	275	500.8305	554.0052	1821.6	0.00%
h	−10%	1.8	500.8305	554.0052	1821.6	0.00%
	−5%	1.9	500.8305	554.0052	1821.6	0.00%
	0%	2	500.8305	554.0052	1821.6	0.00%
	5%	2.1	500.8305	554.0052	1821.6	0.00%
	10%	2.2	500.8305	554.0052	1821.6	0.00%
$\tilde{G}$	−10%	6.12	500.8305	554.0052	1821.6	0.00%
	−5%	6.46	500.8305	554.0052	1821.6	0.00%
	0%	6.8	500.8305	554.0052	1821.6	0.00%
	5%	7.14	500.8305	554.0052	1821.6	0.00%
	10%	7.48	500.8305	554.0052	1821.6	0.00%
γ	−10%	0.81	500.7912	554.0047	1833.2	0.64%
	−5%	0.855	500.8109	554.005	1827.3	0.31%
	0%	0.9	500.8305	554.0052	1821.6	0.00%
	5%	0.945	500.8502	554.0055	1815.8	−0.32%
	10%	0.99	500.8698	554.0057	1810	−0.64%
a	−10%	900	500.8305	554.0052	1821.6	0.00%
	−5%	950	500.8305	554.0052	1821.6	0.00%
	0%	1000	500.8305	554.0052	1821.6	0.00%
	5%	1050	500.8305	554.0052	1821.6	0.00%
	10%	1100	500.8305	554.0052	1821.6	0.00%

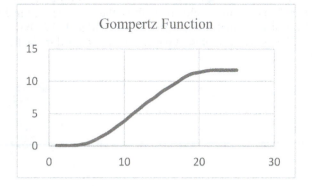

Figure 1.   Gompertz function.

7   CONCLUSION

In the present study we discussed an inventory model of chilled food after reviewing a lot of research articles such as Duun [1], M. de. Giusti [19] etc. we discovered a model for chilled food for a single item vendor system. Such models are popular in the field of inventory models. Chilled food perishable items are very critical to store and supply for a long time. Gompertz model (Figure 1) is used to develop the model because the bacterial growth is similar to Gompertz function. As the temperature rises, these products loss its originality. In our model, the profit increases in the temperature range 5.4 – 6.6 only (Figure 2). Many chilled food products which can be maintained within this rage can be supplied through a single vendor system and the seller, supplier and consumer may be benefitted.

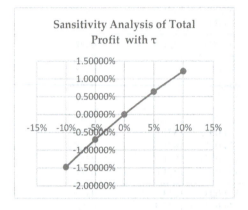

Figure 2.   Sensitivity analysis of total profit with τ.

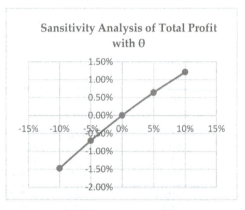

Figure 3.   Sensitivity analysis of total profit with θ.

As far as about the initial bacteria count it may vary from 4.5 – 5.5 colony forming units, the profit remains same in each cycle. If deterioration increases, the profit also increases on the remaining items (Figure 3). When ordering cost increases, then profit remains the same within the same cycle. Variation in inventory holding cost doesn't affect the profit. As deterioration decreases from 0.05 a slight loss is observed in the selling of chilled food (Figure 3). When the index increases from 0.9 a loss is also observed (Figure 4). Optimal profit, optimal value P and optimal value of L altogether are depicted in Figure 5. For the support of model sensitivity, analysis is performed in Table 1. We have given numerical example to validate the model. The present paper can be extended in the field of Agriculture, chemicals, and medicines.

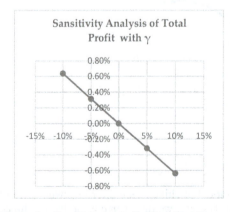

Figure 4.   Sensitivity analysis of total profit with γ.

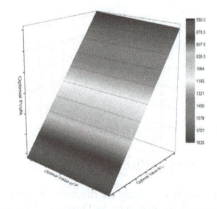

Figure 5.   Optimal profit P/optimal value of cycle L.

REFERENCES

[1] Duun A., Hemmingsen A., Haugland A. & Rustad T. (2008). Quality changes during superchilled storage of pork roast. *LWT-Food science and Technology*, 41(10), pp. 2136–2143.

[2] Montanari R., (2008). Cold Chain Tracking: A Managerial Perspective. *Trends in Food Science & Technology*, 19, pp. 425–431.

[3] Nunes M.C.N., Emond J.P., Rauth M., Dea S. & Chau K.V., (2009). Environmental Conditions Encountered During Typical Consumer Retail Display Affect Fruit and Vegetable Quality and Waste. *Postharvest Biology and Technology*, 51, pp. 232–241.

[4] Parfitt J., Barthel M. & Macnaughton S., (2010). Food Waste Within Food Supply Chains: Quantification and Potential for Change to 2050. *Philosophical Transactions of the Royal Society B: Biological Sciences*, 365, pp. 3065–3081.

[5] Koutsoumanis K., Pavlis A., Nychas G.-J.E. & Xanthiakos K., (2010). Probabilistic model for Listeria Monocytogenes Growth During Distribution, Retail Storage, and Domestic Storage of Pasteurized Milk. *Applied and Environmental Microbiology*, 76, pp. 2181–2191.

[6] Lee K-A., Kim K-T. & Paik H-D. (2011) Physicochemical, Microbial, and Sensory Evaluation of Cookchilled Korean Traditional Rice Cake (Backseolgi) During Storage Via Various Packaging Methods. *Food Sci Biotechnol*, 20, pp. 1069–1074.

[7] Tragethon D. (2011) Vacuum Cooling—the Science and Practice. *International Institute of Ammonia Refrigeration Annual Meeting*, Paper 36, Orlando, FL, USA, pp. 1–22.

[8] Shakila R.J., Raj B.E. & Felix (2012) Quality and Safety of Fish Curry Processed by Sous Vide Cook Chilled and Hot Filled Technology Process During Refrigerated Storage. *Food Sci Technol Int*. 18, pp. 261–269.

[9] Jaczynski J., Tahergorabi R., Hunt A. L. & Park J. W. (2012). Safety and Quality of Frozen Aquatic Food Products. In D. W. Sun (Ed.), *Handbook of Frozen Food Processing and Packaging*. Boca Raton: CRC Press, Taylor& Francis Group. 2nd ed., pp. 343–385.

[10] Islam M. N., Zhang M., Adhikari B., Xinfeng C. & Xu, B. G. (2014). The Effect of Ultrasound-assisted Immersion Freezing on Selected Physicochemical Properties of Mushrooms. *International Journal of Refrigeration*, 42, pp. 121–133.

[11] James S. J., & James C. (2014). Chilling and Freezing. in *Food Safety Management*. Elsevier. pp. 481–510.

[12] Mudgil D., & Barak S. (2017). *Functional Foods: Sources and Health Benefits*. Scientific Publishers, India

[13] Goedhals–Gerber L. L., Stander C., Dyk F. E. V., (2017). Maintaining Cold Chain Integrity: Temperature Breaks Within Fruit Reefer Containers in the Cape Town Container Terminal. *Southern African Business Review*, 21, pp. 362–384.

[14] Rafik H. & Bishara "Cold Chain Management – An Essential Component of the Global Pharmaceutical Supply Chain", *Pharmaceutical Supply Chain*.

[15] Yang M. F., Tseng W. C., (2015). "Deteriorating Inventory Model for chilled Food" *Hindawi Publishing Corporation Mathematical Problems in Engineering*, Article ID 816876, pp. 1–11.

[16] Iao L. C. & Hsiao H. I., (2013). *Application of Continuous Temperature Information on Shelf-Life Prediction and Inventory Control: A case of* 18°C *Ready-to-Eat Food*, National Taiwan Ocean University Research System.

[17] Gibson A. M., Bratchell N., & Roberts T. A., (1989) "The Effect of Sodium Chloride and Temperature on Rate and Extent of Growth of *Clostridium Botulinum* type A Unpasteurized Pork Slurry," *Journal of Applied Bacteriology*, vol. 62, no. 6, pp. 479 – 490.

[18] Qin Y., Wang J., & Wei C., (2014) "Joint Pricing and Inventory Control for Fresh Produce and Foods with Quality and Physical Quantity Deteriorating Simultaneously," *International Journal of Production Economics*, vol. 152, pp. 42–48.

[19] de Giusti M., Aurigemma C. & Marinelli L et al., (2010) "The Evaluation of the Microbial Safety of Fresh Ready-to-Eat Vegetables Produced by Different Technologies in Italy, "*Journal of Applied Microbiology*, 109, 3, pp. 996–1006.

APPENDICES

A. Gompertz Function

$$N(l) = A_0 + \bar{G}e^{-e^{-\Theta(\tau)(l-M)}} \qquad (A.1)$$

$\bar{G}$ is an asymptotic since $\lim\limits_{M \to \infty} \bar{G}e^{-e^{-\Theta(\tau)(l-M)}} = \bar{G}$. l is the time shift along time axis.

$$e^{-e^{-\theta\tau(l-M)}} = \frac{N(l) - A_0}{\bar{G}}$$

$$-e^{-\theta\tau(l-M)} = \ln\left(\frac{N(l) - A_0}{\bar{G}}\right)$$

$$e^{-\theta\tau(l-M)} = \ln\left(\frac{N(l) - A_0}{\bar{G}}\right)^{-1}$$

$$-\theta\tau(l - M) = \ln\ln\left(\frac{N(l) - A_0}{\bar{G}}\right)^{-1}$$

$$l - M = -\frac{1}{\theta\tau}\ln\ln\left(\frac{N(l) - A_0}{\bar{G}}\right)^{-1} \tag{A.2}$$

The time at the half of maximum growth $l = \frac{M}{2}$

$$\frac{M}{2} - M = -\frac{1}{\theta\tau}\ln\ln\left(\frac{N(l) - A_0}{\bar{G}}\right)^{-1}$$

$$\frac{M}{2} = \frac{1}{\theta\tau}\ln\ln\left(\frac{N(l) - A_0}{\bar{G}}\right)^{-1}$$

$$M = \frac{2}{\theta\tau}\ln\ln\left(\frac{N(l) - A_0}{\bar{G}}\right)^{-1} \tag{A.3}$$

B. Level of Inventory with and without Deterioration
Inventory level in the time interval $[0 \le l \le L]$

$$I(l) = \frac{ap^{-\gamma}}{\theta\tau}\left(e^{\theta\tau(L-l)} - 1\right) \tag{B.1}$$

At the beginning

$$I(0) = \frac{ap^{-\gamma}}{\theta\tau}\left(e^{\theta\tau L} - 1\right) \tag{B.2}$$

Suppose $\tilde{I}(l)$ be the inventory level at any time l, $0 \le l \le L$ with no deterioration then

$$\tilde{I}(l) = I(0) - lD(p) \tag{B.3}$$

The loss in inventory stock can be expressed as $\tilde{I}(l) - I(l)$ then,

$$Z(l) = \tilde{I}(l) - I(l)$$

$$Z(l) = \frac{ap^{-\gamma}}{\theta\tau}\left(e^{\theta\tau L} - 1\right) - lap^{-\gamma} - \frac{ap^{-\gamma}}{\theta\tau}\left(e^{\theta\tau(L-l)} - 1\right) \tag{B.4}$$

$$Z(l) = \frac{ap^{-\gamma}}{\theta\tau}\left[e^{\theta\tau L} - e^{\theta\tau(L-l)} - l\theta\tau\right] \tag{B.5}$$

$$Z(L) = \frac{ap^{-\gamma}}{\theta\tau}\left[e^{\theta\tau L} - L\theta\tau - 1\right] \tag{B.6}$$

152

Automation and Computation – Vats et al. (Eds)

A review on lung disease diagnosis using machine learning

Sudhir Kumar Chaurasiya & Nitish Kumar
Goel Institute of Technology & Management, Lucknow, India

Satya Bhushan Verma, Anshita Raj & Shobhit Sinha
Shree Ramswaroop Memorial University, Barabanki, India

ABSTRACT: Over the last two decades, international research in Artificial Intelligence for medical applications has served as a platform for medical trade and the development of a vibrant research community. Artificial Intelligence has been used extensively in medical applications. This paper provides an overview of the history of AI applications in lung disease analysis in research, which includes a wide range of studies in several disciplines to perceive challenging conditions. The analysis of several papers is used to compile a review of research topics and findings relevant to lung disease detection.

1 INTRODUCTION

Machine learning (ML) is a fast-growing subject in healthcare that holds great potential for disease diagnosis, prediction, and management, ultimately enhancing personalized therapy. The potential for machine learning models to identify disease or predict disease trajectories grows as automated data collecting becomes increasingly common in regular care. ML is a kind of artificial intelligence that employs algorithms to find patterns in huge, complicated datasets that traditional statistical methods may struggle to uncover. ML is divided into three types: supervised, unsupervised, and deep learning; each is used depending on the analysis' goal and the data's information. When the severity of a patient's disease is known, supervised approaches create prediction models based on data with annotated outcomes. ANNs are complicated models that employ several interconnected layers of processing units, known as neurons, to extract amounts of information from raw data and provide a set of predictions rules.

Image division is one of the most fundamental and problematic topics in image analysis. Image division is a crucial aspect of image processing. Image division is the process of splitting an image into important sections or components in PC vision. Image division can be used to discover tumors or other pathologies, measure tissue volume, perform PC-assisted medical procedures, plan therapy, investigate anatomical structure, find protests in satellite images, and recognize unique marks, among other things. A division divides a picture into its constituent parts, such as a protest. The qualities of intermittence and similitude [1] are used to classify image division algorithms. Image division is organized as edged based division and area-based division in light of this attribute. Limit or edge-based systems are division strategies that rely on the brokenness feature of pixels. The edge-based division approach attempts to find image division by identifying the edges or pixels between distinct districts with rapid progress in force that are segregated and related to frame shut protest restrictions. The end result is a two-sided picture. There are two basic edge-based division strategies depending on work: dark histogram based and angle-based methodology [2].

DOI: 10.1201/9781003333500-18

2 RELATED WORKS

In 2010 Yudong Zhang *et al.* [9] a wavelet change strategy was utilized in extricating high-lights of a lung disease. In any case, translational variation property in present in discrete wavelet change which will get highlights with two pictures on a similar theme and with little development. To tackle the issue, the paper removes highlight which utilizes SWT rather than DWT. The wavelet coefficients got through SWT are undeniably better as looked at than DWT what's more, we applied SWT to ordinary and unusual cerebrum arrangement. The outcomes exhibit that the classifier based on SWT is more exact than that of DWT. In this paper the prevalence of SWT is talked about which is utilized to acquire highlights. SWT is better when contrasted with discrete wavelet change when we consider its translational invariant property. SWT was applied to get ordinary and unusual cerebrum picture arrangement. The outcomes as gotten in [9] show that grouping of mind picture is productive and precise in ordinary and unusual cerebrum pictures. It is normal that SWT put together elements will be investigated with respect to denoising, combination and pressure.

In 2011 a paper by Yudong Zhang *et al.* expressed that the computerized and precise characterization of MR cerebrum pictures work shows significance for the examination and translation of the pictures and numerous strategies have been proposed in regards to it. They present a neural organisation (NN) based technique for classifying MR mind images as ordinary or odd in this paper. PCA, or head part examination, is used to reduce elements of previously separated highlights created by wavelet transformation. The reduced parts are then sent to a back proliferation (BP) neural organisation, which employs scaled form slope (SCG) to find the neural organization's optimal loads. The creators of this study used this approach on 66 images (18 ordinary, 48 unusual). For both the prepared set and the test set, the calculated computation time for a picture is 0.0451s, and the characterization exactness achieved is 100 percent. The designers of [10] used a cross breed classifier to rank both ordinary and extraordinary images. The creators in [10] have proposed future development with basic focuses as follows: It's first used on LUNG DISEASE images using instruments including T1 weighting, dissemination weighting, and proton-thickness weighting. Second, a high-level wavelet change known as the lift-up wavelet can be used to speed up calculation focuses. Third, research into multiple Lung disease classifications should be possible.

(Deeply and edema. Because of non-obtrusive imaging and delicate tissue difference of lung disease, the mind cancer division procedures are standing out. The creator in [11] states that in the range of twenty years, the PC supported strategies are turning out to be increasingly more redesigned for division of cerebrum cancer which come moving toward the normal applications. The thought process of the creators is to give outline of cerebrum growth division. Initially, a fresh prologue to cerebrum growth division utilizing picture modalities is given. The earlier handling of working and SOTA (best in class) techniques for lung disease-based division is gotten. The consequences of lung disease-based cerebrum growth division is surveyed and approved. At long last, patterns are coordinated for cerebrum lung disease division techniques and future advancements are made. The creators in [11], have given a total layout SOTA (best in class) of lung disease-based cerebrum cancer division. The taking diverse trademark highlights and considering spatial data in a nearby area, a large number of the mind cancer division strategies work lung disease pictures because of the non-obtrusive and great delicate tissue differentiation of lung disease and utilize grouping and bunching techniques.

Around the same time, the writers in [12] have proposed in an article that PC supported discovery/determination (CAD) frameworks can improve the demonstrative abilities of doctors and decrease the time needed for precise analysis. The fundamental reason for the paper is to audit division and order methods and SOTA for the mind LUNG DISEASE. The creators in [12] have expressed why CAD frameworks of human mind can in any case represent an issue. The paper proposes a mixture smart AI method utilizing CAD through which division of mind growth should be possible utilizing LUNG DISEASE. The creators in [12] have utilized input beat coupled neural-network for division, DWT is utilized for include extraction, PCA for dimensionality decrease of wavelet coefficients feed forward back spread neural organization for ordering typical picture and unusual picture. The proposed technique in [12] makes use of the lung disease dataset, which includes 14 common and 87 unusual cerebrum cases. Almost all of the arranging perfection has been achieved. It is significantly more compelling when compared to other ways. It is more powerful, precise, and rapid, according to the results. CAD has become a superior approach as computational knowledge and AI methods have evolved. It has evolved into a crucial procedure in radiology and testing. The papers used in this review are from the middle of 2006 to 2012. The suggested network uses criticism beat coupled organisation to identify the region of interest and picture division, and then uses DWT to extract the pieces. Finally, PCA is used to reduce dimensionality and find more precise classifiers. In conclusion, the decreased highlights are then shipped off back spread feed forward organization to arrange picture as typical or strange. Vigor of the strategy utilized is essential boundary for appraisal. The creators in [12] have understood an enormous number of calculations and contrasted it and the proposed strategy. As indicated by the outcomes, this technique is more productive.

In long term, a distribution by Andrez Larozza *et al.* [13] shows that to foster a grouping model utilizing surface highlights and SVM (support vector machine) interestingly improved T1-weighted pictures to separate between mind metastasis and radiation putrefaction. In 2015, the writers in [14] have expressed in an article that there are many methodologies for exact and programmed grouping of cerebrum lung disease.

In the year 2017, the creators in [15] have expressed that the programmed division [15] of cerebrum growth is the most common way of isolating unusual tissues from typical tissues, White matter (WM), dim matter (GM), and cerebrospinal fluid (CSF) are some of the different types of brain tissue (CSF). The course of division necessitates a great lot of effort due to the changeability in shape, size, and area. The essential data is combined using multimodal imaging technologies for precise cerebrum cancer division. [15] provides an overview of cerebrum growth division tactics using imaging technologies such as Positron Emission Tomography, Magnetic Resonance Imaging, multimodal imaging, and Computer Tomography.

In 2018, in an article distributed by Yanqing Zhang and Jyoti Islam, it was expressed that Alzheimer's infection is a non-reparable, moderate neurological cerebrum issue. Most punctual location of Alzheimer's infection can be utilized to appropriately treat and forestall cerebrum tissue harm. For the conclusion of AD (Alzheimer's disease),many factual and AI models have been utilized by scientists. Investigations of mind lung disease s have been done in clinical review for the detection of AD. Alzheimer's infection location is difficult due to the similarity between AD LUNG DISEASE information and matured individuals standard LUNG DISEASE information. Recently, advanced deep learning processes have been successfully used to demonstrate human level execution in many sectors, including clinical picture examination. The creators proposed in [16] a deep convolutional neural organisation (CNN) for Alzheimer's disease diagnosis based on lung disease data. however, the majority of current solutions use paired arrangement; the model discussed in [16] is used to perceive AD at various levels and has better execution for identifying AD in its early stages. The creators led studies that revealed higher correlation results using various strategies. Using the cerebrum lung disease dataset obtained from OASIS, the authors of [16] demonstrated a viable approach for AD detection. While the majority of the work focuses on double arrangement, [16] authors have made significant progress in multi-class grouping. The proposed technique in [16] could be quite useful for detecting AD in its early stages. The technique provided in [16] is centred on AD and may be applied to many spaces. The proposed approach can also be used to apply CNN to various regions with a limited dataset. The creators plan to apply the proposed methodology to a variety of AD datasets as well as other disease classifications.

In repeating clinical psyche attractive reverberation photographs, Muhammad Febrian Rachmadi et al., (2018) [17], proposed a form of a convolutional neural local area (CNN) plot proposed for fragmenting cerebrum sores with critical mass-impact, to segment white count number hyperintensities (WMH) normal for cerebrums with no or moderate vascular pathology (lung disease). Because of the WMH's small size (i.e. volume) and their similarity to non-neurotic mind tissue, this is a potentially hazardous division. The designers in [17] examine the efficacy of the 2D CNN scheme by comparing its presentation to that obtained from another deep focusing on strategy: Two common gadgets for learning techniques are the Deep Boltzmann Machine (DBM) and the Deep Boltzmann Machine (DBM): Support Vector Machine (SVM) and Random Forest (RF), as well as a publicly accessible tool called Lesion Segmentation Tool (LST), are all considered to be useful for segmenting WMH in LUNG DISEASE. The authors of [17] also provide a method for remembering spatial measures for the WMH division's convolution level of CNN, dubbed total spatial insights (GSI). The perceived relationship between WMH movement, as surveyed using all techniques evaluated, and segment and clinical realities was validated using covariance analysis. Profound acquiring information on calculations outflanks typical contraption dominating calculations through removed from lung disease antiques and diseases that look to be comparable as WMH. Regardless of the local area's settings evaluated, the proposed strategy of fusing GSI accurately assisted CNN in getting higher electronic WMH division. LST-LGA, SVM, RF, DBM, CNN, and CNN-GSI Dice Similarity Coefficient (DSC) values were 0.2963, 0.1194, 0.1633, 0.3264, 0.5359, and 5389, respectively.

Deep convolution neural organisations, or CNNs, finish the top presentation in a slew of PC vision concerns like visual item recognition, location, and division, according to the widely disseminated work [18] by creator Joe Bernal et al. The creators in [18] have expressed that profound convolutional neural organizations are additionally utilized for investigating clinical picture which incorporates sore division, physical division and arrangement. The paper referenced in [18] centers around structures, preprocessing, information planning and post-handling systems [36,38]. The review has three distinct aspects. Convolutional neural organizations (CNNs), an exceptional part of profound learning applications to visual purposes, have procured significant consideration somewhat recently because of its advancement exhibitions in fluctuated PC vision applications, for example, in object acknowledgment, identification and division challenges [19–21], in which they have accomplished astounding exhibitions [22–26].

In the year 2019, Anjali Wadhwa et al., expressed in the paper that the cycle [27] of portioning cancer from lung disease picture of a mind is one of the exceptionally engaged regions locally of clinical science as lung disease is noninvasive imaging. The research includes a writing survey that looks into the methods for mind growth division and cerebrum imaging. The paper discusses state-of-the-art tactics and quantitative end execution of condition-of-workmanship procedures [39,40]. The commitment of various creators as well as diverse strategies for picture division have been discussed. In the paper, it is stated that a work has been done for specialists to investigate new areas of investigation, and it has been discovered that there are other compelling techniques to portion cerebrum cancer, including contingent irregular fields (CRF) with convolutional neural organization (CNN) or CRF with Deep Medic troupe. The study discusses in-depth picture division techniques. For a precise growth conclusion, the clinicians set new headings of exploration as well as quantitative examination having various boundaries help per users among various conditions of workmanship techniques. It has far and wide applications in clinical science, for instance, tissue arrangement, restriction of cancers, growth volume assessment, outline of platelets, careful preparation, map book coordinating, and picture enrollment [28]. Numerical calculations of component extraction, demonstrating and estimation can be taken advantage of in the pictures to identify pathology, an advancement of the infection, or to contrast a typical subject with a strange one [29]. The precise and repeatable evaluation and morphology of growths are made of vital significance for analysis, treatment arranging just as observing of reaction to oncologic treatment for mind cancers [30]. Cerebrum cancer division comprises of isolating the distinctive growth tissues (dynamic growth, edema and rot) from typical mind tissues: GM, WM, and CSF [31].

The creators in [32] have expressed that man-made consciousness and AI has turned into a fact in clinical practice. Somewhat recently numerous man-made consciousness procedures have been utilized which incorporates calculations utilized for determination and mage handling and picture post processing. The creators in [32] have expressed that man-made consciousness has occurred of radiologists and have made imaging work processes. The man-made consciousness method has its own advantages and disadvantages which makes a deterrent in clinical space. The paper examined in [32] has audited computerized reasoning strategies The article also includes a SWOT analysis (Strengths Weaknesses Opportunities Threats).

In view of profound learning design and motion learning, Siyuan Lu *et al.* (2019) [33] thought to consequently identify neurotic mind in attractive reverberation images (LUNG DISEASE). In both academia and industry, deep learning is currently the hot topic. Typically, the amounts of Lung disease datasets required to prepare the entire profound learning system are small. Overfitting is a concern that is addressed during the planning process. As a result, move learning was used to educate the deep brain organisation. First and foremost, the AlexNet structure was purchased. The last three-layer borders were then replaced with loads, and the rests of the boundaries were treated as starting qualities at that point. Finally, the lung disease dataset was used to build the model. The task was completed with 100 percent accuracy. Using AlexNet and motion learning, the creators have presented a method for neurotic mind location. 100 percent exactness was achieved by outperforming five SOTAs (state of the art). AlexNet was retrained as a result of the move realising, which reduced the chances of it being retrained again. Specialists in clinical analysis used the proposed technique in [33]. The authors of [33] provided a novel technique for classifying an example as typical or neurotic, as well as multi-class grouping to separate certain mental infections. The use of obsessive cerebrums plays an important role in clinical treatment, although the creator's plan remains a mystery. The future degree examined in this paper is that the creators plan to utilize more lung disease dataset for their proposed technique and utilize other progressed profound learning structures for obsessive cerebrum discovery.

Muhammad Owais *et al.*, (2019) [34], Medical-picture-basically based visualization is a dreary task, and little injuries in various clinical previews might be ignored with the guide of clinical inspectors because of the limited capacity to focus of the human apparent gadget, which could unfavorably influence clinical therapy. Nonetheless, this issue can be settled by utilizing investigating tantamount cases inside the previous clinical data set through a proficient substance based clinical photograph recovery (CBMIR) contraption. The creators [34] has stated that with the development of many types of clinical imaging modalities, heterogeneous logical imaging knowledge bases have been rapidly developing in recent years. In recent years, a logical clinical specialist has referred to various imaging modalities, including registered tomography (CT), attractive reverberation imaging (LUNG DISEASE), X-beam, and ultrasound, among others, of various organs in order to determine the diagnosis and treatment of a specific ailment. The CBMIR machine's most important role is to precisely classify and restore multimodal clinical imaging reality. Most prior initiatives make use of precisely crafted features for logical photograph classification and recovery, which demonstrate low overall execution across a wide range of multimodal data sets. In spite of the fact that there are a couple of past examinations on the utilization of profound elements for order, the wide assortment of preparing is tiny.

Shigao Huang *et al.*, (2019) [35], Shigao Huang *et al.*, (2019), Shigao Huang *et al.* By using framework learning, 700 malignant growth patients with cerebrum metastases were enlisted and separated into 446 tutoring and 254 looking at companions for this investigation. To analyse the exhibition of most tumours' conclusion for each understanding, seven capacities and seven forecast strategies were chosen. [35] used common data and extreme set with molecule swarm enhancement (MIRSPSO) processes to predict impacted individual's examination with the best accuracy at area under the curve (AUC) = 0.978 with a variance of 0.06. MIRSPSO improved AUC by 1.72 percent, 1.29 percent, and 1.83 percent, respectively, compared to the baseline regular measurable procedure, successive capacity decision (SFS), shared insights with molecule swarm streamlining (MIPSO), and common realities with consecutive component decision (MISFS), separately. Besides, the clinical presentation of the extraordinary determination became better than standard measurement technique in precision, affectability, and particularity. Finally, logical initiatives require recognising the highest grade level AI strategies for the anticipation of conventional persistence in mind metastases.

Jens Kleesiek *et al.*, (2019) [40], Gadolinium-principally based difference retailers (GBCAs) have end up being a fundamental part in every day clinical decision making in the last 3 numerous years. Notwithstanding, there is a broad agreement that GBCAs should be solely utilized if no appraisal detached attractive reverberation imaging (LUNG DISEASE) procedure is accessible to diminish the measure of executed GBCAs in victims. In the current investigate, we assess the chance of anticipating assessment improvement from non-contrast multiparametric mind LUNG DISEASE filters the utilization of a profound dominating (DL) structure.

Mahmoud Mostapha *et al.*, (2019) [39], Deep acquiring information on calculations and specifically convolutional networks have shown top notch satisfaction in clinical picture assessment bundles, albeit exceptionally couple of strategies have been executed to baby lung disease insights due a few intrinsic requesting circumstances, for example, in-homogenous tissue appearance all through the photo, enormous photograph force changeability all through the principal year of presence, and a low sign to commotion setting. The paper gives procedures tending to those requesting circumstances in two chose applications, chiefly minimal one mind tissue division at the isointense degree and pre-indicative issue forecast in neurodevelopmental messes. Relating strategies are explored and as thought about, and open inconveniences are perceived, especially uninformed length limitations, style lopsidedness issues, and lack of understanding of the subsequent profound dominating replies.

3 CONCLUSION

The validations and reliability of machine learning algorithms in biomedical applications for lung disease diagnosis were examined in this paper. It is crucial that models are carefully developed, and that the regulatory, clinical, and ethical frameworks for implementation are taken into account at every stage of the machine learning process, from predevelopment to post-implementation. This is because machine learning can enhance medicinal diagnosis and have a positive impact on clinical care. This is particularly important because lung disease is on the rise and can be accurately identified using machine learning algorithms that are integrated right into clinical care.

REFERENCES

[1] Laurence Germond *et al.*, "A Cooperative Framework for Segmentation of Lung Disease Brain Scans", *Artificial Intelligence in Medicine*, vol. 20, pp. 77–93, 2000, 10.1016/S0933-3657(00)00054-3.

[2] Kapur T. *et al.*, "Segmentation of the Brain Tissue from Magnetic Resonance Images". *Med Image Anal*, vol. 1, issue 2, pp. 109–127, 1996, ISSN 1361–8415.

[3] Teo P.C. *et al.*, "Creating Connected Representations of Cortical Gray Matter for Functional Lung Disease Visualization". *IEEE Trans Med Image*, vol. 16, issue 6, pp. 852–863, 1997, ISSN: 0278–0062.

[4] Warfield S. *et al.*, "Automatic Identification of Gray Matter Structures from Lung Disease to Improve the Segmentation of White Matter Lesions". *J Image Guid Surg*, vol 1, pp. 326–38, 1995.

[5] Ramesh A.N. *et al.* "Artificial Intelligence in Medicine.", *Annals of the Royal College of Surgeons of England*, vol. 86, issue 5, pp. 334–8, 2004.

[6] Olivier Colliot *et al.*, "*Integration of Fuzzy Spatial Relations in Deformable Models – Application to Brainlung Disease Segmentation, Pattern Recognition*, vol 39, issue 8, pp. 1401–1414, 2006, ISSN 0031–3203.

[7] Selvaraj H., "*Lung Disease Slices Classification Using Least Squares Support Vector Machine*", vol. 1, issue 1, pp. 21 of 33, 2007.

[8] El-Dahshan E-S. A. *et al.*, "Hybrid Intelligent Techniques for Lung Disease Brain Images Classification" *Digital Signal Processing*, vol. 20, issue 2, pp. 433–441, 2010, ISSN 1051–2004,

[9] Yudong Zhang *et al.*, "Feature Extraction of Lung Disease By Stationarywavelet Transformand Its Applications", *Journal of Biological Systems*, vol. 18, 2010.

[10] Yudong Zhang *et al.*, "A Hybrid Method for Lung Disease Brain Image Classification", Expert Systems with Applications, vol. 38, issue 8, pp. 10049–10053, 2011, ISSN 0957–4174.

[11] Jin Liu *et al.*, "A Survey of Lung Disease-Based Brain Tumor Segmentation Methods", *Tsinghua Science and Technology*, vol. 19, no 6, 2014.

[12] El-Dahshan E-S. A. *et al.*, "Computer-aided Diagnosis of Human Brain Tumor through Lung Disease: A Survey and a New Algorithm", Expert Systems with Applications, vol. 41, issue 11, pp. 5526–5545, 2014, ISSN 0957–4174,

[13] Andres Larroza *et al.*, "*Support Vector Machine Classification of Brain Metastasis and Radiation NecrosisBased on Texture Analysis in Lung Disease*", *Wiley Periodicals, Inc.* vol. 42, issue 5, 2015, https://doi.org/10.1002/jlungdisease.24913.

[14] Muhammad Nazir *et al.*, "A Simple and Intelligent Approach for Lung Disease Classification", *Journal of Intelligent & Fuzzy Systems*, vol. 28, issue 3, pp. 1127–1135, 2015.

[15] Angulakshmi M. *et al.*, "*Automated Brain Tumor Segmentation Techniques—A Review*", *Wiley Periodicals, Inc*, vol. 27, pp. 66–77, 2017.

[16] Jyoti Islam *et al.*, "Lung Disease Analysis for Alzheimer's Disease Diagnosis using an Ensemble system of Deep Convolutional Neural Networks", *Islam and Zhang Brain Inf*, vol. 2, issue 5, 2018, ISSN 2198–4026UR-https://doi.org/10.1186/s40708-018-0080-3DO

[17] Muhammad Febrian Rachmadi *et al.*, "Segmentation of White Matter Hyperintensities using Convolutional Neural Networks with Global Spatial Information in Routine Clinical Lung disease with None or Mild Vascular Pathology" *Computerized Medical Imaging and Graphics*, vol. 66, pp. 28–43, 2018, ISSN 0895–6111

[18] Jose Bernal *et al.*, "Deep Convolutional Neural Networks for Brain Image Analysis on Magnetic Resonance Imaging: A Review", *Artificial Intelligence in Medicine*, vol. 95, pp. 64–81, 2019, ISSN 0933–3657

[19] Russakovsky O. *et al.*, "Imagenet Large Scale Visual Recognition Challenge", *International Journal of Computer Vision 115*, vol. 252, issue 115, pp. 211–252, 2015, ISSN–1573–1405UR.

[20] Lin T.-Y. *et al.*, "Microsoft coco: Common Objects in Context, In: European Conference on Computer Vision", *Springer*, pp. 740–755, 2014, ISSN 978-3-319-10602-1.

[21] Everingham M. *et al.*, "The Pascal Visual Object Classes Challenge: *A Retrospective*", *International Journal of Computer Vision*, vol. 136, issue 111, pp. 98–136, 2015, ISSN 1573–1405UR.

[22] Krizhevsky A. *et al.*, "Imagenet Classification with Deep Convolutional Neural Networks, In: *Advances in Neuralinformation Processing Systems*", pp. 1097–1105, 2012.

[23] He K. *et al.*, "Deep Residual Learning for Image Recognition", *In: Proceedings of the IEEE Conference on Computer Vision and Pattern Recognition*, pp. 770–778, 2016.

[24] Szegedy C. *et al.*, "Inception-v4, Inception-resnet and the Impact of Residual Connections on Learning, In: *Proceedings of the Thirty-First AAAI Conference on Artificial Intelligence*", pp. 4278–4284, 2017.

[25] Noh H. *et al.*, "Learning Deconvolution Network for Semantic Segmentation, In: *Proceedings of the IEEE InternationalConference on Computer Vision*", 1520–1528, 2015.

[26] Chen L.-C. *et al.*,"Semantic Image Segmentation with Deep Convolutional Nets and Fully Connected CRFs, In: *International Conference on Learning Representations*", 2015, http://arxiv.org/abs/1412.7062

[27] Anjali Wadhwa *et al.*, "*A Review on Brain Tumor Segmentation of Lung Disease Images*", *Magnetic Resonance Imaging*, vol. 61, pp. 247–259, 2019, ISSN 0730–725X

[28] Rabeh A.B. *et al.*, "Segmentation of Lung Disease Using Active Contour Model". *Int J Imaging Syst Technol*, vol. 27, issue 1, pp. 3–11, 2017.

[29] Derraz F. *et al.*, "Application of Active Contour Models in Medical Image Segmentation". *Proc. Int. Conf. Information Technology: Coding and Computing*, vol. 2. pp. 675–81, 2004.

[30] Meier R. *et al.*, "Clinical Evaluation of a Fully-automatic Segmentation Method for Longitudinal Brain Tumor Volumetry". *Sci Rep*, vol. 6, 2016, ISSN 23376.

[31] Gordillo N. *et al.*, "State of the Art Survey on Lung Disease Brain Tumor Segmentation." *Magnetic Resonance Imaging*, vol. 31, issue 8, pp. 1426–38, 2013, ISSN 0730–725X.

[32] Teudoro Martin Noguerol *et al.*, "Strengths,Weaknesses, Opportunities, and Threats Analysis of Artificial Intelligence and Machine Learning Applications in Radiology. "*Journal of the American College of Radiology*, vol. 16, issue 9, pp. 1239–1247, 2019, ISSN 1546–1440.

[33] Siyuan Lu *et al.*, "Pathological Brain Detection based on AlexNet and Transfer Learning", *Journal of Computational Science*", vol. 30, pp. 41–47, 2019, ISSN 1877–7503,

[34] Muhammad Owais *et al.*, "Effective Diagnosis and Treatment through Content-Based Medical Image Retrieval (CBMIR) by Using Artificial Intelligence." *Journal of Clinical Medicine*, vol. 8, issue 4 p. 462, 2019.

[35] Shigao Huang *et al.*, "Mining Prognosis Index of Brain Metastases Using Artificial Intelligence", *Cancers*, vol. 11, issue 8, p. 1140, 2019, ISSN 2072–6694.

[36] Mahmoud Mostapha *et al.*, "Role of Deep Learning in Infant Lung Disease Analysis", *Magnetic Resonance Imaging*, vol. 64, pp. 171–189, 2019, ISSN 0730–725X.

[37] Verma S.B.& Yadav A.K., Detection of Hard Exudates in Retinopathy Images. *ADCAIJ: Advances in Distributed Computing and Artificial Intelligence Journal, Salamanca*, vol. 8, issue 4, 2019, ISSN: 2255–2863.

[38] Verma S. *et al.* Contactless Palmprint Verification System Using 2-D Gabor Filter and Principal Component Analysis, *The International Arab Journal of Information Technology*, vol. 16, issue 1, pp. 23–29, 2019.

[39] Saravanan Chandran and Satya Bhushan Verma, Touchless Palmprint Verification using Shock Filter, SIFT, I-RANSAC, and LPD, *IOSR Journal of Computer Engineering*, vol. 17, issue 3, pp. 01–08, 2015. DOI: 10.9790/0661-17330108

[40] Satya Bhushan Verma and Saravanan Chandran, Analysis of SIFT and SURF Feature Extraction in Palmprint Verification System, *International Conference on Computing, Communication and Control Technology (IC4T), IEEE*, pp. 27–30, 2016.

Automation and Computation – Vats et al. (Eds)
© 2023 the Author(s), ISBN 978-1-032-36723-1

Smart helmet: An innovative solution for timely aid and women safety

Rahul Chauhan, Divyanshu Negi, Ashutosh Uniyal & Himadri Vaidya
Graphic Era Hill University, Uttarakhand, India

ABSTRACT: In road accidents, two wheelers are major contributor. According to a report published in Times of India, a staggering 44.5 percent death involved in road accident is due to two-wheeler. Mostly the accident victim doesn't receive the timely aid and certainly lead to death in various occasion. This project is aimed to develop a solution of timely aid, voice enabled indication and alert generation in case of fall. This project is developed with minimum hardware mainly Raspberry- pi, pressure senor, LEDs, and voice amplifier module. A SMTP (short message transfer protocol) service is being proposed and used for sending the alert emails in case of pressure sensor value crosses the threshold value of 200. Experimental result shows that the system achieved a good accuracy in voice enabled indication control and in alert emails.

1 INTRODUCTION

Two-wheelers are the most preferred medium of transport in today's world due to their compact size, efficiency, low maintenance, mobility for the Indian family, and sheer joy while riding them. In the world, India ranks 2nd position in the production of 2 wheelers. Today cars account for only 13 % of the vehicle population in India and on the other hand two-wheelers account for 70% of the vehicle population in India. Although two-wheelers are the most convenient medium of transport, they are the reason for serious traffic problems. According to the latest data from the NCRB (National Crime Record Bureau) (Ministry of Home Affairs 2020), the rate of accidental death by two-wheelers increased by 43.6% in 2020 compared to 38% in 2019. Though being the covid time there was a 6 % rise in the rate of accidents by two-wheelers. This indicates that in covid time how people shifted to affordable and safe transport mediums. India ranks 1st amongst 199 countries in road accident deaths reported in the World Road Statistics, 2018. According to the report by WHO on Road Safety, 2018,11% of the accident-related deaths in the World are reported in India. According to the report by the world health organization (WHO), about 1.3 million people die every year worldwide as a result of accidents and 30 to 40 million people suffer injuries that result in lifelong disability (Ministry of Road Transport and Highway 2022).

There are various prominent reasons for accidents, a few of them are overspeeding, distracted driving, not wearing helmets, road environment, and many more. The NCRB data showed that speeding was the prominent reason for road accidents, accounting for 75,333 deaths and 2,09,736 were injured (Ministry of Home Affairs 2020). One out of every four two-wheeler riders killed in road accidents in Bangalore during 2018-2020 was not wearing a helmet, according to data from the traffic police. One of the reasons for an accident that are less highlighted is accidents that are caused due to changing course or direction or stopping without indication. Overtaking accounts for around 24.3 % of road accidents. There are numerous fields where for the safety of human beings hamlet is used. The hamlets can be

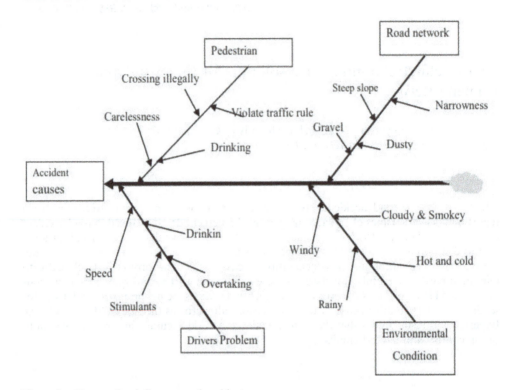

Figure 1.   Factors that influence road accidents.

used for road safety as well as industrial safety. There is various protocol with constructive techniques to enhance the safety standards for the better being of workers. (Chevalier *et al.* 2020; Vaughn *et al.* 2002)

According to the report of the NCRB (National Crime Record Bureau) (Ministry of Home Affairs 2020), there is around 3.75 Lakh (approx. 46%) of accidental death in India were reported in the year 2020 and the majority of which were riders of two-wheelers. Among these, many deaths occur due to a lack of immediate medical attention needed for the injured person.

There are several helmets designed to overcome many problems like these. Sena, the California-based company, maker of Bluetooth communication, now offers smart helmets with an integrated communication facility. Stryker, senas full faces smart helmet with boot mesh intercommunication and Bluetooth intercommunication technology in addition to speakers and a microphone. Stryker provides an intercommunication feature so that users can connect with any sena headset for communication with each other. Mesh Intercom has tough, authentic connections and flexible intercommunication with the advanced Sena's algorithm. Voice commands allow riders to speak directly to their connected helmets. Through this, we can answer the phone call, and start the group's mesh session. Digital assistant access allows the helmet to connect to Google assistant or apple's Siri through which we can perform any task. Altor, India based smart helmet company has some more additional features, including enhanced safety mode and helmet detections, gesture-enabled functions, accepting and rejecting calls, getting notifications, and sending messages with a single swipe. emergency S.O.S. which can be sent with a single click at a time of emergency or accident. Quin spitfire Rossois a full-faced helmet with Intelliquin smart technology, which includes features such as Bluetooth connectivity, crash detection, and an SOS alert.

Intelliquin's crash detection system has smart sensors integrated into it such as an accelerometer, single circuitry, and algorithms for detecting and measuring crashes. Intelliquin S.O.S. beacon provides the live location of motorcyclists. Dual Bluetooth system to optimize crash alerts, listen to music, or make calls. Cross helmet is a helmet-making company that provides some advanced features like a head-up display that provides a crisp and clear image of essential riding information such as route destinations, weather, and time. The Rearview camera, projected onto the HUD, empowers riders with 360-degree vision at a glance. Steel bird an India-based helmet-making company has its SBA-1 helmet which is connected to your mobile phone through an aux cable and can answer calls listen to music and navigate. According to Bengaluru police using headphones for navigation will be a chargeable offense. According to Bengaluru Police, the first violation of the rule will attract a ₹500 fine, while the fine for the second violation will be ₹1,000 (Ministry of Land 2012). Listening to music and receiving calls while riding the two-wheelers can lead to various consequences.

To overcome these problems, we can use the technology we have. IoT (Internet of Things) is the interconnection between things i.e. devices, objects, or any natural or man-made object that can be assigned with an IP address. And which can transfer data over the internet. And later, the data received can be used to process the output. The purpose behind IoT is to have devices that report by themselves in real-time. Today IoT is used in various fields like health care, farming, smart supply food chain, smart city, and many more. Talking about IoT in the healthcare sector contains many varieties of mechanisms like sensors, equipment, big data, telemedicine, information about the patient (Syed et al. 2021), and many more. Taking the help of IoT to solve accidental problems we have come across the solutions for the above-mentioned problems. A smart helmet is a device that serves three major purposes. 1. It consists of voice-enabled indicators present at the front and rear sides of the helmet which indicates the direction in which the user wants to turn. If a person gives the command "left-left" indicators of left will blink and if a person speaks "right-right" indicators of right will blink, hence indicating the direction in which the person wants to turn. 2. In case of an accident, if a person is physically unstable, he can send a pre-programmed message containing his current location through voice command to his family members, friends, nearest police station, and to the nearest hospital through voice commands about him being in danger and needing urgent help. In case the person is unconscious and if his helmet hits the ground at a certain force and if the force received through the force sensor is greater than the threshold limit then the message will be automatically transmitted to family, friends, the police station, and to the nearest hospital. The message will inform the receiver about the danger and provide them with his current location. 3. Women's safety in India is a major concern and therefore should be talked about. Riding through the area if women feel unsafe, she can send a pre-programmed message which includes her current location, to the nearest police station through voice command.

2   HARDWARE SPECIFICATIONS

2.1   *Raspberry-pi*

It is a series of different single-board computers, designed and manufactured by the raspberry pi foundation. In the year 2012 Raspberry pi was launched and there have been various versions since then. The latest model of raspberry pi has a quad-core CPU with a clock frequency of 1.5Hz and 4GB RAM. It is used to learn code and learn to code electronics for physical projects. The Raspberry Pi is a cheap, small card-sized computer that plugs into a monitor or TV and it is capable of performing any task a desktop computer can perform, from surfing the internet or playing video games or any other task that could be performed in a normal computer.

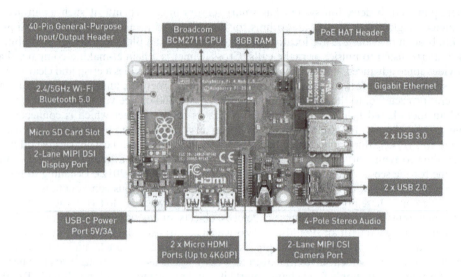

Figure 2.   Raspberry pi.

The raspberry pi board comprises RAM, GPIO pins, processor and graphics chip, Ethernet port, GPU, Xbee socket, CPU, and power source port. And various ports for other external devices such as a mouse, keyboard, and many more. We use an SD flash memory card as a secondary storage device. Raspberry pi board will boot from this SD card similar to that of windows.

2.2   *Voice recognition module*

ELECHOUSE Voice Recognition Module is a small and simple voice recognition module. The board works on 4.5-5.5 voltage with <40mA current. This is a speaker-dependent voice recognition module. Up to 80 voice commands in total are supported by this module. At max 7 voice commands could work at a time. The module is to be trained first by the user to make it usable. This board has 2 controlling ways: Serial Port (full function), and General Input Pins (part of function). General Output Pins on the board could generate waves while the module recognizes the corresponding voice commands and these waves can be further used to perform any operation.

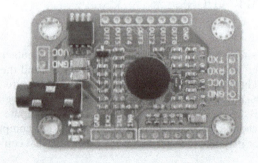

Figure 3.   Voice recognition module v3.

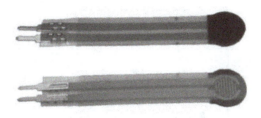

Figure 4.   Force sensor module.

In addition to that, we have to connect a microphone to a port in the voice recognition module, through which we will give commands to the module.

2.3   *Force sensor*

Interlink Electronics FSR^{TM}400series is a single zone Force Sensing Resistors family. Force Sensing Resistors are polymer thick film devices that change the resistance with the change in force that is applied to the surface of the sensor. It is basically a resistor that works on the principle of change in resistive value depending on the pressure applied to it. This force sensitivity is optimized for use in human touch control of electronic devices such as industrial and robotics, electronic devices, and medical systems. The standard sensor is around 7.62mm in diameter. Whereas custom sensor sizes range from 5mm to over 600mm.

2.4   *Battery*

We need power for our raspberry Pi, voice recognition module, pressure sensors, and indicators. so for that, we will be using a pi sugar 2 pro battery which is compact and efficient. A few of its advanced features are RTC: SD3078, Type-c: charging port, with E-mark support, Micro USB: charging port, PH2.0 Battery Plug, Power Switch & Turbo Switch, Programmable Tap Button.

3   PHYSICAL STRUCTURE AND HARDWARE ASSEMBLY

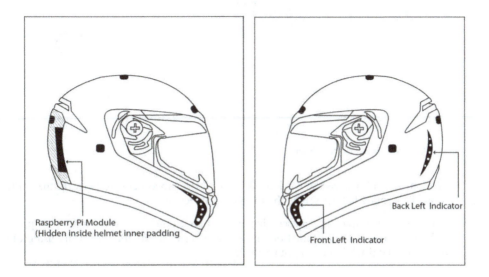

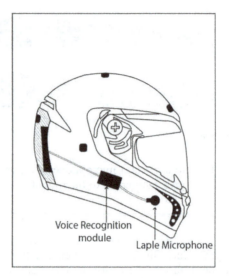

Voice Recognition
module

Laple Microphone

4 METHODOLOGY

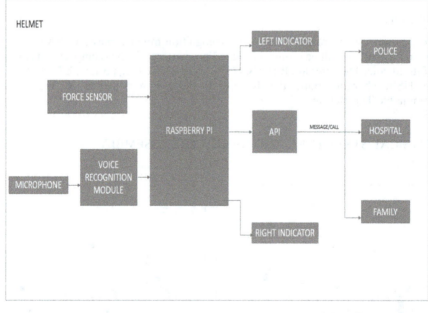

Figure 5.   A workflow of smart helmet.

In this section, we will discuss the working flow of all the sub-systems to perform the entire application. This smart helmet serves the following objectives:

(i)   Avoid accidents that are caused due to changing course or direction or stopping without giving an indication. For this, we can turn on /off the indicators using our voice command while turning our vehicle.

166

(ii) Timely aid, if met with an accident. in case of an accident, a call will be made or a message will be sent to the ambulance, police, or his family members.

(iii) Women's safety: If women feel unsafe in a particular area while traveling, they can send a message or call the police or near ones using voice commands

4.1 Force sensor

This device is used to sense the force applied to the surface. It is a resistor that works on the principle of change in resistive value depending on the pressure applied to it. When there is no pressure, the sensor acts as an open circuit. The harder the pressure on the sensor, the lower the resistance between the two terminals. And if you remove the pressure it will return to its initial value as resistance increases.

We will be using the force sensor in different parts of our helmet which will read the force applied to them. If the force applied to any one of the sensors is above the threshold limit it means that the person wearing that helmet has struck his head somewhere and needs an emergency medication

4.2 Voice recognition module

It is a speaker-dependent voice recognition module. Up to 80 voice commands in total are supported by this module. The module is to be trained first by the user to make it usable. It can be trained with any command. It recognizes the command which we speak through the microphone and work accordingly. If the user speaks "left-left" indicators on left will blink, "right-right" will blink indicators on the right, and "help-help" if women feel unsafe in a particular area," danger-danger" if a person meets with an accident and need immediate medication.

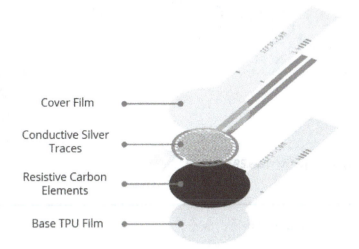

Cover Film

Conductive Silver Traces

Resistive Carbon Elements

Base TPU Film

Figure 6. Construction of a force sensor.

4.3 Raspberry-pi

Raspberry pi act as the heart of the system which acts as the processing unit and medium of communication of the whole system. Connection is made between raspberry pi and the force

167

sensor and voice recognition module; any input provided by the force sensor and voice recognition module is interpreted by the Raspberry Pi and hence performs the task accordingly. It has an inbuilt wifi module which is used to set an internet connection and that will be used to transfer the messages to the respective person.

4.4 *LED indicators*

LEDs are being placed at the front and rear side of the helmet which will perform the task of indicating whether the person is going to turn left or right so that person coming from the front or back be aware. The connection between LEDs and raspberry pi is set and if the instruction of "left-left" is given through the microphone indicators on the left side of the helmet will blink for 5 sec and after 5 sec will automatically turn off, the same for the indicators on another side.

4.5 *SMTP*

Simple Mail Transfer Protocol is a protocol used in sending and receiving e-mail. It is an application layer protocol. SMTP works as a client/server model. First, an email server uses SMTP to send a message from an e-mail client to an e-mail server. Second, SMTP is used to send e-mails to the receiving e-mail server. Third, the server which receives uses an e-mail client to download incoming mail and send it to the inbox of the recipient. We can only send messages with help of SMTP; to make a call we can take the help of API.

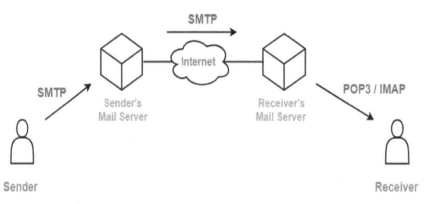

Figure 7. Working of the SMTP protocol.

5 WORKING ALGORITHM

A smart helmet consists of two input devices which are a microphone and a force sensor, a processing unit which is a raspberry pi and uses LEDs as indicators which is an output device.

Any command given to raspberry pi is given either by a microphone that is connected to the raspberry pi through a voice recognition module or by force sensors that are placed on different parts of the helmet which on receiving force above the threshold limit send a signal to raspberry pi to send a message or make a call to the respective person (i.e. hospital, police,

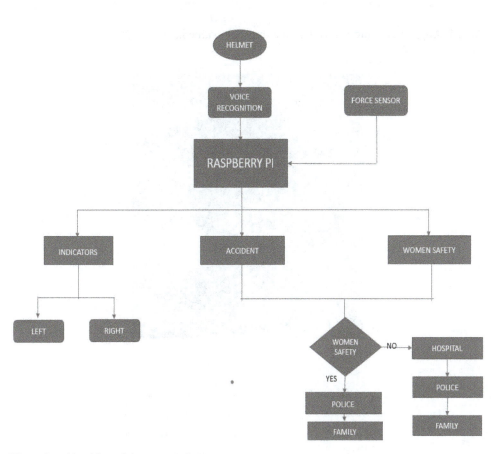

Figure 8. Algorithm of the proposed system.

and home). The sensor keeps on getting the reading continuously and the time when the sensor receives force above a limit (ex 100dB) indicates that the person's head struck somewhere or met with an accident and person needs immediate medication so within a few seconds a message is being transferred which tells the current location of the person and information about the accident, to the nearest hospital, police station and to family members of the person. Any command given through a microphone is interpreted by a voice recognition module which then transfers the corresponding signals according to the commands to the raspberry pi and works according to instruction, If the user speaks "left-left" indicators on left will blink, and "right-right" will blink indicators on the right, and "help-help" if women feel unsafe in a particular area," danger-danger" if a person meets with an accident and need immediate medication. This solution could decrease nationwide accidents and provide immediate medication and reporting.

6 EXPERIMENTAL RESULT

Now after establishing all the connections properly, if we monitor the code we could see in the Figure 12 how on receiving force above the threshold limit mail is being sent to a person. On receiving the voice command such as left-left we can see in Figure 13 how LEDs on the left side are blinking same happens with right indicators.

Case: 1 Applied pressure is less than the threshold value i.e. below 200

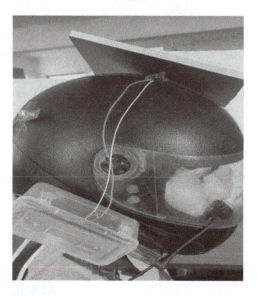

```
Pressure Pad Value: 0              Pressure Pad Value: 79
--------------------------------   ----------------------------------
Pressure Pad Value: 0              Pressure Pad Value: 101
--------------------------------   ----------------------------------
Pressure Pad Value: 0              Pressure Pad Value: 78
--------------------------------   ----------------------------------
Pressure Pad Value: 0              Pressure Pad Value: 12
--------------------------------   ----------------------------------
Pressure Pad Value: 0              Pressure Pad Value: 3
--------------------------------   ----------------------------------
Pressure Pad Value: 0              Pressure Pad Value: 93
--------------------------------   ----------------------------------
Pressure Pad Value: 47             Pressure Pad Value: 0
--------------------------------   ----------------------------------
Pressure Pad Value: 79
--------------------------------
```

Case: 2 Applied pressure is equivalent to the threshold value i.e. approximately 200

```
Pressure Pad Value: 222            Pressure Pad Value: 0
helped called                      ----------------------------------
mailsending...                     Pressure Pad Value: 278
mail send :)                       helped called
--------------------------------   mailsending...
Pressure Pad Value: 0              mail send :)
--------------------------------   ----------------------------------
Pressure Pad Value: 0              Pressure Pad Value: 0
--------------------------------   ----------------------------------
Pressure Pad Value: 258            Pressure Pad Value: 0
helped called                      ----------------------------------
mailsending...                     Pressure Pad Value: 253
mail send :)                       helped called
--------------------------------   mailsending...
                                   mail send :)
```

Case: 3 Speech recognition through the voice module by commands for turning left

```
speak: left left
indicator blinking
speak:
speak:
speak:
speak:
speak:
speak:
speak:
```

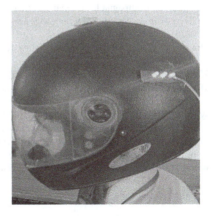

Case: 4 Speech recognition through the voice module by commands for turning right

```
speak: right right
indicator blinking
speak:
speak:
speak:
speak:
speak:
speak:
```

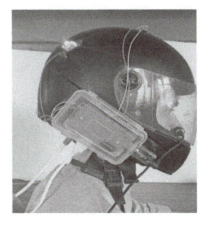

7 CONCLUSIONS

The smart helmet is an IoT device that performs three main operations i.e., voice-enabled indication, timely aid, and women's safety. It makes use of a few electronic components which are raspberry pi, a force sensor, a voice recognition module, and a battery, they are integrated together in a way that we can perform various operations using them, such as turning our left indicators on using command "left-left" same with right indicators if we speak "right-right". If our helmet gets struck and the force experienced by our helmet is above the threshold limit then an emergency is raised by sending a message or call to a family, police station, or hospital. If women feel unsafe in a particular area, they can raise an emergency using the "danger-danger" command. This device will not only decrease the accident caused by improper indication but also will save the lives of people in times of emergencies.

REFERENCES

Chevalier Y., Compagna L., Cuellar J., Drielsma P. H., Mantovani J., Mödersheim S., & Vigneron L. *A High Level Protocol Specification Language for Industrial Security-Sensitive Protocols.* (accessed on 1 November 2020).

Ministry of Home Affairs, National Crime Records Bureau, (2020) *"Accidental Death and Suicide in India 2020".*

Ministry of Land, Transport and Maritime Affairs, Notice No. 2012–945.

Ministry of Road Transport and Highway, Amendment of Motor Vehicles act, 1988 by Motor Vehicles(Fifth Amendment) Act, 2022.

Syed A. S., Sierra-Sosa D., Kumar A., & Elmaghrsby A. (2021). IoT in Smart Cities: A Survey Of Technologies, Practices, and Challenges. *Smart Cities*, 4(2), 429–472.

Vaughn Jr, R. B., Henning R., & Fox K. (2002). An Empirical Study of Industrial Security-Engineering practices. *Journal of Systems and software*, 61(3), 225–232.

World Health Organisation, (2018) "Global Status Report on Road Safety 2018."

Automation and Computation – Vats et al. (Eds)
© 2023 the Author(s), ISBN 978-1-032-36723-1

Computational techniques to recognize Indian stone inscriptions and manuscripts: A review

Sandeep Kaur & Bharat Bhushan Sagar
Department of CSE, Birla Institute of Technology, Mesra, Ranchi, India

ABSTRACT: Many modern scripts and languages can be traced back to various ancient scripts which evolved over the years to their present form and structure. These ancient texts provide significant insight into ancient history and civilization. But deciphering the ancient writings is a complex task as these writings are only available in form of inscriptions on rocks, caves, and palm leaves which are severely degraded due to environmental impact over several years. Preserving these texts and writings is of utmost importance. Digitization of the inscriptions can help in preservation and ease the sharing and searching of the text, like modern-day scripts and writings. Manual digitization is tedious and error-prone, so there is a need for an optical character recognition system for these ancient texts and scripts. Most of the modern-day scripts have various OCR systems implemented and these are in constant enhancement. The purpose of this study is to survey the OCR systems and techniques proposed for the various ancient scripts focusing on ancient Indian scripts.

1 INTRODUCTION

Writing has been the most important means to collect and store information throughout history across different cultures and civilizations. In today's time, paper is the most prominent mode of writing, but in earlier days, before the creation of paper, writing was mostly done on rocks, pillars, slabs, and dried palm leaf. Ancient manuscripts inscribed on pillars, or slabs are called inscriptions, whereas writing documents on palm leaf are called a manuscript. These ancient documents provide valuable information about the civilization's language, literature, and historical, social, spiritual, and political scenarios. Palm leaf writing has traditionally been transmitted from one generation to the next by scholars and scribes. Every time a palm leaf decomposes, it was usual to transfer its contents to brand-new, fresh leaves. And that is how modern civilization received our written ancient literature. Understanding ancient writing materials is a vital way to understand ancient cultures in a better light [1]. Studying these ancient inscriptions is the best way to understand and reveal historical information. These inscriptions have also enabled the preservation of ancient knowledge, history, and culture across centuries, as most of these inscriptions on rocks, and pillars could withstand the impact of nature for centuries. But study and analysis of these ancient scripts require in-depth research by archaeological experts, especially for the inscriptions on rocks and pillars. It requires tedious manual effort as recognizing the characters of scripts is a huge challenge. Most of these very old inscriptions can suffer some damage over time. Also, due to the vast set of ancient languages and changes in scripts/symbols over time, archaeological experts need considerable effort to read inscriptions and identify the symbols. It increases the time required to understand inscriptions and the chances of errors. This leads to the requirement of developing a system to enable automated recognition of symbols from inscriptions. Due to the presence of background noise, the characters in many ancient inscriptions are damaged poses a major challenge in the recognition of ancient scripts.[2] Therefore, there is a need to recognize each character and convert it into readable form. A rigorous development was noticed in Optical Character Recognition (OCR) for modern scripts and languages for digitizing printed documents. Even abundant research has been done on handwritten documents, but it remains a challenge for historical documents and inscriptions [2]. In this paper, we try to address this problem and present a survey of different techniques and methods used to recognize the stone inscriptions of various ancient Indian scripts. This also provides efficient methods for scholars in the humanities to discover and study ancient scripts.

2   ANCIENT INDIAN SCRIPTS

Brahmi is considered a parent script that led to evolving the modern Indian scripts such as Bengali, Devanagari, Kannada, Tamil, Telugu, Gurmukhi, Odia, etc., and some Asian scripts. Figure 1 explains the evolution of Indian scripts from Brahmi. Most inscriptions are in Brahmi scripts and Sanskrit using Brahmi scripts. Palm leaf manuscripts are written in Tamil, Kannada, Sinhala, etc. The oldest surviving Sanskrit Shaivism palm leaf manuscripts date to the 9th century were found in Nepal and are well conserved at the Cambridge University Library.

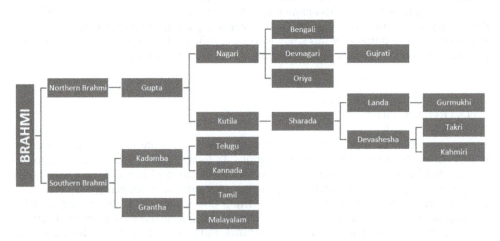

Figure 1.   Indian script evolution [3].

2.1   *Challenges in identifying Ancient Indian script*

Ancient documents are inscribed on stone or pillars, or slabs; the only source of the character-set images are inscription images that scholars and historians have captured from pillars and stones [4]. And some manuscripts are available in libraries, museums, and monasteries in the form of palm leaves manuscripts. But these images are primarily low-resolution images and pose significant challenges as these have a lot of visual defects. Common defects such as cuts and cracks on the inscriptions can intrude on a part of the character in the image [4]. In the case of inscriptions, varied illumination due to sunlight and the angle of capturing the image causes shadows and bright spots to appear. In preprocessing steps like enhancement and binarization, the presence of numerous types of degradations becomes a serious issue [4,5]. The segmentation procedure is impacted by uneven or no white space between words and characters. Recognition of inscription is challenging due to these various types of noise, spots, cracks, and shadows caused by curvature or variable illumination and for manuscripts due to discoloration, fading, and color or brightness intensity changes.

3 OPTICAL CHARACTER RECOGNITION

OCR is a technology that enables automated computer-based text recognition from various sources, including scanned photos, camera images, files (pdf), etc. Unlike photographs, where text cannot be altered or searched through, this makes it possible to store text in a digital format that computers can read and use for searching and editing. OCR can be considered an amalgamation of pattern recognition and AI since, in the simplest words, it seeks to recognize patterns in the input image and classify the patterns into digital data. Various specialized applications have been developed based on OCR. Early OCR systems relied on matching input image with template databases of known images. In 1954, Reader's Digest built the first OCR reading system. Several novel OCR techniques were created due to hardware and software research. Modern OCR systems are more sophisticated and built on machine-learning platforms. To begin with, the OCR system would need to teach the computer using examples of characters from various classes. Once the system has been trained, it can be used to categorize characters from different online or offline data sets and enable character recognition.

OCR systems have a wide range of uses, including processing vast amounts of historical, official, or government documents, processing bank documents like cheques, assisting people with vision impairments in reading, processing, or sorting mail, and natural language processing. Faster data processing and less paper use can result from effective OCR systems [6]. Additionally, the digitization of documents enables seamless sharing and searching of text and helps preserve the document for future reference. A typical ancient character recognition system consists of several constituents like OCR, as shown in Figure 2. The entire process of character recognition can be divided into various stages.

1. The most common input for ancient text is ancient documents or writings on pillars, palm leaves, etc., which are basically inscriptions or ancient text writing on commonly used mediums for writing text.
2. Scanning – This process transforms the data to be digitized from physical to electronic form. An extremely precise digital image of the input data is captured throughout the scanning process. The scanning equipment concentrates light on the image or text. The greyscale or color image is produced based on the light levels in various areas of the supplied text.
3. Pre-processing – The scanned image might have some flaws or noise in it. Also, there may be variations in character sizes and writing styles. Hence, the scanned image needs to be pre-processed to remove this noise and normalize the data through a suitable normalization technique before being sent for pattern recognition. The following describes the pre-processing steps. Historical documents present an additional challenge as most sources are damaged due to wear and tear over several decades and suffer from poor contrast and low resolution. Some of the commonly used pre-processing methods are mentioned below
   a. Skew and slant normalization – The scanned or camera image can introduce a slant in the direction of image text or characters such that the characters in the image text may be slightly curved or tilted. The skewness can also be present due to different writing styles. Detecting and correcting skew in the scanned document is crucial for higher accuracy in OCR.
   b. Image Scaling – A huge image or tiny size image can impact the OCR accuracy. To have better performance in OCR, the image in the entire set needs to be rescaled to a suitable size like 128x128, 256x256, 1024x2096, etc.
   c. Noise reduction/Noise removal – Ancient and historical manuscripts suffer from various physical distortions in the form of cracks, smears, etc. which can significantly

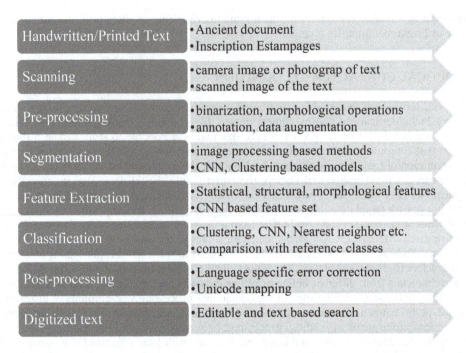

Handwritten/Printed Text	• Ancient document • Inscription Estampages
Scanning	• camera image or photograp of text • scanned image of the text
Pre-processing	• binarization, morphological operations • annotation, data augmentation
Segmentation	• image processing based methods • CNN, Clustering based models
Feature Extraction	• Statistical, structural, morphological features • CNN based feature set
Classification	• Clustering, CNN, Nearest neighbor etc. • comparision with reference classes
Post-processing	• Language specific error correction • Unicode mapping
Digitized text	• Editable and text based search

Figure 2.   Ancient character recognition component.

impact the OCR accuracy. The scanned image may include broken characters lines, gaps in the lines, etc., depending on the condition of the input data (for example, a very old manuscript or photograph) and the resolution of the scanner device. As these flaws can result in poor character recognition, this stage aims to fix these as much as possible using various image processing techniques and enhance the overall image quality. Morphological operations are commonly used to remove stray pixels or to join/fill cracks.

  d. Binarization - Binarization is the process of converting a colored or greyscale image (having a pixel value between 0 to 255) to a binary image (having a pixel value of 0 or 255). The most used method for binarization is Thresholding which includes various variations like Adaptive Thresholding, Otsu Thresholding, etc., to handle scenarios like variable illumination across images.

  e. Thinning or boundary detection – This process involves detecting the exact boundary of the text to be recognized and removing any other dots, lines, etc. which are not part of the text but got introduced during scanning. This is very useful for handwritten text as compared to printed text, as the handwritten text can have different character widths.

4. Segmentation – Rather than implementing an OCR technique over the entire text, the input text is segmented into individual lines, words, and finally characters. Specifically for an ancient script or manuscript not having a standard dataset, developing a character-level OCR is the crucial step. Segmentation can be challenging due to various defects and noise in the inscription image. Various image processing techniques like projection pro-files, contours, etc. can be used for segmentation, and recently advanced techniques like clustering and CNN models have also been employed for image segmentation.

5. Feature extraction – Features are the values or vectors used to identify some specific properties from the raw image such that it can suitably represent the image to enable classification. Feature extraction is a crucial and challenging part of an ancient

176

recognition system. Different types of features or combinations of features need to be evaluated for script identification, depending on the shape and structure of characters in each script. Every text or character has certain qualities or traits that enable us to recognize them. These crucial facets of the character are captured in this stage. This is the most challenging step of the entire OCR process, as we need to identify and capture the minimum set of features which would help classify and identify a character. There are various types of features for OCR -

a. Structural Features – These features are related to the character's structural framework and can handle some level of distortion and variations in style. Strokes like dots, lines, curves, aspect ratio, branch points, etc. are examples of structural features.

b. Topological features – These include features like loops, junctions, convexities, etc., which are invariant to shear deformation.

c. Template features – These are the set of features that can be used to match the character with a predefined template. This technique can be impacted by variations in the size of the character and is also not suitable for recognition from noisy images.

d. Statistical features – These are the features corresponding to the mathematical or statistical dissemination of pixels in the image or a zone/section of the image. There are various statistical features like density features, direction features, variance, height, mean, projection histograms, crossings, distances, etc.

e. Global features – These include features to represent the entire image in a single feature set like Fourier Transform, and moments (Zenrike, central moments). These are mostly scaled, rotation, and translation invariant.

6. Classification – Once the features have been identified and extracted, classification is the next step is to assign the test image to the appropriate character class. Based on the feature set, various classification techniques are used in script identification systems like K-Nearest Neighbor (K-NN) classifier, Support vector machine (SVM), Neural Networks, Quadratic classifier, and Linear Discriminant classifier. The mentioned models are extensively used for big data classification [7–12] but the performance of the classifier depends on various factors like the availability and accuracy of the training set. Like features, multiple types of classifiers or an ensemble of classifiers can be used to improve the overall performance and accuracy. Additionally, the output of one model can be further enhanced and improved by the later models [13–20].

7. Post-processing - Enhancing the effectiveness of language recognizers is one of the post-processing processes. Grouping and error recognition, and error correction are part of this stage. A set of distinct symbols is produced using simple symbol recognition in text. But typically, these symbols do not convey enough information. Words and numbers are made up of these separate symbols that are related to one another. In grouping, strings and text symbols are connected. Post-processors are frequently language-specific and take advantage of the unique characteristics of the language to achieve high accuracy or performance. A dictionary of words with semantic class labels is known as a semantic lexicon. Additionally, it is also applied as a post-processing tool in OCR to find words that have never been seen before.

4   SURVEY IN ANCIENT SCRIPT RECOGNITION

Digitization of ancient script inscriptions like pillars or stone inscriptions or inscriptions on leaves is important to preserve our heritage and easily access and share these ancient writings. But at the same time, it is a challenging task. Optical Character Recognition (OCR) is an emerging field to convert the digitized document into ASCII format text files. Various commercial OCR products have been developed for modern scripts to produce technology-driven applications [21]. Extensive research has been done on modern Indian languages. The latest comprehensive survey presented by Sharma and Kaushik [22] on handwritten offline recognition also discusses the various features and classification techniques used by researchers to recognize modern Indian languages. But for Indian languages or script, dataset is the main concern since the diversity and complexity of scripts and the inadequacy of available public dataset, most researchers construct their own datasets [4]. Kakwani et al. [23] developed an Indian languages monolingual corpus, and data is collected from news crawls. Also, the paper [22], explains the various available datasets available in Indian languages. But for ancient inscriptions, there is no such dataset available. James and Nasseh [24] present a survey on various techniques and tools used to recognize the historical document written in English, Latin etc.

As Brahmi is one of the oldest scripts, Siromoney et al. [25] proposed a method of recognition of characters of the ancient Brahmi script which was found on many Ashoka pillars. Machine-printed characters of the Brahmi scripts were recognized using coded run method. The author claimed any script can be recognized using this technique, which involves manually converting each Brahmi character into a rectangular binary array and transliterating the recognized characters into Roman. Devi et al. [26,27] explained two methodologies Thinning and Thresholding for pre-processing archaeological images of Brahmi script image from rock inscription. Further Warnajoith et al. [28] purposed an algorithm to create a digital depository for Brahmi script characters from the rubbed copy of estampages. The estampages were created by an archaeologist using inscriptions discovered in Sri Lanka on caves, rocks, pillars, and slabs. The author has divided the process into four different stages. Stage 1 generates alphabet fonts from the images/photographs of the Brahmi script. Stage 2 works on identifying the letters based on the fonts identified in stage 1. Stage 3 prepares a database of scripts analyzed in the previous stage. The data is shared on the Web interface with other scholars.

Kumar et al. [29] proposed a method for converting ancient Tamil script text from inscription and palm manuscripts to Tamil Unicode using features based on Affine Invariant Intensity Extrema to extract the features from inscription images and using Tesseract to map these to modern day Tamil characters. Greeba et al. [30] purposed a system to recognize the Vatteluttu, ancient Tamil characters by Combined feature-set constituting Zoning and structural features with a fuzzy logic-based classifier. Merline et al. [31] purposed an ancient Tamil recognition that has been implemented for stone inscription through the process of segmentation using morphological operations, followed by feature generation using corner points, region properties, and scale-invariant features. The recognition is performed using ensemble learning and KNN. Bhuvaneswari et al. [32] proposed a positional feature called INPDM (Image-based) and ZNPDM (Zone based) Normalized Positional Distance Metric along with structural features like horizontal and vertical lines, intersection points, and regional features like orientation, convex area, and eccentricity. Further, nearest neighbor classifier has been used the match the test image with prototype images. Convolutional Neural network structure has been proposed to segment inscription image characters followed by a comparison of the segmented character image with the reference character image dataset to identify the closest match based on Euclidean distance by Giridhar et al. [33]. Further text-to-speech engine has been used to generate the audio output of the recognized text. Histogram-based oriented gradient has been proposed by Manikandan et al. [34] as a feature-set of inscription text followed by a Support Vector machine-based classifier to identify and classify the text of Ancient Tamil characters. Further Manigandan et al. [35] presented a Scale-invariant feature transform based on image features like dots, curves, loops, and number of lines as a feature set for ancient Tamil characters. The recognition is done using the Support Vector machines classifier and Tri-Gram technique for prediction.

Aswatha et al. [36] proposed a HOG feature-based technique to match an estampage section from an inscription image of Kannada, with a predefined template image by calculating the normalized cross correlation between the HOG feature set of the two images. Khan et al. [37] proposed the identification of Kannada stone inscription characters by comparing the mean, Standard Deviation, and the sum of absolute (MSDDA) difference between the test image and database of old Kannada character images.

Narang *et al.* [38] proposed a feature based on the Histogram of oriented gradients and Discrete Cosine Transform for the Devanagari characters followed by evaluation of classification with SVM using RBF kernel, decision tree, and Naïve Bayes. Sachdeva *et al.* [39] purposed a system that uses a feature set for compound Devanagari characters created using Edge Histogram feature extraction. The classification is then evaluated using different systems like SVM, Social Media Optimization techniques, multi-Layer perceptron, and Simple Logistic. Narang *et al.* [40] utilized statistical features including open endpoints, intersection points, centroids, vertical peak extent, and horizontal peak features to generate the feature vector for ancient Devanagari script. Classification was performed and compared with techniques like CNN, NN, MLP, RBF-SVM, and RF.

Ronak Shah *et al.* [41] purposed a system for Sanskrit based on OpenNMT (Open-source neural machine translation) attention-based encoder-decoder model, the author evaluated the recognition accuracy by varying different hyperparameters like the size of batch, number of LSTM layers, iterations, and character size or embedding vector. Dwivedi *et al.* [42] proposed an attention-based LSTM model for the OCR of Sanskrit documents. To improve the overall accuracy data, augmentation and synthetic data have been employed. Sreedevi *et al.* [43] presented a Natural gradient-based flexible ICA (Independent Component Analysis) technique to separate the inscription foreground from the background image. This is followed by Sobel edge detection and dilation to obtain an enhanced foreground image for the characters.

Palaniappan *et al.* [44] proposed a solution consisting of pipeline CNN architecture consisting of the an initial CNN model extracting the features for the key areas, i.e., only text region, only image region, and both text and image region. The output is passed to the second CNN to identify the graphemes from text and image regions. For Brahmi and Hoysala and ancient Kannada scripts, Soumya *et al.* [5] have proposed an identification process employing nearest neighbor clustering for segmentation followed by statistical features like mean, variance, SD, etc. with a Mamdani Fuzzy classifier to identify and display the ancient characters in modern Kannada script.

Table 1 summarizes the various techniques and methodologies to recognize the ancient Indian scripts from the ancient writing material like palm leaf manuscripts and stone inscriptions.

Table 1. Studies on stone inscription/manuscript recognition.

Author	Ancient Language	Proposed Methodology	Result
Siromoney *et al.* [25]	Brahmi Inscription	Coded Run Method	Manually converting each Brahmi character into a rectangular binary array transliterating the recognized characters into Roman. Any script can be recognized using this technique
Devi *et al.* [26]	Brahmi Inscription	ZS thinning, LW thinning algorithm, and WHF thinning algorithm	These algorithms gave considerably good results by preserving the shape of the original image as well.
Anasuya *et al.* [27]	Brahmi Inscription	Global thresholding and local or adaptive thresholding algorithms	Results on Brahmi scripts image from rock inscription obtained by the pixel level processing algorithms.

(continued)

Table 1. Continued

Author	Ancient Language	Proposed Methodology	Result
Warnajith et al. [28]	Brahmi Inscription from Sri Lanka	Modified correlation function method	Modified correlation function method has been proposed which is much more sophisticated than the previous correlation peak method.
Kumar et al. [29]	Ancient Tamil Inscription from Garbarak-shambigai temple	Advanced Maximally Stable Extremal Regions	The author compared the result for the proposed AM-SER method with MSER and MFEAT methods and reported a better recognition rate even with variation in illumination on an inscription image.
Greeba et al. [30]	Ancient Tamil characters from stone inscription	Fuzzy logic-based classifier	The proposed feature set performs well for Tamil characters as it extracts closed regions, stroke convexity, and flat regions from the individual characters. But different language sets would require a different type of feature set
Merline et al. [31]	Tamil Characters from Stone inscriptions and palm leaf from Chola's period	KNN	The segmentation rate of 97.11% and recognition rate of 96.52% has been achieved by the Bagging and KNN ensemble.
Bhuvaneswari et al. [32]	Ancient Tamil Stone inscription images	Normalized Positional Distance Metric (INPDM/ZNPDM) and nearest neighbor classifier.	The recognition result of 84.8% has been achieved for the Tamil script.
Giridhar et al. [33]	Tamil script temple inscriptions between the 7th and 12th centuries.	Convolutional Neural network	Results are presented for various types of inputs e.g., for high-quality printed Tamil characters an accuracy of 91% was achieved, for handwritten characters an accuracy of 70% was obtained and overall accuracy for the system, for various types of inputs, is 77%
Manikandan et al. [34]	Ancient Tamil script from 11th-century stone inscription images	Histogram-based oriented gradient and Support Vector machine.	The results have been presented for the various individual characters such that some of the characters have high recognition accuracy of 100% and some as low as 93.5%. The overall accuracy has been reported as 97.75%

(continued)

180

Table 1. Continued

Author	Ancient Language	Proposed Methodology	Result
Manigandan *et al.* [35]	Tamil Inscriptions images of 9th to 12th century	SIFT (Scale invariant feature transform) with Support Vector machines classifier and Tri-Gram technique for prediction.	The result obtained is not very conclusive and the output displayed shows varying recognition accuracy for the different quality input images.
Aswatha *et al.* [36]	Kannada script inscriptions from Arasikere Temple, Chitradurga Fort, Virupaksha Temple, Shimoga Wooden Palace in Karnataka	HOG feature-based technique by calculating the normalized cross correlation between the HOG feature set of the two images.	The author reported accuracy varying from 83% to 99% for different inscriptions depending on the quality and degradation of the inscription image
Khan *et al.* [37]	Kannada script-based Hoysala, Ganga stone inscriptions in Doddagavanahalli	Mean, Standard Deviation, and sum of absolute (MSDDA)	The author mentioned an accuracy of 98.75% for Hoysala period stone inscription and Ganga period stone inscription characters.
Narang *et al.* [38]	Devanagari-based ancient manuscripts	Feature based on HOG and Cosine Transform with SVM using RBF kernel, decision tree and Naïve Bayes as classfier.	Using an SVM classifier with RBF kernel and DCT zigzag feature of vector size 100, the recognition accuracy of 90.7% has been achieved.
Sachdeva *et al.* [39]	Devanagari	Edge Histogram feature extraction and SVM, Social Media Optimization technique, multi-Layer perceptron, and Simple Logistic classifiers.	Various classification techniques have different accuracy with the highest accuracy of 99.8% with SVM.
Narang *et al.* [40]	Devanagari-based ancient manuscripts	Statistical features with Classifier like CNN, NN, MLP, RBF-SVM, RF.	A recognition accuracy of 88.95% has been achieved using statistical features and based on the majority voting for the classifiers.
Shah *et al.* [41]	Sanskrit manuscript	OpenNMT (Open-source neural machine translation) attention-based encoder-decoder model.	By varying the various hyperparameters accuracy of 99.44% has been mentioned for the Sanskrit test set.
Dwivedi *et al.* [42]	Sanskrit manuscript	Attention-based LSTM model	Using Character Error Rate (CER) and Word Error Rate (WER) for performance evaluation, the proposed LSTM model achieved lower ER as compared to CNN-RNN model, Tesseract, and commercial OCR system Ind.Senz. The proposed system achieved CER of 3.71% and WER of 15.97%

(continued)

181

Table 1. Continued

Author	Ancient Language	Proposed Methodology	Result
Sreedevi et al. [43]	Various scripts e.g., Hampi inscription	Natural gradient-based flexible ICA (Independent Component Analysis) technique with Sobel edge detection	The result of the validation of the proposed method with various inscription images from different historical sites have been presented such the accuracy of 75.5% for recognition of word and 86.7% for character recognition has been achieved
Palaniappan et al. [44]	Indus Script (undeciphered scripts) images from Roja Muthiah Research Library	CNN	The accuracy of graphemes for text regions was reported around 86% and for the image, regions was around 68%
Soumya et al. [5]	Brahmi and Hoysala Script	Statistical features like mean, variance, SD etc. with a Mamdani Fuzzy classifier	Experimental results presented by the author show a segmentation rate of around 90% for both scripts, whereas the recognition rate is better for Brahmi than Hoysala. The recognition success rate is presented using a graphical display for various sample sizes of combined sets of characters for each script

Character recognition technology for handwritten Indian documents is also an emerging field and applications having a high level of accuracy for practical use are very limited. Ancient Indian scripts and regional language recognition are in development but are still at a nascent stage as most of the techniques work well for a specific type of image and may not perform well on another manuscript for the same script.

5 CONCLUSION

OCR systems for modern-day scripts and printed characters are widely available but developing a similar system for ancient text and scripts is a challenging task. Based on the survey, there is no single or combination of features and models which can perform equally well across multiple scripts and different authors have proposed different feature sets and classification systems for various ancient scripts. The survey covers different types of features like statistical, structural, and moments with classifiers ranging from SVMs, clustering models, CNN models, etc. The ancient scripts are also limited by the availability of a standard dataset which is important to efficiently train and developed a classification model. The results are quite variable and significantly depend on the feature set and classification technique used. The finding presented in the survey can be further used across different scripts and languages. Convolution Neural network is an emerging field in image processing which can be utilized for tasks like segmentation and classification of scripts. Moreover, with the advent of new techniques and systems like one-shot learning, transformers, and stable diffusions for image generation systems, Ancient Character Recognition can also overcome major shortcomings of limited dataset such that these techniques can be customized or updated to handle OCR for ancient scripts.

REFERENCES

[1] Sampath V. R. *Quantifying Scribal Behavior: A Novel Approach To Digital Paleography*, 2016. Retrieved from http://hdl.handle.net/10023/9429

[2] Kaur S., & Sagar B.B. Brahmi Character Recognition based on SVM (Support Vector Machine) Classifier Using Image Gradient Features, *Journal of Discrete Mathematical Sciences and Cryptography*, 22(8), 1365–1381, 2019. DOI:10.1080/09720529.2019.1692445.

[3] Pal U., Jayadevan R., & Sharma N. Handwriting Recognition in Indian Regional Scripts: A Survey of Offline Techniques. *ACM Trans. Asian Lang. Inf. Process. (TALIP)* 11(1), 1–35, 2012.

[4] Kaur S., & Sagar B.B. Efficient Scalable Template-matching Technique for Ancient Brahmi Script Image. *Computers, Materials & Continua*, 74(1), 1541–1559, 2023.

[5] Soumya A. & Kumar G. H. Recognition of Ancient Kannada Epigraphs using Fuzzy-based Approach. *2014 International Conference on Contemporary Computing and Informatics (IC3I)*, 657–662, 2014. doi: 10.1109/IC3I.2014.7019645.

[6] Chaudhuri A., Mandaviya K., Badelia P., & Ghosh S. K. (2017). *Optical Character Recognition Systems for Different Languages with Soft Computing*, Vol. 352. https://doi.org/10.1007/978-3-319-50252-6.

[7] Vats S., Singh S., Kala G., Tarar R. & Dhawan S. iDoc-X: An Artificial Intelligence Model for Tuberculosis Diagnosis and Localization, *J. Discret. Math. Sci. Cryptogr.*, 24(5), 1257–1272, 2021.

[8] Vats S., Sagar B.B., Singh K., Ahmadian A. & Pansera B. A., "Performance Evaluation of an Independent Time Optimized Infrastructure for Big Data Analytics that Maintains Symmetry," *Symmetry (Basel).*, 12(8), 2020, doi: 10.3390/SYM12081274.

[9] Vats S. & Sagar B.B., "An Independent Time Optimized Hybrid Infrastructure For Big Data Analytics," *Mod. Phys. Lett. B*, 34(28), 2050311, 2020, doi: 10.1142/S021798492050311X.

[10] Vats S. & Sagar B.B., "Performance Evaluation of K-means Clustering on Hadoop infrastructure," *J. Discret. Math. Sci. Cryptogr.*, 22(8), 2019, doi: 10.1080/09720529.2019.1692444.

[11] Vats S. and Sagar B. B., "Data Lake: A Plausible Big Data Science for Business Intelligence," 2019.

[12] Bhatia M., Sharma V., Singh P. & Masud M., "Multi-level P2P Traffic Classification using Heuristic and Statistical-based Techniques: A Hybrid Approach," *Symmetry (Basel).*, 12(12), 2117, 2020.

[13] Bhati J.P., Tomar D & Vats S., "Examining Big Data Management Techniques for Cloud-based IoT Systems," In *Examining Cloud Computing Technologies Through the Internet of Things, IGI Global*, 164–191, 2018.

[14] Sharma V. et al., "OGAS: Omni-directional Glider Assisted Scheme for Autonomous Deployment of Sensor Nodes in Open Area Wireless Sensor Network," *ISA Trans.*, 2022, doi: 10.1016/j.isatra.2022.08.001.

[15] Agarwal R., Singh S. & Vats S., "Implementation of an Improved Algorithm for Frequent Itemset Mining using Hadoop," In *2016 International Conference on Computing, Communication and Automation (ICCCA)*, 13–18, 2016. doi: 10.1109/CCAA.2016.7813719.

[16] Sharma V., Patel R.B., Bhadauria H.S. & Prasad D., "NADS: Neighbor Assisted Deployment Scheme for Optimal Placement of Sensor Nodes to Achieve Blanket Coverage in Wireless Sensor Network," *Wirel. Pers. Commun.*, 90(4), 1903–1933, 2016.

[17] Agarwal R., Singh S. & Vats S., *Review of Parallel Apriori Algorithm on Mapreduce Framework for Performance Enhancement*, 654. 2018. doi: 10.1007/978-981-10-6620-7_38.

[18] Sharma V., Patel R.B., Bhadauria H.S. & Prasad D., "Policy for Planned Placement of Sensor Nodes in Large Scale Wireless Sensor Network," *KSII Trans. Internet Inf. Syst.*, 10(7), 3213–3230, 2016.

[19] Sharma V., Patel R.B., Bhadauria H.S. & Prasad D., "Deployment Schemes in Wireless Sensor Network to Achieve Blanket Coverage in Large-scale Open Area: A Review," *Egypt. Informatics J.*, 17 (1), 45–56, 2016.

[20] Vikrant S., Patel R.B., Bhadauria H.S. & Prasad D., "Glider Assisted Schemes to Deploy Sensor Nodes in Wireless Sensor Networks," *Rob. Auton. Syst.*, 100, 1–13, 2018.

[21] Ubul K., Tursun G., Aysa A., Impedovo D., Pirlo G., & Yibulayin T. *Script Identification of Multi-Script Documents: A Survey. IEEE Access*, 5, 6546–6559, 2017. https://doi.org/10.110

[22] Sharma R. & Kaushik B. Offline Recognition of Handwritten Indic Scripts: A State-of-the-art Survey and Future Perspectives. *Computer Science Review* 38: 100302, 2020.

[23] Kakwani D., Kunchukuttan A., Golla S., Gokul N. C., Bhattacharyya A., Khapra M. M., & Kumar P. IndicNLPSuite: Monolingual Corpora, Evaluation Benchmarks and Pre-trained Multilingual Language Models for Indian Language. In *Proceedings of the 2020 Conference on Empirical Methods in Natural Language Processing: Findings*, 4948–4961, 2020.

[24] Philips J. & Tabrizi N. Historical Document Processing: Historical Document Processing: *A Survey of Techniques, Tools, and Trends. arXiv*:2002.06300, 2020.

[25] Siromoney G., Chandrasekaran R. & Chandrasekaran M. *Machine Recognition of Brahmi Script, IEEE Transactions on Systems, Man, and Cybernetics*, 4, 648–654, 1983.

[26] Devi H. K. A. *Thinning*: A Preprocessing Technique for an OCR System for the Brahmi Script. *Ancient Asia*, 1, 0, 167–172, 2006a, DOI:http://doi.org/10.5334/aa.06114

[27] Devi H. K. A. Thresholding: A Pixel-Level Image Processing Methodology Preprocessing Technique for an OCR System for the Brahmi Script. *Ancient Asia*, 2006b

[28] Warnajith N., Bandara D., Quarmal S. B., Minato A., & Ozawa S. *Image Processing Approach for Ancient Brahmi Script Analysis. Heritage as Prime Mover in History, Culture and Religion of South and Southeast Asia*, Sixth International Conference of the South and Southeast Asian Association for the Study of Culture and Religion (SSEASR), Center for Asian Studies of the University of Kelaniya, Sri Lanka. (Abstract) 69, 2015.

[29] Kumar A. N. Character Recognition of Ancient South Indian language with conversion of modern language and translation. *A Journal of Composition Theory*. 94–107, 12(10), 2019, ISSN: 0731–6755.

[30] Greeba M.V. Recognition of Ancient Tamil Characters in Stone Inscription Using Improved Feature Extraction. *International Journal of Recent Development in Engineering and Technology*, 2(3), 38–41, 2014, ISSN 2347–6435 (Online).

[31] Merline M. M., & Santhi M. Ancient Tamil Character Recognition from Epigraphical Inscriptions using Image Processing Technique. *Journal of Telecommunication Study*, 4(2), 40–48, 2019.

[32] Bhuvaneswari G., & Bharathi V. S. An Efficient Positional Algorithm for Recognition of Ancient Stone Inscription Characters. *2015 Seventh International Conference on Advanced Computing (ICoAC)*, 1–5, 2015, doi: 10.1109/ICoAC.2015.7562798.

[33] Giridhar L., Dharani A., & Guruviah V. *A Novel Approach to OCR using Image Recognition based Classification for Ancient Tamil Inscriptions in Temples. ArXiv, abs/1907*.04917, 2019.

[34] Bhuvneswari G., & Manikandan G. Recognition of Ancient Stone Inscription Characters Using Histogram of Oriented Gradients. *SSRN Electronic Journal*, 2019. 10.2139/ssrn.3432300.

[35] Manigandan T., Vidhya V., Dhanalakshmi V. & Nirmala B. Tamil Character Recognition from Ancient Epigraphical Inscription using OCR and NLP. 2017 *International Conference on Energy, Communication, Data Analytics and Soft Computing (ICECDS)*, 1008–1011, 2017. doi: 10.1109/ICECDS.2017.8389589.

[36] Aswatha S.M., Talla A.N., Mukhopadhyay J., & Bhowmick P. *A Method for Extracting Text from Stone Inscriptions Using Character Spotting.* In: Jawahar C., Shan S. (eds) Computer Vision - ACCV 2014 Workshops. ACCV 2014. *Lecture Notes in Computer Science*, Springer, Cham, vol. 9009, 2015. https://doi.org/10.1007/978-3-319-16631-5_44.

[37] Khan I., Elizabeth A., Megha S. G., Prakruti K. S., & Gowthami G. R. Read and Recognition of old Kannada Stone Inscriptions Characters using MSDD Algorithm. *International Journal of Engineering Research & Technology (IJERT) RTESIT*, 7(8), 2019.

[38] Narang S., Jindal M. K. & Kumar M. Devanagari *ancient documents recognition using statistical feature extraction techniques. Sādhanā* 44(6), 1–8, 2019.

[39] Sachdeva J., Mittal S., & Yean L. Handwritten Offline Devanagari Compound Character Recognition Using CNN, 2022. 10.1007/978-981-16-6289-8_18.

[40] Narang S. R., Jindal M. K., & Sharma P. Devanagari Ancient Character Recognition using HOG and DCT Features. *2018 Fifth International Conference on Parallel, Distributed and Grid Computing (PDGC)*, 215–220, 2018. doi: 10.1109/PDGC.2018.8745903.

[41] Shah R., Gupta M. K., & Kumar A. (2021). Ancient Sanskrit Line-level OCR using OpenNMT Architecture. *2021 Sixth International Conference on Image Information Processing* (ICIIP), 347–35, 2021. doi: 10.1109/ICIIP53038.2021.9702666.

[42] Dwivedi A., Saluja R., & Sarvadevabhatla R. K. An OCR for Classical Indic Documents Containing Arbitrarily Long Words. 2020 *IEEE/CVF Conference on Computer Vision and Pattern Recognition Workshops (CVPRW)*, 2386–2393, 2020. doi: 10.1109/CVPRW50498.2020.00288.

[43] Sreedevi I., Pandey R., Jayanthi N., Bhola G., & Chaudhury S. NGFICA Based Digitization of Historic Inscription Images. *ISRN Signal Processing*, 2013. 10.1155/2013/735857.

[44] Palaniappan S., & Adhikari R. *Deep Learning the Indus Script*, 2017. Retrieved from http://arxiv.org/abs/1702.00523.

Automation and Computation – Vats et al. (Eds)
© *2023 the Author(s), ISBN 978-1-032-36723-1*

A review of Keras' deep learning models in predicting COVID-19 through chest X-rays

G. Singh, A. Garg, V. Jain & R. Musheer
VIT Bhopal University, Bhopal, Madhya Pradesh, India

ABSTRACT: The advent of the COVID-19 pandemic in 2020 took the world by surprise. The pandemic spread quickly, infecting and killing millions of people. The huge number of infections put immense pressure on practitioners to ensure quick diagnosis and treatment. The healthcare workers stood outnumbered and overwhelmed. While RT-PCR is primarily used for the detection of infection, chest imaging can also be important in making an accurate diagnosis. For making a diagnosis through the chest x-rays (CXR), the presence of a radiologist is essential. This often delays diagnosis and treatment. Artificial Intelligence (AI) and Deep Learning (DL) techniques are growingly used for medical imaging. Convolution Neural Networks (CNN) is a DL technique used for object classification and processing. This paper presents a review of Keras' Convolutional Neural Networks (CNN) that can be used to detect COVID-19.The paper aims to find a model with good testing accuracy. Based on the findings, a model that can be employed to aid medical practitioners and accelerate the process of providing treatment is suggested. Among the found results, Inception V3 has demonstrated the highest training and testing accuracy.

1 INTRODUCTION

John McCarthy introduced the term "Artificial Intelligence" in the mid-1950s. Stated that "It is the science and engineering of making intelligent machines, especially intelligent computer programs" (McCarthy 2004). Humans have come up with ideas and inventions as far back as ~ 400 BC. These ideas have been the stepping stones to bringing AI where it is today. The Greek philosopher Aristotle introduced a method to map logical decision-making and called it "syllogism" (Russell 1995). We can think of artificial intelligence as the ability of machines to reason and make decisions when faced with a problem with the same prowess as humans. The exponential progress of AI in the past two decades can be credited to the massive and readily available datasets that enabled more experimenting. AI has been used for various purposes like processing and mining of big data (Bharti *et al.* 2022; Vats et al. 2019, 2020, 34), making fraud-proof security systems (Vats et al. 2012; Vats et al 2013), improving healthcare (Vats et al. 2021) etc.

In healthcare, (Tran et al. 2019) and (Rong et al. 2020) have noted that from 1999–2018 the number of publications has increased significantly. The pressure on practitioners to diagnose and prevent diseases is constantly increasing. There is an urgent need to find methods that can assist these medical practitioners. In the recent past, Artificial Intelligence (AI) was increasingly being used for image recognition/ object detection tasks. Deep Learning (DL), especially Convolution Neural Networks (CNNs) was extensively used for the detection of disease during medical imaging (Vats et al. 2020).

Since the end of 2019, over 500 million people have been infected and 6 million people have died because of COVID-19 (SARS-CoV2). People infected with SARS-CoV2 often

DOI: 10.1201/9781003333500-21

develop Acute Respiratory Distress syndrome. It can also cause pulmonary fibrosis of the lungs (Sun et al. 2020). The most common method of diagnosing COVID-19 is through laboratory testing through the identification of viral RNA in reverse transcriptase polymerase chain reaction (RT-PCR). When RT-PCR is not available then Chest X-Ray (CXR) has been identified as a suitable way of making a diagnosis. The WHO suggests using chest imaging for diagnosis in the following cases, (1) RT-PCR testing is not available (2) RT-PCR testing is available, but results are delayed, and (3) initial RT-PCR testing is negative but with high clinical suspicion of COVID-19 (WHO 2020).

The pandemic in 2020 taught us that everything could change in the blink of an eye. The CXR is an effective with low-cost and moderate radiation to screen lung abnormalities (Oloko-Oba *et al.* 2022). The world also faced an acute shortage of radiologists during the pandemic. In an article, the Radiology Society of North America reported that Europe has 13 radiologists per 100,000 population and the UK has only 8.5 radiologists per 100,000 people (Hendorson *et al.* 2022). In a study conducted by Hegde *et al.*, a questionnaire was prepared aimed at enquiring about the effect of COVID-19 on radiology practices and radiologists. Close to half of them reported substantial deterioration of their well-being. This could have been because of the increased workload, the anxiety of contracting the disease and further transmission to their families, and uncertainty about the future. More than 60% of the respondents believed that there had been adverse effects on further training and Continuing Professional Development (CPD activities of radiologists (Hegde et al. 2020).

Since it is easy to differentiate between a healthy and an unhealthy CXR, they can be an easy way to detect COVID-19. Due to an acute shortage of radiologists resulting in increased workload, the process of making a diagnosis and providing treatment is also prolonged. This study reviews different pre-trained models with an aim to shorten the duration between taking a CXR of the patient and providing treatment. Moreover, the study also hopes to help healthcare workers who are under immense pressure due to their ever-increasing workload.

2 LITERATURE REVIEW

Several technological advancements have taken place to aid medical practitioners in increasing the efficacy of treatment and diagnosis. One of them includes using deep convolutional neural networks for the diagnosis of diseases through image processing.

Akgül et al. (2022) have used VGG16 and ResNet50 for the diagnosis of COVID-19. They have also proposed an alternative model called IsVoNet. They used a dataset consisting of chest x-rays of 6,157 patients. Since the number of parameters affects the performance and time complexity of the model, in the new model, they have reduced the parameters by at least 5 times. Despite decreasing the number of parameters in IsVoNet, it has demonstrated high accuracy of 99.76% which is comparable to that of other models like VGG16 having an accuracy of 99.92%, and ResNet having an accuracy of 99.78% (Akgül et al. 2022).

Additionally, Appari *et al.* (2022) have employed pre-trained as well as deep learning models for quick diagnosis of COVID-19. VGG16 is used along with XGBoost on a dataset consisting of 1,526 covid and non-covid images. The results depended on the number of images used. An f1 score of 98% is achieved. A deep learning model based on U-Net architecture is also used and using the U-Net-based model, which uses data augmentation, an accuracy of 99.86% is attained (Appari *et al.* 2022).

Similarly, Chauhan et al. (2021) have used transfer learning with DenseNet-121 architecture to enhance the performance of the model to ensure the efficient diagnosis of COVID-19. Here the model is fine-tuned by early stopping. It is done by selecting the model that performs better under certain conditions. The model training is stopped at a low validation loss to achieve high accuracy. Then a comparative analysis is performed among optimizers like Adamax, Adam, AdamW, and stochastic gradient descent along with various loss functions like Cross Entropy, NLLLoss, etc (Chauhan et al. 2021).

Furthermore, Das *et al.* have used deep learning-based models to classify COVID-19. Three pre-trained models are used – ResNet50 V2, DenseNet121, and InceptionV3. CXRs are used from various open-source datasets to train their model. The images are normalized and resized and then split into training, testing, and validation sets. All models are trained for 60 epochs. The model with the use of Adam optimizer and Stochastic Gradient Descent was able to yield an accuracy of 91.62%. Their prototype needs a GUI to help medical practitioners make easy computer-based analyses (Das et al. 2021).

Fan *et al.* have used DenseNet for the diagnosis of COVID-19 using chest x-rays. They evaluated the reliability of the results obtained from a deep learning model. The Score-CAM method has been used to get rid of any dependence on the gradient. The activation mask obtained from Score-CAM is normalized and multiplied by the original image. The final result is obtained by the linear combination of weights and an activation graph. They have combined ProtoPN*et along* with DenseNet for final classification. It has been concluded that deep learning methods are reliable for making diagnoses when there is enough, clean and diverse data to train the model (Fan *et al.* 2020).

Kaur, Harnal, *et al.* have proposed a system that uses a deep learning model that uses Inception V4 and Multiclass SVM classifier for detection of COVID-19. The classifier detects and classifies the infection into 4 different classes. The classes are "normal", "viral pneumonia", "COVID-19" and "bacterial pneumonia". The images used for training the model are from a publicly accessible database. In comparison, the proposed model C19-Net performs better than Inception V4. C19-Net has managed to attain an accuracy of 96.24% (Kaur 2021).

Shazia, Anis *et al.* have performed a comparative analysis on various pre-trained deep learning models to find the best CNN for the detection of COVID-19. The models VGG16, VGG19, DenseNet121, Inception-ResNet-V2, Inception V3, ResNet50, and Xception have been used. On taking the loss and accuracy into account, It has been observed that DenseNet and ResNet have the highest accuracy with the lowest validation loss (Shazia et al. 2021).

The next sections of this paper include the Methods Used that have the workflow, a brief discussion of the 5 pre-trained models, and the dataset used. It is followed by Result and Discussion that compares the training and testing accuracies of the 5 models used. The Conclusion mentions the findings of the study.

3   METHODS USED

For the diagnosis of COVID-19, the model uses neural networks and deep learning models to scan chest x-rays. The sigmoidal activation function is then used to give a binary output. The architecture of a neural network is inspired by the smallest unit of the nervous system-the neuron. Neural networks aid in learning and optimization. Deep learning is a toolbox of methods that primarily uses neural networks to develop models that can learn independently. Neural networks learn by changing the weight associated with different input values. The weight can be thought of as the output's sensitivity to a given parameter. The higher the weight associated with a parameter, the greater its importance in determining the output. Neural networks are able to learn these weights and bring the error to zero using gradient descent. In the training phase, the model is first fed to a dataset that has both input and output. This allows the network to find optimal weights that minimize error. The model is then 'tested' on a separate dataset to see if the model is making accurate predictions after it has been trained. A neural network is able to make accurate predictions by virtue of its ability to learn itself using huge databases.

ANNs are highly susceptible to overfitting i.e. instead of learning parameters from the data that help in the classification, they learn the data itself.

A convolution neural network is robust to the problem of overfitting by decreasing the number of parameters. It contains an input and an output layer with several hidden layers in between. As shown in Figure 1, the hidden layers consist of convolution and pooling layers with multiple fully connected layers at the end. TensorFlow is a deep learning framework used to decrease code complexity and increase runtime performance. Keras is a library of convoluted neural networks that run with the aid of the TensorFlow framework.5 CNNs from the Keras library are used to find out which neural network is best for the prediction of COVID-19.

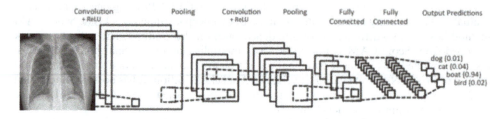

Figure 1. Architecture of a CNN.

3.1 *Models used*

VGG is one of the most popular CNNs used for image classification. It has an accuracy of about 91.8%. It has a convolution filter of 3x3 with a stride of 1. This has significantly increased the depth of the model. Its maxpool layer is 2x2 with a stride of 2. It has been trained (Simonyan *et al.* 2014) on The ReLu activation function is used in each hidden layer. VGG16 has been trained on the ImageNet database that contains 12 subtrees with 5247 synsets and 3.2 million images in total (Deng 2009). The model has 13 convolution layers, 5 max pooling layers, and 3 dense layers. All of this adds to 21 layers. But it only has 16 weighted (containing trainable parameters) layers, hence it is called VGG16 (Gopalakrishnan et al. 2017).

ResNet50 is a deep convolution layer network that is 50 layers deep. ResNet stands for Resolution Networks and is extensively used in tasks relating to computer vision. Akin to VGG16, it also has pretrained weights on the ImageNet database. As the CNN gets deeper, there is a problem with the gradient of the picture diminishing with every subsequent layer. ResNet50 addresses this by use of shortcut connections as shown in the figure. Shortcut connections skip multiple layers and perform identity mapping (He et al. 2016). It involves skipping over some layers of the network to add the input to the output of the CNN. This does not increase computing complexity or introduce new parameters.

DenseNet is another CNN made to address the problem of vanishing gradient with an increase in the depth of the network. While ResNet addresses this problem by connecting subsequent layers, DenseNet connects each and every layer of the network with all other layers present. By doing this, every successive layer receives inputs from all the layers preceding it. While ResNet adds (+) the output of the previous layers, DenseNet concatenates (.) it. The availability of all output gradients from previous layers and original input ensures high trainability by improving the flow of information.

Uniformly increasing the size of the neural network to increase its performance comes with drawbacks. One of them includes the dramatic increase in computational resources. It also increases the chances of overfitting because of the increase in the number of parameters (Szegedy et al. 2015).

InceptionV3 uses Factorized Convolutions to reduce the cost of computation by reducing the number of parameters and using asymmetric convolutions (Szegedy et al. 2016). It uses smaller convolution to decrease training time (Szegedy et al. 2016). It also uses the Auxiliary Classifier as a regularize and uses a grid of the architecture.

This was designed to scale up ConvNets which can achieve better accuracy and efficiency (Tan *et al.* 2019). It has been done by scaling all the parameters of image resolution, depth, and width. This is done by finding a scaling coefficient by performing a grid search. The scaling coefficient is then applied to the baseline network to achieve desired computational expense. A new baseline network was designed using AutoML MNAS framework to further improve the goal of increasing accuracy and efficiency (Tan 2019).

3.2 *Dataset*

The Covid-19 Radiograph Dataset from Kaggle used in this study has been compiled by a team of researchers from Qatar University, Doha, Qatar, and the University of Dhaka, Bangladesh along with collaborators from Pakistan and Malaysia (Covid-19 Radiograph Dataset).

From the dataset, we have used a total of 7,625 images. 4,013 CXR images are of healthy lungs and 3,621 images are of lungs with COVID-19. Out of all the images, 70% of the images are used to train the model, 10% to test the model, and 20% to validate it. All the CXRs are verified to ensure that only high-quality and clear scans are included in the database used to train the model. The COVID-19 and normal images have also been verified by experienced medical practitioners.

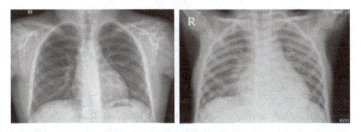

Figure 2.   (a) Healthy lungs; (b) lungs of COVID-19 patients.

3.3 *Workflow and equations*

Keras is an open-source Python library for developing and evaluating deep learning models. It is a collection of deep learning models, consisting of pre-trained weights that automatically get downloaded when installing a model. Before training, the images are synthetically augmented by applying various image transformation techniques such as rotation with 20, shifting (width and height with 0.1), zoom range set to 0.1, and shearing set to 0.1.

A Deep Learning computer model learns to perform a set of tasks using an input such as images, sound, or text. Its general architecture consists of an input layer, hidden layers, and an output layer. The input layer contains the image data. The image is represented in form of a three-dimensional matrix. The matrix is reshaped into a single column before feeding it to the neural network.

The hidden layers consist of the Convo Layer, the Pooling Layer, and the Fully Connected layer. In the Convo layer, the convolution operation is performed to calculate the dot product between the receptive field and the filter. The filter slides over the input matrix by a stride until the whole input is covered. This output becomes the input for the next layer. The ReLU activation function makes all the negative values zero. The pooling layer makes the computation economic by reducing the volume of the input after the Convo layer. The Fully Connected layer contains neurons, weights and biases to sort images into different classes. The activation function is at the end of the Fully Connected layer. This is a binary classification problem since the CXRs need to be classified as COVID-19 positive or COVID-19 negative. Binary classification problems classify the output into one of the two classes. The Sigmoid function is ideal since its range lies between (0, 1).

Equation 1 is that of the Sigmoidal Activation Function.

$$\sigma(x) = \frac{1}{1 + e^{-x}} \tag{1}$$

Equation 2 shows the derivative of Equation 1.

$$\sigma'(x) = \frac{e^{-x}}{(1 + e^{-x})^2} \tag{2}$$

Equation 3 shows a more simplified form of Equation 2 that can be used in Machine Learning.

$$\sigma'(x) = \frac{1}{(1 + e^{-x})} \cdot \left(1 - \frac{1}{1 + e^{-x}}\right) \tag{3}$$

Equation is derived on rewriting Equation 3 with refence to Equation 1.

$$\sigma'(x) = \sigma(x)(1 - \sigma(x)) \tag{4}$$

The Output Layer contains labels that are assigned to the input at the end of the classification process.

4 RESULTS AND DISCUSSION

5 different CNNs are used, namely – VGG16, DenseNet, ResNet50, Inception v3, and EfficientNet B0 from Keras. The models are run on a system having 16GB RAM, with a Windows 10 operating system having an intel i7 processor. Experiments are performed for each model separately to find the training accuracy, testing accuracy, loss, precision, and recall as shown in Table 1.

Table 1. Compilation of results.

Model	Training Accuracy (%)	Testing Accuracy (%)	Precision (%)	Recall (%)	Training Time (s)	Testing Time (s)	Parameters (in Millions)
VGG16	92.59	82.75	83.59	84.50	46885	7.33	138 M
ResNet 50	90.33	83.23	82.00	81.30	36223	6.83	23 M
DenseNet	93.79	87.13	84.91	85.00	36237	6.02	7.1M
Inception V3	96.12	89.46	89.74	87.50	27486	4.34	41.2 M
EfficeintNet B0	94.85	85.50	86.79	86.21	16860	4.83	11 M

The dataset used to run the experiments are divided into three parts – training, testing, and validation. Each model is run for 200 epochs. From the experiments and as shown in Table 1, it is evident that inception v3 has the highest training and testing accuracy at 96.12% and

89.46% respectively. Even though Inception v3 has a fraction of parameters as compared to VGG 16, it performs significantly better. DenseNet is close behind InceptionV3 with a testing accuracy of 87.13%. ResNet has performed with training and testing accuracy of 90.33% and 83.23%. EfficientNet achieved testing accuracy comparable to that of DenseNet within a significantly smaller amount of time and lesser parameters as shown in Table 1.

Figure 3 below shows the comparison between the testing accuracy, training accuracy, precision, and recall of the 5 CNNs used in this study.

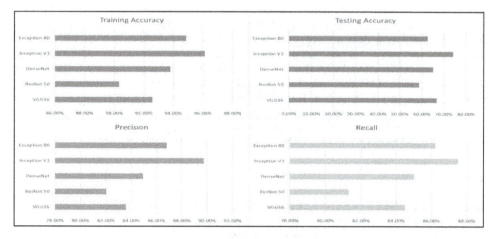

Figure 3.

Below, Figure 4 shows the confusion matrices for all 5 models used in this study.

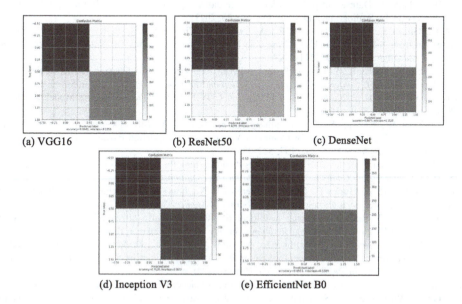

(a) VGG16   (b) ResNet50   (c) DenseNet

(d) Inception V3   (e) EfficientNet B0

Figure 4. Confusion matrices.

From both the ROC and Confusion matrices, it is clear that Inception V3's performance stands out in diagnosing COVID-19 out of the selected models. This model can be used to aid medical practitioners by fast-tracking the process of making a diagnosis and providing early treatment to patients.

5 CONCLUSION

The COVID-19 pandemic devastated the world and overwhelmed health sectors across the world. It was most likely to be contracted by people in close proximity to an infected person. At its peak, the virus became airborne (Zhang et al. 2020). This created the need for rapid diagnosis, and due to this radiologists faced an exponential increase in their workload. In this paper, experiments are performed to find a more efficient way to make a diagnosis. Five deep learning convolutional neural networks (CNN) are utilized to diagnose COVID-19 in chest radiographs. Different pre-trained models are selected for evaluation including VGG16, Inception V4, DenseNet, ResNet 50, and EfficentNet B0 with 16, 48, 121, 50, and 237 layers each. From the experiments, Inception V3 has demonstrated the highest accuracy during both training and testing. It obtained a testing score of 89.46%. Inception v3 is therefore a good model to make a COVID-19 diagnosis using chest x-rays.

REFERENCES

Akgül I., Kaya V., Ünver E., Karavaş E., Baran A., & Tuncer S. 2022. Covid-19 Detection on x-Ray Images Using a Deep Learning Architecture. *Journal of Engineering Research*. 1–16.

Appari N. V. L., & Mahendra G.K. 2022. Soft Computing and Image Processing Techniques for COVID-19 Prediction in Lung CT Scan Images. *International Journal of Hybrid Intelligent Systems* 18(1–2):111–131.

Bhati J. P., Tomar D., & Vats S. 2018. Examining Big Data Management Techniques for Cloud-based IoT Systems. *Examining Cloud Computing Technologies Through the Internet of Things, IGI Global* 164–191.

Chauhan T., Palivela H., & Tiwari S. 2021. Optimization and Fine-Tuning of DenseNet Model for Classification of COVID-19 Cases in Medical Imaging. *International Journal of Information Management Data Insights* 1(2):100020.

Covid-19 Radiograph Dataset. https://www.kaggle.com/datasets/preetviradiya/covid19-radiography-dataset? select=COVID-19_Radiography_Dataset (October 25, 2022).

Das A.K., Ghosh S., Thunder S., Dutta R., Agarwal S., & Chakrabarti A. 2021. Automatic COVID-19 Detection from X-Ray Images Using Ensemble Learning with Convolutional Neural Network. *Pattern Analysis and Applications* 24(3):1111–1124.

Deng J., Dong W., Socher R., Li L. J., Li K., & Fei-Fei L. 2009. Imagenet: A Large-scale Hierarchical Image Database. *2009 IEEE Conference on Computer Vision and Pattern Recognition*. 248–255.

Fan B. B., and Yang H. 2020. Analysis of Identifying COVID-19 with Deep Learning Model. *Journal of Physics: Conference Series* 1601(5).

Gopalakrishnan K., Khaitan S. K., Choudhary A., & Agrawal A. 2017. Deep Convolutional Neural Networks with Transfer Learning for Computer Vision-based Data-driven Pavement Distress Detection. *Construction and Building Materials*, 157, 322–330.

He K., Zhang X., Ren S., & Sun J. 2016. Deep Residual Learning for Image Recognition. In *Proceedings of the IEEE Conference on Computer Vision and Pattern Recognition*. 770–778.

Hegde G., Azzopardi C., Hurley P., Gupta H., Vemuri N. V., James S., & Botchu R. 2020. Impact of the COVID-19 Pandemic on Radiologists. *Indian Journal of Medical Sciences*, 72(3), 177.

Hendorson M. 2022. Radiology Facing a Global Shortage, Specialty affected by COVID-19, Aging Population and Demand for Imaging. *Radiology Society of North America*. Illinois: RSNA.

Kaur P., Harnal S., Tiwari R., Alharithi F.S., Almulihi A.H., Noya I.D. & Goyal N. 2021. A Hybrid Convolutional Neural Network Model for Diagnosis of Covid-19 Using Chest x-Ray Images. *International Journal of Environmental Research and Public Health* 18(22).

McCarthy J. 2004. *What is Artificial Intelligence?* Computer Science Department, Stanford University.

Oloko-Oba M, Viriri S. 2022. A Systematic Review of Deep Learning Techniques for Tuberculosis Detection From Chest Radiograph. *Front Med (Lausanne)*. 10;9:830515.

Rong G., Mendez A., Assi E. B., Zhao B., & Sawan M. 2020. Artificial intelligence in healthcare: review and prediction case studies. *Engineering*, 6(3), 291–301.

Russell S. J., & Norvig P. 1995. *Artificial Intelligence: A Modern Approach*. Englewood Cliffs, New Jersey: Prentice Hall.

Simonyan K. and Zisserman A., 2014. Very Deep Convolutional Networks for Large-scale Image Recognition. arXiv preprint arXiv:1409.1556.

Shazia A., Xuan T.Z., Chuah J.H., Usman J., Qian P. & Lai K.W. 2021. A Comparative Study of Multiple Neural Network for Detection of COVID-19 on Chest X-Ray. *Eurasip Journal on Advances in Signal Processing* 2021(1):1–16.

Sun J., He WT., Wang L., Lai A., Ji X., Zhai X., Li G., Suchard MA., Tian J., Zhou J., Veit M., Su S. 2020. COVID-19: Epidemiology, Evolution, and Cross-Disciplinary Perspectives. *Trends Mol Med.* 26(5):483–495.

Szegedy C., Liu W., Jia Y., Sermanet P., Reed S., Anguelov D., ... & Rabinovich A. 2015. Going deeper with convolutions. In *Proceedings of the IEEE Conference on Computer Vision and Pattern Recognition*. 1–9.

Szegedy C., Vanhoucke V., Ioffe S., Shlens J., & Wojna Z. 2016. Rethinking the Inception Architecture for Computer Vision. In *Proceedings of the IEEE Conference on Computer Vision and Pattern Recognition*. 2818–2826.

Tan M., & Le Q. 2019. Efficientnet: Rethinking Model Scaling for Convolutional Neural Networks. *International Conference on Machine Learning*. 6105–6114.

Tan M. 2019. EfficientNet: Improving Accuracy and Efficiency through AutoML and Model Scaling. *Google Researc.* May, 2019 [Blog]. Available at https://ai.googleblog.com/2019/05/efficientnet-improving-accuracy-and.html (Accessed October 25, 2022).

Tran B. X., Vu G. T., Ha G. H., Vuong Q. H., Ho M. T., Vuong T. T. & Ho R. C. 2019. Global Evolution of Research in Artificial Intelligence in Health and Medicine: A Bibliometric Study. *Journal of Clinical Medicine*, 8(3), 360.

Vats S., Dubey S.K. & Pandey N.K. 2012. *Criminal Face Identification System*. Koln: LAP Lambert Academic Publishing.

Vats S., Dubey S.K. & Pandey N.K. 2013. Genetic Algorithms for Credit Card Fraud Detection. *2013 International Conference on Education and Educational Technologies*. 42–53.

Vats S. & Sagar B. B. 2019. Data lake: A Plausible Big Data Science for Business Intelligence. *Communication and Computing System*. CRC Press.

Vats S. & Sagar B.B. 2019. Performance evaluation of K-means Clustering on Hadoop Infrastructure. *Journal of Discrete Mathematical Sciences and Cryptography*. vol. 22, no. 8.

Vats S. & Sagar B.B. 2020. An Independent Time Optimized Hybrid Infrastructure for Big Data Analytics. *Modern. Physics Letters B*. 2050311, Oct. 2020.

Vats S., Sagar B. B., Singh K., Ahmadian A., & Pansera B.A. 2020. Performance Evaluation of an Independent Time Optimized Infrastructure for Big Data Analytics That Maintains Symmetry. *Symmetry (Basel)*, vol. 12, no. 8.

Vats S. Singh G. Kala R. Tarar, and S. Dhawan. 2021. iDoc-X: An Artificial Intelligence Model for Tuberculosis Diagnosis and Localization. *Journal of Discrete Mathematical Sciences and Cryptography*. 1257–1272.

World Health Organisation. 2020. *Use of Chest Imaging in COVID-19: A Rapid Advice Guide*. Geneva: World Health Organisation.

Zhang R., Li Y., Zhang A.L., Wang Y., Molina M.J. 2020. Identifying Airborne Transmission as the Dominant Route for the Spread of COVID-19. In *Proceedings of the National Academy of Sciences*. 14857–14863.

Automation and Computation – Vats et al. (Eds)
© 2023 the Author(s), ISBN 978-1-032-36723-1

Encryption and decryption technique in optically transformed color images

Satya Bhushan Verma
Shree Ramswaroop Memorial University, Barabanki, India

Manish Maurya
Goel Institute of Technology and Management, Lucknow, India

Nidhi Tiwari & Kartikesh Tiwari
Shree Ramswaroop Memorial University, Barabanki, India

Sandesh
ATME College of Engineering, Mysuru, India

ABSTRACT: We now need quick and reliable security solutions due to the growth in digital communication and multimedia data, such as digital photographs and videos. One technique to secure our data from unauthorized users and achieve high security is via encryption. A mix of picture encryption and image stitching methods is used in this paper's suggested approach to secure the photos. An innovative cryptographic technique based on the combination of multiple optical transformations is suggested in order to address the issue of the low complexity and security of digital picture encryption algorithms. Plain-text pictures and the random matrix are subjected to coding and basic procedures. The enhanced logistic map generates the random matrix. The random sequence generated by the hyper-chaotic system dynamically provides the ways of coding and operation.

1 INTRODUCTION

The digital era has only lately begun to take shape as a result of the enormous advances in communication technology. The dissemination of photos has gained popularity. The images are digital in 70% of the cases. information that was sent online on the other side, modern computer processors have made it trivial to get unwanted access to data sent through the Internet. Image transmission is no longer just used in the everyday lives of common people. In addition to the everyday lives of regular people, image transmission has uses in the military, medical, and industrial sectors.

In these applications, image security against diverse attacks is crucial. The most crucial method of delivering picture security is most often image encryption. Contrarily, the amount of time it takes for the picture to travel from the transmitter to the receiver is crucial because, if the delay exceeds the threshold, irreparable damage might result. Therefore, the encryption mechanism used in such applications must provide high security and minimal operating time in order to meet the mentioned criteria. Therefore, none of the picture encryption techniques in use today are appropriate for these uses. Several common encryption techniques have been suggested for text encryption. These algorithms have poor security and a lengthy encryption time due to the massive volume, correlation between close pixels, and visual data

DOI: 10.1201/9781003333500-22

redundancy. The setting for connection and data exchange, which play a big role in human existence, is correspondence networks.

There are several methods, including cryptography, steganography, and encryption, to assure information security, which is of the utmost importance [20]. To prevent unauthorised access, many approaches are used in cryptography [7]. When using steganography techniques, the main goal is to hide a mystifying message in the majority of covers. The most important computerised information that has ever been used was probably pictures. We need to use a variety of tactics to protect photographs from exploitative exploitation. One of the famous methods that may be used to protect most types of information is encryption. We presented a technique in which picture parameters are dependent upon certain independent factors in order to increase the image security [3,8]. In this research, we presented an optical transformation-based approach to digital colour picture encryption [1].

1.1 Digital image security

Every cell in our body performs a certain function. Ordinary cell division occurs in normal cells. When they get worn out or injured, they pass away, and new cells replace them. When cells begin to proliferate unchecked, this is cancer. The cancer cells continue to multiply and create new cells. They push out healthy cells. The area of the body where the cancer first manifested is affected by this [20].

Other bodily areas may potentially get infected with cancerous cells. For instance, lung cancer cells might go to the bones and develop there. The term "metastasis" refers to the spread of cancer cells. Lung cancer is still referred to as such after it has spread to the bones. For medical professionals, cancer cells in the bones resemble lung cancer cells exactly. Without a bone origin, it is not referred to as bone cancer. Mesothelioma is a cancerous tumour that develops in the lining of the lungs, abdomen, or heart and is brought on by inhaling asbestos fibres. Chest discomfort and shortness of breath are some symptoms. Most mesothelioma patients may expect to live for around a year following their diagnosis. Radiation, chemotherapy, and surgery are all forms of treatment that may improve prognosis. Mesothelioma is an asbestos-related cancer that usually attacks the lining of the lungs. It is incurable. However, mesothelioma tumours may grow in the heart's or abdomen's lining [15].

1.2 Digital image

Pixels are the visual components that make up digital pictures. Ordinarily, pixels are arranged in a rectangular array that is ordered. The dimensions of this pixel array define the size of a picture. The array's column and row counts are represented by the image's width and height, respectively [18]. The pixel array is a matrix with M columns and N rows as a result. We establish a pixel's coordinates at x and y in order to refer to a particular pixel inside the picture matrix [17]. Image matrices' coordinate system specifies that x increases from left to right and y increases from top to bottom.

1.3 Types of an image

1. BINARY IMAGE–As its name implies, a binary image consists of only two pixel components, 0 and 1, where 0 denotes black and 1 denotes white. Monochrome is another name for this picture [16].
2. BLACK AND WHITE IMAGE – A black-and-white image is one that solely uses these two colours.
3. 8 bit COLOR FORMAT The most well-known picture format is the 8 bit colour format. It is sometimes referred to as a grayscale image and has 256 distinct colour tones. In this format, 0 denotes black, 255 denotes white, and 127 denotes grey.

4. 16 bit COLOR FORMAT: This is a format for colour images. It comes in 65,536 distinct colours. High Color Format is another name for it. The distribution of colour in this format differs from that in a grayscale picture.

Red, Green, and Blue are the other three formats that make up a 16-bit format. the well-known RGB format.

1.4 *Encryption of digital images*

A procedure called encryption transforms an initial message, known as plaintext, into cypher text, which is the message's encrypted version, using a finite set of instructions called an algorithm. To encrypt or decode data, cryptographic techniques often need a key, which is a collection of characters. We can encrypt or decrypt plaintext into cypher text and then convert cypher text back to plaintext with the use of a key and an algorithm.

Similar to how software encrypts words, an image may also be encrypted. Encryption software modifies the values of the numbers in a picture in a predictable manner by applying a series of mathematical operations, referred to as an algorithm, to the binary data that makes up the image. The encryption code can only be decrypted using a software key, which is generated by the same programme that scrambles the image. To reduce the likelihood that both may be intercepted by a hacker, the receiver receives the encrypted picture and the key separately. The encoded picture is decoded using the software key, which is often a form of password, which is entered into decryption software. How challenging it is to decrypt the encrypted data determines how secure the encryption is.

Encrypted data is decrypted to restore its original form. Encryption is typically reversed. Decoding the information uses a secret key or password that can only be accessed by trusted individuals. Privacy is one of the reasons for using an encryption-decryption system. It's important to carefully examine any access from unapproved groups or people as information goes across the Internet.

2   RELATED WORK

Image security has grown more crucial in many contemporary application domains as the necessity of information security has risen. Picture encryption, image concealing, and image watermarking are all included in the research of image security [21]. The use of picture encryption technology has been extensively used in a variety of application domains, including quantum-secured imaging [20], 3D image encryption, data monitoring, data tracking, and secret data transfer in the military and medical industries. Image security that fully exploits optical parallel features has grown in importance as a research area in recent years, and we have also shown that optical image encryption and hiding are technically possible. These techniques could be useful in the implementation of all-optical systems in the future19.

However, one of the biggest obstacles to optical image security has been the size of the data volume needed to store or transmit holograms. Numerous holographic compression approaches have been proposed in recent years to address this issue [19]. However, due to the development of hologram laser speckling [8] and the common use of electrical methods for hologram compression, their efficiency has been shown to be restricted [11]. A novel technological method for hologram compression in the optical domain is provided by the recently discovered theory of compressive sensing (CS), which captures the non-adaptive linear projections of compressible signals at a rate that is much slower than the Nyquist rate [9,10]. Then, using an optimization procedure, these signals are rebuilt from these projections. In addition, CS is integrated with other specific techniques.

Wider applications for imaging techniques are being sought after, as shown in quantum imaging [10], photon counting imaging, coherent imaging of various wavelengths, and the measurement of electric fields. The Internet of Things (IoTneeds)'s for large data processing and information security may also be effectively addressed by these features. Recently, a number of compressive sensing-based picture encryption techniques have been developed, including parallel image encryption, image encryption using an Arnold transform [13], and colour image encryption. These techniques, however, only apply to digital picture encryption; entirely optical image encryption techniques based on compressive sensing have not been covered.

Guo *et al.*'s [3] proposal for a shading image encryption method based on Arnold change and discontinuous fragmentary arbitrary change in IHS shading space was in Arnold change space. While the tint and immersion portions are encoded using Arnold change, the force component is encoded using discrete partial arbitrary change. As encryption keys, partial orders of discrete partial arbitrary change, irregular grids of discrete fragmentary arbitrary change, and the number of Arnold change emphases are also used. A shading image encryption method was presented by Liu *et al.* [10] in light of Arnold modification and Discrete Cosine Transform (DCT). Arnold change is used to combine the image's various shades. As encryption keys, arbitrary Arnold change bounds and points are used. A two-step image encryption method employing Arnold change and discrete partial precise change was presented by Liu *et al.* [17]. Arnold modification is used to scramble the pixels in an unanticipated way at a local level. It uses a discrete partial rakish alteration to jumble up the altered complicated function. As additional keys for enhancing security, the bounds of the discrete fragmentary rakish change and the Arnold change are used. The Fractional Fourier Transform (FrFT) has been used in the literature to offer several image encoding computations [5]. The FrFT, also known as the FrFT boundary, is a theory of the Fourier Transform that provides the additional advantage of partial request. Additionally, Liu [12] suggested an irregular FrFT that may be used for image encoding and interpretation by randomizing the traditional FrFT. A photo encryption approach involving arbitrary movement in the FrFT region was presented by Hennelly and Sheridan [13]. Zhang *et al.*'s [22] developed a technique for encoding pictures that uses the borders of the FrFT as the encoding's keys. By using various random stages in wavelet sub bands, Chen and Zhao [15] present a different method of encoding colouring pictures. Here, the keys for encryption include wavelet bundle channels, irregular phases, and partial orders of the FrFT. Singh and others [16] presented an optical encryption technique using grid or lattice-added graphics in a two-stage arbitrary encoding framework. Another image encryption technique using multichannel and multistage partial Fourier area filtering was suggested by Liu *et al.* [17].

3 PROPOSED METHODOLOGY

3.1 *Data pre-processing*

Images are pre-processed at the lowest level of abstraction, also referred to as pre-processing. A matrix of image function values (brightness) is used to represent an intensity image, the same as the original data captured by the sensor. As a result of pre-processing, unwanted distortions in the image are suppressed or certain features of the image are enhanced before they are further processed (e.g. rotation, scaling, translation). In pre-processing steps involved following sub stages:

- Read image
- Resize image
- Remove Noise (De-noise)
- Segmentation
- Morphology (smoothing edges)

3.2 *Image encryption procedure*

Figure 1 corresponds to the comprehensive explanation of the suggested encryption technique. In the encryption method, a coloured original picture of size M N is first divided into Red (R), Green (G), and Blue (B) colour integrals (B). Arnold is applied to each of the R, G, and B planes separately, followed by the application of DWT to all three parts of Arnold, which creates four subbands for each plane: LL, LH, HL, and HH. Next, FWHT is applied to each part, SVD is applied to the decomposition part, and the singular values Si of each subband of the R, G, and B are multiplied by the parameters bi. For I = 1; 2; 3; 4: In order to rebuild all sub bands, fresh singular values Si are used, where Si = Si bi. The encoded picture is created by combining all three colour planes.

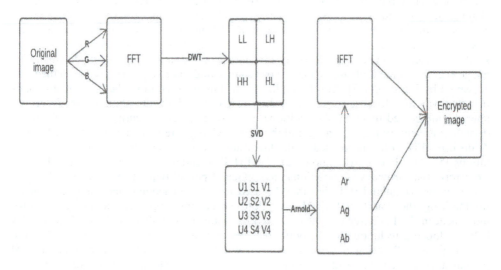

Figure 1.   Encryption procedure.

3.3 *Image decryption procedure*

Figure 2 corresponds to the comprehensive explanation of the suggested decryption system. The encrypted colour picture is divided into three colour planes—Red, Green, and Blue—in the decryption method. Apply DWT to each colour plane and SVD to each of the DWT's sub bands. Divide the parameters bi by the diagonal matrix of singular values Si.

Reconstruct each section using changed singular values Si then apply IFWHT to the reconstruction portion where I = 1, 2, 3, 4,.............12 in the proper sequence. Reconstruct the colour planes using the appropriate subband arrangement using inverse DWT (i.e. LL, LH, HL, HH). Combining all three colours is one of the completely correlated characteristics will provide a properly encrypted picture after applying inverse Arnold to the results.

If even one of the following pieces of information is absent, it will be challenging to recover the input picture from the encrypted one: the proper breakdown of the final encrypted picture, the accurate understanding of DCST bands, the rearrangement of MSVD sub-bands, the organisation of decomposed sub images produced from MSVD sub-bands, and values of matrix U and its application in different sub images.

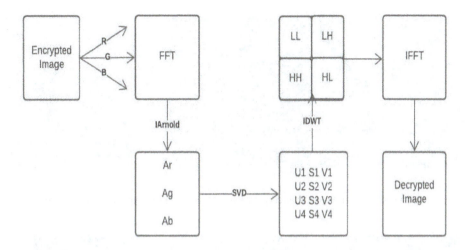

Figure 2.   Decryption procedure.

4   RESULT AND ANALYSIS

4.1   *Image encoding and decoding*

The simulation results presented in this part serve to demonstrate the effectiveness, unpredictability, and resilience of the suggested encryption approach. We utilise several common photos and tests to examine the proposed approach. The procedures for encrypting and decrypting As illustrated in Figure 3, the suggested approach is applied to various test photos, each measuring 512X512 pixels. The suggested work's encrypted version of the Lena picture is shown in Figure 4 (b). Figure 4(c) displays the properly encrypted picture with the proper parameters and organisation.

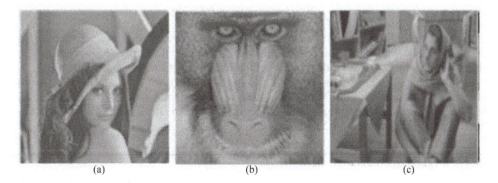

Figure 3.   Various test images: (a) Lena; (b) Baboon; (c) Barbara.

Squared Mean Error
The commutative square error (MSE) is the discrepancy between the authentic and successfully decoded pictures. MSE values are shown in Figure 3 for each test picture. The values between the original and decrypted photos of the proposed work are less than the

Figure 4.   (a) Leena input image, (b) encrypted Leena Image, (c) Decrypted Leena. Image.

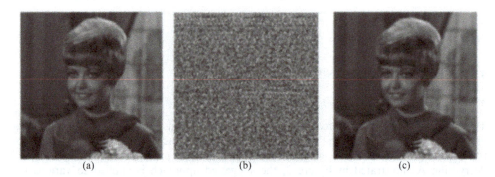

Figure 5.   (a) girl image, (b) encrypted girl image, (c) decrypted girl image.

MSE values of the work published in [12] which indicates that the suggested work can rebuild the images more effectively and with less information loss.

Ratio of Peak to Signal (PSNR)

The peak signal-to-noise ratio between two pictures, measured in decibels, is computed by the PSNR block. This ratio is used to compare the original and compressed images' quality. The quality of the compressed or rebuilt picture improves with increasing PSNR [2].

The peak signal-to-noise ratio (PSNR) and mean-square error are used to compare the quality of image compression (MSE). The MSE shows the total squared error between the original picture and the compressed one. The PSNR shows the peak error. The error goes in the opposite direction of the MSE value [4].

SSIM, or Structural Similarity Index Metric

It is a way to measure how much the quality of an image has changed because of things like data compression or data loss during transmission. PSNR is the most common way to measure image quality because it is easy to understand and has a clear physical meaning. However, it does not correlate well with how good an image looks. It is a full reference metric that needs two images from the same image capture: a reference image and a processed image. The picture that has been processed is usually compressed. For example, it could be found by looking at a JPEG picture saved as a reference (at any quality level). SSIM has a few uses in still photography [14].

4.2 Mean Squared Error

MSE is the commutative square error between the original (I) and correctly decrypted images. MSE values for all the test images shown in Figure 2 are given in Table 1. It can be analysed from Table 1 that the proposed work can reconstruct the images more efficiently with less information loss as the MSE values between original and correctly decrypted images.

Table 1. MSE value of test image.

S.no.	Input image	Input color images	Previously developed methods	Proposed work
1	Barbara	Red	$9.2770 \times 10-30$	0
		Green	$5.6062 \times 10-30$	0
		Blue	$6.5952 \times 10-30$	0
2	Lena	Red	$3.0591 \times 10-29$	0
		Green	$7.5336 \times 10-30$	0
		Blue	$8.9521 \times 10-30$	0
3	Baboon	Red	$2.2753 \times 10-29$	0
		Green	$8.1118 \times 10-30$	0
		Blue	$6.3833 \times 10-30$	0

Table 2. PSNR values of test images.

S.no.	Input image	Input color images	Proposed work
1	Barbara	Red	5.0558
		Green	7.1063
		Blue	7.4949
2	Lena	Red	2.7411
		Green	7.2000
		Blue	7.2830
3	Baboon	Red	4.7603
		Green	5.4300
		Blue	5.9825

4.3 Peak to Signal Ratio (PSNR)

The PSNR block computes the peak signal-to-noise ratio, in decibels, between two images. This ratio is used as a quality measurement between the original and a compressed image. The higher the PSNR, the better the quality of the compressed, or reconstructed image.

5 CONCLUSION

With the help of Arnold, DWT, SVD, and FWHT as well as certain predefined parameters and keys, the colour picture encryption system. The encrypted picture cannot be effectively decoded without knowledge of the right keys and the perfect arrangement of rippling sub bands. Additionally, we have provided a comparison between previously presented approach and our proposed technique. When compared to the old system, the MSE and PSNR values of each three-color channel of the correctly decoded picture exploitation projected scheme are low, demonstrating the usefulness of the suggested technique. Low MSE and PSNR values of a correctly encrypted picture show that no significant data will be lost during the decryption process.

REFERENCES

[1] Bishop, C. A., Humble, T. S., Bennink, R. S. & Williams, B. P. *Quantum-Secured Surveillance Based on Mach-Zehnder Interferometry*. arXiv Preprint arXiv:1303.6701 (2013).

[2] Candes, E. J., Romberg, J. & Tao, T. Robust Uncertainty Principles: Exact Signal Reconstruction from Highly Incomplete Frequency Information. IEEE T. Inform. Theory. 52, 489–509 (2006).

[3] Clemente, P., Durán, V., Tajahuerce, E. & Lancis, J. Optical Encryption based on Computational Ghost Imaging. Opt. Lett. 35, 2391–2393 (2010)

[4] Clemente, P. *et al.* Compressive Holography with a Single-pixel Detector. Opt. Lett. 38, 2524–2527 (2013).

[5] Donoho, D. L. Compressed Sensing. IEEE T. Inform. Theory. 52, 1289–1306 (2006).

[6] Howland, G. A., Lum, D. J., Ware, M. R. & Howell, J. C. Photon Counting Compressive Depth Mapping. Opt. Express. 21, 23822–23837 (2013).

[7] Petitcolas, F. A. P., Anderson, R. J. & Kuhn, M. G. Information Hiding – A survey. Proc. IEEE. 87, 1062–1078 (1999).

[8] Malik, M., Magana-Loaiza, O. S. & Boyd, R. W. Quantum-secured Imaging. Appl Phys Lett. 101, 241103 (2012).

[9] Magana-Loaiza, O. S., Howland, G. A., Malik, M., Howell, J. C. & Boyd, R. W. Compressive Object Tracking Using Entangled Photons. Appl. Phys. Lett. 102, 231104 (2013).

[10] Sun, B. et al. 3D Computational Imaging with Single-Pixel Detectors. Science. 340, 844–847 (2013).

[11] Jun, L., Yuping, W., Rong, L. & Yaqin, L. Coherent Single-detector 3D Imaging System. *Proceedings of the SPIE – The International Society for Opt Eng.* 8913, 891303 (2013).

[12] Li, J. et al. Two-step Holographic Imaging Method based on Single-pixel Compressive Imaging. J Opt Soc Korea. 18, 146–150 (2014).

[13] Lu, P., Xu, Z., Lu, X. & Liu, X. Digital Image Information Encryption based on Compressive Sensing and Double Random-phase Encoding Technique. Optik. 124, 2514–2518 (2013).

[14] Liu, X., Cao, Y., Lu, P., Lu, X. & Li, Y. Optical Image Encryption Technique based on Compressed Sensing and Arnold Transformation. Optik. 124, 6590–6593 (2013).

[15] Satya Bhushan Verma, Abhay Kumar Yadav (2019) Detection of Hard Exudates in Retinopathy Images. *ADCAIJ: Advances in Distributed Computing and Artificial Intelligence Journal* (ISSN: 2255-2863), Salamanca, v. 8, n. 4

[16] Verma, S. et al. 2019. Contactless Palmprint Verification System Using 2-D Gabor Filter and Principal Component Analysis, The International Arab Journal of Information Technology. 16(1), pp 23–29.

[17] Saravanan Chandran and Satya Bhushan Verma, Touchless Palmprint Verification using Shock Filter, SIFT, I-RANSAC, and LPD, IOSR Journal of Computer Engineering, Volume 17, Issue 3, PP 01–08, 2015. DOI:10.9790/0661-17330108

[18] Satya Bhushan Verma and Saravanan Chandran, Analysis of SIFT and SURF Feature Extraction in Palmprint Verification System, International Conference on Computing, Communication and Control Technology (IC4T), *IEEE*, pp. 27–30, 11–12 November, 2016

[19] Tajahuerce, E. & Javidi, B. Encrypting Three-dimensional Information with Digital Holography. Appl Opt. 39, 6595–6601 (2000).

[20] Verma, Satya Bhushan, and Abhay Kumar Yadav. "Hard Exudates Detection: A Review." Emerging Technologies in Data Mining and Information Security 117–124 (2021).

[21] Tanha, M., Kheradmand, R. & Ahmadi-Kandjani, S. Gray-scale and Color Optical Encryption based on Computational Ghost Imaging. Appl. Phys. Lett. 101, 101108 (2012).

[22] Wu, L., Zhang, J., Deng, W., & He, D. Arnold Transformation Algorithm and Anti-Arnold Transformation Algorithm. In 2009 *First International Conference on Information Science and Engineering* (pp. 1164–1167). IEEE, (2009, December).

Automation and Computation – Vats et al. (Eds)
© 2023 the Author(s), ISBN 978-1-032-36723-1

Recent innovations in building smart irrigation systems

Ila Kaushik*, Nupur Prakash* & Anurag Jain*
University School of ICT, GGSIP University, Delhi, India

ABSTRACT: India is an agricultural land where a large number of populations are engaged in farming. Water is a basic need for farming. Shortage of clean water all over the world has initiated the urge for optimizing natural resources. The use of water makes it possible to raise livestock and adds to the economic growth of the country by producing a high-yielding variety of food grains. Proper management of irrigation is important to promote plant growth. It is used for more substantial properties that are required for managing the volume, rate, and timings of water applications to match the compatibility of the farm. The paper presents the effective use of water to enhance the productivity of the field and to the nurture development of plants by managing the right moisture levels in the soil. The second important aspect is to have backup facilities required to sustain the farms when water levels are low in case of natural calamities. Different types of irrigation schemes are introduced which gives an outline of which crop is suited for which type of irrigation practice. The working mechanism of the Internet of Things (IoT) based network for smart agriculture is viewed as dealing with different parameters required in farming. Management strategies lay their foundation in providing optimal production and high yield of crops. The later section of the paper describes different innovative schemes for setting up smart irrigation-based systems along with recent technologies perspectives in achieving smart water-saving mechanisms in the farms.

Keywords: Internet of Things, Smart Irrigation, Innovative Schemes, Water Monitoring System, Agriculture

1 INTRODUCTION

Water is a basic necessity for setting up any agriculture system. In order to provide a good yield of crops, an irrigation facility is considered as one of the most important attributes as water is the basic necessity for nurturing the crops [1]. Most small-scale farmers are totally dependent on rain for the water supply required for the crops [2]. Technological advancement in agriculture provides a response to varying climatic conditions. As the number of food crops is at an alarming rate, in order to increase productivity smart irrigation systems are used which helps in fulfilling the energy requirements from renewable energy sources [3]. Farmers can control the water flow according to the requirement of the crops based on surrounding temperature, and moisture content. Depending upon the level of water, it needs to get pumped out of the system which is monitored and turned off on an alarm alert [4]. The scarcity of water resources around the World offered a requisite to optimize the water resources. IoT-based smart irrigation system helps in managing the optimum use of water

*Corresponding Authors: ila.kaushik.8.10@gmail.com; nupurprakash1@ gmail.com and anurag@ipu. ac.in

DOI: 10.1201/9781003333500-23

resources. As the population increases, water requirements for freshwater are at their high peak which is to be addressed. The major tranche of water intake by the agriculture industry. Due to the unavailability of a cost-effective irrigation system, developing countries are utilizing more water in agriculture as compared to developed countries for the same yield. Therefore, there is an urge to build smart irrigation systems based on smart strategies to overcome the problem of water scarcity [5]. The sensors used in smart farming collect soil information which effectively uses an irrigation facility to water the crops by observing the moisture content in the soil, so over-watering and under-watering of plants can be rectified which protects the crop from being damaged. IoT-based smart irrigation system in the area of smart farming plays a very important role in increasing water productivity along with reducing manpower, use of fertilizers, and water requirements [6]. Smart irrigation system proves an efficient system during water crises or shortage of annual rainfall as many farmers are still dependent on rainfall for watering crops. A farmer owning more than one farm can remotely access other parts of the farm using mobile apps which give an alert message regarding the water level required for nurturing the crops [7]. Both agriculture and irrigation practices mainly focus on the quality of crops and environmental factors required for precise control. This enables IoT-based smart irrigation systems to be of utmost priority. IoT-based sensors provide real-time updations and even collect information when the fluctuations in the moisture and soil content are very less. Managing water expenses is one of the crucial factors needed in the quality and health of the crop. Irrigation management controls the quantity of water needed by the crop and the right quantity of water at a specific timeline [8]. IoT makes it very simple by connecting different sensors where the processed information collected by the sensors can be accessed using mobile apps. Wastage of water which affects crop health and leads to soil damage can be prevented by using water waste prevention strategies that are inbuilt into the irrigation system. Ever increasing demand for the crop is at its peak. In order to increase the productivity of demands, farmers are using optimized techniques based on smart systems to produce more food crops in a short span of time [9]. IoT-based analytics provides real-time updates notifying you about recent market trends, and opportunities and a tracking system for observing environmental conditions like humidity, temperature, and rainfall [10]. Some of the key points highlighted in using smart irrigation are: (i) Increased farmland productivity, (ii) Exemption of manpower, (iii) Reduction in soil erosion, (iv) Increased quality of food products, and (v) Efficient water utilization.

The rest of the paper is organized as follows: Section 2 discussed the literature review; Section 3 described different types of Irrigation Systems; Section 4 explained the Architecture for Smart Irrigation Systems; Section 5 illustrated the Recent Innovations carried out in Smart Irrigation followed by the Conclusion in Section 6.

2 LITERATURE REVIEW

India is an Agrian country. Almost 60% of its population depends on farming. It adds to the economic growth and development of the country. Earlier farmers are only and only dependent on monsoon rain for crop productivity but as the demand increases towards producing crops in large quantities, there is a shift towards modern farming techniques. The gathered information from different sensors that monitors the water, soil, and environmental conditions are used to determine the state of water required and the necessity for requiring fresh water. Countries raised with higher funds are implementing water management systems in smart farming to minimize the utilization of water resources and to decrease the environmental impact of utilizing a large amount of water. Systems formulated for irrigation and agriculture are quite exorbitant, making it tough for small farmers to execute smart irrigation-based systems. In order to overcome large costs inculcated in irrigation systems, developers are using low-cost sensors linked to the nodes in the system meant for agriculture monitoring and irrigation management. Different low-cost sensors such as a water turbidity sensor made with colored and infrared led emitters and receptors [11], a leaf water stress monitoring sensor [12], a water salinity monitoring sensor made with copper coils [13], a multi-level soil moisture sensor comprised of copper rings placed along a PVC pipe [14] are used. Much research has been done on cloud-based irrigation solutions which gather the information from the sensors and transfer the content to the cloud for the irrigation system. Research scholars at Colorado University have designed an evaporation-based irrigation system that uses soil water balancing and information-related queries from Colorado Agricultural Meteorological Network (CoAg-Met) and Northern Colorado Water Conservation District (NCWCD) weather stations [15]. A smartphone app is used for user interaction with the interface to know about the moisture content, weather measurements, and quantity of water needed for irrigation purposes. In [16], authors have suggested a communication system based on a wireless sensor network that is deployed in real scenarios. This system is built to monitor plants' need for water using sensors and actuators. The Strawberry smart irrigation system [17] measures the crop transpiration of two commonly used strawberry cultivators which evaluate irrigation efficiencies, crop yielding, and water productivity. The android application was evolved for promoting irrigation scheduling in the strawberry sector. Different algorithms such as principal components analysis, and k-means of the time series of the agro-climatic variables are used for enabling methodology in the system. John R. Dela Cruz et al [18] gave water usage optimization of smart farm automated irrigation system as a solution for redesigning the water usage in smart farm automated irrigation system. Pushkar Singh et al. [27] introduced Arduino-based smart irrigation using a soil moisture sensor, temperature, and Wi-Fi module. The Arduino-based model has a simple workflow toward execution and cost. Riadh Zaier et al. [19] designed and implemented smart irrigation for groundwater use at small farms. Data is gathered at a single node collection on each farm. All center points are considered slave nodes. One node act as the master node and fuses sensors with a solenoid valve. The master node is responsible for the positioning of water as and when required. Joaquín Gutiérrez et al. [20] proposed a smartphone irrigation sensor created using optical sensors and cell phones to control the field. L. Garcia Paucar et al. [21] proposed decision system support by using wireless distributed sensors to remotely arrange sensors in the system. Table 1 shows the findings of the different methodologies used in smart irrigation.

Table 1. Findings of different methodologies in smart irrigation.

Author's	Methodology	Findings
Parmenter et al. [22]	Ethernet, USB	To explore the availability of wireless connectivity.
Kumar et al. [23]	Moisture sensors; Xbee communication;	Installing low-effort sensors perform well in detecting the dampness of the ground.
Gutiérrez et. al. [20]	Optical sensor, smartphone, Android App, wireless sensor Network	Cell phone-based water sensor network with the use of Arduino board.
Jain et al. [24]	Arduino microcontroller	Optimized utilization of water resources and reduction in labor costs in agricultural systems.
Paucar et al. [21]	Wireless sensor network, decision support system	Decision support by using wireless distributed sensors to remotely arrange sensors in the system.

(continued)

Table 1. Continued

Author's	Methodology	Findings
Zaier et al. [19]	TCP/IP protocol and XBee network	• Designed and implemented smart irrigation for groundwater use on a small farm. Data is gathered at a single node collection on each farm. • All center points are considered slave nodes. • One node act as the master node and fuses sensors with a solenoid valve. The master node is responsible for the arrangement of water as and when required.
Bandara et al. [25]	Green Roof	The technique is used for choosing water measurement in view of evapotranspiration.
Ghosh et al. [26]	Cloud, Android And Data Mining	Sensor diagrams will be shown on smartphones and PC for effective monitoring of the system.
Singh et al. [27]	Arduino	• Arduino-based smart irrigation using soil moisture sensor, temperature, and Wi-Fi module. • Arduino-based model has a simple workflow towards execution and cost.
Kodali et al. [28]	MQTT protocol.	Simple water pump controller used in the smart farming model.

3 TYPES OF IRRIGATION SYSTEMS

The implementation of smart irrigation depends upon the variety of crops, and environmental conditions like temperature, rainfall, and moisture. The various types of irrigation systems that are connected to IoT controllers and also used according to the variety of crops are as explained follows.

3.1 Flood irrigation system

It is the oldest form of irrigation technique. In this type of irrigation system, the flow of water is directed by small, thin parallel channels used to carry water for irrigation. The supply of water is done using an underground pipe. Only half of the water is used by the crops other half is wasted in form of evaporation, infiltration of uncultivated land, and transpiration through weeds leaves. This is not a much-suited irrigation technique.

3.2 Drip irrigation system

It is the most systematic nutrient-delivery system required for growing crops. In a real-time frame, the right amount of water and nutrients are provided to the crops needed for the growth of the plants. High yielding variety of crops is produced with optimum use of water, fertilizers, and crop protection products. With the use of dripper lines, water and minerals are supplied to the crops in the entire field. Smaller units of drippers also work on the fields which emit water droplets resulting in the uniformity of water and minerals across the entire field.

3.3 Micro irrigation system

This is one of the modern types of irrigation system, in which water is irrigated with the help of drippers, sprinklers, foggers, and other emitters. Major components involved in this type of irrigation system include sources of water, pumping device, valves, fertigation equipment, filters, control valves, emitters, etc. In this technique, water is applied drop by drop at each root of the crop. Drippers are fixed in between the spacing provided between two crops.

3.4 *Sprinkler irrigation system*

This type of irrigation system allows the flow of water under high pressure with the help of a pump. A small nozzle is fixed in the pipe from where the water releases from a small hole. A system of pipes is connected from where water is equally distributed, sprayed, and irrigated on the land. This type of irrigation is mostly suited for field and tree crops and water is sprayed over the crop canopy. For treating sensitive crops, this technique is not suited as the flow of water can leads to crop damage. Table 2 describes the advantages, disadvantages, and different suitable crops in different types of irrigation systems.

Table 2.   Different types of irrigation systems.

S. No.	Irrigation System	Advantages	Disadvantages	Suitable Crops
1.	Flood Irrigation System	1. Minimum land wastage. 2. Low cost incurred to purchase equipment. 3. Reduction in chemical leaching 4. Minimum cost of pumped water. 5. Higher yielding of crops.	1. Unequal water distribution. 2. Water loss through evaporation. 3. More cost in leveling of land. 4. Possibility of removal of the top layer of soil. 5. Uneven distribution of water.	Wheat, Maize, Legume
2.	Drip Irrigation System	1. Irrigation can be done under low pressure at a low cost. 2. Reduction in weed growth and soil erosion. 3. Field leveling 4. Compatible with all soil types. 5. Chemigation and Fertigation are done in easy steps.	1. High initial investment. 2. Damage to system components. 3. Wastage of water if drippers are not properly installed. 4. Cloning of tubes 5. Tube breakage because of excessive sun heat.	Grapes, Banana, Tomato
3.	Micro Irrigation System	1. Reduced surface crusting. 2. Reduction in pests. 3. Joint system for irrigation and fertilization. 4. Protection of small horticulture crops. 5. Equal distribution of water.	1. More time is required for installation. 2. Loopholes in tubes. 3. Reduction in the fertility of the soil. 4. Wastage of time and water if not installed properly. 5. More cost than initial system cost.	Sunflower, Groundnut, Cashew-nut.
4.	Sprinkler Irrigation System	1. Equal distribution of water. 2. Easy installation of setup. 3. No terracing is required. 4. Lower labor requirements in comparison to the traditional system 5. Adding fertilizers and pesticides in a feasible way.	1. Wind sensitivity resulting in evaporation losses. 2. Deposition of sediments resulting in nozzle breakage. 3. Constant water supply. 4. High cost 5. Problem while supplying saline water.	Cotton, Gram, Jowar, Fenugreek

4 ARCHITECTURE FOR SMART IRRIGATION SYSTEM

The architecture for building a smart irrigation system comprises the following components and is also shown in Figure 1. The components are further grouped into layers discussed as follows.

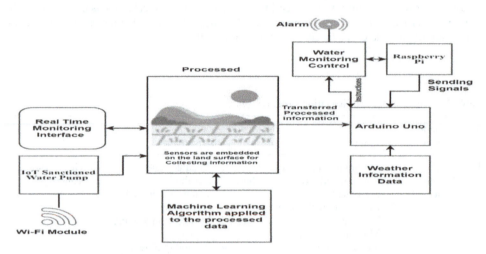

Figure 1. Architecture for smart irrigation system.

4.1 *Data collection*

On the basis of the field's requirement, sensors are deployed on the field or individual sensors are installed for gathering information about soil, temperature, and humidity. The collected information is then processed and transferred to the Arduino board for further processing.

4.2 *Weather data collection*

This unit works for the online collection of weather forecast data. Different weather forecasting information like cloudiness, temperature, and humidity of various portals are aggregated. The captured information is kept in the log file for future reference purposes.

4.3 *Soil moisture prediction algorithm*

Different algorithms like k mean clustering and regression techniques are applied to the data collected by sensors working on soil moisture content. In order to have effective irrigation which is according to the requirement of the crop, data gathered by soil sensors are analyzed to keep a track record of the moisture content in the soil so that according to the previously studied pattern, water can be supplied in the right time frame with right quantity.

4.4 *Water motor control*

This module works with the functioning of the motor. The application can be accessed with the help of Raspberry PI which sends the signal to Arduino which operates on the opening and closing of the motor. An alarm is also attached to the motor, when water reaches to the prescribed level of water an alarm is initiated to stop the motor so that there is no unnecessary wastage of water.

4.5 Real-time monitoring interface

This module works with monitoring moisture content in the soil in the real timeline and accordingly irrigation facilities can be provided to particular crops whose moisture content is low. Irrigation can be provided on a scheduling basis so that crops cannot get damaged due to excessive spraying of water.

4.6 IoT sanctioned water pump

This module enables a switch that is supervised by a Wi-fi enabling node. With the help of this module, farmers can remotely access their fields and on time shifts each field is provided by irrigation facilities.

5 RECENT INNOVATIONS IN SMART IRRIGATION

Constant monitoring of irrigation systems is required in order to produce a high-yielding variety of crops, especially in remote areas where there is a problem of water scarcity. In response to varying climatic conditions, advancement in technology helps in providing a high yield of crops. According to the requirement of crops, water flow is controlled to crops so that excessive water spraying does not damage the crops. A number of different innovative ideas are introduced to make irrigation more powerful for better crop productivity. Some of them are explained as follows-

5.1 IoT-based solar energy-based irrigation system

Depletion of natural resources results in a negative impact on environmental sources. Therefore, to restore natural resources, many developments and innovations are done in this regard. More than 70% of water comes from natural resources like lakes, rivers, and underground systems to support irrigation in farms and feed livestock. It is an important factor to maximize the productivity of crops, reduction in water wastage, and efficient energy consumption. Sun is the major source of solar energy and it is most widely used in an irrigation system for producing clean energy for the environment. This feature was adopted in smart irrigation to automate the irrigation process in smart agriculture. Irrigation system based on solar energy generates electricity from the photovoltaic cell through solar technology. For minimal usage of power, the backup battery is also installed in the system which is used in power shortage scenarios. The whole performance of the system is managed using an Arduino-based microcontroller. The climatic conditions based on temperature, humidity, and rainfall are collected with the help of sensors and are used for monitoring real-time data. The electricity produced by solar radiation helps the water pump to supply water in the field. An alarm-based system is also installed which gives a signal to the users when to off the system so that there is no unnecessary wastage of water in the field and based on data collected by the sensors, the right quantity and at the right timings, water is provided to the crops so that they are not getting damaged by excessive water supply [1].

5.2 IoT-based sensor water system

With the help of the IoT framework, there is depth knowledge about the interdependency of energy, water, and food resources using wireless connectivity. With real-time scenarios, farmers can optimize water resources using smart irrigation control strategies and precisely knowing when to harvest the crops resulting in reducing labor input and energy. The drip irrigation system is most widely used as the water reaches the crops by dripping near plant roots with the help of drippers. This technique improves productivity by reducing water usage and is relatively cheaper in cost. The system design is fully automated using real-time data from wireless nodes to schedule the irrigation events periodically. The system integrates a switch, water pump, solenoid valves, relay, and Arduino board. Sensors are linked to the microcontroller board and the relay activates the solenoid valve to the set threshold value. Different environment and agriculture parameters are measured with the help of a sensor. The water level in the leaf and moisture in the soil are updated at each interval. Wireless units sense these collected data and transmit it to a gateway. All these functionalities such as sensing, data collection, and communication are programmed in an open-source environment. To inbuilt power saving mechanism, the microcontroller can inbuilt real-time clock to only awaken the device at the time of requirement. For the power-saving purpose, the device operates in sleep mode and collects data at each hour [2].

5.3 Weather monitoring system

For irrigation and the performance of crops, weather conditions are considered key factors. Air temperature, luminosity, precipitation, ultraviolet radiations, wind speed, and atmospheric pressure are the most monitored parameters needed in an irrigation system [3]. The amount of precipitation gives the idea that additional water is needed for plants. Temperature monitoring is the most important factor in weather monitoring. DHT11 (Adafruit Industries, New York, USA), DHT22 (Adafruit Industries, New York, USA), and the LM35 (Texas Instruments, Dallas, USA) are the most utilized temperature sensors [4]. For monitoring purposes, embedded devices enable self-protection of the environment. Deployment of sensors in the environment brings real-life object interaction with other objects through the network. Analyzed results through information gathered by sensors are visible to end users with the help of Wi-Fi connectivity across remote access. The system collects the information with the help of sensors and transmits it over the cloud by passing through the gateway. The gathered data is also helpful in the future for developing countries and industrial areas for weather monitoring. By continuous monitoring of weather conditions, this system helps users from pollution content in the air. The gathered information by the sensors is very helpful in predicting weather trends like if rainfall is expected on any particular day, water can be saved by not irrigating during at day by the irrigation system. The purpose of watering of plants can be fulfilled by rainfall as predicted by the system. An additional parameter like an alarm can be added to the functionality of the system so that users can be notified at regular intervals about excess smoke, humidity, and temperature conditions [5].

5.4 Integrated mobile and IoT monitoring system

For effective utilization of water needed for agricultural crops, an automated irrigation system is designed. Sensors capturing data about temperature and soil moisture are placed in the root of plants. The amount of water required for plants depends upon the prescribed threshold value according to which nurturing of plants is done [7]. The system design takes into consideration of raspberry pi, moisture and temperature sensors, water pump, etc. The communication unit comprises the smartphone. Different varieties of crops are monitored starting from the beginning when the seed is sown in the soil. Water levels required for nurturing the plants are recorded at each level for future reference. Plants are then watered at different stages according to the threshold value recorded at each stage of plant development. Sensors collecting data about water levels are connected to main irrigation canals. These sensors are connected to a gateway operating on wireless technology which periodically sends the data to a web server. The database keeps all the track records based on past history for monitoring the water level needed for different varieties of crops. The sensors used to automate irrigation facilities help in improving water usage efficiency. Raspberry pi is a low-cost, small-sized computer that plugs into a computer and monitors the process.

Information gathered from the sensors is transferred to the microcontroller board where different algorithms are applied for the monitoring process. Mobile applications are used for giving an alert message to farmers owning more than one farm so that they can remotely monitor their farms [6].

5.5   *IoT and LoRaWAN technologies for irrigation system*

Low Power Wide Area Networks are mainly designed for long-range communication sites and low power consumption. This network is best suited for farms situated far away from villages and towns where energy supply is a major issue. So, this type of network deals with providing low energy consumption. Wireless connectivity is required for the deployment of these types of networks and the transfer of data can be done from both sides. It is an open standard and infrastructure setup can be installed according to the requirement of the farm. This design helps in monitoring hydrants which are used as active agents to control the flow of water consumed by each crop. Earlier use of hydrants does not give periodic check updates about the watering of plants to owners. An additional cost is required by the farmers to install their own meters to check for the consumption of water for their crops. Additional features such as talk bots and voice assistants in order to design the system more user-friendly, easy to use, and interactive for people finding technology use difficult. A data logic module can be added to this system for making logical decisions. Enhancement can be done by using machine learning and artificial-based technologies for making decisions with high accuracy. System knowledge can be increased by training large data set from various systems' requirement [9].

5.6   *Prototype design model of the irrigation controller*

Irrigation techniques provide water to irrigation fields for maintaining healthy crops. Different irrigation techniques are applied to different crops depending on their varieties. Atmospheric parameters such as humidity, temperature, and rainfall give assistance in calculating the next irrigation practice. On observing the previous trend of different parameters, new practices can be used in the farms in increasing the productivity of food grains. Weather, soil, and crops are considered as three pillars of agriculture. Changes in weather cannot be traced, but based on crop information additional supplements can be provided to the field in order to attain the required yield. The technique used behind in designing an irrigation controller is to have a more efficient water management technique. There are already many irrigation systems like real-time feedback systems, volume-based systems, time-based systems, open-loop systems, and closed-loop systems in practice but the best suited is the controller-based system. In recent use technologies, high-tech systems making use of controllers, Arduino boards, and sensors are used for monitoring the farms [10].

6   CONCLUSION

The increase in World's population and need for freshwater leads to the problem of water paucity. The agricultural sector is the major consumer of fresh water. As there is no availability of cost-effective irrigation systems in developing countries, a large amount of water is being consumed as compared to developed countries. There is an urgent requirement for setting up of smart irrigation system dealing with advanced technologies for the effective utilization of water resources. Automation helps farmers to remotely access their fields and make their work easier. Different sensors and technologies help farmers to know moisture content, temperature, humidity, and amount of water required for nurturing the crops. Different types of irrigation systems along with their advantages and disadvantages are discussed along with compatibility with different plant species. Due to the uneven use of natural resources, the level of freshwater is declining which is a major issue to be addressed. As agriculture is the main source of freshwater consumption, therefore different recent innovations are discussed in the later section of the paper. This offers an understanding of different recent technologies operating in the field of smart irrigation. The work can further be extended by exploring new technologies for smart irrigation to provide sustainability, increase profit and reduce the depletion of natural resources in the environment.

REFERENCES

[1] Al-Ali, A. R., *et al.* "IoT-solar Energy Powered Smart Farm Irrigation System." *Journal of Electronic Science and Technology* 17.4 (2019): 100017.

[2] Mekonnen, Yemeserach, *et al.* "IoT Sensor Network Approach for Smart Farming: An Application in Food, Energy and Water System." 2018 *IEEE Global Humanitarian Technology Conference* (GHTC). IEEE, 2018.

[3] García, Laura, *et al.* "IoT-based Smart Irrigation Systems: An Overview on the Recent Trends on Sensors and IoT Systems for Irrigation in Precision Agriculture." *Sensors* 20.4 (2020): 1042.

[4] González-Amarillo, Carlos Andrés, *et al.* "An IoT-based Traceability System for Greenhouse Seedling Crops." *IEEE Access* 6 (2018): 67528–67535.

[5] Girija, C., and Shires Andreanna Grace. "Internet of Things (IoT) Based Weather Monitoring System." *International Journal of Engineering Research & Technology (IJERT)*.

[6] Vaishali, S., *et al.* "Mobile Integrated Smart Irrigation Management and Monitoring System Using IOT." *2017 International Conference On Communication and Signal Processing (ICCSP)*. IEEE, 2017.

[7] Seenu, N., Manju Mohan, and V. S. Jeevanath. "Android Based Intelligent Irrigation System." *International Journal of Pure and Applied Mathematics* 119 (2018): 67–71.

[8] Matilla, Diego Mateos, *et al.* "Low-cost Edge Computing Devices and Novel User Interfaces for Monitoring Pivot Irrigation Systems Based on Internet of Things and LoRaWAN technologies." *Biosystems Engineering* (2021).

[9] Davcev, Danco, *et al.* "IoT Agriculture System Based on LoRaWAN." *2018 14th IEEE International Workshop on Factory Communication Systems (WFCS)*. IEEE, 2018.

[10] Poyen, Faruk Bin, *et al.* "Prototype Model Design of Automatic Irrigation Controller." *IEEE Transactions on Instrumentation and Measurement* 70 (2020): 1–17.

[11] Daskalakis, S.N.; Goussetis, G.; Assimonis, S.D.; Tenzeris, M.M.; Georgiadis, A. A uW Backscatter-morse_leaf Sensor for Low-power Agricultural Wireless Sensor Networks. *IEEE Sens. J.* 2018, 18, 7889–7898.

[12] Guruprasadh, J.P.; Harshananda, A.; Keerthana, I.K.; Krishnan, K.Y.; Rangarajan, M.; Sathyadevan, S. Intelligent Soil Quiality Monitoring System for Judicious Irrigation. In *Proceedings of the 2017 International Conference on Advances in Computing, Communications and Informatics (ICACCI)*, Udupi, India, 13–16 September 2017.

[13] Parra, L.; Ortuño, V.; Sendra, S.; Lloret, J. Low-Cost Conductivity Sensor based on Two Coils. In *Proceedings of the First International Conference on Computational Science and Engineering*, Valencia, Spain, 6–8 August 2013.

[14] Sendra, S.; Parra, L.; Ortuño, V.; Lloret, L. A Low Cost Turbidity Sensor Development. In *Proceedings of the Seventh International Conference on Sensor Technologies and Applications*, Barcelona, Spain, 25–31 August 2013.

[15] Bartlett A., Andales A., Arabi M., Bauder T., A smartphone app to extend use of a cloud-based irrigation scheduling tool, *Comput. Electron. Agric.* 111 (2015) 127–130, doi:10.1016/j.compag.2014.12.021.

[16] Sales N., Remdios O., Arsenio A., Wireless sensor and actuator system for smart irrigation on the cloud, In: *2015 IEEE 2nd World Forum on Internet of Things (WF-IoT)*, 2015, pp. 693–698, doi:10.1109/WF-IoT.2015.7389138.

[17] Londhe G., Galande S.G., Automated Irrigation System By Using ARM Processor, *International Journal of Scientific Research Engineering & Technology (IJSRET)*, ISSN 2278-0882, Vol. 3, No. 2, May 2014.

[18] De la Cruz, John R., Renan G. Baldovino, Argel A. Bandala, and Elmer P. Dadios. "Water usage optimization of Smart Farm Automated Irrigation System using the artificial neural network." In *5th*

International Conference on Information and Communication Technology (ICoIC7), pp. 1–5. IEEE, 2017.

[19]   Zaier, R., Zekri S., Jayasuriya H., Teirab A., Hamza N., and Al-Busaidi H. "Design and implementation of smart irrigation system for groundwater use at the farm scale." In *7th International Conference on Modelling, Identification, and Control (ICMIC)*, pp. 1–6. IEEE, 2015.

[20]   Jagüey, J.G., Villa-Medina J.F., López-Guzmán A., and Ángel Porta-Gándara M. "Smartphone irrigation sensor." *IEEE Sensors Journal*, Vol. 15, No. 9, pp. 5122–5127, 2015.

[21]   Paucar, L. Garcia, A. Ramirez Diaz, Federico Viani, Fabrizio Robol, Alessandro Polo, and Andrea Massa. "Decision support for smart irrigation by means of wireless distributed sensors." In *IEEE 15th Mediterranean Conference on Microwave Symposium (MMS)*, 2015, pp. 1–4. IEEE, 2015.

[22]   Parmenter, Jason, Alex N. Jensen, and Steve Chiu. "Smart irrigation controller." In *International Conference on Electro/Information Technology (EIT)*, pp. 394–398, 2014.

[23]   Kumar, Akash, Khurram Kamal, Mohammad Omer Arshad, Senthan Mathavan, and Tanabata Vadamala. "Smart irrigation using low-cost moisture sensors and XBee-based communication." In *IEEE International Conference on Global Humanitarian Technology Conference (GHTC)*, pp. 333–337. IEEE, 2014.

[24]   Jain, Prateek, Prakash Kumar, and D. K. Palwal. "Irrigation management system with microcontroller application." In *1st International Conference on Electronics, Materials Engineering and Nano-Technology (Genentech)*, pp. 1–6. IEEE, 2017.

[25]   Bandara, A. G. N., B. M. A. N. Balasooriya, H. G. I. W. Bandara, K. S. Buddha Sri, M. A. V. J. Muthugala, A. G. B. P. Jayasekara, and D. P. Chandima. "Smart Irrigation Controlling System for Green Roofs Based On Predicted Evapotranspiration." In *IEEE International Conference on Electrical Engineering Conference (EECon)*, pp. 31–36, 2016.

[26]   Ghosh, Subhashree, Sumaiya Sayyed, Kanchan Wani, Mrunal Mhatre, and Hyder Ali Hingoliwala. "Smart Irrigation: A Smart Drip Irrigation System Using Cloud, Android And Data Mining." In *IEEE International Conference on Advances in Electronics, Communication and Computer Technology (ICAECCT)*, pp. 236–239. IEEE, 2016.

[27]   Singh, Pushkar, and Sanghamitra Saikia. "Arduino-based smart irrigation using water flow sensor, soil moisture sensor, temperature sensor and ESP8266 WiFi module." In *IEEE International Conference on Humanitarian Technology Conference (R10-HTC)*, pp. 1–4, 2016.

[28]   Kodali, Ravi Kishore, and Archana Sahu. "An IoT Based Soil Moisture Monitoring on the Losant platform." In *2nd International Conference on Contemporary Computing and Informatics (IC3I)*, pp. 764–768. IEEE, 2016.

Automation and Computation – Vats et al. (Eds)
© 2023 the Author(s), ISBN 978-1-032-36723-1

Performance evaluation of pre-trained models on Tuberculosis dataset

Chakshu Grover*, Satvik Vats* & Vikrant Sharma*
Department of Computer Science and Engineering, Graphic Era Hill University, Dehradun, India

ABSTRACT: Witnessing the mass influence of Tuberculosis on the world population, threatening the world security and challenging healthcare and pharmaceutical services. unprecedented emergence of pandemics like Covid-19 delayed the research and studies on these globally affected diseases, and increased the chances of incidences of cases worldwide, collectively disrupting and underutilization the essential healthcare resources and services. Due to the passive nature of tuberculosis bacteria, Laboratories, radiologist, and necessary time is being underutilized because of the traditional methods of Diagnosis and treatments, which can be optimized using the computerized method like advance automated computer systems, statistics, and artificial intelligence, and Big Data. Methodologies and tools like Deep Learning architectures when integrated with Transfer learning can provide an efficient and fast Data-fed systems that can quantifiably reduce the engagement of these important radiological and diagnostic capital that were previously were underutilized. The performance evaluation of such architectural models inferred the VGG-16 as the best performing model with performance metrics like precision of 93% and recall of 92%, resulting in a F-1 Score of 92.5% and overall accuracy of 90.3%. The best architecture can practically lessen the classification result calculation time from weeks to not more than 200ms, with it being approx. 90 percent accurate.

Keywords: Tuberculosis, VGG16, Resnet50, MobileNet, InceptionNet, Transfer Learning

1 INTRODUCTION

In some of the last decades, Science, research, and technology witnessed the exponential growth in practically unquantifiable bytes of Data [1]. Computer science and research have seen phenomenal growth in the last five to six decades. The studies, development, and research are now opening multiple gateways for the new and advanced methodologies in the field of agriculture, infrastructure, automation, business, Health, and pharmaceuticals, and so on. The emerging technologies of the last some decades such IoT, Cloud Services, Big Data Analytics, and wireless Networks had been a reason for the emergence of numerous advanced technologies and disciplines [2–6]. Wireless network systems and sensor system is providing the humongous amount of Data with broad spectrum of domain and fields, and the rise of interconnected geographical data exchange systems led to the worldwide public access to the unseen data & the need for essential automation led to the foundation of artificial intelligence now we have machine learning and deep learning [7–10]. This has been the core idea of several breakthrough day-to-day resources as well as a human substantial

---

*Corresponding Authors: Chakshugrover22502@gmail.com; svats@gehu.ac.in and vsharma@gehu.ac.in

                                                    DOI: 10.1201/9781003333500-24

development project [11–13]. This research survey is based on such application of Machine learning and artificial intelligence in the field of healthcare and pharmaceuticals, where the research intends to portraits the idea of developing a computerized Neural network-based automated patient tuberculosis polarity classification using deep learning's transfer learning approach [14,15]. The inspiration for this survey is the attention gained by the tuberculosis in the last some decades, being infamous for its infectiously deadly and hard to diagnose nature. The World Health Organization (WHO) has been concerned over this massive health and security danger because of its worldwide spread [16].

2   LITERATURE REVIEW

Called Phthisis by Greeks, and Tuberculosis by the modern science, Tuberculosis is among the 10 deadliest and contagious diseases, which has been at large since a while [17]. Since its first formal medical discovery in 1882 by Dr. Robert Koch to today, Tuberculosis has been among humans silently and showing serious complications from then [18]. As per W.H.O.'s report, worldwide, about 10 million positive cases were confirmed, with 1.5 million deaths were confirmed in 2020, primarily because of Tuberculosis, making it a contender next to COVID in terms of the severity in the infectious disease [19].

The Covid emergency has significantly delayed the research and development we had in past some years [19], and boosted the growth of tuberculosis cases because of its correlation

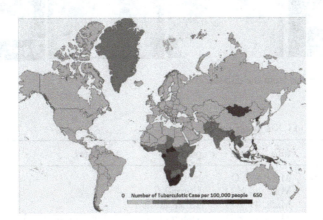

Figure 1.   Visualisation to report of confirmed cases of Tuberculosis worldwide in 2020.

of their features [20,21]. Tuberculosis is a pathogen-based infectious disease, called *myco-bacterium tuberculosis*, which causes enzyme formation in the region of impact, resulting in rotting and tearing up of the tissues in the lung. The mode of transmission of the pathogen is Airborne and is hard to detect and diagnose. 90% of the cases are not detected by the traditional means of treatment, but the most optimal result was given by the CXR reports, which can detect the possible area of impact even without the premature symptoms [22]. Hence, the CXRs can be used as the first diagnostic procedure for detection. The traditional mean of diagnosis requires resources like trained radiologists with years of training and expertise, Laboratory occupancy for the test as infrastructure and time to test and frame report to the patients. This inefficiency can bring underutilization of resource, resulting in taking up to 3 weeks for the final reports, which is an important duration as the pathogens may mature significantly. The visualisation Figure shows the active estimated presence of Tuberculosis cases, with its presence in Indo–Pacific and lower African subcontinent.

One of the major inspirations of this research is that the pathogen responsible for the TB, mycobacterium tuberculosis is difficult to detect and does not show any presence in the pulmonary tissues [23]. The pathogen's thick cell wall protects it from the immunity barrier of the lung tissues and can let bacteria to settle into deeper parts of the lungs. Since the human immune system Is sufficient to tackle the pathogen, some variants and other biological states like HIV, pregnancy, weaker immunity, and the after-effects of pandemics like CoVid-19 can work like catalysts as they can directly weaken the immune efficiency [24]. With coexistence of chronic respiratory disease, tuberculosis increases possibility of infection by 2.46 times (C.I. of 1.76–3.44, with 95% likelihood) [25]. The "Figure 2 – CXR images from the Dataset" looks at some of the examples of the Chest X-Rays sample from the dataset, images 1 and image 2 are normal samples whereas, images 3 and image 4 have confirmed tuberculosis, most likely to be the area surrounded by a red outline. "Table 1 – Metadata of CXR image dataset" shows the metadata about the images shown below in respective order [26].

The traditional modalities used in the tuberculotic test were the microbiological, biochemical, and biopsy tests, where several biological tests like surveillance of body fluids,

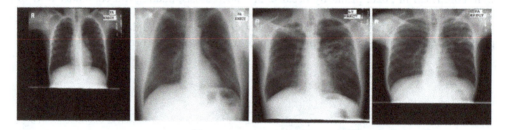

Figure 2.   CXR images from the dataset.

Table 1.   Metadata of CXR image dataset.

Image	Info	polarity	Attributes
MCUCXR_0042_0	Male, 9 Y.O.	Normal	–
MCUCXR_0089_0	Male, 44 Y.O.	Normal	–
MCUCXR_0166_1	Male, 49 Y.O.	Tuberculosis	–
MCUCXR_0264_1	Male, 15 Y.O.	Tuberculosis	Cavitary infiltrate LUL, Non-Consolidate

glucose, microorganisms, or other types of tests by which the labs which may take several hours or maybe weeks in some cases [27,28]. Other advanced computerized tests like radiographical tests are done by radiologists with certain expertise, years of training but still, manual tasks are done by humans and still have some degree of biases and errors.

This inefficiency of biological test alone can underutilize the most crucial time of the patient, as well as the important share of time of the labs, radiologists, and Healthcare capitals [24]. The introduction of an automated computer system is possible. Here is a Possibility to build some automated, highly precise computerized methods which will reduce the occupant weeks of resources to a couple of seconds [29]. Here, we introduce the use cases of data science and deep learning for image processing. Artificial Intelligence and its subsets have shown a tremendous amount of improvement in decision-making throughout multiple domains. For instance, Business, Engineering, Computer science, Space Research, agriculture have been benefited from technology [30]. The modern technique called Deep

Learning has been gathering a lot of applause for its replicational qualities of a human brain and being the better, cheaper, and faster solution to human cognitive-based problems [10]. We believe that the problem of time and errors while the detection of tuberculosis can be solved by the application of Deep Learning by teaching a virtual model from the pre-determined results with the help of medical organizations.

3   METHODOLOGY

Deep Learning as commonly acknowledged as "a subset of machine learning where artificial neural networks, algorithms inspired by the human brain, learn from large amounts of data" [31]. Deep learning is the concept where we have data that requires so deep understanding and the learning process is so deep that traditional machine learning and Artificial intelligence could not catch. This Algorithm explores the depth of the data and gets insightful information for every possible bit of data and significantly increases the performance. Deep Learning's core principle is the Artificial Neural Networks, where the mathematical functions mimic human neurons and gather information, aiding the decision-making process. Deep learning's reason for success is unlike other algorithms, deep learning can have any possible number of computational layers and each layer can have multiple numbers of neurons or functions for the analysis, but each layer of neuron sets uses some calculation capabilities of the resource.

This research used the deep learning Convolutional neural network, which is renowned for its dynamicity, agility, and effectiveness over unstructured data like imagery and audio input. The Convolutional neural networks are a set of different layers of Convolutional layers, fully connected layers, Pooling layers, and Dense layers, where we mainly divide the complete architecture into 3 parts: Input Layer, Hidden Layer, and Output Layer.

The most optimal results were induced by the VGG16 architecture which is used as the base architecture for the research, proposed by Andrew Zisserman and Karen Simonyan in

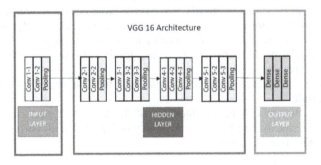

Figure 3.   Architectural design of VGG-16 model.

2014 [32,33]. The architecture's name was deduced from the Visual Geometry Group lab of Oxford, hence VGG. The VGG architecture has 2 models, VGG16 and VGG19, numbers denoting the number of layers in the model. VGG16 model is pretrained on the ImageNet Dataset, where 15 million high-resolution images belonging to nearly about 22 thousand categories, train it's near about 138 million weights [33]. This pre-trained model can further be trained as per the local data for the tuning of weights, and then can give a significantly high accuracy on the testing set. Figure 3 shows the VGG 16 Architecture. The model achieved to enter the top 5 test accuracy with a 92.7% score on the ILSVRC – 2014, an international competition where the dataset comprised 1.2 million datasets, with 50k validation images from 1000 categories.

4 DATASET

Two of the most renowned public datasets, Montgomery and Shenzhen were chosen, which provided by the National Library of Medicine, near about 4.08 GB with an average image size of 4MB [26,27,34]. This provides High-resolution Chest X-ray images, classified into 2 main classes, Normal and Tb. The complete data was divided into 70%, 20%, and 10% as training, validation, and testing data respectively. The Montgomery data is provided by the tuberculosis control program of the Department of Health and Human Services of Montgomery County, MD, USA. Whereas the Shenzhen data is provided by the Shenzhen No.3 Hospital in Shenzhen, Guangdong providence, China [26]. In addition, this dataset provided the masked images of the lung, and the metadata of all the images providing the SEX, AGE, POLARITY, and FEATURES of the image.

5 MODEL EVALUATION

The experimental results of the research will be based on the model testing accuracy over the untrained set of images, called the test dataset. The test dataset comprises 10% percent of the entire dataset. The comparative analysis is generated between multiple pretrained transfer learning CNN architecture such as VGG16 [32], Mobilenet [35], densenet [36], resnet [37], & inceptionNet [38], and will be compared by the testing parameters like F1 score, recall and precision and ROC curve.

A model has once trained, showing different combinations of the results, which may be the permutations of the possible chances of the inference. Since the research is using a binary classification approach of Tuberculosis polarity, the possible status of a case can be either positive or negative. While the model's predictions will also have 2 chances to deduce: positive or negative. A confusion matrix is a type of table layout, which shows all the possible chances of combinations of results that a model can infer, by classifying the outcome to each category of the result. In a binary classifier model, the possible outcomes are:

True positive (TP): predicts true for a positive result *False-positive (FP)*: when predicts false for a positive result

False-negative (FN): predicts false for a negative result *True-negative (TN)*: predicts true for a negative result

This confusion matrix aids in the inference of the important performance metrics like accuracy, precision, recall, and F1 score. Accuracy is the ratio of all correct inferences to all prediction outcomes (see Equation 1)

$$Accuracy = \frac{correct\ prediction}{All\ predictions} = \frac{TP + TN}{TP + FP + TN + FN} \tag{1}$$

The precision is ratio of correctly predicted positive predictions to the total predicted positive outcomes, while recall, or sensitivity, is the ratio of correct predictions to all positive predictions. These two metrics help us to determine the success of the model when the data has an imbalance in classes (see Equation 2)

$$Precision = \frac{TP}{TP + FP}(i) \quad Recall = \frac{TP}{TP + FN}(ii) \tag{2}$$

To evaluate both of the above metrics, we use the F1-score (see Equation 3), which is the twice product of precision and recall, divided by the sum of precision and recall. F1 score significance is in the presentation of the predictive performance by combining both the above metrics.

$$\text{F1} - \text{Score} = \frac{2(precision * recall)}{precision + recall} = \frac{2 * TP}{2 * TP + FP + FN} \tag{3}$$

The confusion matrix also helps us determine some of the additional performance metrics like sensitivity and specificity. Sensitivity is the rate of actual positive predictions out of all positive predicted value, whereas the specificity Is the rate of the actual negative predictions out of every negative predicted values (see Equation 4).

$$\text{Sensitivity} = \frac{TP}{TP + FP} \quad \text{Specificity} = \frac{TN}{TN + FN} \tag{4}$$

6 OBSERVATION

The Comparative analysis using different neural network architecture bassed on the model performance upon the Dataset inferred that the VGG-16 and ResNet50 model provided the least error with accuracy of 92.3% and 91.7% respectively. The Precision and recall differ by

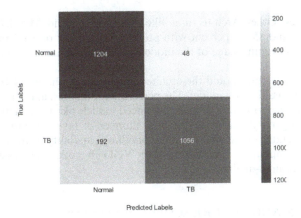

Figure 4. Epoch chart of VGG-16.

	precision	recall	f1-score	support
NORMAL	0.86	0.96	0.91	52
TB	0.96	0.85	0.90	52
accuracy			0.90	104
macro avg	0.91	0.90	0.90	104
weighted avg	0.91	0.90	0.90	104

Figure 5. Confusion matrix.

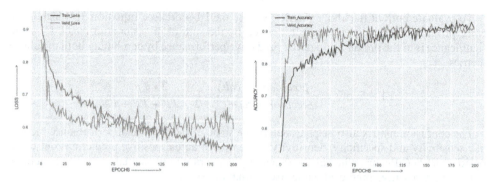

Figure 6.   (a) Performance metrics for the VGG-16. (b) Accuracy-epoch chart of VGG-16.

Table 2.   Performance observations of neural network architectures.

Model	Total accuracy	Precision	Recall	Sensitivity	specificity	F1 Score
VGG-16	90.3	93	92	86.21	95.656	92.49
ResNet	89.0	90.1	90.7	87.02	86.23	90.39
XceptionNet	88.5	89.8	89.3	89.45	87.23	89.54
MobileNet	85.8	86.2	86	87.45	88.23	86.09

$\pm1\%$ in both cases. Other Architectures like XceptionNet and MobileNet delivered the accuracy of 89.5% and 85.8% percent with precision and recall of $\pm1\%$ in these cases also. The Hyperparameter tuning stage of the model will further enhance the performance of the model.

As per the overall accuracy, and the architectural strength, the VGG-16 model is chosen to represent as the classification model. The performance metrics, and analysis of the dataset suggests that the causes of failure of more advanced models like ResNet50 is its incapability of handling the larger dataset, since the high number of layers significantly increase the training and computation time, and the error handling capability of the model decreases in the scenario. The computational versatility of the VGG 16 model and its assurance to handle the large amount of high-quality data.

7   CONCLUSION AND FUTURE SCOPE

The research was intended to put down the computational advancements and computational capabilities to reduce the pathological resources using computer vision and deep learning. The research successfully develops a high accuracy model to detect the polarity of the CXR report for the Tuberculotic patient using transfer learning. The comparative analysis of the most famous deep learning architecture was done based on their confusion matrix, F1 score and overall accuracy on the unseen Dataset, and deduced the VGG-16 architecture to be the best performing architecture. The research paved the way to more technologically advance and complex system to not just identify, but to segment and localize the infection onto a designated area on the Chest. The future research will be focussed on the increasing the dataset and its availability, developing better architecture for the computation and more precise calculation of the results, intentions to create the multiclassification system to differentiate the infected area and the type of tuberculotic infection, and last but not the least the features of tuberculosis in the patient and segmentation of those feature independently.

REFERENCES

[1] Vats S. and Sagar B.B., "Data Lake: A Plausible Big Data Science for Business Intelligence," *Undefined*, pp. 442–448, 2019, doi: 10.1201/9780429444272-70.

[2] Vats S., Sagar B.B., Singh K., Ahmadian A. and Pansera B.A. "Performance Evaluation of an Independent Time Optimized Infrastructure for Big Data Analytics that Maintains Symmetry," *Undefined*, vol. 12, no. 8, Aug. 2020, doi: 10.3390/SYM12081274.

[3] Vats S. and Sagar B.B., "An Independent Time Optimized Hybrid Infrastructure for Big Data Analytics," *Undefined*, vol. 34, no. 28, Oct. 2020, doi: 10.1142/S021798492050311X.

[4] Bhati J.P., Tomar D. and Vats S., "Examining Big Data Management Techniques for Cloud-Based IoT Systems," *Undefined*, pp. 164–191, Oct. 2018, doi: 10.4018/978-1-5225-3445-7.CH009.

[5] Sharma V. et al., "OGAS: Omni-directional Glider Assisted Scheme for Autonomous Deployment of Sensor Nodes in Open Area Wireless Sensor Network," *ISA Trans*, 2022, doi: 10.1016/J. ISATRA.2022.08.001.

[6] Sharma V., Patel R.B., Bhadauria H.S. and Prasad D., "Pneumatic Launcher Based Precise Placement Model for Large-Scale Deployment in Wireless Sensor Networks," *International Journal of Advanced Computer Science and Applications*, vol. 6, no. 12, 2015, doi: 10.14569/IJACSA.2015.061222.

[7] Kumar A., Sharma V. and Prasad D., "Distributed Deployment Scheme for Homogeneous Distribution of Randomly Deployed Mobile Sensor Nodes in Wireless Sensor Network," *International Journal of Advanced Computer Science and Applications*, vol. 4, no. 4, 2013, doi: 10.14569/ IJACSA.2013.040422.

[8] Vikrant S., Patel R.B., Bhadauria H.S. and Prasad D., "Policy for Planned Placement of Sensor Nodes in Large Scale Wireless Sensor Network," *KSII Transactions on Internet and Information Systems*, vol. 10, no. 7, pp. 3213–3230, Jul. 2016, doi: 10.3837/TIIS.2016.07.019.

[9] Sharma D., Patel R.B., Bhadauria H.S. and Prasad D., "NADS: Neighbor Assisted Deployment Scheme for Optimal Placement of Sensor Nodes to Achieve Blanket Coverage in Wireless Sensor Network," *Wirel Pers Commun*, vol. 90, no. 4, pp. 1903–1933, Oct. 2016, doi: 10.1007/S11277-016-3430-6.

[10] Vats S. and Sagar B.B., "Performance Evaluation of K-means Clustering on Hadoop Infrastructure," https://doi.org/10.1080/09720529.2019.1692444, vol. 22, no. 8, pp. 1349–1363, Nov. 2020, doi: 10.1080/ 09720529.2019.1692444.

[11] Sharma V., Patel R.B., Bhadauria H.S. and Prasad D., "Deployment Schemes in Wireless Sensor Network to Achieve Blanket Coverage in Large-scale Open Area: A Review," *Undefined*, vol. 17, no. 1, pp. 45–56, Mar. 2016, doi: 10.1016/J.EIJ.2015.08.003.

[12] Bhatia M., Sharma V., Singh P. and Masud M., "Multi-level p2p Traffic Classification Using Heuristic and Statistical-based Techniques: *A Hybrid Approach*," *Symmetry (Basel)*, vol. 12, no. 12, pp. 1–22, Dec. 2020, doi: 10.3390/SYM12122117.

[13] Vikrant S., Patel R.B., Bhadauria H.S. and Prasad D., "Glider Assisted Schemes to Deploy Sensor Nodes in Wireless Sensor Networks," *Undefined*, vol. 100, pp. 1–13, Feb. 2018, doi: 10.1016/J. ROBOT.2017.10.015.

[14] Agarwal R., Singh S. and Vats S., "Review of Parallel Apriori Algorithm on Mapreduce Framework for Performance Enhancement," *Advances in Intelligent Systems and Computing*, vol. 654, pp. 403–411, 2018, doi: 10.1007/978-981-10-6620-7_38/COVER.

[15] Agarwal R., Singh S. and Vats S. "Implementation of an Improved Algorithm for Frequent Itemset Mining Using Hadoop," *Proceeding – IEEE International Conference on Computing, Communication and Automation, ICCCA 2016*, pp.13–18, Jan. 2017, doi: 10.1109/CCAA.2016.7813719.

[16] *"Tuberculosis."* https://www.who.int/health-topics/tuberculosis#tab=tab_1 (accessed Nov. 17, 2022).

[17] *"The Top 10 Deadliest Diseases in the World."* https://www.healthline.com/health/top-10-deadliest-diseases#diarrhea (accessed Nov. 17, 2022).

[18] *"History | World TB Day | TB | CDC."* https://www.cdc.gov/tb/worldtbday/history.htm (accessed Nov. 17, 2022).

[19] *"Global Tuberculosis* Report 2021." https://www.who.int/teams/global-tuberculosis-programme/tb-reports/global-tuberculosis-report-2021 (accessed Nov. 17, 2022).

[20] "COVID-19 Causes Increase in Tuberculosis Deaths: Global TB Report 2021 Communitymedicine4all." https://communitymedicine4all.com/2021/10/17/covid-19-causes-increase-in-tuberculosis-deaths-global-tb-report-2021/ (accessed Nov. 17, 2022).

[21] Udwadia Z.F., Vora A., Tripathi A.R., Malu K.N., Lange C. and Sara Raju R., "COVID-19-Tuberculosis Interactions: When Dark Forces Collide," *Indian Journal of Tuberculosis*, vol. 67, no. 4, pp. S155–S162, Dec. 2020, doi: 10.1016/J.IJTB.2020.07.003.

[22] Zaidi S.Z.Y., Akram M.U., Jameel A. and Alghamdi N.S., "A Deep Learning Approach for the Classification of TB from NIH CXR Dataset," *IET Image Process*, vol. 16, no. 3, pp. 787–796, Feb. 2022, doi: 10.1049/IPR2.12385.

[23] "TB Online – The Path to Diagnosing TB." https://www.tbonline.info/posts/2016/3/31/path-diagnosing-tb-1/ (accessed Nov. 17, 2022).

[24] Pogatchnik B.P., Swenson K.E., Sharifi H., Bedi H., Berry G. J. and Guo H.H., "Radiology-pathology Correlation Demonstrating Organizing Pneumonia in a Patient Who Recovered from COVID-19," *Am J Respir Crit Care Med*, vol. 202, no. 4, pp. 598–599, Aug. 2020, doi: 10.1164/RCCM.202004-1278IM/SUPPL_FILE/DISCLOSURES.PDF.

[25] Yang J. *et al.*, "Prevalence of Comorbidities and its Effects in Patients Infected with SARS-CoV-2: A Systematic Review and Meta-analysis," *International Journal of Infectious Diseases*, vol. 94, pp. 91–95, May 2020, doi: 10.1016/J.IJID.2020.03.017.

[26] "Index of public/Tuberculosis-Chest-X-ray-Datasets/Shenzhen-Hospital-CXR-Set/." https://data.lhncbc.nlm.nih.gov/public/Tuberculosis-Chest-X-ray-Datasets/Shenzhen-Hospital-CXR-Set/index.html (accessed Nov. 17, 2022).

[27] "Index of public/Tuberculosis-Chest-X-ray-Datasets/Montgomery-County-CXR-Set/MontgomerySet/." https://data.lhncbc.nlm.nih.gov/public/Tuberculosis-Chest-X-ray-Datasets/Montgomery-County-CXR-Set/MontgomerySet/index.html (accessed Nov. 17, 2022).

[28] Arora D., Singh A., Sharma V., Bhaduria H.S. and Patel R.B., "HgsDb: Haplogroups Database to Understand Migration and Molecular Risk Assessment," *Bioinformation*, vol. 11, no. 6, pp. 272–275, Jun. 2015, doi: 10.6026/97320630011272.

[29] Vats S., Singh S., Kala G., Tarar R. and Dhawan S., "iDoc-X: An Artificial Intelligence Model for Tuberculosis Diagnosis and Localization," *Undefined*, vol. 24, no. 5, pp. 1257–1272, Jul. 2021, doi: 10.1080/09720529.2021.1932910.

[30] Dilek S., Cakır H. and Aydın M., "Applications of Artificial Intelligence Techniques to Combating Cyber Crimes: A Review," *International Journal of Artificial Intelligence & Applications*, vol. 6, no. 1, pp. 21–39, Jan. 2015, doi: 10.5121/ijaia.2015.6102.

[31] Liu W., Wang Z., Liu X., Zeng N., Liu Y. and Alsaadi F.E., "A Survey of Deep Neural Network Architectures and their Applications," *Neurocomputing*, vol. 234, pp. 11–26, Apr. 2017, doi: 10.1016/J.NEUCOM.2016.12.038.

[32] Simonyan K. and Zisserman A. "Very Deep Convolutional Networks for Large-Scale Image Recognition," 2015, Accessed: Nov. 17, 2022. [Online]. Available: http://www.robots.ox.ac.uk

[33] "[PDF] Very Deep Convolutional Networks for Large-Scale Image Recognition Semantic Scholar." https://www.semanticscholar.org/paper/Very-Deep-Convolutional-Networks-for-Large-Scale-Simonyan-Zisserman/eb42cf88027de515750f230b23b1a057dc782108 (accessed Nov. 17, 2022).

[34] Jaeger S., Candemir S., Antani S., Wáng Y.-X.J., Lu P.-X. and Thoma G., "Two Public Chest X-ray Datasets for Computer-aided Screening of Pulmonary Diseases", *Quant Imaging Med Surg*, vol. 4, no. 6, p. 475, Dec. 2014, doi: 10.3978/J.ISSN.2223-4292.2014.11.20.

[35] Howard A. G. *et al.*, "MobileNets: Efficient Convolutional Neural Networks for Mobile Vision Applications", Apr. 2017, doi: 10.48550/arxiv.1704.04861.

[36] Huang G., Liu Z., van der Maaten L., and Weinberger K.Q., "Densely Connected Convolutional Networks," *Proceedings 30th IEEE Conference on Computer Vision and Pattern Recognition, CVPR 2017*, vol. 2017-January, pp. 2261–2269, Nov. 2017, doi: 10.1109/CVPR.2017.243.

[37] Jian K.H.X.Z.S.R., "Deep Residual Learning for Image Recognition arXiv:1512.03385v1". *Enzyme Microb Technol*, vol. 19, no. 2, pp. 107–117, 1996, Accessed: Nov. 17, 2022. [Online]. Available: http://image-net.org/challenges/LSVRC/2015

[38] Szegedy C. *et al.*, "Going Deeper with Convolutions," *Proceedings of the IEEE Computer Society Conference on Computer Vision and Pattern Recognition*, vol. 07-12-June-2015, pp. 1–9, Sep. 2014, doi: 10.48550/arxiv.1409.4842.

Automation and Computation – Vats et al. (Eds)
© *2023 the Author(s), ISBN 978-1-032-36723-1*

URL-based technique to detect phishing websites

R. Lakshman Naik & Sourabh Jain
Department of CSE, IIIT, Sonepat, Haryana, India

ABSTRACT: Phishing website means a type of cyber threat to steal confidential and sensitive data like login credentials of a person or a company through uniform resource locators (URLs) and webpages, email, telephone or by sending a fraudulent message. The objective of this work is to predict phishing websites by training classification algorithms on the dataset. This paper, mostly concentrated on main eight features, which are used to predict phishing URL or legitimate URL in an effective way. As the proposed system is based on training by classification techniques; the performance level of each technique is measured its metrics and compared. Then the proposed system achieved 86.6% accuracy.

1   INTRODUCTION

Phishing means fishing for confidential or private information. In cyber security, phishing is playing a major role. In phishing, hackers trap the victim by sending a familiar website login page and leads the victim to enter confidential information like username and password.

The attackers get the victims credential information or other financial data for financial gain, then the attackers get satisfied. Otherwise, attacker sends phishing mails to grasp the employee login information and other details to attack an organization. Phishing uses cyber-crime attacks like advanced persistent threats (APTs), Advanced Persistent Threats (APTs) and ransomware.

Many researchers published articles on predicting phishing websites by using data mining techniques. However, there are no proper training datasets published due to very less literature made on features to identify phishing websites. It is also very difficult to prepare a dataset to cover all possible features, which are used to predict phishing websites. In the cyber world among all the cyber security attacks, phishing is an important security threat and it is causing huge amount of financial loss to industries as well as to individual persons [1]. Compare to all the existing cyber security attacks, phishing attack needs a special attention to avoid effect on economy [2–4].

In the year 2016, according to APWG report, around the world 12, 20,523 phishing attacks identified and it is also identified that 65% of phishing attack is increased compared over 2015 [5]. Over one decade 57.53% of phishing attacks increased per month. The main aim of phishing attack is not only to get sensitive credentials, but also became an important method to send malicious software's like ransomware [6,22].

A graph representation given after the survey done regarding phishing attacks in different fields like financial, payments, retail & services, social networking and internet service provider from the year 2015 to 2022. It clearly explains about the percentage of phishing attacks clone every year and states whether the attacks are increased or decreased is as shown in Figure 1. In the year 2015 and 2020 most of phishing attack is done on Internet Service Providers (ISP). When it comes to the year 2016 most of the phishing attack happened on Retail & Service sectors that is on the financial transactions only. It is clearly observed that

DOI: 10.1201/9781003333500-25

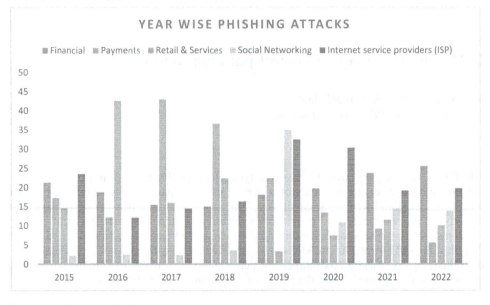

Figure 1. Year wise phishing attacks.

in the year 2017 and 2018 most of the phishing attacks on payments. Most of the Phishing attacks happened in the year 2019. In the year 2021 and 2022 phishing attacks concentrated on social networks.

Developing an effective Anti-Phishing framework is a challenging task. Most of the researchers and developers proposed various phishing detection solutions. Most of these solutions Like Heuristic Approach and Visual Comparison Approach produces mostly False Positive Occurrence but not capable to deal with Zero-Day attack. The Blacklist Based Detection approach have Quick Access Time, but not capable to identify Zero-Hour attack. Because of this, the phishing problem still available and proposing anti-phishing solution became a challenging task [21].

Therefore, it is very important to design a framework, which can classify phishing websites efficiently. Recently, the researchers are using Machine Learning Models to detect phishing attacks.

The said classification methods take some characteristics as inputs to identify a legitimate website or a phishing website [7]. The training data, is used to specify that, the proper feature selection set, and classification algorithm used to detect how effective methods are used [8].

This paper organised as; Section 2 describes the type of phishing attacks. Section 3 discuss about the Architecture and algorithm of proposed approach. Section 4 shows the implementation and evaluation details. In the section 5 paper is concluded with future research work in this field.

2   TYPES OF PHISHING

Phishing is a fraud that uses both technological and social engineering fraud to steal customers' financial account information and personal identity information.

In the Social engineering cyber security, the hacker uses fake email accounts and email messages to target naive victims by creating a belief that they are working in a trusted and legitimate entity. Using these techniques customers are directed to phoney websites that deceive users into disclosing financial information such passwords and usernames.

224

In order to catch their prey from the "sea" of online users, these hackers utilised emails as "hooks" (Kay 2004) [9]. There are several recognised phishing attack methods in use today, as shown in Figure 2.

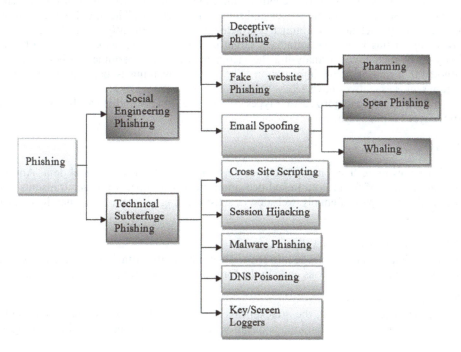

Figure 2.   Type of phishing attacks.

2.1   *Social engineering*

Social engineering is the practise of deceiving someone in order to achieve goals that may be malicious and detrimental to the victim. These methods aim to get access to certain systems or obtain information about a person or group in order to benefit financially [10].

2.1.1   *Deceptive phishing*
It is a most difficult kind of phishing. In this technique the attacker tricks the victims to provide confidential information. After that attackers use this data to commit financial crimes or carry out other types of offences. Deceptive phishing is when someone sends you a fake bank email asking you to click a link and confirm your account information.

2.1.2   *Fake website phishing*
Some phishing tricks make use of JavaScript commands to change the website's address bar. To accomplish this, either cover the address bar with an image of a real URL or close the existing bar and open a new one that contains the real URL. These are copied websites.
   Pharming: Like phishing, pharming directs consumers to a fake website that looks official. In this technique, customers are brought to the fake website without even click on a bad link. Even if the user types in the right URL, attackers can hijack either the user's computer or the website's DNS server and route them to a phoney website.

2.1.3   *Email spoofing*
The sending of email messages with a falsified sender address is known as email spoofing. Since the fundamental email protocols lack an authentication mechanism, spam and

phishing emails frequently use spoofing to trick or outright lie to recipients about the message's origin. Two types are being sorted.

Spear phishing: this technique targets particular individuals rather than concentrating on huge population. Attackers mostly conduct online and offline research on their targets. Then they personalise their communications and sound more genuine. The initial step used to get past a company's defences and launch a targeted assault is frequently spear phishing.

Whaling: in this technique, attackers target a "big fish," such as CEO. here attackers mostly spent their time in analysing the victim in order to determine the best time and method for acquiring login information. The issue of whaling is particularly important because senior executives have access to many corporate data.

There are mainly categorized two types of phishing attacks and it can be divided in to various types explained below.

2.2 Technical subterfuge

A phisher uses more than simply social engineering tricks to steal personal data. Another common method of fraud is technical deception, in which either a phisher transmits malicious malware as part of emails, websites, or self-executing programmes (generally crack of any software).

2.2.1 Cross-Site Scripting (XSS)
This attach injects malicious code into trusted websites. in this attack, the attacker uses web application and sends malicious code in the form of client-side script to all the end users.

2.2.2 Session hijacking
In a session hijacking attack, the attacker takes control of the user session. When you log into a service, such as a user banking application, a session begins and ends with your logout.

2.2.3 Malware phishing
In malware-based phishing, credentials are stored on the victim's computer by malware and sent to the owner, or the phisher.

2.2.4 DNS poisoning
In an attempt to lure the client into communicating with a bogus DNS server, the phisher uses DNS poisoning. Once the victim establishes a connection, they may be taken to dangerous websites or even have malware installed on their devices.

2.2.5 Key/screen loggers
With the advent of screen logging software, key loggers have become completely undetectable, rendering virtual keyboards useless. These take screenshots of the screen and record mouse movements, which are then forwarded to the phisher who is located in remote.

3 PROPOSED WORK

There are many cybercrimes happening regarding banking or financial sectors. It is still evident that the primary source of cyberattacks is phishing emails sent by hackers and other online criminals. We anticipate seeing firms continue to use authentication technology to safeguard against unauthorised and fraudulent senders as more businesses become aware of and address email problems. Phishing is also a kind of cybercrime, which sent a link or URL to the users. The unaware users open it and fill all the credentials needed. By this easily the common people are becoming the victims. To overcome such problems, we proposed a frame work.

The proposed approach's system architecture is shown in Figure 3. Our method extracts and examines a variety of features from dubious websites in order to successfully identify widespread phishing attacks. The main contribution of this study is the identification of excellent feature set. To increase the effectiveness of phishing website detection, we suggested eight new features. The domain names in URL are matched with the queried webpage URL using Patter Mathing Algorithm. Here the website URL features are automatically gathered by web crawler.

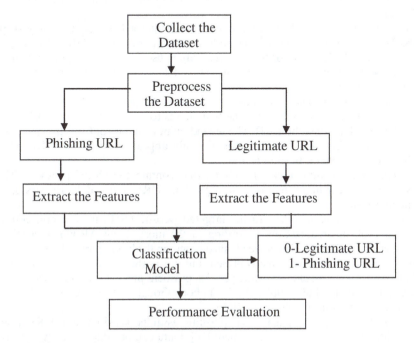

Figure 3.   Architecture of proposed system.

The system architecture of the proposed work is shown in above Figure 3. The proposed work collected the Phishing URL dataset from [17] and legitimate URLs dataset from [19] for training the classification model. The suggested work analyses and extracts a variety of information from dubious websites in order to successfully identify phishing attacks. The chosen aspects are the work's standout qualities and main contribution. To increase the effectiveness of phishing website detection, we suggested eight features.

We employed pattern-matching techniques to compare the domain name in the web address to the domain name of the requested webpage URL. The proposed algorithm draws its features from website URLs. In our method, we primarily make use of the client-side particular URL characteristics. The eight URL-based qualities that we have selected as URL counterfeits are explained in the section below.:

- *Number of dots:* This feature checks the number of dots (.) in the URL. In the URL of the legitimate website, it contains three dots, but phishing URL may contain more than three dots and more sub-domains to confuse users. The said subdomains are separated by the dot symbol. if any URL contains more than three dot symbols, then it is considered as phishing [21].

- *Presence of special symbol:* it determines special symbol like "@" and "-" (dash symbol) in the URL address. If "@" symbol exists in the URL path, then it is considered as real domain for retrieving the website and treat it as legitimate URL. If dash "-" symbol used in the URL path, then it is considered as phishing URLs. Mostly dash "-" symbol is used by the attacker resemble the fake URL as original URL.
- *URL length:* The legitimate web URLs are very short, rememberable and very significant. But phishing web URLs are longer, not a meaningful domain, because the phisher hide the brand or enterprise name and hides the longer URL redirection information. In the proposed work if we found a longer URL that is greater than 72 characters, then it is decided as phishing website.
- *Suspicious words in URL:* This function checks whether a URL contains any suspicious words. To establish trust, the phisher adds dubious terms to the URL. In the phishing URLs, we found nine keywords that were often used: security, login, sign in, bank, account, update, include, webs, and online. The URL is considered phishing if any of these keywords are included in it.
- *Position of top-level domain:* The top-level domains (TLDs) in the URL are examined in this feature in terms of two factors. The first step is to look at where the TLD is located within the base domain section [21]. The second point is that it confirms the URL contains more than one TLD. The top-level domain only appears once in the URL of a valid website; the base domain portion is absent.
- *http count:* This function tracks the URLs that contain the "http" protocol. The http protocol may appear more than once in a phishing URL, but it only appears once on a legitimate website.
- *Enterprise brand name in URL:* In the targeted website URL, phisher concentrated on brand name of an organization. according to the most recent APWG data [17] in the phishing websites, 45.96% URLs are phished based on the brand name. This feature marks a site as phishing if a brand name is present but not in the proper location. The top 500-phishing targets we have chosen include banks, payment processors, and other institutions. USAA, Amazon, PayPal, Apple, AOL, Yahoo, Google, Dropbox and other popular names can be found in the phishing URLs.
- *Data URI:* These days, it appears that phishing assaults based on data URIs (universal resource identifiers) are the most prevalent [18]. Data can be added directly to web pages using the Data URI protocol, just like external resources. Instead of using numerous HTTP requests, this method enables the retrieval of various elements including HTML, pictures, and JavaScript in a single HTTP request. The following is a list of data URI syntax:

  data: [<media type>] [; base64], <data>

  Without hosting the actual data online, media contents can be displayed in a web browser using the data URI technique. Actually, the phishing web pages are not at all hosted anywhere on the internet, conventional anti-phishing techniques are unable to identify it. In this attack, people can be tricked without having to interact with a server.

After extracting the features. the data is split into 100-50 i.e., 10000 training samples & 5000 testing samples. This is an obvious supervised machine learning challenge based on the dataset. In Supervised Machine Learning, regression and classification are two major issues. This data collection has a classification issue because the input URL might be either legal or phishing (1). (0). XGBoost, Random Forest, Naive Bayes, Support Vector Machines, and Decision Tree are the supervised machine learning models taken into consideration to train the dataset in this work. All of these models were developed using the dataset, and the test dataset was used to assess the models.

Input: dataset which contains URLs information
  Output: Features ∈ [0,1], 1 indicate phishing link; 0 indicate legitimate link
  Begin

1. Collect the Dataset
     wget command is used to collect csv file which contains phishing URLs
     1.1 loading the phishing URLs data from "onlin-valid.csv" file
            data1 = read(onlin-valid.csv)

     1.2 Loading the legitimate URLs data from "Bening_list_big_final.csv" file
            data2 = read (Bening_list_big_final.csv)

2. Select the margin value of 10000 phishing URLs & 5000 legitimate URLs.
     2.1 Collect 10000 Phishing URLs data randomly.
            P_url = random. Sample (range (0, data1), 10000)

     2.2 Collect 5000 Legitimate URLs randomly.
            L_url = random. Sample (range (0, data2), 5000)

3. Extract the features from the URLs dataset
     3.1 checking No. of dots in URL
            If dots in URL >= 3, then
            set F1=1
            else set F1=0

     3.2 Presence of special symbol in URL
            If URL having '@ '|| '– 'symbol then
            set F2=1
            else set F2=0

     3.3 URL length
            If URL length >72, then set F3=1
            else set F3=0

     3.4 Suspicious words in URL
            If URL having suspicious word, then set F4=1
            Else set F4=0

     3.5 Position of the top-level domain
            If top level domain name presents in base domain one time or more than one
            time, then set F5=1.
            Else set F5=0

     3.6 http count in URL
            if http count in URL greater than 1, then set F6=1
            else set F6=0

     3.7 Brand name in URL
            if a popular band's name appears in the URL incorrectly, then Set F7=1
            Else set F7=0

     3.8 Data URI
            if Data URI present, then set F8=1
            else set F8=0

```
  4. Function to extract features
       feature Extraction (url, label)
     f_list= [ ]
       f_list.append(F1(url)
       f_list.append(F2(url))
       f_list.append(F3(url))
       f_list.append(F4(url))
       f_list.append(F5(url))
       f_list.append(F6(url))
       f_list.append(F7(url))
       f_list.append(F8(url))

    4.1 Legitimate URLs
         Extracting the features from data1 & storing them in a list
         converting the list to data frame

    4.2 Phishing URLs
         Extracting the features from data2 & storing them in a list
         converting the list to data frame

    4.3 Concatenating the data frames into one
         Storing the data in CSV file

  5. The classification models for supervised machine learning that were thought to be
     trained on the dataset
         Apply classification algorithms to identify legitimate URLs and phishing URLs.
         Analyse the models and forecast the results in terms of accuracy

     End
```

This phishing algorithm helps the users to be identify the website, weather it is phishing URL or Legitimate URL. The attackers these days are sending links which looks as an official link. The attacker or the hacker sends a URL request or a link to the user to capture the confidential information. People unaware of such attackers, imagine that they might have got this information from the official company or bank and give their details or clicks the URLs.

4   PERFORMANCE EVALUATION

The suggested method creates a true or false classifier based on the eight features, correctly classifying legitimate and phishing websites. The performance matrix and training and testing dataset used in our suggested approach are described in this section.

4.1   *Dataset training and testing*

the dataset is split into 10000 training samples & 5000 testing samples. We have collected Phishing URL dataset from Phishtank [19] source and Legitimate URLs dataset from University of New Brunswick [20] sources. this data set comes under classification problem, as the input URL is classified as phishing (data1) or legitimate (data2). The classification technique considered to train the dataset such as Decision Tree, Random Forest, XGBoost, Support Vector Machines, Naive Bayes. All of these models were developed using the dataset, and the test dataset was used to assess the models.

4.2 Environment set-up

Here we used the laptop machine with the specification of i5 intel processor with 3.0 GHz clock speed and 8.810 GB RAM to the proposed anti-phishing framework. Python programming is employed to carry out our suggested strategy. Python has extensive library support and has a fast compilation time. In order to identify phishing websites, we first determined the pertinent and practical features. We next created a dataset by extracting features from both legal and phishing websites. The random forest classifier is trained using the labelled dataset.

For each feature, a separate function has been developed. More libraries are needed to extract the website features. These libraries can be downloaded and extracted from the original websites. The approach is made more dependable and effective by feature extraction from the webpage's source code, which helps to decrease processing and response times.

4.3 Performance metric

To assess the effectiveness of the suggested anti-phishing technique, we calculated the accuracy, true positive rate (TPR), false positive rate (FPR), true negative rate (TNR), and false negative rate (FNR). These are the common measures used to evaluate anti-phishing strategies. NL and NP indicate the total number of genuine and phishing websites, respectively.

In the Table 1, different classification algorithms False Positive Rate (FPR), True Positive Rate (TPR) and accuracy rate is compared. Among this XGBoost algorithm have less False Positive Rate (2.43%) and greater True Pasitive Rate (89.41%). It is also showing greater accuracy rate (98.25%).

Table 1. Performance of proposed system.

Algorithms	FPR (%)	TPR (%)	Accuracy (%)
SVM	5.25	97.32	97.26
Naive Bayes	3.49	96.46	96.89
Decision Tree	2.29	97.53	97.15
XGBoost	2.43	98.41	98.25
Random forest	1.35	97.89	98.19

The same results also shown in the below graph.

Table 2. Comparison of classification algorithm in terms of accuracy.

Algorithm	Training Accuracy (%)	Testing Accuracy (%)
SVM	79.8	81.8
Naive Bayes	85.3	86.3
Decision Tree	81.4	83.4
XGBoost	86.6	86.4
Random forest	81.4	83.4

The dataset generated using proposed algorithm is trained and tested with different classification techniques as show in above table and evaluated to find out the accuracy among them. Figure 5 shows the training and test dataset accuracy by the respective classification techniques. It's clearly states that XGBoost classifier achieves greater accuracy than the other classification.

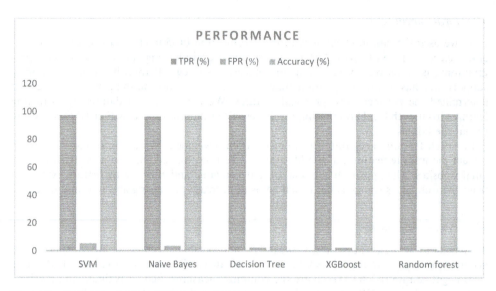

Figure 4.   Performance of classification techniques.

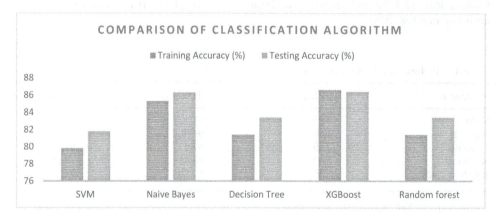

Figure 5.   Comparison of classification algorithm.

5   CONCLUSION

This paper presents a new URL based technique to identify the phishing websites at client-side URL. The classification technique is used to classify inputted URL to legitimate or phishing. The main contribution of this work is to identify the various features to detect the URL is phishing URL or Legitimate URL. With the proposed system we met the accuracy of 86.6%. for further future work we have to integrate more feature to increase the accuracy and to detect the phishing attack more accurate.

The phishing detection algorithm is entirely depending upon the set of features, which are used to detect the URL is phishing URL or legitimate URL. Therefore, the detection of phishing URL has to extend to consider the more features, content based, domain-based and other techniques is the aim of our future scope.

REFERENCES

[1]   Jain A. K., & Gupta B. B. Detection of Phishing Attacks in Financial and E-Banking Websites Using Link and Visual Similarity Relation. *International Journal of Information and Computer Security, Inderscience*, 2017 (Forthcoming Articles).

[2]   Gupta S., & Gupta B.B. Detection, Avoidance, and Attack Pattern Mechanisms in Modern Web Application Vulnerabilities: Present and Future Challenges. *International Journal of Cloud Applications and Computing*, 7(3), 1–43.

[3]   Almomani A., et al. A Survey of Phishing Email Filtering Techniques. *IEEE Communications Surveys & Tutorials*, 15.4, 2070–2090.

[4]   Gupta B.B., et al. Fighting Against Phishing Attacks: State of the Art and Future Challenges. *Neural Computing and Applications*, 28(12), 3629–3654.

[5]   APWG Q4 2016 Report available at: http://docs.apwg.org/reports/pwg_trends_report_q4_2016.pdf. Last accessed on September 22, 2017.

[6]   Razorthorn Phishing Report, Available at: http://www.razorthorn.co.uk/wp-content/uploads/2017/01/Phishing-Stats-2016.pdf. Last accessed on September 22, 2017.

[7]   Gowtham R., & Krishnamurthi I. (2014). A Comprehensive and Efficacious Architecture for Detecting Phishing Webpages. *Computers & Security*, 40, 23–37.

[8]   Aboudi N. E. & Benhlima L. Parallel and Distributed Population-based Feature Selection Framework for Health Monitoring. *International Journal of Cloud Applications and Computing*, 7(1), 57–71.

[9]   Kay S. Globalization, Power, and Security. *Security Dialogue*, 35(1), 9–25. https://doi.org/10.1177/0967010604042533.

[10]   Psannis K., *Defending Against Phishing Attacks: Taxonomy of Methods, Current Issues and Future Directions.*

[11]   Baykara M. & Gurel Z.Z. "Detection of Phishing Attacks", *2018 6th International Symposium on Digital Forensic and Security* (ISDFS), 2018.

[12]   Hussah A., Rasheed A. & Eyas E., *Reverse of E-Mail Spam Filtering Algorithms to Maintain E-Mail Deliverability.* 297–300. 10.1109/DICTAP.2014.6821699.

[13]   Montazer G. A., ArabYarmohammadi S. Detection of Phishing Attacks in Iranian E-banking Using a Fuzzy-rough Hybrid System. *Applied Soft Computing*, 35, 482–492.

[14]   Xiang G., Hong J., Rose C. P., & Cranor L. Cantina+: A Feature-rich Machine Learning Framework for Detecting Phishing Web sites. *ACM Transactions on Information and System Security* (TISSEC), 14 (2), 21.

[15]   El-Alfy E. S. M. Detection of Phishing Websites Based on Probabilistic Neural Networks and K-medoids clustering. *The Computer Journal*. https://doi.org/10.1093/comjnl/bxx035.

[16]   Zhang W., Jiang Q., Chen L. & Li C. Two-stage ELM for Phishing Web pages Detection using hybrid features. *World Wide Web*, 20(4), 797–813.

[17]   Verified Phishing URL, Available at: https://www.phishtank.com. Last accessed on November 9, 2022.

[18]   Phishing dataset available at: https://www.openphish.com/. Last accessed on November 9, 2022.

[19]   URL dataset (ISCX-URL2016) at: The University of New Brunswick, https://www.unb.ca/cic/datasets/url-2016.html. 2016.

[20]   Dataurization of URLs for a More Effective Phishing Campaign. Available at: https://thehackerblog.com/dataurization-of-urls-fora-more-effective-phishing-campaign/index.html. Last accessed on November 9, 2022.

[21]   Jain A.K., Gupta B.B. Towards detection of phishing websites on client-side using machine learning based approach. *Telecommun System* 68, 687–700, 2018. https://doi.org/10.1007/s11235-017-0414-0

[22]   Aljofey A., Jiang Q., Rasool A. *et al.* An Effective Detection Approach for Phishing Websites using URL and HTML features. *Sci Rep* 12, 8842, 2022. https://doi.org/10.1038/s41598-022-10841-5

Automation and Computation – Vats et al. (Eds)
© *2023 the Author(s), ISBN 978-1-032-36723-1*

Current and future roadmap for artificial intelligence applications in software engineering

Isha Sharma
Assistant Professor, Chandigarh University, Chandigarh, India

Rahul Kumar Singh
Assistant Professor, UPES, Dehradun, India

Pardeep Singh
Assistant Professor, GEHU, Dehradun, India

ABSTRACT: To improve the development process of software in the current scenario there is a need for applying Artificial intelligence techniques. There are number of techniques that can be used and advocated by researchers like knowledge-based systems, fuzzy logic, neural networks, and data mining. Software, because of its dynamic nature in the context of operation, needs improvement in the quality. Traditionally the expertise of the developer, with the same kind of projects and his knowledge is the parameter for improving the quality but the dynamic nature and the volatile environment today stresses upon in exploring the potential of AI for making all the improvements in the development phases of software being developed. The objective of this paper is to bring out the application of AI in software engineering and suggesting the best method of AI that will help in the overall improvement in software development phases. This paper covers the various AI methods like neural networks, genetic algorithms, fuzzy logic, and natural language processing techniques for software engineering and provides an organized survey of current and future aspects of AI in software engineering.

1 INTRODUCTION

The software engineering industry is witnessing a software crisis since a half-century ago. A large number of methodologies, techniques, and environments have been unitized and implemented to solve the software crisis issue. Though the changing scenario of the market and the complex nature of the software to be developed along with the dynamic state of software had hit hard in the software industry and is halting the progress. This changing scenario is leading to more chances of software failures. Various software fails before the actual development work starts. This makes the forecasting of the estimated development time or the delivery time taken by the project a difficult step. Sometimes there might be the case that due to changing constraints about the software, the deadlines need to be shifted and thus impacting the overall cost by increasing the budget for the development (Fox, *et al.* 2005). This adds more risk to the risk analysis module that needs to be used before the start of the software project to predict the delivery dates.

Researchers have seen a perceptible rise in the area of Artificial Intelligence (AI) that enforces and recommends a change in software engineering (Booch 1986). The research paper is an endeavor that focuses on the applicability of AI in improving software engineering. An attempt has been made to look for the perspective in using AI methods for the

DOI: 10.1201/9781003333500-26

development of software projects that are more prone to risks to save them for immature fallouts due to changing market scenarios (Aggarwal *et al.* 2021).

This review article differs from the previous article in such aspects a (1) classified the current studies based on various AI applications utilized in the software development life cycle. (2) the article emphasized the use of AI applications in the different phases of SDLC. (3) summarized the available and used techniques of AI in software development.

The major objective of this survey paper is to understand the use of AI methods in the field of software development life cycle and how can we enhance the current system. The paper also brought out the future prospects of using AI in software development.

The structure of the paper as follows: Section 2 defines the project planning based on AI. Requirement specification and software design based on AI is described in section 3. Software coding and testing phrase-based AI techniques are defined in section 4. Section 5 summarizes the AI techniques used in SDLC. At last, conclude the article in section 6.

2 ARTIFICIAL INTELLIGENCE BASED PROJECT PLANNING

Before the actual development of the project, there are number of factors that need to be planned off before time. Planning these in advance is required for the success in development of software projects. These factors not only include planning the project completion time well in advance and total cost for the project, but it also involves the staffing requirements and dependencies between various activities in the project. These are mere estimates that will be guiding the correct path to be followed in the project development.

The research is now focusing on the application of Artificial intelligence in the area of software engineering. Different AI approaches have been proposed that allow the use of AI in various phase of software development life cycle model. Figure 1. demonstrates the various AI techniques like genetic algorithms (Singh *et al.* 2022), case-based reasoning, knowledge-based systems, and neural networks in the planning of the software project and its usefulness.

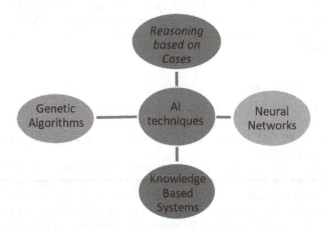

Figure 1. AI techniques in the planning of the software project.

2.1 *Genetic algorithms*

The researches in the application of Genetic algorithms (GA) to scheduling the project activities are very large in number in various areas (Tewari et al. 2013) (Ammar et al. 2012) (Harman 2012) (Jain 2011). An investigation of their appliance in manufacturing and operations management can be found in (Meziane *et al.* 2009) (Raza 2009). In GA terminology the project plan is viewed as a constraint satisfaction problem with an objective. The objective is to optimized and transformed into the optimal problem with the aid of GA.

In (Srivastava *et al.* 2009), the author proposed evolving functions using a Genetic Algorithm for calculating the total effort required to develop the software project well in advance time. A wide range of mathematical functions and control statements are used that are applied on target grammars. As per the statistics, more than 400 software projects are used as training set data that is tested for more than 30 features. The major features are being tested using the approach for example the requirements for the projects, changing market scenario, type of application. The output is collected and compared with the other approaches and was found to be better. The GA approach uses non-linear team size as compared to traditional estimates used for calculating total effort for the software project.

Assigning the staff to activities is crucial for any software project that had a direct impact on the schedule of the activities too. In (Hooshyar et al. 2008), the author also uses GA to produce optimal time and effort estimates. The research proposes chromosomal representation for assigning people to the activities and maintaining the precedence relationship among the activities so that the child produced after cross-over operation ensures a valid timeline. The outcome of this approach validates the usage of GA in the field of software development and more in software planning. A future proposal of finding the usage of GA when the project deadlines got shifted due to uncertain or certain reasons is still required.

It has been observed that while developing the software projects, quality's focused more as a result improving the quality of project might result in increased cost of the project and also changing the completion time. This further increases user satisfaction. It has been also observed that higher quality is also achievable by adding more resources to the project, thereby helping in decreased overall cost though the activity level cost is high and early development of the project (Pankwar *et al.* 2022).

In (Bertolino 2007), the author advocated the application of GA to deal with these variations. In their experiment, the crossover and mutation operations are used to represent the schedule of the project activity. A fitness function is there that considers both cost and time factors. The experiment is implemented on a large number of projects having numbers of activities not more than 30. The experiment favored the use of GA In (Sharma *et al.* 2018), the author suggested a Meta-Heuristic Approach for Software Project Risk Schedule Analysis using GA. The research illustrates a technique to shorten the time duration using the structured method.

2.2 *Reasoning based on cases*

Experiences from the past of the development of projects are one of the factors that affect the success of the software projects. The more bad experiences the developer has gained, the more chances is there that there is less probability of repeating the same experience in developing similar projects.

A less number of researches are there that support the use of CASE-BASED REASONING (CBR) for planning the development of the software. In (Kobbacy et al. 2007), the author proposed the use of CBR in planning the software project. HCA (Hierarchical Criteria Architecture) is used for representing the projects in terms of cases where the case is describing the project in terms of requirements of the client, resources that the project needs. HCA uses varying weights allowing variations to be tested and provide more flexibility to the project team leader. With the new project, similar cases are extracted using data mining. This aids project planning. The author (Yang *et al.* 2009) used more than 40 projects as an input set to show the applicability of CBR and data mining to help in finding out optimal total completion time of the project with more probability percentage.

2.3   Knowledge based systems

Experience in working with similar projects in past helps in planning the new projects opti-mally. Several studies have been made that support this analogy and use this experience in Knowledge based Systems (KBS). In (Reddy, et al., 1985), the author supported the use of KBS in the planning of the projects by providing a well-defined representation for the same. They proposed new semantics to support successful project planning. A frame-based lan-guage is used for developing this representation. The research supports the use of the lan-guage for characterizing the project milestones, targets, tasks, and duration. Other researches point out the usage of associative networks and production rules supporting KBS approach to be implemented (Shradhanand *et al.* 2007). However, the adoption of KBS for representing knowledge is a costly affair. In (Aljebory *et al.* 2019), the author focused on using an automated Software Modelling information software tool that will assist in project planning in easy manner using the concept of expert systems.

2.4   Neural networks

Neural networks have been extensively and effectively used for problems that require clas-sification. Neural networks are found to be perfect for the software engineering domain thereby dealing with the situations where there is a requirement of making a prediction for example about the risks in the maintenance module (Yong et al. 2006), analyzing the risks (Wappler *et al.* 2006) and for fault prediction using matrices (Ge *et al.* 2006). In (Hu et al. 2006), the authors proposed new research in this domain. They used the approach of iden-tifying the risk assessment features based on old classifications by taking the input by interviewing the project managers. A total of 39 risk factors are found in the study that is further classified into 5 risk categories. The area of software engineering, project manage-ment, teamwork, cooperation, and project complexity is found to be more risky in nature and the classification of risk is made according to these parameters. Further using the PCA (Principal components analysis) approach, the 39 risk factors are dependent on each other are reduced to 19 independent factors. As per the study made, and experiments are per-formed, the Neural network approach was found to be the best approach for forecasting the risk. A total of 50 examples are used for this study. 35 examples out of the total 50 examples are used for training and 15 examples are used for testing purposes. A genetic algorithm trained Neural Network was used for the prediction and the result are grouped under suc-cessful projects, partially failed projects, and failed projects. Although the confidence in the research that Neural Network is better the approach was made on the small data set and more number of projects needed to be tested to have permanent confidence in the research made that calls for future extension in the research work made.

3   REQUIREMENTS ENGINEERING AND SOFTWARE DESIGN

Requirements engineering is the first phase of any software development life cycle model. This phase is part of every life cycle model and is the most important phase too. The phase is actually not a single step of execution but has multiple areas defined in it that seek attention (Briand et al. 2005). Requirement engineering helps the companies in defining the require-ments of the clients but due to different perceptions of different people involved in software development like company members and clients, there is need to express requirements in such a language that is natural in understanding and need to be documented (Baresel et al. 2004). As we already know the requirement engineering phase is not single activity rather it involves sub-activities like gathering of the needs for the project and analysis of the same. The valid document achieved after requirement engineering is the valid set of requirements that have been added in the contract between the client and the company as the requirements to be achieved. The requirement specification document is then further formulated to be base for designing the software. Though it sounds so easy that gathering the requirements is not a tedious task but it is actually the most important and deciding task in software development that form the basis for effective design. Therefore it is essential to look for the gaps in the requirement gathering and analysis step do that before actual implementation of the system in form of designing it , the same will be resolved and not carried forward thereby making a loop of problems and errors in the software process. The following gaps are identified during this phase:

- The natural language expression of the requirements are often found to be ambiguous thereby making the requirements invalid and inconsistent. [33]
- Requirements are mostly the confusing needs of the customers that is maximum times not sure about what they demand from the software process being developed d by the organization. this incompleteness and fuzzy nature of the customer need calls for another gap in requirement engineering.
- Requirement are dynamic in nature and hence keeps on changing and improving over the time and time need to be devoted to incorporate them in the software system
- The most exigent task in this phase is perceptive the requirements of the customer.
- The management of requirements is also a difficult task. The success of this task will further decide the success of other recurring tasks that follows requirement engineering.

Artificial intelligence is involved in this phase of engineering requirements in software projects as follows:

- By developing the tools that are able to extract and transform natural language requirements in no ambiguous form.
- By providing knowledge-based systems and ontologies.
- Combining intelligence at computation level to deal with problems coming up with the requirement engineering phase

4 SOFTWARE CODING, TESTING AND ARTIFICIAL INTELLIGENCE

4.1 *Coding*

A number of AI techniques are there that will help in easing the task of coding for the real world software projects. With the help of AI it is possible to mechanize the entire process of software coding. The basic idea behind this mechanized task is to involve the programmer in developing the algorithmic solution for the problem, and then to use AI techniques that will generate automatic functions or the complete program as per the programmer specification. There are various AI methods that can be applied. The below is the list of such methods contributing computational intelligence in software domain:

- Object oriented method (Gupta et al. 2004)
- Case based Reasoning analogies
- Constraints prior programming approach
- Genetic Algorithms
- Heuristic software engineering based on search optimization

In (Husain et al. 2019), the author devised Artificial Intelligence for semantic-based code searching in which the required code is extracted by querying the software system with the application of NLP. The AI based software supports a huge number of functions that will ease the querying based on natural language from the software.

4.2 *Testing*

The important phase of developing the software is to check for the effective working by means of testing the software against anomalies. The task of testing the software being developed is a lengthy procedure. It will be taking more time if done manually by team of testers, therefore, is requiring the need of computerizing the whole task of testing and there AI is an important component that will help in this domain. In 1991, DeMasie and Muratore has drafted the work in this direction by using the concept of constraint-based testing (DeMasie *et al.* 1991). In (Bering *et al.* 1988), the authors suggested the use of knowledge based expert systems to automate the process of testing. In the mid of 1990, the concept of planning testing using AI has blossomed by the development of the planner supporting the generation of test cases and executing the same to find the loopholes in the software being developed. In (Wallace *et al.* 2004), the author supported the use of genetic algorithms for simulating the task of software testing and supports the generation of optimal cases for testing.

In 2019–2020 (King et al. 2019) (Liang et al. 2020) recommended the use of Deep neural networks that will help in finding out the complex anomalies software code. Deep Neural Networks have the advantage of adaptation that will help the software development team in identifying the rare errors and thus improves the overall software quality. AI-based fuzzy systems are found to be far accurate than the other available hybrid or manual versions (Takanen et al. 2018).

AI eases the process of software based testing and because of the automation profile, it is helping in saving the manpower hours in executing the tests and documentation overheads as well. The software development cost and the Time to market costs are also found to be reduced significantly (Hourani et al. 2019).

5 ARTIFICIAL INTELLIGENCE TECHNIQUES USED IN SOFTWARE ENGINEERING

This section focuses on the available and used techniques from the AI domain that are applicable in software development. Figure 2 demonstrates the AI techniques in SDLC.

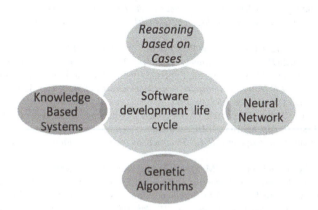

Figure 2.   AI technique used in SDLC.

Table 1 describes the available and used techniques of AI in the SDLC.

Table 1. AI techniques for software development.

Techniques of AI	Where to apply in software engineering
Neural networks	Risk analysis, software Maintenance
Genetic Algorithms	Software testing
Fuzzy systems	Software requirements and engineering, software testing
Knowledge-based systems	Software design, planning
Natural Language processing	Requirement engineering
Expert system	Software risk management
Case based Reasoning analogies	Software coding
Software engineering based on search optimization	Software coding

6 CONCLUSION AND FUTURE SCOPE

This paper aims to find the application of artificial intelligence and software engineering. The introduction of AI was targeted in the area of software engineering to solve the real-time problems software development life cycle for developing the projects faced. The application of computational intelligence in the software domain has seen major decrease in the development time of the software development and decrease in cost for the production of the same and hence resulted in better quality projects.

REFERENCES

Aggarwal Alankrita [*et al.*] An Implementation Analysis of Risk Mitigation in Software Reusability using Matrix Approach. *International Conference on Emerging Technologies: AI, IoT, and CPS for Science & Technology Applications*. NITTTR, Chandigarh: *CEUR Workshop Proceedings*, 2021. Vol. 3058.

Aljebory K M and QaisIssam M Developing AI Based Scheme for Project Planning by Expert Merging Revit and Primavera Software. In *2019 16th International Multi-Conference on Systems, Signals & Devices (SSD)* (pp. 404–412). IEEE. 2019.

Ammar H H, W Abdelmoez and Hamdi M S Software Engineering Using Artificial Intelligence Techniques: Current State and Open Problems. *In Proceedings of the First Taibah University International Conference on Computing and Information Technology (ICCIT 2012)*, Al-Madinah Al-Munawwarah, Saudi Arabia (Vol. 52). 2012.

Baresel A [*et al.*] Evolutionary Testing in the Presence of Loop-assigned Flags: A Testability Transformation Approach. *ACM SIGSOFT Software Engineering Notes*. 2004. Vol. 29 (4). pp. 108–118.

Bering C A and Crawford M W Using an Expert System to test a Logistics Information System. In *Proceedings of the IEEE National Aerospace and Electronics Conference* (pp. 1363–1368). Dayton, OH. Washington DC: IEEE Computer Society. 1988.

Bertolino A Software Testing Research: Achievements, Challenges, Dreams. In *Future of Software Engineering* (FOSE'07) (pp. 85–103). IEEE. 2007.

Booch G Object-oriented Development. *IEEE Transactions on Software Engineering*. 1986. Vol. 2. pp. 211–221.

Briand L C, Labiche Y and Shousha M Stress Testing real-time Systems with Genetic Algorithms. *In Proceedings of the 7th Annual Conference on Genetic and Evolutionary Computation* (pp. 1021–1028). 2005.

DeMasie M P and Muratore J F Artificial Intelligence and Expert Systems In-flight Software Testing. In *Proceedings of the Tenth IEEE Conference on Digital Avionics Systems Conference* (pp. 416–419), Los Angeles, CA, Washington DC: IEEE Compute. 1991.

Fox T L and Spence J W The Effect of Decision Style on the use of a Project Management Tool: An Empirical Laboratory Study. *ACM SIGMIS Database: The Database for Advances in Information Systems*. 2005. Vol. 36(2). pp. 28–42.

Ge Y and Chang C Capability-based Project Scheduling with Genetic Algorithms. In *2006 International Conference on Computational Inteligence for Modelling Control and Automation and International Conference on Intelligent Agents web Technologies and International Commerce (CIMCA'06)* (pp. 161–161). IEEE. 2006.

Gupta M [*et al.*] Automated test Data Generation Using MEA-graph Planning. In *16th IEEE International Conference on Tools with Artificial Intelligence* (pp. 174–182). IEEE. 2004.

Harman M The Role of Artificial Intelligence in Software Engineering. In *2012 First International Workshop on Realizing AI Synergies in Software Engineering (RAISE)* (pp. 1–6). *IEEE.* 2012.

Hooshyar B, Tahmani A and Shenasa M A Genetic Algorithm to Time-Cost Trade off in project scheduling. In *2008 IEEE Congress on Evolutionary Computation (IEEE World Congress on Computational Intelligence)* (pp. 3081–3086). IEEE. 2008.

Hourani H, Hammad A and Lafi M The Impact of Artificial Intelligence on Software Testing. In 2019 *IEEE Jordan International Joint Conference on Electrical Engineering and Information Technology (JEEIT)* (pp. 565–570). IEEE. 2019.

Hu Y [*et al.*] A Neural Networks Approach for Software Risk Analysis. In *Proceedings of the Sixth IEEE International Conference on Data Mining Workshops* (pp. 722–725), Hong Kong. Washington DC: IEEE Computer Society. 2006.

Husain H [*et al.*] *Codesearchnet challenge: Evaluating the State of semantic code search.* arXiv preprint arXiv:1909.09436. 2019.

Jain P Interaction Between Software Engineering and Artificial Intelligence-a Review. *International Journal on Computer Science and Engineering.* 2011.Vol. 3(12). p. 3774.

King T M [*et al.*] AI for Testing Today and Tomorrow: Industry Perspectives. In *2019 IEEE International Conference On Artificial Intelligence Testing (AITest)* (pp. 81–88). IEEE. 2019.

Kobbacy K A, Vadera S and Rasmy M H AI and OR in Management of Operations: History and Trends. *Journal of the Operational Research Society.* 2007. Vol. 58(1). pp. 10–28.

Liang H [*et al.*] Sequence Directed Hybrid Fuzzing. In *2020 IEEE 27th International Conference on Software Analysis, Evolution and Reengineering (SANER)* (pp. 127–137). IEEE. 2020.

Meziane F and Vadera S Artificial Intelligence Applications for Improved Software Engineering Development: New Prospects. *IGI Global.* 2009.

Pankwar Deepak [*et al.*] Firefly Optimization Technique for Software Quality Prediction. Soft Computing: Theories and Applications. *Lecture Notes in Networks and Systems*, Springer, 2022. Vol. 425.

Raza F N Artificial Intelligence Techniques in Software Engineering (AITSE). In *International MultiConference of Engineers and Computer Scientists (IMECS2009).* 2009. Vol. 1.

Reddy Y R [*et al.*] KBS: A Knowledge Based Simulation System. *IEEE Software.* Camegie Mellon University, 1985: Vol. 3(2). p. 26.

Sharma I and Chhabra D Meta-Heuristic Approach for Software Project Risk Schedule Analysis. In *Analyzing the Role of Risk Mitigation and Monitoring in Software Development IGI Global.* 2018. pp. 136–149.

Shradhanand A K and Jain D S Use of Fuzzy Logic in Software Development. *Issues in Information Systems.* 2007. Vol. 8(2). pp. 238–244.

Singh Pardeep [*et al.*] *Knowledge Application to Crossover Operators in Genetic Algorithm for Solving the Traveling Salesman Problem.* IGI Global, 2022. Vol. 10(1).

Srivastava P R and Kim T H Application of Genetic Algorithm in Software Testing. *International Journal of software Engineering and its Applications.* 2009. Vol. 3(4). pp. 87–96.

Takanen A [*et al.*] *Fuzzing for Software Security Testing and Quality Assurance.* Artech House, 2018.

Tewari J, Arya S and Singh P N Approach of Intelligent Software Agents in Future Development. *Int. J. Adv. Res. Comput. Sci. Softw. Eng.* http://ijarcsse. com/Before_August_2017/docs/pap ers/Volume_3/5_ May2013/V3I4-0319. pdf. 2013.

Wallace L and M Keil Software Project Risks and their Effect on Outcomes. *Communications of the ACM.* 2004. Vol. 47(4). pp. 68–73.

Wappler S and Wegener J Evolutionary Unit Testing of Object-oriented Software Using Strongly-typed Genetic Programming. In *Proceedings of the 8th annual conference on Genetic and evolutionary computation* (pp. 1925–1932). 2006.

Yang H L and Wang C S Recommender System for Software Project Planning One Application of Revised CBR Algorithm. *Expert Systems with Applications.* 2009. Vol. 36(5). pp. 8938–8945.

Yong H [*et al.*] A Neural Networks Approach for Software Risk Analysis. *In Sixth IEEE International Conference on Data Mining-Workshops (ICDMW'06)* (pp. 722–725). IEEE. 2006.

Automation and Computation – Vats et al. (Eds)
© 2023 the Author(s), ISBN 978-1-032-36723-1

An efficient Artificial Intelligence approach for identification of rice diseases

Richa Gupta & Amit Gupta
Graphic Era Hill University, Dehradun, Uttarakhand, India

ABSTRACT: In India rice is the primary Kharif crop which is cultivated in the autumn season and consumed more or less in every state of India however the crop may be vandalized by numerous diseases. In agriculture domain the automatic identification of rice disease is essential as it influence the economy as well as the food supply of the nation. Modern AI methods are appropriate choice for detection of rice diseases as it has the capabilities to extract the features which are important for the classification and remove the undesirable and destructive features. This paper presents an intelligent method for classification of rice disease among three classes. It is demonstrated in the paper that Deep Neural Network can predict the diseases more efficiently than other AI methods like SVM. The classification accuracy of 98.3%, with area under the curve 99.9% achieved by the model.

Keywords: Artificial Intelligence, CNN, Machine Learning, SqueezeNet, Brown Spot, Leaf Blast, Hispa

1 INTRODUCTION

In India and some other countries, rice is important and basic crop which it is cultivated in almost all Asian countries [1]. In India specially the Southern area of the country heavily depends on the rice for their food [2]. The quality and the quantity both decreases due to any disease in the crop [3,12]. Early and accurate diagnosis of the crop disease is very important as farmers depend on the crop production economically [4]. If using various ML techniques, some of the diseases like brown spot, hispa and leaf blast in rice crops can be forecast at the early stage then the loss can be minimized So that the crop can be protected by taking the safety measures like pesticides. The detection or diagnosis of these particular disease can be done with the help of expert advice of experienced persons but sometime it take too much time that the crop is compromised, thus developing some automated models can be used for the early detection of the diseases. With the technological and conceptional of AI many automated models have been developed yet for early detection of the crop disease.

In this digital era (data generated exponentially), AI methods like machine learning algorithms, Neural networks and computer vision plays a great role in implementation of these types of models [17–31]. These methods are utilized for the important feature extraction from the data set of the images of the crops [2]. AI methods enable the model to detect the diseases in timely manner so that preventive measures can be used to reduce the loss. This improves the scope of AI in the field of agriculture. Different CNN models are available like ResNet, VGG-19, VGG-16, Inception V3 mobileNet, GoogleNet & SqueezeNet.

In the proposed work the CNN model SqueezeNet is utilized for the feature extraction. These CNN models have multiple layers within the artificial neural networks [4]. Multiple layers have multiple features which can be trained so that the prediction will be accurate. Squeezenet has 1.2 million learnable parameters.

  DOI: 10.1201/9781003333500-27

Further, paper is organized as section-2 described the work previously done by the researchers, section-3 discussed the proposed methodology with the SqeezeNet architecture. Results and Discussion part is done in the section-4 and the last section is the conclusion of the paper.

2 RELATED WORK

CNN with inceptionv3 and VGG16 is applied to find out the disease in the rice crop [1]. Five CNN models called VGG-16, mobile net, Nas Net, inception V3, squeezenet and simple CNN, are utilized for the experiment by the authors and three types of training namely: Baseline Training, Transfer Learning & Fine Tuning and is applied on all five CNN models. VGG-16 outperforms among all the models with 97.12% with fine tuning learning.

CNN with the ReLu give better results for classification problems [2]. In the implementation of the model two types of CNN architecture i.e., GoogleNet and Cifar10 is applied on the 9 kinds of maize leaf diseases and achieve the accuracy of 98.9 and 98.8% respectively.

Dense-net is a deep CNN which is called Dense-Incep is applied on rice data set and the model performance is outstanding for the classification of disease [3]. Different CNN models VGG-16, VGG-19, ResNet, DenseNet, InceptionV3 with the proposed model is utilized for the experiment and the proposed model outperforms with 90% accuracy.

With CNN feature extraction approach and Transfer learning is applied to find the better results [4] on the millet crop to identify the mildew diseases. VGG-16 used with feature extraction which is an approach of transfer learning. VGG-16 model used the weights from imageNet. The performance of the model give the accuracy of 95%.

Transfer learning of CNN is applied on rice disease with four classes for classification [5] which also give above 90% accuracy with 80–20 deviation of dataset. The authors use Convolutional Neural Network as feature extractor and Support Vector Machine as classifier. AlexNet model is used which is pretrained on ImageNet.

Another new approach for weather-based prediction of plant disease with Support Vector Machine is introduced [6]. Six weather variables are used as prediction features. Two models called cross location and cross year are utilized. Web server for the prediction is also developed by the authors.

CNN is also compared with other machine learning models where CNN won the battle with 95% accuracy [7]. 10 rice diseases are predicted with the CNN model while PCA is used for feature extraction.

Image segmentation with soft computing technique is applied to find out the disease in the rice plants [8]. first the image segmentation is applied on the images and after segmentation the boundary detection method is applied. After the boundary is detected the spot detection is applied. For classification self-organizing map (SOM) neural network is used.

In another approach deep neural network with Jaya algorithm is implemented for classification of rice disease [9]. In this method authors first convert the RGB images into HSV images then the diseased part is extracted by saturation parts.

Support vector machine with deep features is also applied by some other authors and they find that it works better than some small CNN with 98.38% accuracy [10]. 13 other CNN models applied but ResNet50 with SVM performs outstanding. CNN models are used for feature extraction and after this SVM is applied for classification.

This indicate that a much research is carried out on the rice disease yet. In a survey paper [11] researchers find out the different data set of rice diseases and the machine learning technique used for classification. This paper demonstrates that that different image processing methods are used for identification of diseases with different machine learning techniques. Thus, it has been proved that Machine Learning which is a method of AI is widely adopted by researchers to build the models to find out the anomaly in the plants. A detailed survey is depicted in Table 1 for the previous contributions of the researchers.

Table 1. Analysis of different methods for different crop diseases.

S. No.	Authors and Year	Crops	Diseases Identified	Method Used	Accuracy
1	Rakesh Kaundal et al. [6] 2006	Rice	Rice Blast	SVM	
2	Santanu Phadikar et al. [8] 2008	Rice	Leaf Blast and Brown Spot	SOM Neural Network	92%
3	Haiguang Wang et al. [12] 2012	Wheat	Wheat stripe & leaf rust	principal component analysis (PCA) and backpropagation (BP) SOM	100%
4	Haiguang Wang et al. [12] 2012	Grape	Grape downy mildew & Grape Powdery Mildew	Generalized Regression & Probabilistic Neural Networks	94.29%
5	Yang Lu et al. [7] 2017	Rice	Rice Bacterial, leaf Blight Sheath Rot, Rice False Smut, Rice Blast, Rice Bakanae, Rice Brown Spot	CNN	95.48%
6	XIHAI ZHANG et al. [2] 2018	Maize	Dwarf Mosaic, Grey Leaf Spot, Curvularia, Brown Spot, Northern Leaf Blight, Round Spot & Southern Leaf Blight	Google Net	98.9%
7	Solemane Coulibaly et al. [4] 2019	Millet	Plant dead, yellowing, Malformation of ear, Plantule, Partial green ear	VGG-16	95%
8	Vimal K. Shrivastava et al. [5] 2019	Rice	Rice Blast, Sheath Blight, Heavy Leaf & Bacterial Leaf	CNN	91.37%
9	S. Ramesh et al. [9] 2019	Rice	Brown Spot, Bacterial Blight, Leaf Blast, Sheath Rot	DNN_JOA	98.9%
10	Junde Chen et al. [3] 2020	Rice	Rice White Tip, Rice Leaf Scald, Rice Leaf Smut, Rice Stack Burn, Rice Bacterial Leaf Streak	DENS-INCEP	92.86%
11	Chowdhury R. Rahman et al. [1] 2020	Rice	Bacterial Leaf Blast, False Smut, Brown Plant Hopper, Sheath Blight, Hipsa, Brown Spot	CNN-VGG-16	99%
12	Prabira Kumar Sethy et al. [10] 2020	Rice	Brown Spot, Bacterial Blight, Tungro, Blast	ResNet50	98.38%
13	Yibin Wang et al. 2021 [13]	Rice	Brown spot, Hispa, Blast	Attention based depth wise separable neural network with Bayesian optimization (ADSNN-BO)	94.65%
14	Vaibhav Tiwari et al. 2021 [14]	Six different crops	Different Diseases	Dense Neural Network	99.58%

(continued)

Table 1.   Continued

S. No.	Authors and Year	Crops	Diseases Identified	Method Used	Accuracy
15	Santosh Km. Upad-hyay et al. 2022 [15]	Rice	Leaf smut, Brown spot and leaf blight	CNN	99.7%
16	Dhiman Mondal et al. 2022 [16]	Tomato	Nine diseases	R-CNN	99.62%

3 PROPOSED METHODOLOGY

In this section the data set, System specification and the model which is utilised for the classification is described.

The entire set of images consist 3355 which is a large dataset for classification. In which 595 images of Brown Spot, 1416 images of healthy leaves, 565 images of Hispa and 779 of leaf blast. The data set contains four labelled class 1. Brown spot, 2. Hispa, 3. Healthy, 4. Leaf blast thus the multiclass classification is performed on the dataset. The division of the data set is in 70–30 ratio for training and testing which is the standard deviation. The model is applied on the jpeg images of rice leaves. The sample of raw images which are used as input to the model are given in Figure 1.

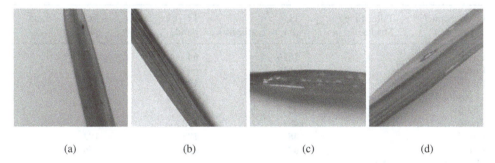

(a)   (b)   (c)   (d)

Figure 1.   Sample input images of four classes (a)Brown spot (b)Healthy (c)Hispa (d)Leaf blast.

A personal computer is used for the experiment with 4GB RAM, 2TB HDD with i5 processor.

In the proposed model CNN with Sqeeznet embedder is utilized with activation function ReLu. It displays accurate and precise prediction for the rice diseases classes. The dominant strength of this model is the incremented performance although the data is even noisy. First the CNN-SqueezeNet architecture is explained then the proposed model is explained in detail.

3.1   CNN-Squeeze Net

CNN used back-propagation for the extraction of features from the various image features. There are multiple layers in CNN like pooling, fully connected & Softmax layers. The architecture of CNN with SqueezeNet is given in Figure 2. SqueezeNet is a compressed architecture as it has relatively small number of parameters. In squeeze net CNN the

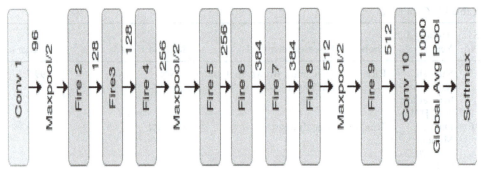

Figure 2.   Architecture of CNN-SqueezeNet.

squeezed convolution 1x1 layers are feed into the expand layers with 1x1 convolutional layers and 3x3 convolutional layers by reducing the number of filters. So the number of parameters are automatically reduced. SqueezeNet reduces the model size with higher accuracy. The connections in SqueezeNet are simple bypass which boost the accuracy. The SqueezeNet model layer architecture is shown in Table 2, in this Table all the details of filter size and stride size is illustrated. The interesting thing about the architecture of the SqueezeNet is the lack of fully connected layer and the extra feature which is added to this architecture is the "Fire modules". The advantage of the Squeezenet model is the reduction in the size of the model by other CNN architectures. Another advantage is the deep compression.

Table 2.   Layer architecture of the SqueezeNet model.

Layer Name/ Type	Filter Size / Strides	1x1 (#1 × 1 squeeze)	1x1(#1 × 1 expand)	3x3(#3 × 3 expand)
Fire 2		1616	64	64
Fire 3		1616	64	64
Fire 4		3232	128	128
Maxpool 4	3X3/2			
Fire 5		32	128	128
Fire 6		48	192	192
Fire 7		48	192	192
Fire 8		64	256	256
Maxpool 8	3X3/2			
Fire 9		64	256	256

Fire Module-It is the mixture of 1x1 and 3x3 convolution filters. A squeeze convolution layer of 1x1 is fed into this mixture. In the fire module 3 hyper parameters are there, first one is s1x1 is number of squeeze layer, second one is e1x1 is the number of expanded 1x1 layer and e3x3 is the number of expanded 3x3 layer. The number of s1x1 is always lesser than the number of e1x1 and e3x3 so that the limit of input channel can be controlled.

The proposed method works in four steps as illustrated in Figure 3, in the first phase the input images are fed into he CNN, then these images are passed to the squeezed convolutional layer. In second part the output of the first layer is fed into the expanded 1x1convolutional layer and 3x3 convolutional layer by reducing the number of layers. In third phase the output of this expanded layer is concatenated into one parameter, in last phase the output images are classified into different categories. 10-fold cross validation is used with learning rate 0.0001.

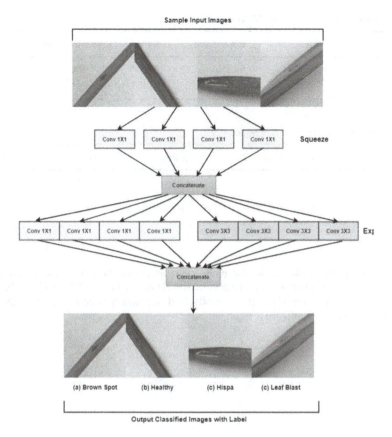

Figure 3.   Architecture of proposed CNN SqueezeNet model.

With SqueezeNet ReLu (Rectified Linear unit) activation function is used in the proposed method. It is mostly adopted function as it does not activate all the neurons at the same time, so this function is efficient than other function. The function is given by equation (1).

$$f(x) = max(0, x) \tag{1}$$

where x is the input variable.

4   RESULTS AND DISCUSSION

The model proposed in this paper is CNN SqueezeNet which performs well overall, for demonstrating its improved exhibition SVM is additionally utilized on a similar informational index. SVM doesn't give such exact outcome as CNN, this can be illustrated in the Table 3. In this Table it is find that CNN give 98.3% accuracy and F1, Precision and Recall estimations of CNN are additionally 98.3% while SVM accuracy is just 47.8%. So for this kind of dataset CNN model is a decent decision as it requires some investment to prepared and productively arranged the illnesses in rice leaves.

CNN SqueezeNet performance can also be visualized with the confusion matrix which is demonstrated in Table 4. Here for Brown spot disease the accuracy is above 99%, for Healthy plants it is almost 99%, for Hispa 98% and for Leaf Blast 96% . So overall accuracy of the model is 98%.

Table 3. Comparison of CNN and SVM performance.

Model	AUC	CA	F1	precision	Recall
SqueezeNet	**99.9**	**98.3%**	**98.3**	**98.3**	**98.3**
SVM	75.3	47.8%	47.0	48.3	47.8

Table 4. Confusion matrix of CNN SqueezeNet.

	Brown Spot	Healthy	Hispa	Leaf Blast
Brown Spot	99.5%	0.5%	0.0%	0.0%
Healthy	0.5%	98.9%	0.6%	0.1%
Hispa	0.2%	1.6%	98.2%	0.0%
Leaf Blast	0.9%	2.7%	0.1%	96.3%

For the evaluation of the model the ROC curve is chosen which an effective method of evaluation is. In the Figure 4 visualization of the ROC curve of SVM and CNN for each disease is represented, from where it can easily find out that for each disease CNN covers a large area which show its effectiveness.

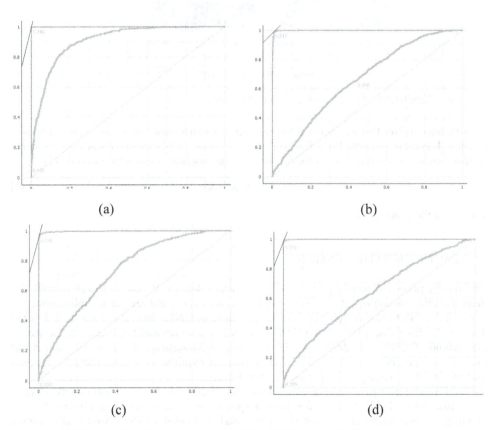

Figure 4. ROC curve for (a) Brown spot, (b) healthy, (c) Hispa and (d) Leaf blast.

5 CONCLUSION

The proposed model deployed CNN SqueezeNet for arrangement of rice infection which is a proficient strategy for AI. The model gives higher exactness than the past embraced models as it foresees with 98.3% precision and for the assessment of the model the ROC bend which is a famous assessment strategy. This model may help the farmers for ahead of schedule and exact forecast of the yield illness so the quality and amount can be improved or saved. In this manner the end is that in the field of horticulture AI is an insightful and productive methodology for the early and precise determining of a maladies.

In future new sort of data sets can be utilized on similar model for order of various ailments.

REFERENCES

[1] Rahman C. R., Arko P. S., Ali M. E., Khan M. A. I., Apon S. H., Nowrin F., & Wasif A. Identification and Recognition of Rice Diseases and Pests Using Convolutional Neural Networks. *Biosystems Engineering*, *194*, 112–120, 2020.

[2] Zhang X., Qiao Y., Meng F., Fan C., & Zhang M. Identification of Maize Leaf Diseases using Improved Deep Convolutional Neural Networks. *IEEE Access*, 6, 30370–30377, 2018.

[3] Chen J., Zhang D., Nanehkaran Y. A., & Li D. Detection of Rice Plant Diseases Based on Deep Transfer Learning. *Journal of the Science of Food and Agriculture*, *100*(7), 3246–3256, 2020.

[4] Coulibaly S., Kamsu-Foguem B., Kamissoko D., & Traore D. Deep Neural Networks with Transfer Learning in Millet Crop Images. *Computers in Industry*, *108*, 115–120, 2019.

[5] Shrivastava V. K., Pradhan M. K., Minz S., & Thakur M. P. Rice Plant Disease Classification Using Transfer Learning of Deep Convolution Neural Network. *International Archives of the Photogrammetry, Remote Sensing & Spatial Information Sciences,* 2019.

[6] Kaundal R., Kapoor A. S., & Raghava G. P. Machine Learning Techniques in Disease Forecasting: A Case Study on Rice Blast Prediction. *BMC Bioinformatics*, *7*(1), 1–16, 2006.

[7] Lu Y., Yi S., Zeng N., Liu Y., & Zhang Y. Identification of Rice Diseases using Deep Convolutional Neural Networks. *Neurocomputing*, *267*, 378–384, 2017.

[8] Phadikar S., & Sil J. Rice Disease Identification Using Pattern Recognition Techniques. In *2008 11th International Conference on Computer and Information Technology*. IEEE, 420–423, 2008.

[9] Ramesh S., & Vydeki D. Recognition and Classification of Paddy Leaf Diseases Using Optimized Deep Neural network with Jaya algorithm. *Information Processing in Agriculture*, *7*(2), 249–260, 2020.

[10] Sethy P. K., Barpanda N. K., Rath A. K., & Behera S. K. Deep feature Based Rice Leaf Disease Identification Using Support Vector Machine. *Computers and Electronics in Agriculture*, *175*, 105527, 2020.

[11] Shah J. P., Prajapati H. B. & Dabhi V. K. A Survey on Detection and Classification of Rice Plant Diseases. In *2016 IEEE International Conference on Current Trends in Advanced Computing (ICCTAC)*, 1–8. IEEE, 2016.

[12] Wang H., Li G., Ma Z., & Li X. Image Recognition of Plant Diseases Based on Principal Component Analysis and Neural Networks. In *2012 8th International Conference on Natural Computation*. IEEE, 246–251, 2012.

[13] Wang Y., Wang H., & Peng Z. Rice Diseases Detection and Classification Using Attention Based Neural Network and Bayesian Optimization. *Expert Systems with Applications*, *178*, 114770, 2021.

[14] Tiwari V., Joshi R. C., & Dutta M. K. Dense Convolutional Neural Networks Based Multiclass Plant Disease Detection and Classification Using Leaf Images. *Ecological Informatics*, *63*, 101289, 2021.

[15] Upadhyay S. K., & Kumar A. A Novel Approach for Rice Plant Diseases Classification with Deep Convolutional Neural Network. *International Journal of Information Technology*, *14*(1), 185–199, 2022.

[16] Mondal D., Roy K., Pal D., & Kole D. K. Deep Learning-Based Approach to Detect and Classify Signs of Crop Leaf Diseases and Pest Damage. *SN Computer Science*, *3*(6), 1–12, 2022.

[17] Gupta R., Tripathi V. & Gupta A. An Efficient Model for Detection and Classification of Internal Eye Diseases using Deep Learning. *International Conference on Computational Performance Evaluation (ComPE)*, 045–053, 2021, doi: 10.1109/ComPE53109.2021.9752188.

[18] Vats S., Singh S., Kala G., Tarar R. & Dhawan S., "iDoc-X: An Artificial Intelligence Model for Tuberculosis Diagnosis and Localization," *J. Discret. Math. Sci. Cryptogr.*, 24, 5, 1257–1272, 2021.

[19] Vats S., Sagar B. B., Singh K., Ahmadian A. & Pansera B. A., "Performance Evaluation of an Independent Time Optimized Infrastructure for Big Data Analytics That Maintains Symmetry," *Symmetry (Basel).*, 12, 8, 2020, doi: 10.3390/SYM12081274.

[20] Vats S. & Sagar B. B., "An Independent Time Optimized Hybrid Infrastructure for Big Data Analytics," *Mod. Phys. Lett. B*, 34, 28, 2050311, 2020, doi: 10.1142/S021798492050311X.

[21] Vats S. & Sagar B. B., "Performance Evaluation of K-means Clustering on Hadoop Infrastructure," *J. Discret. Math. Sci. Cryptogr.*, 22, 8, 2019, doi: 10.1080/09720529.2019.1692444.

[22] Agarwal R., Singh S. & Vats S., "Implementation of an Improved Algorithm for Frequent Itemset Mining Using Hadoop," In 2016 *International Conference on Computing, Communication and Automation (ICCCA)*, 13–18, 2016. doi: 10.1109/CCAA.2016.7813719.

[23] Agarwal R., Singh S. & Vats S., Review of Parallel Apriori Algorithm on Mapreduce Framework for Performance Enhancement, 654, 2018. doi: 10.1007/978-981-10-6620-7_38.

[24] Bhati J. P., Tomar D. & Vats S., "Examining Big Data Management Techniques for Cloud-based IoT Systems," In *Examining Cloud Computing Technologies Through the Internet of Things, IGI Global*, 2018, 164–191.

[25] Sharma V. *et al.*, "OGAS: Omni-directional Glider Assisted Scheme for Autonomous Deployment of Sensor Nodes in Open Area Wireless Sensor Network," *ISA Trans.*, 2022, doi: 10.1016/j.isatra.2022.08.001.

[26] Sharma V., Patel R. B., Bhadauria H. S. and Prasad D., "NADS: Neighbor Assisted Deployment Scheme for Optimal Placement of Sensor Nodes to Achieve Blanket Coverage in Wireless Sensor Network," *Wirel. Pers. Commun.*, 90, 4, 1903–1933, 2016.

[27] Sharma V., Patel R. B., Bhadauria H. S. & Prasad D., "Policy for Planned Placement of Sensor Nodes in Large Scale Wireless Sensor Network," *KSII Trans. Internet Inf. Syst.*, 10, 7, 3213–3230, 2016.

[28] Sharma V., Patel R. B., Bhadauria H. S. & Prasad D., "Deployment Schemes in Wireless Sensor Network to Achieve Blanket Coverage in Large-scale Open Area: A Review," *Egypt. Informatics J.*, 17, 1, 45–56, 2016.

[29] Bhatia M., Sharma V., Singh P. and Masud M., "Multi-level P2P Traffic Classification Using Heuristic and Statistical-based Techniques: A Hybrid Approach," *Symmetry (Basel).*, 12, 12, 2117, 2020.

[30] Vikrant S., Patel R. B., Bhadauria H. S. and Prasad D., "Glider Assisted Schemes to Deploy Sensor Nodes in Wireless Sensor Networks," *Rob. Auton. Syst.*, 100, 1–13, 2018.

[31] Vats S. & Sagar B. B., 2019. Data Lake: A Plausible Big Data Science for Business Intelligence. In *Communication and Computing Systems*. CRC Press, 442–448.

Automation and Computation – Vats et al. (Eds)
© 2023 the Author(s), ISBN 978-1-032-36723-1

An IoT based smart shopping cart

Manjeet Singh Pangtey, Sunita Tiwari, Keshav Rawat & Mandeep Kumar
G B Pant DSEU Okhla 1 Campus, New Delhi, India (Formerly G B Pant Engineering College)

ABSTRACT: In today's world, having a fast-growing population with a wide range of demands, shopping for different products in the supermarkets demands a lot of time and patience from the customers in coordinating themselves for a successful shopping outcome. We need to address this problem by efficiently using our technologies. Hence, this product Smart Shopping Cart (SSC) is being proposed by our team to streamline the shopping experience of the customers. Currently technologies like this are rare in the market and this is our attempt at making this technology more feasible and attractive while also cutting the cost. Our model of Smart Shopping Cart system uses RFID (Radio frequency Identification) to detect the products put inside the cart. It also uses a weight sensor which helps in detecting any anomalies during the shopping. It then sends the scanned products data to the Realtime Database which is further fetched and shown in the user interface. Our proposed system will also have various functions which will assist the customers during shopping like cart to communication, automatic shopping list and various other functions which can be provided to the customer by using the mobile interface.

Keywords: Internet of Things (IoT), RFID, Smart Shopping Cart, Cart to Cart communication

1   INTRODUCTION

In our cities, we witness tremendous amount of surge at the shopping complexes and it reaches to another level during the holidays and weekends which results in long queues at the billing counter. The cashier prepares the bill using a barcode reader, which can be a time-consuming process. The one alternative one could think of is online shopping but it also suffers from various issues like inability to see products, late deliveries, online scams etc. Customers who shop physically can get a feel for the goods and avoid the inconvenience of returning it, as many customers find the return process for online shopping to be overly complicated or time consuming.

Shopping is also about having fun and recreation. Physically shopping for products also works as a mental therapy. Many people enjoy shopping after some stressed out week days. But this experience is ruined by the everlasting queue at the billing counter. According to Phononic's "Store of the Future report 2019", out of 1118 consumers, 51% believe that supermarkets needs to "enter the modern age" or else they may find other ways to shop Consulting (2019).

One of the most infamous reasons behind customer dissatisfaction is to queue up for the billing process, and if any problem occurs at the time of checkout, be it non- functioning of the card or to suddenly remember an item and to leave the queue for that only adds up to the problem. Also, there are other issues that arise in traditionalshopping such as communicating with the co-shopper, if more than one persons are shopping with each one having their own shopping cart it becomes very easy to shop for duplicate items and only realizing it

DOI: 10.1201/9781003333500-28

during checkout. Going over-budget, following the shopping list or making sure every required product is shopped for etc. are also some of the problems faced during shopping. With the rising population, there's a need for a product in the shopping complexes which cuts down the time spent in queues, provides convenience during shopping, and overall streamlines the shopping process. We are proposing a solution to resolve this issue by using a Smart Shopping Cart that can be integrated into supermarkets, department stores, or any small-scale shopping mart. With this product, the customers can shop hassle-free without having to spend time standing in the queue for billing. It will also help in creating a superior shopping experience for the customers by giving various facilities to the user like cart to cart communication, automatic shopping list and providing shopping related metrics like shopping history, categories of items shopped etc.

Today IoT is being used in different domains such as smart homes, agriculture, healthcare, industries etc. The Internet of Things (IoT) is a vast network architecture of many types of devices that connect physical and virtual items through communication and data collecting. Castillejo (2013)-Xia (2012). In our paper we are focusing on developing a model of Smart Shopping Cart which is based on IoT and uses RFID technology to scan the RFID tags present on the products Mitton (2012). The RFID system is made up of tags that act as transmitters and responders, as well as RFID readers that operate as transmitters and receivers. They communicate with the help of radio waves which makes them readable even when they are not in the line of sight Finkenzeller (2010). The proposed design of the Smart Shopping cart will eliminate the need of the traditional checkout/billing process as the bill would be automati- cally calculated by the cart. A load cell is also employed which verifies the weight of the scanned products. After all this, the scanned products are displayed on the user interface using a backend Server and a Realtime database. The user interface will provide information regarding the shopped products and will also act as a medium to checkout and pay for the shopped items. The user interface provides customers with features like Cart to Cart Communication which will allow customers to share their cart details (like the items they have in their cart) with their partners so that they can shop together easily. The customers can also create a shopping list on the interface prior to shopping which will assist them during shopping by automatically removing the scanned/shopped products from the list, it will also provide shopping history and other useful metrics to the user.

The overall objective of this paper is to introduce a smart shopping system, which is capable of identifying product using a RFID scanner attached on the smart cart and verify the weight of the scanned products using a load cell. All the scanned products will be displayed on the interface which most likely would be a smartphone. Through the interface, customers can easily pay for the shopped items. Also features like cart to cart communication, automatic shopping list and other shopping related metrics will be provided to the customer with the help of this user interface.

The remainder of this paper is organized as follows: Section II gives the summary of related works done on this topic, the architecture and the design of the proposed smart shopping cart system is discussed in Section III, the development process and the final outcome is discussed in Section IV. Finally, Section V gives our conclusion and the future works related to smart Shopping cart.

2 RELATED WORK

A lot of research work has been done in the field of IoT applications but the concept of smart shopping System using IoT is something which is still very new. Although there are some good research works related to this topic which are published in recent years. A considerable amount of study has been done on smart shopping system which uses RFID technology as the medium to detect products and Zigbee for establish- ing communication between the smart cart and the backend server Kumar (2013)- Chandrasekar (2014). In 2016, Ruinian Li *et al.* presented a smart shopping system which has a secure communication protocol between the server and the cart. They also presented security analysis and performance evaluations Li (2016). Sakorn M. in 2020 proposed the idea of a smart shopping system which can notify the shoppers of the location of various products on the shelves inside the shopping complex, this will assist the customers in finding the desired products Mekruksavanich (2020). In 2019 Ponnalagu R. N. and Sudipta R.S. proposed a secure processing system to pay the bill after shopping. They proposed a system where the customer will be first authenticated using their UID (Universal Identification Number) like aadhar in India. After this the payment can be initiated using UPI (an instant real-time payment system in India) Subudhi (2019). Ragesh N *et al.* designed and proposed a system which uses Deep Learning for object detection. This system is useful for edible objects like fruits or vegetables on which applying RFID tag is not feasible. It also uses a load cell for calculating the weight of the product Ragesh (2019). Bindhu R, Chandana S, Pranathi M, Shilpa V B, SK Khadar Basha included a load cell in the smart shopping system which can detect larceny. All the items after getting scanned will be verified with the help of a load cell Basha (2020). Also A. Yewatkar, F. Inamdar, R. Singh, A. Bandal *et al.* proposed a system in which the smart shopping system will give recommendations to the customers during shopping Yewatkar (2016).

3 ARCHITECTURE OF SMART SHOPPING CART

3.1 *Design goals*

The overall goal of our proposed Smart Shopping system is as follows:

- Product Identification: The smart cart should be able to identify the product. Each product will have a RFID tag which stores information about the product and the RFID scanner present on the cart will read the tags.
- Product Verification: The load cell should alert the customer if any discrepancies are found in the weight of the scanned product. This feature makes the cart more user friendly by alerting the customers about any possible error while shopping (e.g. if an item which is not scanned is put inside the cart the load cell will detect this and alert the customer with the help of a buzzer).
- Sending the scanned products data to the database and the server so that it can be displayed on the user interface (this interface could be mobile application on a smartphone).
- Verification of shopped products at the end of shopping using RFID scanners. We propose installing RFID readers before the exit door, which can scan all the items present in the smart cart and check whether every item is paid for.

Apart from these goals many additional features like cart to cart communication, automatic shopping list, shopping history etc. will be implemented which will assist the customers while shopping. Using the cart to cart communication feature, the cart will be able to connect to other carts using the user interface which will allow all connected users to see the products present on the all the other carts. A shopping list can be made prior to shopping by searching for items in the interface and adding them to the list. During shopping, the items will be automatically removed from the list when they are scanned and put inside the cart. The user will also be able to see the shopping history and other data like the item categories he/she has shopped for through the interface.

3.2 *Components*

The smart Shopping System consists of the following components:

- Server and Database: All the registered user data, product data will be stored on the Database. The product data will include the product ID, product name, manufacturing date etc. The Smart Shopping Cart itself will communicate with the server whenever a user updates the cart by placing an item.
- Smart Shopping Cart: The specific description of each component is present in Table 1.

Table 1.   Specification of hardware.

Sr. No.	Components (*t*)	Description (*t*)
1	Arduino Uno micro-controller	Arduino Uno is a microcontroller based on the ATmega382P. It has 14 digital I/O pins. It has 32 KB of flash memory.
2	MFRC522 RFID module	The RC522 is a 13.56MHz RFID Reader/Writer Module that is based on MFRC522 IC by NXP semiconductors. It is a slave device that operates on 3.3V DC power supply.
3	Load cell	A transducer that works on the Wheatstone Bridge Principle and converts the force applied on it into a measurable electrical signal.
4	HX711 amplifier module	Signal Amplifier; Analog-to-Digital Converter;
5	ESP8266 Wi-Fi module	A low-cost Wi-Fi module that is used to connect with a Wi-Fi network.
6	Piezo Buzzer	Audio signaling device that converts electrical audio signals into sound.

The smart shopping cart will be equipped with these components.:

- o Microcontroller: The cart will be equipped with an Arduino UNO micro-controller which will control the RFID scanner, load cell, Wi-Fi Module to successfully perform the intended function.
- o RFID scanner: This will be used to read the RFID tags present on the items.
- o RFID tags: These tags will store the information about the products like weight, price, product status etc. RFID tags can easily be updated using a RFID scanner.
- o Load Cell: This will be used to calculate the weight of the items which are put inside the cart. As the load cell generates very weak signal, it will also require a signal amplifying module to increase the magnitude of signal so that the signal can be interpreted by the Arduino UNO microcontroller.
- o Wi-Fi module: The updates from the smart cart's microcontroller will be sent to the database using this module.

- User Interface: This is the interface on which all the information related to the shopped items will be displayed. Preferably, this interface would be a mobile application present on the customer's smartphone.

3.3   *System design*

To design our proposed system, the cart would first need to fetch product information by scanning the RFID tag present on the item using the RFID Scanner. The RFID tag stores information about the respective product. Once an item is scanned it will be placed on the cart where the load cell is present which will verify the weight of the item with the weight information fetched from the RFID tag. The RFID tag also contains a field named "product status" which is originally set to "available" as shown in Figure 1. By scanning the tag, the product status stored on the RFID tag will change to "Sold" which means the item has been scanned, this is shown in Figure 1. To remove an item from the cart, the item is again placed near the RFID scanner. This time the RFID scanner updates the product status on the RFID tag to "available". In this way the user can add or delete an item from the cart.

a	Product_ID: exampleID, Product_weight: exampleWeight, Product_status: **available**	b	Product_ID: exampleID, Product_weight: exampleWeight, Product_status: **sold**

Figure 1.   (a) RFID tag before scanning; (b) RFID tag after scanning.

If an item which is not scanned is placed on the cart, the load cell will detect the difference in weight and the cart will starts beeping, alerting the customer about it. After the item which was not scanned is removed the load cell will again verify the weight and will stop beeping. If the customer puts a different item instead of the scanned item the cart will detect the difference in weight using the product weight data fetched from the RFID tag of that product and will alert the customer. The cart will halt all other process until the discrepancies are removed. After the scanning and weight verification, the information fetched using the RFID scanner is sent to the server by using the ESP-01 Wi-Fi module. The information sent contains product ID along with other data. The database will contain all the information about the products and the product ID will act as a key to fetch the respective product data from the database so that it can be displayed on the interface. If the user scans the shopped item again, the product ID again reaches the database but this time it already finds the product with the same product ID in the cart and it removes it from the interface Figure 2 shows a basic diagram of the smart shopping system process.

Figure 2.   Basic architecture of smart shopping system.

To start shopping the user will first have to register him/her to the store's application. After the registration he/she can simply log in into his/her account and connect a cart to its interface by entering the Cart Number. This can also be changed to a QR code and the customer can simply connect its interface to the cart by scanning the QR code on the cart. We also propose a screen on the smart shopping cart which can display randomly generated QR code or numbers. Once the cart is connected he/she is ready to start the shopping process.

After the basic system is ready. Other proposed features can be implemented like the cart to cart communication, automatic shopping list etc. To implement cart to cart communication a user will need to connect his/her cart to other carts. This can be done with the help of the user interface by following these steps:

• The user will connect a cart to its interface to start the shopping.
• One of the two users will generate a pairing code and the other user will use that code to pair the cart.
• After entering the code, the user who generated the code will get a pairing request and depending on the user's choice the request can be accepted or rejected.

After all this, the items present in all the connected carts will be visible to both the users. To differentiate between the items shopped by different users, the items would be color coded depending on which user has shopped for them. Similarly, a user can create a shopping list using the interface. The shopping list will be created prior to shopping by searching for the desired item using a search bar. The database will contain the information off all the products available. The user will just need to search and add the item to the list. Once the list is ready the user can start the shopping and the items present on the list will get automatically removed from the list once they are added to the cart. By using the previous shopping history useful metrics can be given to the user such as categories of items shopped, most frequently shopped items etc.

After the user has completely shopped for all the items. He can press the checkout button to finish the shopping. Since the database has fetched details of all the shopped product including the product status it will generate a bill for all items for which product status is set to "sold". Then by some payment mechanism the customer can pay for the items. After that the user will have to pass through a checkout scanner to exit the store. The checkout scanner will consist of multiple RFID's. If there is any product inside the cart which is not scanned then the product status field of that item on the RFID tag would be set to "available". The RFID will check all the products inside the cart and if the product status field of any product is set to "available" then the exit scanner will start beeping. This final checking process will ensure that all the products have been paid for.

4   IMPLEMENTATION

4.1   *Basic implementation*

Based on the above design, a prototype of smart shopping cart was developed which successfully performed the above-mentioned functionalities. A system using RFID scanner and load cell is made which is able to scan the RFID tags present on products to add and remove items from the cart. A load cell is employed for weight verification process. Further, ESP-01 Wi-Fi module is used which is used to send the product ID of the scanned item to the server. The circuit diagram of the smart shopping cart is shown in Figure 3.

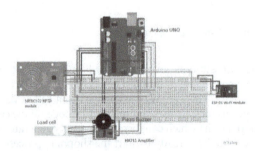

Figure 3.   Connection diagram of the hardware components.

A basic prototype of user Interface was also made. In Figure 4 it can be seen that the cart is empty. When a user scans a product's RFID tag and places the item on the load cell. After verification the ESP-01 Wi-Fi module sends the product ID to the server which then fetches information about the product from the database and displays it on the interface as shown in Figure 5.

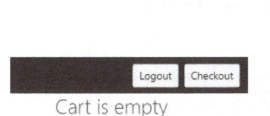

#	Item	Qty	Total
1	Lays Chips	3	90
2	Sunscreen	1	320
3	Dark Chocolate	5	300
4	Pasta	2	100
5	Oats	2	400

Smart Cart — Swati

Total ₹ 1210

Figure 4.   Interface when the cart is empty.

Figure 5.   Interface when items are added in the cart.

4.2 *Cart features*

The cart to cart communication feature was also implemented. The user will first generate a pairing code using the interface as shown in Figure 6. After this, the other users can use this pairing code to connect with the cart. The items shopped by all the connected user will be displayed to everyone and they would be color coded depending on which user has shopped the item as shown in Figure 7.

Pair Cart

Your pairing code

764379

Or

Enter pairing code here [Go]

[Close]

| 3 | Maggie Noodles | 4 | 100 |

Total ₹ 340

Smart Cart — Both — Karan — Swati

#	Item	Qty	Total
1	Pepsi	1	60
2	Olive Oil	1	180
3	Maggie Noodles	4	100
4	Lays Chips	3	90
5	Sunscreen	1	320
6	Dark Chocolate	5	300
7	Pasta	2	100
8	Oats	2	400

Total ₹ 1550

Figure 6.   Pairing code generation.

Figure 7.   Display of color coded items.

5 CONCLUSION AND FUTURE WORKS

We have proposed a design of a smart shopping system which uses RFID technology to scan items and a load cell to detect anomalies. This system can enhance the shopping experience by cutting down the time spent in queue for billing. We also discussed about the additional features of the smart shopping system like cart to cart communication, automatic shopping list etc. which helps in creating a superior shopping experience for the customers.

There is tremendous scope of innovation in this new piece of tech. The work that we have done in this project is to cover the most basic aspects of a Smart Shopping Cart. One of the major developments that could happen is Indoor Navigation which will help the customer in locating a product in the store. Using Indoor Navigation, the cart can show the optimal route the customer can shop through based on the shopping list prepared by the customer.

Recommending products to the customer during shopping on the basis of currently shopped items and the items shopped in the past. A concept of payment mechanism can also be thought of which may use a balance system or some other method.

REFERENCES

Ali Z. and Sonkusare R. (2013) "Rfid based Smart Shopping and Billing," *International Journal of Advanced Research in Computer and Communication Engineering* **2** (12) 4696–4699.

B.R, C.S, P.M, S.V.B and Basha S.K.S.K. (2020) "Smarttrolley to Avoid Larceny Using load cell," *International Research Journal of Engineering and Technology (IRJET)* 07(08) 1708–1711.

Consulting R. (2019) *Store of the Future*. [Online]. Available: https://content.phononic.com/2019/store-of-the-future?utmsource=PRutmmedium=PressReleaseutmcampaign=SOTF2019.

Castillejo P., Martinez, J.-F., Rodríguez-Molina, J. and Cuerva, A. (2013). Integration of Wearable Devices in a Wireless Sensor Network for an E-health Application. *Wireless Communications, IEEE.* **20** 38–49.

Chandrasekar P. and Sangeetha T. (2014) "Smart Shopping Cart with Automatic Billing System through Rfid and Zigbee," *Information Communication and Embedded Systems (ICICES)*.

Finkenzeller K. and Muller D. (2010) *"RFID Handbook: Fundamentals and Applications in Contactless Smart Cards, Radio Frequency Identification and Near-Field Communication,"* Wiley.

Gupta S., Kaur A., Garg A., Verma A., Bansal A. and Singh A. (2013) "Arduino based Smart Cart," *International Journal of Advanced Research in Computer Engineering & Technology*, **2** (12).

Kumar R., Gopalakrishna K. and Ramesha K. (2013) "Intelligent shopping cart." *International Journal of Engineering Science and Innovative technology* **2** (4) 499–507.

Li R., Song T., Capurso N., Yu X. J. and Cheng, (2016) "IoT Applications on Secure Smart Shopping," *IEEE International Conference on Identificaton, Information, and Knowledge in the Internet of Things* 238–243.

Mitton N., Papavassiliou S., Puliafito A. and Trived K. S. (2012) "Combining cloud and sensors in a smart city environment," *EURASIP journal on Wireless Communications and Networking* **1** 1.

Mekruksavanich S. (2020) "Supermarket Shopping System using RFID as the IoT Application," *Joint International Conference on Digital Arts*, Media and Technology with ECTI-Northern Section Conference 83–86.

Ragesh N., Giridhar B., Lingeshwaran D., Siddharth P. and Peeyush K.P. (2019) "Deep Learning based Automated Billing Cart," *International Conference on Communication and Signal Processing* 0779–0782.

Subudhi S.R. and Ponnalagu R. N., (2019) "An Intelligent Shopping Cart with Automatic Product Detection and Secure Payment System," *IEEE 16th India Council International Conference (INDICON)*, Rajkot, India.

Tianyi S., Ruinian L., Bo M., Jiguo Y., Xiaoshuang X. and Xiuzhen C. (2017). "A Privacy Preserving Communication Protocol for IoT Applications in Smart Homes." *IEEE Internet of Things Journal* 1–1.

Xia F., Yang L. T., Wang L. and Vinel A., (2012) "Internet of Things." *International Journal of Communication Systems* **25** (9) 1101.

Yewatkar A., Inamdar F., Singh R.and Bandal A., (2016) "Smart Cart with Automatic Billing, Product Information, Product Recommendation Using Rfid & Zigbee With Anti-Thef," *Procedia Computer Science* **79** 793–800.

Automation and Computation – Vats et al. (Eds)
© 2023 the Author(s), ISBN 978-1-032-36723-1

Pixel-wise binary visual cryptography for black and white images

Amit Singh*, Apoorv Srivastava & M. Sweety Reddy
School of Computer Science, University of Petroleum and Energy Studies, Dehradun, India

ABSTRACT: Encrypting data has become a crucial component of our secure transmission as a result of the rise in cybercrime. This includes audio, text, and image data. One of the effective methods for encrypting data that blends images to ensure secure information transfer is called visual cryptography (VC). Binary images, sometimes known as hidden images, can be sent into VC as input, which further divides it into two or more segments. Shares or transparency are the common names for these parts. At the receiving end, these shares are combined to produce the covert image. A significant amount of computing power is typically needed for encryption and decryption. But among picture encryptions, VC is favored since it uses very little processing power. The visual encryption method that ensures the security of data is most frequently composed of pixel expansion and contrast. This work proposes a non-expansion-based method for the (K, N)-visual secret sharing scheme. The proposed algorithm reconstructs the image without enlarging it, producing the shares. The suggested approach keeps the size of the original image impaired. In each sharing, the primary (i.e., secret) image is encoded in 4-pixel blocks to 4-pixel blocks in each share. Finally, binary and black-and-white images are used to validate the suggested algorithm. In addition, colored picture encryption is also shown to work.

1 INTRODUCTION

We now belong to the internet, a global network that is rapidly expanding. Given the size of the network, maintaining data privacy and confidentiality has become a significant challenge especially when profits, data copyrights, etc. are at stake. Worldwide, several techniques and algorithms have been created to protect data and personal information. Although each of them has specific advantages and disadvantages. Since the development of cybercrime, picture, text, and audio encryption has become a crucial component of our security framework. Visual cryptography was created as a unique method of concealing images that could subsequently be decrypted using only the keys. Similar methods were used in this study project.

By mixing and secretly communicating segmented images, Visual Cryptography (VC) [1] is one of the effective encryption techniques that achieve potent secrecy [2–4]. As was previously discussed, VC accepts a binary image as input that has been partitioned into two or more shares. At the receiving end, these shares are superimposed to create a hidden image. Its non-computational resource required for decryption is a distinctive feature that categorically sets VC apart from other related techniques [5]. Additionally, the sole human visual recognition system is used for decryption (the human eye). A Quality of Service component of the visual cryptography scheme under discussion is as follows:

(i) Pixel expansion
(ii) Contrast

---

*Corresponding Author

DOI: 10.1201/9781003333500-29

This research paper demonstrates the non-expansion algorithm for the (K, N)-visual hiding and sharing system, also known as Visual Secret Sharing (VSS) [6–9]. The reconstructed picture formed during the decryption phase and the shares created during the encryption phase are both the same size. Each share is further divided into 4-pixel blocks once the original image has been separated into 4-pixel blocks. This system's goal is to protect data so that only the customer and other authorized users can view confidential or secret information [10,11]. Additionally, image-based information is safer because only the intended recipients can reveal the actual image. The suggested visual cryptography methods are designed to exchange (i.e., encrypt) data, particularly images, without the need for intricate calculations. Additionally, the decryption process is dependent on human vision.

A GUI interface allows users to choose images from their local system and turn them into encrypted form after pressing the "ok" button and choosing the portion of the image to be encrypted. It also creates a Share Combiner that allows users to decrypt the encrypted version of the selected image. A class of users who need to secure their data can utilize this type of encryption-decryption program. The layperson, technical/non-technical, and expert technical users can all benefit from the application as it is presented here.

A human-detected interactive media element is the image. The smallest component of a digital image is a pixel, though. Each pixel in a computer image with 32 bits has four components: Alpha, Red, Green, and Blue. The size of each piece is 8 bits [12,13].

The first component, Alpha, represents the level of transparency. At its "off chance", all of Alpha's bits are set to "0". It implies that the image will be entirely transparent. Humans' visual systems perform an "OR" operation. The result is transparent when two transparent objects are stacked on top of one another. The overall effect will be nontransparent in all other circumstances (which means that any item is not transparent) [14,15]. It resembles "OR" behavior in the following ways:

(a) *0 OR 0 = 0*, It stands for transparency.
(b) *1 OR 0 =1*, It stands for non-transparency.
(c) *0 OR 1 =1*, It stands for non-transparency.
(d) *1 OR 1=1*, It stands for non-transparency.

Figure 1 shows a diagrammatic illustration of an OR gate in action.

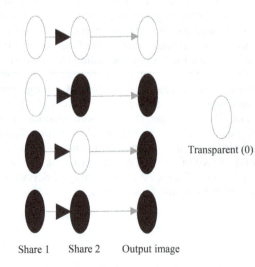

Transparent (0)   Non-Transparent (1)

Share 1   Share 2   Output image

Figure 1.   An *"OR"* function, the human visual system.

The following traits are crucial in figuring out how effective a visual encryption method is. The first is pixel expansion, which enlarges the size of the image, and the second is contrast. Pixel expansion is the term used to describe the expected number of sub-pixels needed to address each pixel in the original image. The contrast between a white pixel and a dark pixel in the original distinct image.

It is suggested to use a Novel and Simple Non-Expansion-based Algorithm (NSNEA) to translate the pixels of the Secret Image (SI) to transparency [10]. These transparencies are currently a bit-block of size 4 bits. Since the transparencies are the same size as the concealed image, the recovered image is the same size as the original secret image [11].

Let's assume that the data (image) D is divided into n shares. Any k shares out of n shares can be used to create D. Complete data on (k-1) shares reveals no information on D. In this procedure, k out of n shares are needed to retrieve concealed information from the secret image.

a.) Plain   b.) Encrypted   c.) Encrypted   d.) Decrypting using
Text   share 1   share 2   share 1 and share 2

Figure 2.   Restoration of an encrypted image using 2 shares.

The paper is divided into sections for organizing. Section II includes a summary of the existing literature on the subject of visual cryptography coverage. In Section III, the suggested algorithm NSNEA is described and illustrated. In Section IV, the results are discussed and presented. Finally, Section V is concluded the insightful of the proposed work and suggests new directions.

2 RELATED WORK

One of the crucial steps in visual cryptography is pixel expansion, therefore several solutions have been offered to deal with this problem. A probabilistic method for binary photo encryption is described by Ito *et al.* [16]. Yang [8] also presents a straightforward implementation-based non-expanded technique based on conventional VSS. In this way, the original image's pixels are represented in the sharing as either black or white pixels. If the original pixel is white, a randomly chosen column from an easily accessible n→m matrix S0 is used; if the original pixel is black, a randomly chosen column from an easily accessible n→m matrix S1 is used.

By Chen *et al.* [17], a size invariant approach is provided. For multilevel-VSS, Chen *et al.* propose a block-based encoding technique. In this method, the original image is divided into various blocks. Additionally, the encoding is done in blocks rather than one pixel at a time. Each block creates a share with an equal number of white and black pixels (i.e., s/2 white and s/2 black pixels), and is then encoded into k-blocks on k shares. The size of the reconstructed image is similar to that of the original.

For instance, Pooja and Lalitha [18] employ a comparable strategy. The authors used halftone photographs to show the applicability. The final image is around the same size as the hidden image's original size. The method is founded on a cutting-edge (2, 2) visual cryptography strategy. The technique makes use of the ideas of phony randomization and pixel reversal for the demonstration and validation of the results.

Ching *et al.* [19] highlight a method based on two covert images. With this method, there is no pixel expansion seen. Huang and Chang [20] proposed a solution that combines the non-expanded system with the ability to conceal sensitive data during communication in order to prevent data identification. This method divides the secret image into four parts to achieve encryption. The block encoding method is then used to generate the share blocks without expanding them. To improve the effectiveness of VC for image encryption and decryption, a few improvements are also suggested [21].

3 PROPOSED WORK

The accompanying Figure 3 illustrates the proposed approach executing of encryption and decryption process.

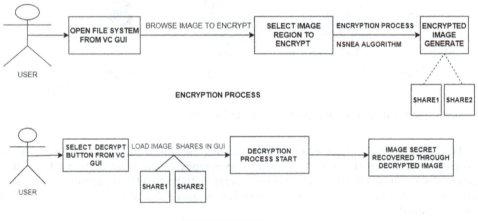

Figure 3. Proposed workflow of encryption-decryption process.

The activity diagram shows the progression of the encryption and decryption procedures as well as the creation of the encrypted image's share. The user can access the GUI-based VC system and load the image in the first stage. The user can also choose which part or area of an image to encrypt. The user can select values for K and N (ranges between 2 to 4). An image's share count is broken into N. However, the user must provide the share count needed to recreate the image at the destination end (i.e. K).

In order to decrypt the image and retrieve the desired secret information, shares of the relevant Secret Image (SI) must be put into a share combiner (at least 2 shares).

3.1 *Novel and Simple Non-Expansion-based Algorithm (NSNEA)*

The proposed algorithm is described as follows:

- The process begins by making the transparencies and mapping the hidden image's pixels into 4-bit blocks. The decryption approach does, however, restore the same size image because the transparencies and the original secret images are of comparable size. To calculate the original image size as $x- > Y$ initially, NSNEA checks the x (*i.e. width*) and y (*i.e. height*) of a secret image **if** $x * y\%2 = 0$ **else** encoding starts on $x0 -> y0$ of SI where $x0 = x - x\%2$ and $y0 = y - y\%2$. The last column (*in the case where* $x\%2! = 0$) and the last row (*in the case where* $y\%2! = 0$) are processed separately.

- In accordance with the quantity of accessible white pixels, each 4-bit block f is then encoded into a 4-cycle block for each share (transparency). Here, the bottom right square pixel's direction is $(x + 1, y + 1)$ while the upper-left pixel's direction is (x, y).
- Verify whether or not adding white pixels to f makes it equal to four. In the unlikely event that "yes" f is perceived as white. The encoding maps the main column to share1 and the second line to share2, depending on the erroneous identification of a framework C0.
- Check to see if f has any white pixels and if so, whether it equals three. If "yes," depending on where the black pixel is situated, f will be seen as being half white and half black. The encoding square will be placed so that two dark pixels surround it (one black other than the first one). It verifies that the recovered image accurately depicts the original pixel.
- The encoding block created by the previous steps is based on a randomly selected matrix framework C01, or if the black pixel is the first in b (b = 1000), or C02, or if the black pixel is second in b, and so on. The subsets of C0 are C01, C02, C03, and C04.

The algorithm described below serves as a demonstration for all the processes presented previously in this work. The algorithm creates shares that are nothing more than the communication of information's picture encryption.

NSNEA algorithm

Input: Secret Image *(SI), C0, C1*
Output: Encrypted Images / Shares *S1, S2*
start
1. *f* is a four-pixel block.
2. *fc* is a four-pixel block in the last column;
3. *fr* is a four-pixel block in the last row;
4. $n0(f)$ is the quantity of *0's* in *f*;
5. $width = SI_{width} - (SI_{width})\%2$;
6. $height = SI_{height} - (SI_{height})\%2$
7. **for** $\forall f$ in *SI (width*height)* do
8.   **if** $n0(f) == 4$ then
     encodeWhite (f);   /* f is white */
9.   **else if** $n0(f) == 3$ then /* *f* is seen as half-white, half-black */
10.   *switch (f)*
11.   *case 1000:*
12.   *encodeC0I (f, 1)*

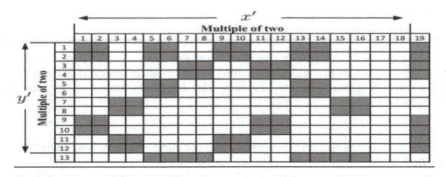

Figure 4. Encoding block using random selection matrix.

13.   *case 0100:*
14.   *encodeC0I (f, 2)*
15.   *case 0010:*
16.   *encodeC0I (f, 3)*
17.   *case 0001:*
18.   *encodeC0I (f, 4)*
19.   *else if* $(n0(f))$ then
20.   *if* $(f = 1001$ *ORf* $= 0110)$ then
21.   *encodeC0I (f, 5)*
22.   *else encodeBlack (f)*
23.   *else encodeBlack (f)*
24. *if* $(SI_{width})\%2 = 1$
25.   *encodeTheBlockLastColumn(fc);*
26. *if* $(SI_{tallness})\%2 = 1$
27.   *encodeTheBlockLastRow (fr);*
28. *if* $(SI_{width})\%4 \,!=0$ *OR* $SI_{tallness})\%4 \,!=0$
29.   *encodePixelWise(SI_{width}, SI_{height});*
end
encodeWhite(f){
    randomly select M0 from C0;
    encode *f* by the 1st row of M0 to share 1;
    encode *f* by the 2nd row of M0 to share 2;}
encodeC0I(f, i){
    randomly select M0i from C0i;
    encode *f* by the first row of M0i to share 1;
    encode *f* by the second row of M0i to share 2;}
encodeBlack(f){
    randomly select M1 from C1;
    encode *f* by the first row of M1 to share 1;
    encode *f* by the second row of M1 to share 2;}
encodeTheBlockLastColumn(fc){
    encode *fc* using the same criteria as *f*, but taking into account the different
    forms of *fc*;}
encodeBlockLastRow(fr){
    encode *fr* according to the same principles as encoding *f*, but take into ac
    count the different forms of *fr*;}*encodePixelWise(SI_{width}, SI_{height})*{
    /* encoding of $(SI_{width})\%4$ pixels
in the last row and $(SI_{height})\%4$ pixels
    *in the last column. */*
    for $\forall$ pixels do
    if white then
        randomly encode by 0 or 1 to both shares;
    else
        randomly encode by 0 or 1 to share 1 and the complement to share 2; }
After the data is sent to the desired location, the decryption procedure is carried out by
stacking shares to produce the original data. The decryption procedure is demonstrated
below:

step I:
    Enter each offer's height (h), width (w), and the number of approved offers (k).
step II:
    Make a two-layered cluster share[k][w*h] to hold the pixel upsides of each offer.*step III:*

```
for i = 0 to k − 1 {
    Specify the name of the ith image to be shot
    for j = 0 to (w * h − 1) {
        Filter each pixel in the ith image share and
        save the value in hare[i][j] }
    }
step IV:
for i = 0    to    (k − 1) {
        for j = 0 to (w * h −    1) {
        final[j]=final[j] | share[i][j] } /*| is bitwise OR */}
step V:
    Make an image using final[w * h].
    //secret information (SI) can be recovered from the decrypted image output
```

Below, in Figure 5, is a demonstration of the encryption and decryption procedure using a plain text secret image (SunShine) as input. Additionally, at the encryption stage, two shares, share1 and share2, are formed. Additionally, both shares are layered throughout the decryption process to recreate the original image at the receiver end.

All of the shares created during the encryption step are necessary to regenerate the original picture at the target end.

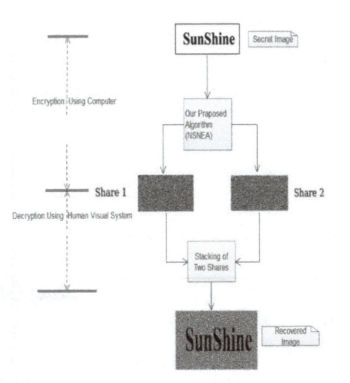

Figure 5.   A working demonstration of the proposed work.

4 RESULTS AND DISCUSSION

4.1 *User interface*

Figure 6 below illustrates a user-friendly GUI-based interface.

The user has control over which image will be encrypted, as well as which area inside the image. It is possible to set the parametric values of K and N and create shares for transmission. The original data [that can be used to construct the Secret Image] can, however, be rebuilt using K or more shares throughout the decryption process. Surprisingly, if there are fewer shares than K, the original information cannot be disclosed.

4.2 *Simple text image encryption-decryption*

The simple text image containing the word "sunshine" is encrypted, as seen above in Figure 5, and then decryption is carried out. As shown in Figure 7, two shares have been created. Here, K and N are both given values of 2 to illustrate the results.

4.3 *Black & white image encryption-decryption*

As indicated in the following Figure 8, the value of K and N for the black-and-white image is taken into consideration in this result analysis to be 2.

If there is a data breach when the information is being transmitted from source to destination, the shares created during the encryption stage are unpredictable. Visual

Figure 6.   GUI-based user interface for encryption-decryption.

a.) Plain pixel-wise   b.) Encrypted   c.) Encrypted   d.) Decrypting using
   binary image   share 1   share 2   share 1 and share 2
                                                          simple text image

Figure 7.   Simple text image encryption-decryption.

a.) Plain black &   b.) Encrypted   c.) Encrypted   d.) Decrypting using
white images   share 1   share 2   share 1 and share 2
                                                             black & white image

Figure 8.   Black & white image encryption-decryption.

A → 1000001   A →   ▬▬▬▬   ▬▬▬▬   A → 1000001
B → 1000010   B → 1000010   B → 1000010
a.) Plain text   b.) Encrypted   c.) Encrypted   d.) Decrypting using
  image   share 1   share 2   share 1 and share 2

Figure 9.   Pixel-wise binary image encryption-decryption.

cryptography provides the protection necessary for any hacker to recreate an original image that is 80 to 80 percent accurate.

4.4   *Pixel-wise binary image encryption-decryption*

Pixel-wise refers to the processing stage where each pixel is taken into account. The values of K and N are taken into consideration in the following result demonstration, which is displayed in Figure 9.

It is important to note that the precision of the decrypted image is inversely correlated with the volume of shares generated during encryption. Only two shares are taken into account to be stacked in this outcome during the image regeneration. As can be observed in Figure 9 (c), the quality of the regenerated image is somewhat low as a result.

5   CONCLUSION AND FUTURE SCOPE

Visual cryptography encompasses a broad range of academic areas, including subjects relating to protecting images, data concealment, multimedia, color imaging, and others. It also addresses issues with file formats, cybercrime, and other related issues.

The suggested scheme's guiding principle is:

1. The secret image should be encoded in a 4-pixel block. Each block is given a 4-pixel block every two shares based on the proportion of white pixels in the block.
2. The algorithm does not enlarge the shares (transparencies) it produces or the rebuilt image. These all closely correspond to the dimensions of the original (secret) image.
3. A few scenarios have been explored using the suggested method (NSNEA), and the results have been encouraging and reliable. With a reconstructed image that is between 80% and 85% accurate, the proposed method has been verified for correctness.

The reconstructed image in the proposed effort is anticipated to receive a little additional refinement. It can also be expanded to include colorful images. Although the suggested VC-based encryption-decryption is only capable of working with simple text, binary data organized by pixels, and monochrome images. The work can be expanded, nevertheless, to include real-world applications including document and copyright protection, biometric authentication, secret communication, and data storage.

REFERENCES

[1] Weir J. and Yan W.Q. *A Comprehensive Study of Visual Cryptography.* Transactions on DHMS V, LNCS, 6010, 2010.

[2] Abbas T. and Beiji Z. A Novel Non-expansion Visual Secret Sharing Scheme for Binary Image. *International Journal of Digital Content Technology and its Applications (JDCTA)*, 4(6):106–114, 2010.

[3] Kumar S., Jana M.B., Hait G. Survey on Size Invariant Visual Cryptography. *International Journal of Computer Science and Information Technologies(IJCSIT)*, 5(3):3985–3990, 2014.

[4] Hou Y.C. Visual Cryptography for Color Images. *Pattern Recognition*, 36, 2003.

[5] Yang C. N. New Visual Secret Sharing Schemes Using Probabilistic Method. *Pattern Recognition Letter*, 25(4):481–494, 2004.

[6] Supraja, A. and Anil Kumar K. "Analysis on Hybrid Approach for (k, n) Secret Sharing in Visual Cryptography." *2019 International Conference on Data Science and Communication (IconDSC)*. IEEE, 2019.

[7] Kandar, S. and Chandra Dhara B. "Kn Secret Sharing Visual Cryptography Scheme on Color Image Using Random Sequence." *International Journal of Computer Applications* 975 (2011): 8887.

[8] Verheul, E.R. and Van Tilborg H.C.A. "Constructions and Properties of k Out of n Visual Secret Sharing Schemes." *Designs, Codes and Cryptography* 11.2 (1997): 179–196.

[9] Hofmeister, T., Krause M. and Simon H.U. "Contrast-optimal k out of n Secret Sharing Schemes in Visual Cryptography." *International Computing and Combinatorics Conference*. Springer, Berlin, Heidelberg, 1997.

[10] Lakshmi S.B. and Sree Lakshmi G. "A Novel Cryptographic Technique Under Visual Secret Sharing Scheme for Binary Images." *International Journal of Engineering Science and Technology* 2.5 (2010): 1473–1484.

[11] Xiaotian W., Weng J. and Yan W.Q.. "Adopting Secret Sharing for Reversible Data Hiding in Encrypted Images." *Signal Processing* 143 (2018): 269–281.

[12] Chaturvedi R.N., Thepade S.D. and Ahirrao S.N. "Quality Enhancement of Visual Cryptography for Secret Sharing of Binary, Gray And Color Images." *2018 Fourth International Conference on Computing Communication Control and Automation (ICCUBEA)*. IEEE, 2018.

[13] Yu, B., Xu X. and Fang L. "Multi-secret Sharing Threshold Visual Cryptography Scheme." *2007 International Conference on Computational Intelligence and Security Workshops (CISW 2007)*. IEEE, 2007.

[14] Lukac, R. and Plataniotis K.N. "Bit-level Based Secret Sharing for Image Encryption." *Pattern Recognition* 38.5 (2005): 767–772.

[15] Yu, B., Fu Z. and Fang L. "A Modified Multi-secret Sharing Visual Cryptography Scheme." 2008 *International Conference on Computational Intelligence and Security*. Vol. 2. IEEE, 2008.

[16] Ito R., Kuwakado H. and Tanaka H. Image Size Invariant Visual Cryptograph. *IEICE Trans. Fundam. Elect. Commun. Comput. Sci.*, E82-A(10):2172–2177, 1999.

[17] Chen Y.F., Chan Y.K., Huang C.C., Tsai M.H. and Chu Y.P. A Multiple-level Visual Secret-sharing Scheme Without Image Size Expansion. *Information Sciences*, 177(21):4696–4710, November 2007.

[18] Pooja and Lalitha Y. S. Non Expanded Visual Cryptography for Color Images Using Pseudo-randomized Authentication. *International Journal of Engineering Research and Development*, 10(6):01–08, June 2014.

[19] Wang C.-L., Wang C.-T. and Chiang M.-L. The Image Multiple Sharing Schemes Without Pixel Expansion. In *International Conference on Machine Learning and Cybernetics*, volume 4, Guilin, 10–13 July 2011.

[20] Huang Y.-J. and Chang J.-D. Non-expanded Visual Cryptography Scheme with Authentication. *IEEE 2nd International Symposium on Next-Generation Electronics (ISNE)*, February 25–26 2013.

[21] Naor M. and Shamir A. Visual cryptography. In Proceeding of Advances in Cryptology EUROCRYPT'94, *Lecture Notes in Computer Science*, volume 950, pages 1–12. Springer-Verlag, 1995.

Automation and Computation – Vats et al. (Eds)
© 2023 the Author(s), ISBN 978-1-032-36723-1

A survey of open source and free software engineering tools for academia and developer community

Sugandha Sharma & Kaushik Ghosh
UPES, Dehradun, India

Manika Manwal & Sonali Gupta
Graphic Era Hill University, Dehradun, India

ABSTRACT: Software engineering is a prominent and indispensable area of computer science and engineering education. The role of software engineering in industry too is well established and it is needless to say that it plays a pivotal role in managing software projects. Over the years the field of software engineering has evolved and as a result, it demands a host of tools for dealing with tasks like designing, bug fixing and cost estimation of different software projects. At present although numerous state of the art tools are present for the said purposes, yet it has to be kept in mind that many of them are proprietary in nature. Considering the costs of such proprietary tools, it becomes difficult for both the academia and small software companies to purchase them for the purpose of learning and development. Open source tools on the other hand come free of cost and give the provision of customized feature addition. Usage of open source software engineering tools therefore will enable such companies to provide cost effective solution to their clients and thereby remain relevant in the industry space. Academia too will be benefited by using such tools in their resource-constrained environment. Moreover, the outbreak of Covid 19 has led to transition of many industries, including education, from offline to online mode. As a result of this operational shift, both the industry and academia became dependent on different software tools and platforms, like never before. In this paper we have therefore presented a group of open source software tools which may be used by both industry and academia, and to make students learn, appreciate and apply good software engineering techniques in their professional life. A detailed classification of the discussed tools and platforms has been done too, depending upon their utility and domain of usage.

1 INTRODUCTION

The domain of software engineering has evolved to a large extent since the flourishing of IT industries and computer education throughout the world. The past two decades have seen a tremendous rise in the growth and adoption of application softwares in different industries. This is due to the fact that massive digitization has taken place throughout the world in the form of automation [1]. The jobs and processes performed manually erstwhile are now automated by different softwares. As a result, a huge demand for software engineers was created cutting across the barriers of different industries. Computer education nowadays cannot therefore do away with software engineering course. The pedagogy style for this particular course demands considerable hours of hands-on sessions with the help of tools that meet the industry benchmark and allows the student community to adopt to good software engineering practices [1]. Moreover, the outbreak of the pandemic of Covid-19 has

DOI: 10.1201/9781003333500-30

changed the mode of operation for many industries, including education. Classroom teaching experienced a major operational shift in the form of online teaching. During this period, platforms like Zoom, Google meet and Microsoft team were used even by schoolteachers for delivering lectures – a fraternity not accustomed with conducting web meetings as a part of their routine job profile. The challenges in higher education, particularly technical education, were even greater [2]. Other than theory classes, laboratory sessions for different subjects too needed to be conducted which in turn generated the demand for novel and indigenous platforms [2]. Just like academia, this transition from offline to online mode affected the industry as well. New platforms needed to be envisioned for allowing geographically dispersed teams and team-members to collaborate and contribute to their assigned projects [3]. The tools and platforms developed for software development enabled the developer fraternity and the under-graduate and post-graduate student community to develop and deliver their IT projects and products within deadlines. It is a well-known fact that any software product goes through the following stages during its software development lifecycle: requirement analysis, design, development, testing, deployment and maintenance. Numerous tools are available in market for each of the above mentioned phases [3]. These tools increase the efficiency, productivity and QoS of the delivered software products. However, it is to be kept in mind that many of the state-of-the-art tools present in the market are proprietary in nature. This makes it difficult for the academia and start-ups to leverage the benefit of these tools [4]. Open source and free software tools on the other hand address this issue by providing a cost-effective solution. In this paper, we have discussed about a handful of open source and free software product development and management tools which allow the programming community to plan, develop and test quality code through collaboration. Along with this, we have categorized the tools and platforms into different genres as per their usage and utility. A software development project management approach is generally divided into four phases. These four phases encompass six major classes of management tasks related to software development projects [5]. Each of these tasks correspond to a management process, and each achieves a significant and crucial role in the course of software development project management. Figure 1 represents how the corresponding processes fit into this framework.

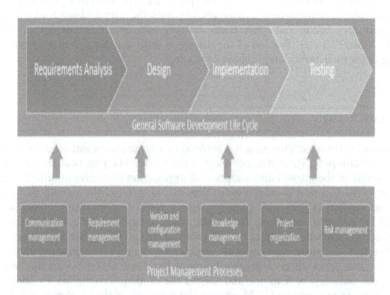

Figure 1.   Overview of processes involved in software development project management.

A project manager's workload can often become very exhausting because of the various roles and responsibilities. There are a plethora of software project management tools available to assist; they range from individual to mid-sized and high-end systems [6][7]. In order to identify and select the best-fit tool, it is important to first determine the team size, the nature of software projects to be developed, and the nature of development methodology to be followed [6][7]. Figure 2 presents some of the features that support the planning, execution, and control of each phase of the SDLC.

Project estimation	Permissions and user roles
Scope definition, verification & control	workflow processes
Project Scheduling /Calendar	Portfolio management
Resource Mangement	Change management
Risk Management, Planning & Control	Configuration management
Subtasks	Product roadmap
Task Management with priorities and categories	Release management
Budget management & forecasting	Version control
Time and expense management	Bug Tracking
Project monitoring & control	Document management & file storage
Team collaboration & communication	Team collaboration & communication
Kanban boards, Gantt charts, and PERT diagrams	Analysis,reports,KPIs,Objectives,key results
Milestone tracking	Client portals for communication, UAT & approvals

Figure 2.   Features to consider while choosing software project management tools.

The rest of the paper is organized as follows: in section 2 we have identified and presented few popular open source and free productivity tools for software engineering. This is followed by conclusion in section 3.

2   SURVEY

In view of the Covid-19 pandemic situation where teams worked remotely, the software development community too had to adopt social collaboration tools [8]. There exists various distinct kinds of project management solutions and tools, varying from personal to-do task list to sophisticated project and portfolio management (PPM) platforms [8]. Figure 3 presents the various categories of software project development and software project management tools into broad categories.

The following section details different open source and free tools that belong to different categories mentioned in Figure 3 along with their features and benefits for the academia and industry.

Redmine is an open source platform that provides both Project Management and Project Communication [9]. Some of the prominent features of Redmine are Multiple projects support, Flexible role based access control, Flexible issue tracking system and a central repository of project information. Here, each project is capable of information sharing and distribution with the help of the following built-in features: Wiki, Forums, News, Issue

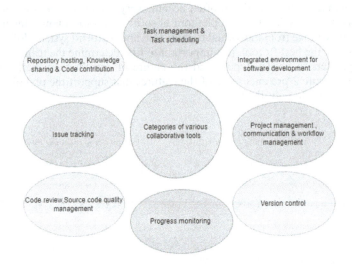

Figure 3. Categories of various collaborative tools for software development and software project management.

tracking with associated scheduling, Document/Files, time tracking, versioning, Gantt charts and calender [9]. This way it becomes possible to provide permissions and access control to the product, assign issues to specified versions of the product, monitoring projects and providing feedback. A development issue can be moved from the "New status" into "Development" and from "Development" into "Integration Test". However, an issue can be closed only by the "Integration Test" team and that too after conducting a test. For intra-team communication, an asynchronous communication path is provided by Redmine Forums, Wikis, and Issues, to all members, independent of space and time. Moreover, the History feature present here provides the current project status [9].Project decisions too can be captured and revisited as and when needed. The "state" of Redmine and the Subversion repository are used both as a regular scheduled turn-in, as well as for keeping the project documents updated. For a geographically scattered team, Redmine has provision for the members to participate and contribute to the team. For project presentation, one can use team's Redmine and Subversion sites directly [9].

Turnkey offers more than 45 stacks that are ready-to-use and are open source software based [10]. For example Redmine, MySQL database, a LAMP stack, Tomcat on Apache, Moodle, Bugzilla and so on. Few of the prime features of these appliances are that, they are free, pre-configured as well as pre-tuned and have a standard set of pre-installed tools. Each of these appliances can be backed up to Amazon's S3 Cloud Service at a nominal cost. Also the instructor has the permission to deploy the whole appliance in Amazon EC2 cloud service, in a single touch. This extends the capability of the student teams to scale up the system along with the bandwidth as per their requirement [10].

Eclipse IDE of Eclipse Foundation offers a multi-language and integrated environment for software development that is widely used in the industry [11]. It comes with an extensible plug-in system that also provides support for a wide range of SDKs. It also provides access to quality code, library, and IDE tutorials. Apache Subversion that is open source provides version control system for each project. The integration between Eclipse and distinct team's Subversion source-code repository is achieved by the open-source Subclipse Eclipse plug-in [12].

GitHub is a prevalent service for hosting repository where the github community can explore open-source code [13]. They can upload private as well as public projects. The GitHub community has established itself into a giant and diverse group of developers who utilize this platform to explore resources, share code, work in a collaborative manner, and develop quality software. They offer free plans for basics for individuals and organizations [13].

Stack Overflow is a large and active programmer community comprising of beginners as well as experienced developers that utilize this platform to acquire coding skills, contribute, and build their careers in software development [14]. It allows knowledge sharing and collaboration. This community comprises of professionals including front-end developers, back-end developers, DevOps engineers, mobile app developers, full-stack developers and database administrators as well as the student community. This forumis popularly used to get one's technical queries and doubts solved. They have free plans for teams comprising upto 50 members [14].

Git is a free version control system which enables developers to manage their projects rapidly and efficiently [15]. Being easy to learn and an open-source distributed system, it is broadly used even by the beginners in coding. A developer can easily modify the code along along with tracking the changes because Git saves all the modifications till the last version of the code. Anybody can collaborate, contribute as well as retrieve a copy of any developer's code [15].

GitLab is a popular web-based tool that offers an unified solution integrating features like issue tracking, reviewing code, monitoring, version control, CI/CD along with security for software development and DevOps lifecycle management[16]. GitLab is an easy-to-install platform that facilitates developer community in accelerating their projects. It offers free plan for individuals [16].

Jenkins os is an automation server that possesses orchestration capabilities required to deploy applications [17]. This Apache software is open-source and functions in servlet containers. It allows development, testing, deployment as well as continuous integration. It is also used for monitoring the CI/CD pipeline as a dependable tool [17].

Docker is an open source containerization platform that caters to developers and system administrators [18]. Docker containers allow programmers to package software as a file system, dependencies and related libraries. It eases the process of application development and helps in code-building, code-shipping, and executing distributed apps [18]. It further assists the isolation along with security to execute multiple containers concurrently on a virtual machine or any host [18].

IntelliJ IDEA is an Open Source Java IDE, which the professional developers utilize to build commercial as well as non-commercial products [19]. This Platform also integrates developer tools and plugins that are language-aware. Text editor, user interface framework, virtual file system, debugger and testing feature are the core components of this project [19].

Trello is a very popular project management tool in software engineering. Being a Kanban-style application, it represents a project as a board or a list. Every list comes with movable cards that have drag-and-drop functionality [20]. Trello allows setting of due dates, adding attachments, inserting colored labels, comment-writing, checklist preparation and integration with other apps. Trello offers a **free plan** for individual users and small teams [20].

CodeProject is an active coding and learning software engineer community [21]. It provides access to tutorials and free source code pertaining to software development, programming languages, web-development etc. Developers can find significant information through variety of resources available on this platform. The objective of Code Project Open License (CPOL) allows the developers willing to share their code, along with a license to interested users along with the policy of usage [21].

Chrome DevTools is a free comprehensive toolkit for web-development built directly into Google Chrome [22]. DevTools enables to create websites, web-authoring, debugging, and editing pages in real-time, diagnosing problems easily. This set of software engineering tools is very useful for improving productivity [22].

Restya's community edition is an open source license which is free and is suitable for testing and development activities [23]. The Restya board has features like trello but also allow starred board listing, addition of board with predefined templates as well as obtain board statistics. One can import board from trello, Taiga, Asana, Wekan and so on.The Restya Core is a tool for managing work flows effectively [23].

ProofHub is a free and open source collaboration software that is preferred by software developers as well as beginners and non-technical community [24]. It offers useful features like team-communication, workflow management, project-centralization, custom project reports, boards and timesheets [24].

BitBucket is a web-based tool that is built for revision control and for hosting repository. It allows the coding community to collaborate with pull requests as well as inline comments [25]. Owned by Atlassian, it integrates well with other tools like Confluence and Jira. Bitbucket Cloud provides unlimited collaborators and unlimited private repositories at no-cost for academia [25].

OpenProject is an open source web-based tool for project management and team colla-boration [26]. Its key features include task-management, planning, collaboration, scheduling, roadmap planning, release-planning, agile, scrum, Time tracking, bug tracing and tracking, costing, reporting and budgeting, meeting agendas and meeting minutes, forums and wikis [26].

SonarQube is an open source platform for managing source-code quality used by devel-opment teams. The major objective is to allow teams to manage code quality with nominal effort. It offers services like code analyzers, debugging modules, reporting tools etc. It also offers a plugin mechanism to expand the functionality [27]. SonarQube offers language support for Java, JavaScript, PHP, PL/SQL and Cobol through open source as well as commercial plugins. It enables to cover quality on 7 axes and further report on duplication of code, coding standards, unit tests, code-complexity, finding bugs, comments, design and architecture [27]. It has been designed to support global strategies for continuous improve-ment in code quality in an organization and therefore can also be leveraged as a shared central repository for code quality management [27].

StarUML is a freely available software modeler that aims to support agile modeling. The target users are Agile and small development teams [28].

Creately is a SaaS visual collaboration tool that offers extensive diagramming and design capabilities that comes in two versions: an online cloud edition and a downloadable offline edition [29].

Feedly is an open-source software that offers a a basic interface by providing feed stories [30]. It is essential for software engineer community to operate in sync with the emerging technologies and trends in their domain. Feedly allows one to create a list of one's preferred publications, favourite news topics, YouTube channels and blogs. It is essentially a cloud-based aggregator service which organizes one's feed according to one's priorities [30].

3 CONCLUSION

Free and open source software tools are a massive support for academia and startups as they help to reduce the development cost of a software product, without compromising the QoS. The survey done here in this paper provides the reader with some of the very popular tools of software engineering, used by academia and industry. The survey will provide the readers with an outline of the different tools that are extensively used in the domain of software engineering.

REFERENCES

[1] Raibulet C., Fontana F.A. and Pigazzini I. (2019) "Teaching Software Engineering Tools to Undergraduate Students," *Proceedings of the 2019 11th International Conference on Education Technology and Computers* [Preprint]. Available at: https://doi.org/10.1145/3369255.3369300.

[2] Fitoussi R. and Chassidim H., "Teaching Software Engineering During Covid-19 Constraint or Opportunity?," in *2021 IEEE Global Engineering Education Conference (EDUCON)*, 2021.

[3]   Ralph P. et al., "Pandemic Programming: How COVID-19 Affects Software Developers and How their Organizations Can Help: How COVID-19 Affects Software Developers and How Their Organizations Can Help," *Empir. Softw. Eng.*, vol. 25, no. 6, pp. 4927–4961, 2020.

[4]   Sarasa-Cabezuelo A. and Rodrigo C., "Development of an Educational Application for Software Engineering Learning," *Computers*, vol. 10, no. 9, p. 106, 2021.

[5]   Schreiber R. R. and Zylka M. P., "Social Network Analysis in Software Development Projects: A Systematic Literature Review," *Int. J. Softw. Eng. Knowl. Eng.*, vol. 30, no. 03, pp. 321–362, 2020.

[6]   Fitoussi R. and Chassidim H., "Teaching Software Engineering During Covid-19 Constraint or Opportunity?," in *2021 IEEE Global Engineering Education Conference (EDUCON)*, 2021.

[7]   Santos J. M. D., "Best project Management Software (2022) - Features & Tools," *Project-Management.Com*, 15-Nov-2021.[Online]. Available: https://project-management.com/top-10-project-management-software/. [Accessed: 28-Oct-2022].

[8]   O'Loughlin E. "Not Every Project Needs Project Management Software," *Top Business Software Resources for Buyers – 2022 Software Advice*, 03-Mar-2017. [Online]. Available:https://www.softwareadvice.com/resources/social-collaboration-tools/. [Accessed: 28-Oct-2022].

[9]   Rahim M. S., Chowdhury A. E., Nandi D. and Rahman M., "Issue starvation in software development: A Case Study on the Redmine Issue Tracking System Dataset," *Journal of Telecommunication, Electronic and Computer Engineering (JTEC)*, vol. 9, no. 3–3, pp. 185–189, 2017.

[10]   *Have you Checked our the Turnkey Hub Yet.* (2022, January 14). TurnKey www.turnkeylinux.org

[11]   *Enterprise-class Centralized Version Control for the Masses.* (2022, January 14). Apache Subversion. http://subversion.apache.org/

[12]   *Desktop IDEs.* (2022, January 14). Eclipse. https://www.eclipse.org/ide/

[13]   *Where the World Builds Software.* (2022, January 14). GitHub. https://github. com/

[14]   *Where Developers Learn, Share, & Build Careers.* (2022, January 14). Stack Overflow. https://stackoverflow.com/

[15]   *Branching and Merging.* (2022, January 14). Git. https://git-scm.com/about

[16]   *The DevOps Platform has arrived.* (2022, January 14). GitLab. https://about.gitlab.com/

[17]   *Build Great things at any Scale.* (2022, January 14). Jenkins. https://www.jenkins.io/

[18]   *Developers Love Docker.* (2022, January 14). Docker. https://www.docker.com/

[19]   *Why IntelliJ IDEA.* (2022, January 14). IntelliJ IDEA. https://www.jetbrains.com/idea/

[20]   *trello helps teams move work forward.* (2022, January 14). Trello. https://trello.com/

[21]   *The Code Project Open License (CPOL) 1.02.* (2022, January 14). CodeProject. https://www.codeproject.com/info/cpol10.aspx

[22]   *Documentation.* (2022, January 14). Chrome DevTools. https://developer.chrome.com/docs/devtools

[23]   *Restya Products.* (2022, January 14). Restya. https://restya.com/

[24]   *The One Place for all Your Projects and Team Collaboration.* (2022, January 14). Proof\hub. https://www.proofhub.com/

[25]   *Built for Professional Teams.* (2022, January 14). Bitbucket. https://bitbucket.org/

[26]   *Open Source Project Management.* (2022, January 14). OpenProject. https://www.openproject.org/

[27]   *Code Quality and Code Security.* (2022, January 14). SonarQube. https://www.sonarqube.org/

[28]   *A Sophisticated Software Modeler for Agile and Concise Modeling.* (2022, January 14).Star UML. https://staruml.io/

[29]   Software Design Toolset. (2022, January 14). Creately. https://creately.com/lp/uml-diagram-tool/

[30]   *Track Insi ghts Across the Web Without Having to Read Everything.* (2022, January 14). Feedly. https://feedly.com/

[31]   Teel S., Schweitzer D., and Fulton S., "Teaching Undergraduate Software Engineering Using Open Source Development Tools," *Issues Informing Sci. Inf. Technol.*, vol. 9, pp. 063–073, 2012.

Automation and Computation – Vats et al. (Eds)
© *2023 the Author(s), ISBN 978-1-032-36723-1*

Detection of pulmonary tuberculosis with thoracic radiograph on ensemble deep learning model

Abdul Karim Siddiqui
School of Computer Science & Applications, LPU-Punjab, India

Vijay Kumar Garg
School of Computer Science & Engineering, LPU-Punjab, India

Vikrant Sharma
Department of Computer Science and Engineering, Graphic Era Hill University, Dehradun, Uttarakhand, India

ABSTRACT: The scarcity of abundant information for effective diagnosis and incomplete or unnoticed follow-up may trouble pulmonary tuberculosis patient. Any delay in diagnosing TB may lead to interrupting governments' campaigns to eliminate national TB control programs. Improper planning and lengthy treatment may affect your bills. The wrong prediction may deteriorate an individual's health. All such cases need to be revised with an absolute and fast intelligent system of prediction of pulmonary Tuberculosis. Pulmonary Tuberculosis as a critical contagious issue requires opting for thoracic images to be applied with efficient Deep-learning, so a low-cost system can be developed for everyone in society. Through this paper, an effective deep learning model is presented to predict and classify pulmonary Tuberculosis among other pulmonary issues. The efficacy of the model relies on pre processing of dataset inputs. An accuracy of 96% is recorded on VGG16 whereas InceptionV3 resulted in 93% of overall accuracy.

1 INTRODUCTION

Being a contagious disease Mycobacterium tuberculosis spread through air medium to a normal person. The pathogenic bacteria are the causative agent named Mycobacterium tuberculosis.

Patient having similar coughing issues or loss of weight with mucus formation or patients having HIV positive may be prone to positive Tuberculosis. There may be a chance of infecting other pulmonary diseases such as pneumonia, pleural effusion or chronic obstructive pulmonary disease. In the last three years, the cases of Covid-19 have worsened the situation. It is now important to design an efficient model that gives quick information for categorizing and analyzing thoracic urgencies. The mortality of a huge population by pulmonary tuberculosis has alarmed the situation. World health organization presents every year reports on TB and other contagious diseases; also it earmarks provisions for the prevention of pulmonary tuberculosis. Slowing down of TB diagnosis and lack of improved interpretation during the Covid period has impacted the increase in TB mortality [20], [21]. The worldwide pandemic situation has turned back the direction of assistance in availing urgent TB predictions and prevention campaigns. Among all the major nations that accounted for 93% of this drop-down in newly, TB diagnosed are India, Indonesia, and the Philippines. Worsening trends suggest that the negative impacts on TB mortality will remain to continue in 2022 [WHO]. Thoracic images are common to find early remarks on pulmonary tuberculosis. The suspected patient may have other types of Lung issues which when not traced clearly, may divert the treatment. Altered diagnosis may cause health. As chest X-rays are used everywhere as one of the finest tools to diagnose TB undoubtedly, preprocessing of image input may boost classification and segmentation.

DOI: 10.1201/9781003333500-31

2 RELATED WORK

TB is a hot topic among researchers and there a great number of papers are available to discuss on the topic. Detailed literature reviews on qualitative data inputs and machine learning models have shown a good impression. Despite the training of systems are sometimes compromised to generate an accurate model. In this paper features using DL models and uniting them with features of texture, shape, etc. are evaluated. CNN is a noble way to describe image inputs in the classification model [1]. Transfer learning was applied to CNN models to detect Covid positive cases with more than 90% sensitivity [2]. Pneumonia confirming X-rays were processed using Resnet, ImageNet, Xception, and Inception by Chhikara et al. [3]. He applied preprocessing such as filtering and gamma correction. Abbas et al. [4] presented transfer learning modified CNN termed as DeTraC that works on the imbalanced dataset. Variations in parameters of DL layered CN networks are done to detect PTB-positive cases. S. Vats et al. presented iDoc-X model to diagnose TB so that medical practitioners can better predict thoracic problems [6]. Hooda et al. [5] apply DL approach to investigate TB and normal cases with an accuracy of 82%. A CAD approach to intelligent pattern recognition presneted by Evalgelista and Guedes by using CNN with an accuracy of 88% [7]. Author [8] compared a pre-trained model,-DenseNet on Shenzhen and Montgomery County repositories. Meraj et al. [9] worked with VGG16, VGG19, GoogLeNet and RestNet50. Yadav et al. [10] sketched out TB detection using transfer learning and claimed 94% accuracy. Rahman et a. applied 09 deep CNNs in transfer learning with 3 separate experiments were carried out for segmentation and x-ray radiographs classification. He claimed accuracy of 96% [14]. Dasanayaka et al. ensembled 02 DCCNs to classify abnormal TB radiographs with is 0.971 avg. accuracy [15]. Munadi et al. use 03 image enhancement algo. feeds the images into ResNet with some other DL models for transfer learning and claims 89% of accuracy [16].

2.1 Deep learning

Being associated with machine learning deep learning is a subset of it. Essentially, it is a neural network with many layers. Neural networks learn to simulate human behaviour as he learns from brain and makes deep learning capable to learn from huge data. A single neural network layer can make approximate predictions. On other hand, additional hidden layers help to optimize so that refined and accurate outcomes can be gained. Conventional machine learning algorithms such as SVM, Decision Tree, and Bayesian Networks, etc. require pre processing of raw data before they are fed into the system. Any impurity remained before passing it into DL may lead to incorrect learning model and thus will not classify them clearly. Figure 1 shows how an intelligent model grows during development phase.

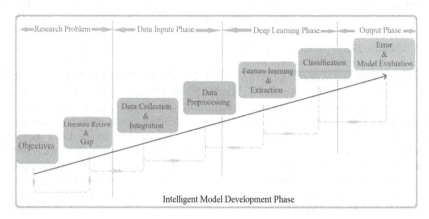

Figure 1.   Intelligent model development phase.

277

Feature learning & extraction is a complex phase. DL algorithms collect high-level features from the dataset in an incremental manner. It also combines the steps of feature extraction and classification. The bigger advantage of the deep learning algorithm is that it may be dealt smoothly with extensive dataset input.

2.2 *Convolutional networks*

Convolutional networks have been proved the landmark in the image recognition, classification and image segmentation. In the field of medical sciences and research, CNNs have become emergent in image classification. CNN architecture has got prominent place to extract and passing features to next layers. The two main modules of CNNs are feature extraction module and the classifier module. Feature extraction module reads relevant features through convolution and pooling. The classifier module classifies the extracted features. Figure 2 shows a simplified structure of convolutional neural networks. An input image is fed into the convolution stack to extract the necessary features. Extracted features are then pooled to reduced map size. It is finally mapped to vector classes for the problem through dense layer.

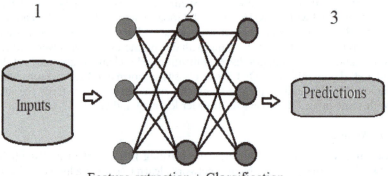

Feature extraction + Classification

Figure 2.   Basic steps involved in deep learning problem.

2.3 *CNN classifier*

The emergence of VGGNets made the researcher more reliance on deeper network structure. VGG has different variants such as VGG16, and VGG19. The issue comes with degradation as the depth increases, it causes saturation. This happens because of complexity. The degradation problem is solved by ResNets residual structure. A residual besides mapping between x to H(x) could better optimize. This also ease skipping the connections in front and back layer to facilitate Back propagation of gradient in training phase. And thus ResNets train deeper network and referred as multi-branch model construction method. Residual Network was designed to handle vanishing Gradient and degradation. It has multiple variants based on number of layers such as ResNet18, ResNet50, ResNet101 and ResNet152. ResNet is effectively trending in clinical image classification in association with transfer learning.

In DenseNet class of neural networks each layer receives inputs from layer behind it and forwards its own feature map to next subsequent layers. Here every layer acquires the knowledge from preceding layers in a feed-forward manner. Inception network is composed of a composite architecture. It may act as perfect model to apply in classification problems. For improved result most of the CNN models prefers extra number of layers in between them. Inception model takes fewer number of layers in classifying problems, resulting less amount of time and cost. Different versions of InceptionNet have emerged viz. InceptionNet V1, V2, V3.1, V4 and Inception-ResNet.

2.4 *Transfer learning*

In Transfer learning the model trained for a purpose is reused again as an initial point to train another model. The dilemma of inadequacy in practicing data is resolved by Transfer learning. The mechanism is to shift the information from the origin domain to the last domain by reducing the premise that the practice data and the test data need to be comparable. The Transfer learning process is described in following Figure 3. The objective of deep transfer learning is to acquire knowledge from other domain. And thus source domain and the target domain need not to be equivalent identically. Due to shortage of available dataset Transfer learning is suitably applicable in CNN application. Transfer learning is equally used in manufacturing and medical field[11]. This overcomes the need of large dataset input and saves the long training period [12].

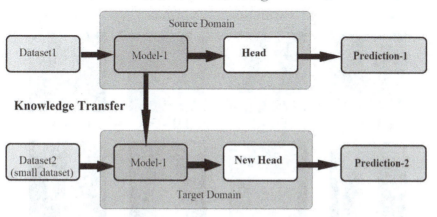

Figure 3.   Transfer learning process.

3   METHODS AND MATERIALS

3.1 *Data set*

The online dataset was collected from Montgomery County CXR Set[18], Shenzhen Hospital CXR Set[19], and NIAID TB dataset[17]. It contained more than 3000 TB positive chest X-ray images listed on the National Library of Medicine-USA. Image dataset such as CXR has a high sensitivity for pulmonary TB. It is one of the most popular tools to identify TB in the differential diagnosis of pulmonary TB. As per WHO, the Standard characteristics of CXR associated with active TB disease cover infiltration, pneumonia, atelectasis, mass, nodule, and effusion.

3.2 *Data preprocessing*

The thoracic radiographs having equal dimension of the standard digital picture were required. The accessed databases had variations in resolution. Thus, they were first converted to an acceptable size by Resize operation, and then normalized. It is beneficial in training of CNN. The normalized image input can emphasize the accuracy of image processing to get a more precise classification. Wireless Sensor Networks are a vital systems of an automated machine and are preferably used in computer assisted diagnosis to monitor imaging textures and cavity marks [13]. Acquiring x ray images from different imaging machines may produce artifacts and low contrast or patient movement during imaging or reflection caused by lighting. All such variations should be addressed before proceeding to

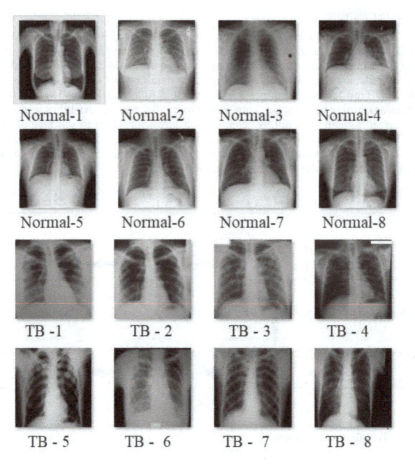

Figure 4.   Enhanced X-ray images of normal and tuberculosis.

Machine Learning. Preprocessing is a key task in biomedical imaging transformations. We need to extract the region of interest also called ROI from the chest X-ray and then pre-process it until better image features come for subsequent processing. Some of the relevant Preprocessing techniques involve Unsharp Masking, High-frequency Emphasis filtering and Contrast Limited AHE. The Enhanced thoracic data input are shown in given Figure 4.

3.3   Experimental dataset & environment

Datasets are equally divided into 3 classes for training inputs as Normal, TB and other Lungs diseases so that class imbalancement issue can be overcome. The division of the training and testing dataset are mentioned below:

Table 1.   Dataset – training.

| DATASET | Class | | |
	(Normal)	(PTB)	Other Lung Issue
No.	1000	1000	1000

Table 2.   Dataset – testing.

DATASET	Class		
	(Normal)	(PTB)	Other Lung Issue
No.	300	300	300

3.4   Dataset partition

We have taken a total of 3900 images of Pulmonary diseases Vs normal thoracic radiographs including pulmonary Tuberculosis and other lung related complications. We divided these images as 77% for training and 23% for testing datasets.

3.5   Modeling

The pretrained architecture contains neural networks viz. VGG16, InceptionV3 and ResNet50 described earlier. Training of weights in a multilayer feed-forward neural environment is done by backpropagation algorithm. VGG is framed on less convolutional filters. There are 3 convolutional layers and 3 fully connected. X-Ray images are further molded into size of 224X224X3. First two layers have Filter size of 3 X 3 with "same" padding. Feature map with size 64 comes out. It turns to a max pooling with 2 convolution layers. Max pool layer has stride (2, 2). Next the two convolution layers with dimension (112, 112) and filter of size (3, 3) exist. Another max pooling layer follows the same criterion. A set of convolutional layers with size (56,56) and filter size (3, 3) is followed which produces feature map of 256. Next it has 3 convolution layers followed by a max pooling layer. A feature map of 512 is received. Most familiar Activation function ReLU is applied in hidden layers to ease computations. InceptionV3 is efficient in parameters count and memory utilized with 48 deep layers whereas ResNet50 has total 48 convolution layers. Additionally, the proposed model is composed of one max and average pool layer.

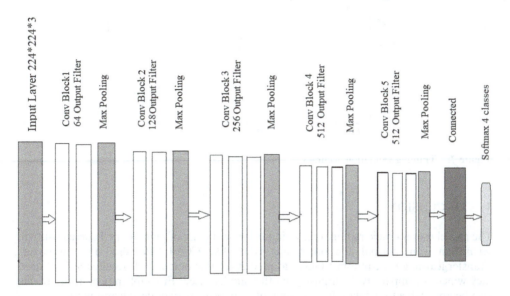

Figure 5.   Proposed CNN architecture.

3.6 System architecture

The training input are split into a batch of 32 and trained for 50 epochs. Further Adam optimizer used to give better accuracy. Over fitting issue is addressed through Custom call back and Model Checkpoint. Transfer learning technique is applied to support InceptionV3 and ResNet50 over VGG16. More than 95% testing accuracy is achieved on VGG16. The proposed system architecture was trained on GPU. X-ray images require greater processing space so that proportional computation power is important. Here two GPUs viz. Tesla K80 on Google Colab, and Tesla K100, were used.

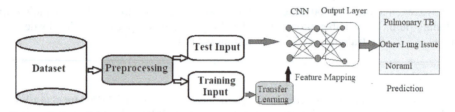

Figure 6. Proposed system architecture.

3.7 Experimental results and analyses

Initially Training accuracy is low on epoch. Later it achieved stability which ranges from 0.95 - 0.97. Some spikes were found with validation accuracy of range 0.90 - 0.96. Comparing result achieved from Table 3 Visual Geometry Group has better accuracy with 96%. InceptionV3 has also relevancy by accuracy of 93%.

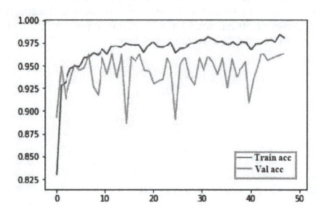

Figure 7. Training validation accuracy.

4 CONCLUSION

The model uses thoracic radiographs as input for predicting of pulmonary Tuberculosis with separation from normal and other lung issues like COVID-19, Viral Pneumonia etc. Transfer learning is helpful in VGG16 to predict reliable outcome. It resulted 96% of accuracy which is comparatively inspiring to other models used. Ensemble model in future may also be applicable to increase the accuracy of classification over deep learning models.

REFERENCES

[1] Krizhevsky A., Sutskever I. and Hinton G. E. "Imagenet Classification with deep Convolutional Neural Networks," In *Proc. Adv. Neural Inf. Process. Syst. (NIPS)*, 2012, pp. 1097–1105.

[2] Tahir A., Qiblawey Y., Khandakar A., Rahman T., Khurshid U., Musharavati F., Islam M. T., Kiranyaz S. and Chowdhury M. E. H., "Coronavirus: Comparing COVID-19, SARS and MERS in the eyes of AI," 2020, arXiv:2005.11524. [Online]. Available: http://arxiv.org/abs/ 2005.11524

[3] Chhikara P., Singh P., Gupta P. and Bhatia T., "Deep Convolutional Neural Network with Transfer Learning for Detecting Pneumonia on Chest X-rays," In *Advances in Bioinformatics, Multimedia, and Electronics Circuits and Signals (Advances in Intelligent Systems and Computing)*, vol. 1064, L. Jain, M. Virvou, V. Piuri, and V. Balas, Eds. *Singapore*: *Springer*, 2020, doi: 10.1007/978-981-15-0339-9-13.

[4] Abbas A., Abdelsamea M. M. and Gaber M. M., "DeTrac: Transfer Learning of Class Decomposed Medical Images in Convolutional Neural Networks," *IEEE Access*, vol. 8, pp. 74901–74913, 2020.

[5] Hooda R., Sofat S., Kaur S., Mittal A. and Meriaudeau F., "Deep-Learning: A Potential Method for Tuberculosis Detection using Chest Radiography," In *Proc. IEEE Int. Conf. Signal Image Process. Appl. (ICSIPA)*, Kuching, Malaysia, Sep. 2017, pp. 497–502.

[6] Vats S., Singh S., Kala G., Tarar R. and Dhawan S., "iDoc-X: An Artificial Intelligence Model for Tuberculosis Diagnosis and Localization," *J. Discret. Math. Sci. Cryptogr.*, vol. 24, no. 5, pp. 1257–1272, 2021.

[7] Evalgelista L. G. C. and Guedes E. B., "Computer-aided Tuberculosis Detection from Chest X-ray Images with Convolutional Neural Networks," In *Proc. Anais do XV Encontro Nacional de Inteligência Artif. e Computacional (ENIAC)*, Oct. 2018, pp. 518–527.

[8] Nguyen Q. H., Nguyen B. P., Dao S. D., Unnikrishnan B., Dhingra R., Ravichandran S. R., Satpathy S., Raja P. N. and Chua M. C. H., "Deep Learning Models for Tuberculosis Detection from Chest X-ray Images," In *Proc. 26th Int. Conf. Telecommun. (ICT)*, Apr. 2019, pp. 381–386.

[9] Meraj S. S., Yaakob R., Azman A., Rum S. N., Shahrel A., Nazri A. and Zakaria N. F., "Detection of Pulmonary Tuberculosis Manifestation in Chest X-rays using DIFFERENT Convolutional Neural network (CNN) Models," *Int. J. Eng. Adv. Technol. (IJEAT)*, vol. 9, no. 1, pp. 2270–2275, Oct. 2019.

[10] Yadav O., Passi K. and Jain C. K., "Using Deep Learning to Classify X-ray Images of Potential Tuberculosis Patients," *In Proc. IEEE Int. Conf. Bioinf. Biomed. (BIBM)*, Dec. 2018, pp. 2368–2375.

[11] Christodoulidis S., Anthimopoulos M., Ebner L., Christe A. and Mougiakakou S., "Multisource Transfer Learning with Convolutional Neural Networks for Lung Pattern Analysis," *IEEE J. Biomed. Health Inform.*, vol. 21, no. 1, pp. 76–84, Jan. 2017.

[12] Akçay S., Kundegorski M. E., Devereux M. and Breckon T. P., "Transfer Learning Using Convolutional Neural Networks for Object Classification Within X-ray Baggage Security Imagery," In *Proc. IEEE Int. Conf. Image Process. (ICIP)*, 2016, pp. 1057–1061.

[13] Sharma V., Patel R. B., Bhadauria H. S. and Prasad D., "Deployment Schemes in Wireless Sensor Network to Achieve Blanket Coverage in Large-scale Open Area: A Review," *Egypt. Informatics J.*, vol. 17, no. 1, pp. 45–56, 2016.

[14] Rahman T. *et al.*, "Reliable Tuberculosis Detection Using Chest X-Ray With Deep Learning, Segmentation and Visualization," in *IEEE Access*, vol. 8, pp. 191586–191601, 2020, doi: 10.1109/ ACCESS.2020.3031384.

[15] Dasanayaka C. and Dissanayake M. B. (2020): Deep Learning Methods for Screening Pulmonary Tuberculosis Using Chest X-rays, *Computer Methods in Biomechanics and Biomedical Engineering: Imaging & Visualization*, DOI: 10.1080/21681163.2020.1808532

[16] Munadi K., Muchtar K., Maulina N. and Pradhan B., "Image Enhancement for Tuberculosis Detection Using Deep Learning," in *IEEE Access*, vol. 8, pp. 217897–217907, 2020, doi: 10.1109/ ACCESS.2020.3041867.

[17] Rosenthal A, Gabrielian A, Engle E, Hurt DE, Alexandru S and Crudu V *et al.* The TB Portals: an OpenAccess, Web-Based Platform for Global Drug-Resistant-Tuberculosis Data Sharing and Analysis. *J Clin Microbiol.* 2017;55(11):3267–3282. Dataset must be cited.

[18] Index of public/Tuberculosis-Chest-X-ray-Datasets/Montgomery-County-CXR-Set/MontgomerySet/ CXR_png/ (nih.gov)

[19] Index of public/Tuberculosis-Chest-X-ray-Datasets/Shenzhen-Hospital-CXR-Set/CXR_png/ (nih.gov)

[20] https://www.who.int/publications-detail-redirect/9789240037021

[21] <https://tbcindia.gov.in/showfile.php?lid=3659

Automation and Computation – Vats et al. (Eds)
© 2023 the Author(s), ISBN 978-1-032-36723-1

A study of the challenges faced by the hotel sector with regards to cyber security

Aniruddha Prabhu B.P.
Assistant Professor, Department of Computer Science, Graphic Era Hill University, Dehradun, India

Rakesh Dani
Associate Professor, Department of Hospitality Management, Graphic Era Deemed to be University, Dehradun, India

Chandradeep Bhatt
Assistant Professor, Department of Computer Science, Graphic Era Hill University, Dehradun, India

ABSTRACT: This research was carried out with the intention of drawing attention to the vital role that cyber security plays in the hotel business. This research identifies and evaluates various prevalent network dangers, and it suggests effective security policies and strategies that hotels may use to avoid cyber assaults. This research, which makes use of up-to-date, relevant data from the hospitality sector, offers a wealth of knowledge that the IT department may use to tighten up the hotel industry's already stringent regulations on the protection of electronic data. This study employs a novel blend of the qualitative and review research methods to learn about the fundamental problems that arise in the hospitality industry and the cutting-edge technical and practical answers that hotels have come up with to address these problems. The findings of this study suggest that the current measures hotels take to ward off cyber-attacks are, for the most part, inadequate and dated. The hospitality industry is exposed to cyber threats and attacks, but a survey shows that most hotel employees lack the knowledge to deal with them. At the conclusion of the study, various consequences and advices are presented to the decision-makers of hotels, with the goal of assisting in the prevention of security breaches affecting both the hotels and their visitors.

Keywords: Hotels, Cyber security, Information, Guests

1 INTRODUCTION

E-marketing, international connectivity and increased transaction volumes are all driving elements in the hotel sector's technical evolution [1]. Using IT to improve customer service, bookings, F&B management, sales, catering, maintenance, security, and hotel accounting is a possibility [2]. These days, hotels and other lodging businesses have more alternatives for the quick, tailored, and localized delivery of services thanks to the Internet of Things [3]. Thermostats, motion detectors, and ambient light sensors are all examples of Internet of Things (IoT) devices that might be utilized to save money on energy bills by adjusting restaurant temperatures and lights automatically depending on customer usage patterns. It is possible that edge/fog computing will be used by the hospitality industry to provide location-based services [4]. The incorporation of technology has improved the guest experience by reshaping the delivery of services, but it has also created new issues, the most pressing of which is the need to ensure the cyber security of these technologies in the hospitality business [5].

DOI: 10.1201/9781003333500-32

Information about visitors is frequently acquired by hotels using technology, and this may lead to a data breach and the theft of personal information [6]. Organizations keep an eye out for a wide range of security threats in order to avoid monetary losses due to things like computer-aided fraud and espionage, sabotage, vandalism, hacking, system failures, fires, and floods [7]. Organizations typically try to conceal data breaches and cyber-attacks on their computer systems in the hotel industry, since consumer loyalty and trust quickly translate into cash, to preserve public trust and dissuade copycat hackers [8]. That's why this paper addresses the challenges and hazards to hospitality's cyber security, as well as new technologies and tactics that may be utilized to fight cyber assaults. It also presents consequences and advice for hotels to better safeguard the data of both their own customers as well as visitors.

2 PURPOSE AND QUESTIONS OF THE RESEARCH

This study will evaluate and advise critical security procedures and processes associated to electronic information and network systems to prevent cyber-attacks in hotels.

This study's novel method answered the following research questions.

1. How do hotels secure their computer networks and data?
2. What threatens hotel computer security?
3. What security measures does the hotel have in place to protect its computer network?
4. How important is network security in hotels?
5. How do hotels safeguard the personal and financial information of their clients on their websites?
6. How can hotels protect the security of their computer networks and online logins?

3 BACKGROUND AND COMPREHENSIVE REVIEW

In this section, we will begin with a review of some important literature on hotel cyber security, followed by a discussion of some background information. This subject covers a wide range of topics, including information security technologies and strategies, cyber risks, hazards, and challenges within the hospitality industry, as well as techniques for combating cyber assaults. This research investigates a range of methods that may be used to protect hotel businesses from cyber assaults. The hospitality industry is plagued by a number of cyber security issues.

3.1 *Hardware and software applied within the hospitality sector*

The most important functions of information technology are data storage and transmission [9]. Businesses that are dependent on information technology have a responsibility to protect and secure any electronic data that they manage and are in charge of [10]. Administrative managers have the responsibility of protecting the assets and information of the organization. The information technology systems used in hotels, similar to those used in other types of organizations, consist of both software and hardware components [11]. The property management system (PMS), the point-of-sale system (POS), the call accounting system (CAS), and the hotel accounting system [12] are all crucial pieces of hotel software. Hardware in hotels can refer to a wide range of electronic components, such as front-desk PCs, back-office POS terminals and servers, cameras and printing equipment, routers, switches, network cables, and other Internet protocol (IP) devices. Network connections link the POS terminals, back- and front-office computers, and printers to the routers and switches [13].

There is a good chance that the hotel you are staying at will provide you with access to a local area network (LAN). The local area network (LAN) at the hotel is connected to the Internet as well as other networks via the use of routers. Firewalls prevent unauthorized users from connecting to the hotel's network from the outside [14]. In the hotel industry, the point-of-sale (POS) system and the property management system (PMS) are used to avoid having two separate reservations for the same day and time.

3.2 *Instruments and methods for the information security*

Malware is just one of many risks that can affect an organization's IT infrastructure [15]. Viruses, internal attackers, stolen laptops or other electronic devices, spoofing, uncontrolled internal access, unauthorized external assaults, and denial of service attacks are among the most typical sorts of threats [16]. The purpose of information security is to reduce the amount of potential damage caused by security lapses as much as possible, with the end objective of increasing both the profitability of organizations and investments [17]. Most information security systems are built with the three pillars of security in mind: privacy, reliability, and accessibility. The basic goal of information security systems is to prevent sensitive data from falling into the wrong hands. This level of security can be achieved using a combination of technical and non-technical safeguards such as the locking down of computers and other valuables, user ID and password management systems, biometric access control systems, and firewalls [18]. In the following, we will describe some of the information security terminologies, as well as some of the tools and techniques:

Digital Identifiers, or IDs, are the modern replacements for paper documents including driver's licenses, passports, and membership cards (IDs) [19]. Digital identifiers typically consist of two parts: a user name and a password. In asymmetric cryptography, which can be thought of as a type of information security technology, public and private keys are utilized as identifiers in the same way. In asymmetric cryptography, the validation of the legitimacy of public keys and identifiers is accomplished via the use of digital certificates. When a certificate digitally signs both the key of a user or public system and the ID of the user or system, the certificate creates a connection between the ID and the public key of the user or system [20].

Second, an intrusion detection system watches the activities that are going on inside a computer system or network in order to discover any attacks or intrusions that may be taking place [21]. The term "firewall" is commonly used to describe this type of security system. Threats to security service availability, integrity, and confidentiality may result from an intrusion. One of the most common forms of cyber-attacks perpetrated by bad actors is the distributed denial of service attack, or DDoS attack. Intrusions may be caused by unauthorized users, approved users, or even authorized users who misuse and abuse their powers [22].

When we discuss the concept of physical security, we are referring to the process of ensuring that a company's computer network and associated hardware are stored in a secure location [23].

A firewall, which may be either hardware, software, or a combination of the two, can be used to monitor the devices and/or networks that are connected to it [24]. A software firewall, as opposed to a hardware firewall, is used to monitor the flow of traffic over a network. A hardware firewall is physically connected to a network and installed on devices that are connected to the network (such as computers, tablets, and smartphones, for example). Through the use of the firewall, malicious packets have the potential to be prevented from entering or exiting a network [25].

It is necessary to encrypt the information in such a way that it is either impossible or very difficult to decode in order to protect its confidentiality. Encryption's main purpose is to provide an additional layer of security in addition to secrecy. The purpose of encryption is to prevent anybody other than the data's lawful owners from accessing it [26].

Speech recognition and gait analysis are both examples of biometric technology, and they are only two of the many sorts of biometric identification techniques available. The use of biometrics as an identity verification method might prove to be extremely helpful in the future. (The identify of a person may be verified by using biometric technologies, which include taking measures of a person's physical characteristics and comparing those measurements to data that has been gathered in the past [27]. Techniques for limiting system resources to authorized users/processes are referred to as "Access Control." Verifying a user's identity and granting permissions are the two most common components of an access control scheme [28].

Assessment of Vulnerabilities Scanning the system for possible vulnerabilities, Scan informs the system administrator of these flaws so that the system may be protected against them [29].

3.3 *The hospitality industry faces a wide range of cyber security challenges*

It has always been possible to conduct a crime online, ever since computers were invented; yet, the nature of the assaults and crimes committed online evolves as technology advances. The most common sorts of security attacks include hacking, technology theft, and fraud; however, other types of security attacks are always possible [30]. Most hacking activities are done with the goal of illegally gaining important information (for example, the financial information of banking accounts or the information associated with user accounts). One type of this crime is when an attacker knowingly establishes a connection to a computer in order to steal proprietary data or source code. Without proper authorization, theft of trade secrets happens when an individual or group uses confidential business information for financial gain [31].

To commit fraud, an attacker must gain unauthorized access to a computer system or assume the identity of a genuine user of the system [32].

Cybercrime is expected to cost roughly $7.51 billion by the end of this year [33], prompting firms to increase expenditure on cyber security to guard against the loss of sensitive information. However, this figure only takes into account the measurable, direct expenses associated with the data breach. Organizational data breaches typically cost millions of dollars, but that doesn't include any other expenses that may arise as a result of the breach [34]. The true cost of a data breach to a corporation is far higher when the long-term effects on the organization's prospects and the effects on its guests and staff are factored in [35].

Whereas the hotel sector strives to provide visitor with a level of acceptance and relaxation, accomplishing this goal has become increasingly difficult as a result of the proliferation of potential dangers. These dangers aim to undermine not only the credibility of hotels but also the faith placed in them by customers by stealing and misusing the guests' private and financial information [36]. Hotel owners need to be aware that these kinds of dangers are always present, and they have a responsibility to accept liability for any data loss that occurs on their properties. Hotel owners can protect data loss and security breaches from occurring by putting in place appropriate preventative measures [37].

The prevalence of data breaches in organizations serves as a constant reminder to businesses of the importance of implementing cyber security technology and tactics. A hotel's guests and employees need to be familiar with basic cyber security concepts and practices, in contrast to the establishment of cyber security tools and procedures. Being careful and vigilant about the sources from which you obtain information, and understanding the distinction between secure websites and emails, is one of the most critical things you can do to protect your data [38].

If the personnel at your hotel visits an unsafe website or opens a link in an unsolicited email, it could cause serious issues for the business. At addition to the laptops in hotels, the Wi-Fi networks themselves are a potential target for cybercriminals. Nowadays, free Wi-Fi is a standard amenity at hotels, available in public areas like the lobby and meeting rooms. Unfortunately, most modern hotels' Wi-Fi networks are not encrypted, making guests vulnerable to identity theft. Hackers can use hotels' Wi-Fi to spread malware-infected updates of popular programs like Adobe Reader and Flash Player, encouraging users to download and install them [39]. The malware in those updates is used by cybercriminals to steal sensitive information from users' PCs and mobile devices.Of note is the fact that hacker and cyber criminals often utilize ancient software, some of which is more than a decade old. Because of the carelessness and negligence of many hotels and the absence of system upgrades, hackers may still use even these ancient software applications to steal hotel guests' sensitive information [40].

A company's internal and external attackers might come from the same source. There is a greater risk of insider attacks in the hotel and tourist business because of the huge turnover [41]. To avoid insider attacks and offer infrastructure to assure just one sign-on by a worker, certain firmsadopt measures. When an employee leaves or resigns, the system simply has to be cleared of one record, ensuring that the former staff can no longer access it.

One of the most popular ways for hackers to steal guests' personal information at hotels is through a practice known as "Fake Reservation" [42] in this scam, the hackers establish a website that appears and acts just like the official hotel website, complete with the same domain name. This results in a flood of potential guests visiting the infected website, some of whom may even book their accommodations through it, so providing the hackers with their personal and financial details [43].

3.4 *Strategies for preventing hospitality sector cyber attacks*

Even though there is no security programs, virus protection, or possibly other technology that can guarantee that a hotel or other company will be safe from cyber threats one hundred percent of the time, hotels are obligated to utilize the most up-to-date and effective technologies and strategies to protect themselves from such threats [44]. When it comes to information security, hotels often divide their procedures into three distinct phases: prepare and protect, defend and detect, respond and recover [45].

Web application firewall is one approach that hotels may adopt to protect themselves from the risk of a data breach (WAF). When it comes to problems with the security of online applications like SQL injections, buffer overflows, and security misconfigurations, a WAF may filter the content of particular web apps and, as a result, help prevent assaults [46]. In addition, WAF solutions are efficient in detecting and preventing data theft since they can identify and block the database of a credit card [47]. This makes WAF solutions useful in both detecting and preventing data theft.

The usage of digital certificates is another approach that hotels may use in order to safeguard the sensitive data of their clients. The use of digital certificates enables the provision of non-repudiation security services when appropriate. The use of digital certificates in the hospitality industry may help prevent customers or company owners from making false claims, which may be challenged in court and the truth established. Due to the fact that it has digital certificates that attest to its authenticity, a hotel's website may be relied upon [48].

Hotels may wish to consider obtaining cyber security insurance as an extra method of safeguarding themselves and the personal information of their visitors in the event that a data breach occurs [49].

The author Butler recommends that proprietors of hotels get themselves some kind of cyber security insurance. Because general liability policies do not cover data breaches or cyber claims, it is even more important for hotel owners to give some thought to and maybe invest in cyber insurance in case their establishment is the target of a cyber-assault [50]. These insurers will offer coverage for both the first party and third parties in the event that a data breach or cyber-attack occurs. Third parties may take several forms, including customers and governmental agencies.

4 METHODOLOGY

In order to demonstrate how vital it is for the hospitality industry to have solid cyber security, this research takes use of two distinct research methodologies. An in-depth analysis of all academic and professional papers that are now available in the subject of hospitality cyber security will be one of the tactics that will be employed. This will be done as one of the ways that the problem will be approached. The writers of this piece have presented a summary of the most pertinent findings and issues associated to cyber security and dangers affecting the hotel industry in second section of this research. The second method of investigation was qualitative in nature, and the authors conducted sixty interviews with persons who were involved in the hospitality business, academics, and hotel guests. For the purpose of this study, interviews were conducted with a total of sixty individuals: six of them were hospitality professors who led a class titled "Hotel IT;" fourteen of them were hotel managers who oversaw establishments of varying sizes; twenty of them were hotel staff members (front desk clerks); and the final twenty individuals were hotel guests.

Each interview took an average of twenty minutes to complete; around five to seven minutes were spent talking to visitors, and fifteen minutes were spent talking to staff and management. There was a striking similarity between the queries that were posed by the management and the academics and the research questions of this study. On the other hand, the queries that were posed by both the staff members working at the hotel and the visitors were, for the most part, basic inquiries about the security of computers, emails, and websites.

5 FINDINGS AND RESULTS

Based on interviews with front desk staff, guests, managers, and academics, the findings and conclusions of this study show that many hotels employ core tools and software like antiviruses to prevent data intrusion and loss. These results and discoveries were uncovered as a direct result of this study. Finding out that a large percentage of medium- to small-sized hotels do not employ an information technology manager or a competent computer security specialist was a bit of a surprise. Some of these hotels, surprisingly, do not have a contract in place with any information technology business to deal with cyber security issues. In the event that these hotels have problems with their computers, they will call unrelated IT professionals working for other organizations in order to resolve the difficulties. Since of this, there is a considerable risk to the hotel's safety because the unidentified information technology specialist may constitute a threat to the hotel's computer system and network.

Three managers interviewed for this research said they had witnessed a data breach in the past five years. One of them speculated that the perpetrator was likely a former colleague who had a personal issue with the IT manager and wished to take revenge. Another management who participated in this study's interviews also said their organization has not been the victim of a data breach in the previous five years. Both companies' upper management declined an opportunity to discuss the scope of the data breach because of the delicate nature of the topic. Furthermore, the most of hotel staff reported not receiving any suspicious or insecure emails in which the sender requested personal information be provided by clicking on a link provided in the email. These emails are thought to be insecure because they ask the recipient to provide information that could be used to steal the recipient's identity. The vast majority of the staff members at the hotel confirmed this information. In point of fact, the majority of the hotel's staff as well as the hotel's visitors who were questioned lacked considerable experience about cyber security. This was found to be the case when the hotel customers were questioned. In the course of the interview with one of the guests, divulged that she had previously been a victim of having her personal data taken through the usage of a hotel network system. This information was provided by the guests while the interview was in progress. The guest said that she was a participant in the hotel's loyalty program and that she had seen some strange transactions on her credit card. After then, the news of the data breach that occurred at the hotel spread all over the world. A guest who stayed at one of the nearby hotels related his experience with what he called a "Fake Booking" to the other guests. He said that he used a well-known hotel chain's website to make his online reservation for a hotel room; but, when he arrived at the hotel, the personnel at the front desk were unable to find his reservation. In addition, he said that this took place despite the well-known status of the hotel brand. Next, he presented the hotel personnel with the written verification of his booking, at which point they saw that the website's name was misspelt and that the website in question was not the official website for the hotel. When asked if they use the hotel's Wi-Fi, the vast majority of guests admitted they did, despite the fact that hotel Wi-Fi is notoriously insecure and leaves guests wide open to cyber-attacks. Due to the unsecured nature of the hotel's Wi-Fi network, thieves may potentially access and use guests' personal information, including financial account details and login credentials, without the visitors' knowledge or consent. When using a hotel's Wi-Fi connection or entering sensitive information on a website while staying at that hotel, the authors recommend that guests utilize a virtual private network (VPN).

It has been brought to the attention of the media by management that there is currently no official training in place for employees in the areas of data privacy and cyber security. The management is the source of this information. Because of the high turnover rate in the hospitality business, management were concerned that it would not be cost-effective to teach staff members on "marginal" issues. This was due to the fact that employees frequently left their jobs.They said that this churn made it unlikely that employees would retain the information they were taught. It was unsettling to hear this from a hotel manager who should have been aware of the possible damages that may result from a data breach. These losses could be far more than the cost of training staff members. During our discussion with the professors, not only did we discuss the theoretical aspects of cyber security, but the professors also provided hotels with critical guidance on how to better safeguard the privacy and security of their visitors. This article touches on a few of these concepts throughout its readings.

It has been discovered that the majority of hotel guests and staff members (including managers and front desk personnel, for example) demonstrate a lack of knowledge and carelessness. Another finding was that there was a lack of training for staff members, as well as the usage of effective tools and procedures, to protect the hotel's computer systems and network. According to the findings of this research, the lack of IT personnel in hotels is connected to data security risks in hotels.

The final finding of this study was that hotels are susceptible to security breaches because they do not often change their passwords, do not update their software, and do not use strong passwords.

6 FINAL THOUGHTS, SUGGESTIONS, AND IMPLICATIONS

Cyber security is an extremely important concern for the hospitality industry. The hotel industry may be able to better protect itself against potential cyber-attacks by making use of the tools and strategies that are outlined in this paper. The authors of this research were limited in their ability to draw conclusions about other aspects of hotel information security based on interviews because of the sensitive and confidential issue under investigation. While this study's findings do shed light on a few of the most pressing reasons for security gaps in the hospitality sector, the researchers seem unable to identify many other elements that might compromise the information security of hotels.

After conducting an in-depth analysis using a variety of academic and professional materials, we were able to identify five of the most important threats and challenges that hotels have experienced to this point. According to Hiller [9], the following is an outline of these five challenges:

Credit card fraud and identity theft have been responsible for a significant number of data breaches and the theft of information from hotel network systems.

Only two examples of cybercrime tactics that hotels utilize to escape discovery during quiet intrusions are social engineering (also known as phishing) and advanced persistent threats (APTs), which have been more common in recent years.

The vast majority of hotels either do not undertake security audits at all or they do it on a timetable that is less regular, placing investors and customers both in peril.

Hotels have the risk of being targeted by terrorist acts as well as other types of violent crimes, such as abduction.

Hotels lose their opportunity to compete and have a bleak outlook as a direct consequence of the dangers posed by cyber security breaches.

An assault on a computer network might take any one of the following three forms:

It is possible that the intruder will get unauthorized access to the network.

It's possible that the invader may corrupt, delete, or otherwise mess up the data in some way.

It is possible for the intruder to gain a user account on the phone and then proceed to carry out a malicious procedure that results in the device malfunctioning, hanging, or rebooting itself.

This article focuses on a few of the various advantages and recommendations that users in the hospitality industry may make use of to their advantage. These advantages and recommendations include:

When users submit personal and financial information on websites, they should exercise extreme caution and verify both the website's name and its secure socket layer (SSL) certification before submitting the information. Users should also exercise extreme caution when dealing with websites that do not have SSL certification. For the convenience of our visitors, we have compiled a collection of travel and hotel-related hints and information. Customers should refrain from the following activities when utilizing public computers or public Wi-Fi, as recommended by our company: (i) using online banking; (ii) accessing email inboxes while travelling; (iii) disabling automatic connection to unknown Wi-Fi networks; (iv) using remote desktop applications rather than saving sensitive information on laptops or smartphones while travelling; and (v) connecting to a VPN network to browse the web and enter personal information.

According to the findings of a number of studies, hotels that provide loyalty programmes expose themselves to a greater possibility of experiencing a data breach due to the greater quantity of patron information that they own. To put it another way, the personal information of hotel guests and customers is at a greater risk as a result of this. The following are some of the measures that hotel management may take to protect the personal information of repeat customers:

Notifying customers and educating them on the risks associated with having their personal information taken by hackers is an important step.

As an added precaution, business owners should warn their clients to avoid using the same password on many websites. Clients may also be reminded to keep an eye on their accounts more often. Customers that are security conscious might get rewards from their managers.

In the event that a consumer's password is changed or their account is accessed, an automated email and notice are sent to alert them so that they may report any suspected abuse right away.

If the system is set up to use two-factor identification, guests may need to submit not only their user name and password, but also the pin number that will be delivered to the same email address or phone number that they gave when they registered for the profile. Cyber – criminals attempting a masquerade attack won't be able to gain access to a user's account if they can't reach them via the contact information they provided (such as an email address or phone number).

There are a variety of ways in which security weaknesses in the computer system and network can be investigated. Depending on the hotel's budget, several methods could be used to secure visitors' data and personal information. To ensure the security of their IT infrastructure, hotels should have contracts in place with reputable IT companies or professional IT managers. The hotel should also set up policies and procedures for the use of the hotel's network and cyber infrastructure. In addition, hotels should provide their staff with cyber security training to safeguard their use of electronic communication channels like email and social media. Guests can reserve rooms and services online without worrying about being hacked or misused if the hotel has a secure and certified website that employs extended validation or at least domain validation. Hotels should purchase cyber insurance to safeguard themselves financially and legally in the event of a data breach or other cyber-attack.

REFERENCES

[1]   Freeman, R., & Glazer, K. (2015). Services Marketing. *Introduction to Tourism and Hospitality in BC.*
[2]   Law, R., & Jogaratnam, G. (2005). A Study of Hotel Information Technology Applications. *International Journal of Contemporary Hospitality Management.*
[3]   Kansakar, P., Munir, A., & Shabani, N. (2019). Technology in the Hospitality Industry: Prospects and Challenges. *IEEE Consumer Electronics Magazine, 8*(3), 60–65.
[4]   Shabani, N., & Munir, A. (2020, July). A Review of Cyber Security Issues in Hospitality Industry. In *Science and Information Conference* (pp. 482–493). Springer, Cham.
[5]   Domanski, M. (2020). The Concept of a Smart Hotel and its Impact on Guests' Satisfaction, Privacy and the Perception of the Service Quality.
[6]   Shabani, N. (2017). A study of Cyber Security in Hospitality Industry-threats and Countermeasures: Case Study in Reno, Nevada. *arXiv Preprint arXiv:1705.02749.*
[7]   Scherf, J. A. (1974). *Computer and Data Security: A Comprehensive Annotated Bibliography.* Massachusetts Inst of Tech Cambridge Project Mac.
[8]   Chen, H. S., & Jai, T. M. C. (2019). Cyber Alarm: Determining the Impacts of Hotel's Data Breach Messages. *International Journal of Hospitality Management, 82,* 326–334.
[9]   Smelser, N. J., & Baltes, P. B. (Eds.). (2001). *International Encyclopedia of the Social & Behavioral Sciences* (Vol. 11). Amsterdam: Elsevier.
[10] Martin, K., Shilton, K., & Smith, J. (2019). Business and the Ethical Implications of Technology: Introduction to the Symposium. *Journal of Business Ethics, 160*(2), 307–317.
[11] Khatri, I. (2019). Information Technology in Tourism & Hospitality Industry: A Review of Ten Years' Publications. *Journal of Tourism and Hospitality Education, 9,* 74–87.
[12] Regmi, K. K., & Thapa, B. (2010). Software in Tourism Industry: *A Study On Emerging New Niches Of Software In Hotel Industry.*
[13] Chiang, L. C. (2000). Strategies for Safety and Security in Tourism: A Conceptual Framework for the Singapore Hotel Industry. *Journal of Tourism Studies, 11*(2), 44–52.
[14] Damon, E., Mache, J., Weiss, R., Ganz, K., Humbeutel, C., & Crabill, M. (2014). Cyber Security Education: The Merits of Firewall Exercises. In *Emerging Trends in ICT Security* (pp. 507–516). Morgan Kaufmann.
[15] Jang-Jaccard, J., & Nepal, S. (2014). A Survey of Emerging Threats in Cybersecurity. *Journal of Computer and System Sciences, 80*(5), 973–993.
[16] Alkhalil, Z., Hewage, C., Nawaf, L., & Khan, I. (2021). Phishing Attacks: A Recent Comprehensive Study and a New Anatomy. *Frontiers in Computer Science, 3,* 563060.
[17] Kuzminykh, I., Ghita, B., Sokolov, V., & Bakhshi, T. (2021). Information Security Risk Assessment. *Encyclopedia, 1*(3), 602–617.
[18] Cankaya, E. C. (2020). Security and Privacy in Three States of Information. In *Security and Privacy From a Legal, Ethical, and Technical Perspective*. IntechOpen.

[19] Kuperberg, M., Kemper, S., & Durak, C. (2019, June). Blockchain Usage for Government-issued Electronic IDs: A survey. In *International Conference on Advanced Information Systems Engineering* (pp. 155–167). Springer, Cham.

[20] Herzberg, A., Mass, Y., Mihaeli, J., Naor, D., & Ravid, Y. (2000, May). Access Control Meets Public Key Infrastructure, or: Assigning Roles to Strangers. In *Proceeding 2000 IEEE Symposium on Security and Privacy. S&P 2000* (pp. 2–14). IEEE.

[21] Ellison, C., & Schneier, B. (2000). Ten risks of PKI: What You're Not Being Told About Public Key Infrastructure. *ComputSecur J, 16*(1), 1–7.

[22] Clarke, N., Karatzouni, S., & Furnell, S. (2009, May). Flexible and Transparent User Authentication for Mobile Devices. In *IFIP International Information Security Conference* (pp. 1–12). Springer, Berlin, Heidelberg.

[23] Garcia, M. L. (2007). *Design and Evaluation of Physical Protection Systems*. Elsevier.

[24] Gao, J., Liu, J., Rajan, B., Nori, R., Fu, B., Xiao, Y., ...& Philip Chen, C. L. (2014). SCADA Communication and Security Issues. *Security and Communication Networks, 7*(1), 175–194.

[25] Rizvi, S., Willett, J., Perino, D., Vasbinder, T., & Marasco, S. (2017, March). Protecting an Automobile Network Using Distributed Firewall System. In *Proceedings of the Second International Conference on Internet of things, Data and Cloud Computing* (pp. 1–6).

[26] Lafuente, G. (2015). The Big Data Security Challenge. *Network security, 2015*(1), 12–14.

[27] Rui, Z., & Yan, Z. (2018). A Survey on Biometric Authentication: Toward Secure and Privacy-Preserving Identification. *IEEE Access, 7*, 5994–6009.

[28] Hu, V. C., Kuhn, R., & Yaga, D. (2017). Verification and test Methods for Access Control Policies/Models. *NIST Special Publication, 800*, 192.

[29] Holm, H., Sommestad, T., Almroth, J., & Persson, M. (2011). A Quantitative Evaluation of Vulnerability Scanning. *Information Management & Computer Security*.

[30] Mohammed, I. A. (2020). Artificial Intelligence for Cybersecurity: A Systematic Mapping Of Literature. *International Journal of Innovations In Engineering Research And Technology [IJIERT], 7*(9).

[31] Hannah, D. R. (2005). Should I keep a Secret? The Effects of Trade Secret Protection Procedures on Employees' Obligations to Protect Trade Secrets. *Organization Science, 16*(1), 71–84.

[32] Heartfield, R., & Loukas, G. (2018). Detecting Semantic Social Engineering Attacks with the Weakest Link: Implementation and Empirical Evaluation of a Human-as-a-security-sensor framework. *Computers & Security, 76*, 101–127.

[33] Dobrynin, V., Mastroianni, M., & Sheveleva, O. (2021, December). A New Structured Model for ICT Competencies Assessment Through Data Warehousing Software. In *International Conference on Innovations in Bio-Inspired Computing and Applications* (pp. 435–446). Springer, Cham.

[34] Mansfield-Devine, S. (2016). Securing Small and Medium-size Businesses. *Network Security, 2016*(7), 14–20.

[35] Huang, H. H., & Wang, C. (2021). Do Banks Price Firms' Data Breaches?. *The Accounting Review, 96*(3), 261–286.

[36] Stankov, U., & Gretzel, U. (2020). Tourism 4.0 Technologies and Tourist Experiences: A Human-centered Design Perspective. *Information Technology & Tourism, 22*(3), 477–488.

[37] Coburn, A., Leverett, E., & Woo, G. (2018). *Solving Cyber Risk: Protecting Your Company and Society*. John Wiley & Sons.

[38] Abawajy, J. (2014). User Preference of Cyber Security Awareness Delivery Methods. *Behaviour& Information Technology, 33*(3), 237–248.

[39] Lee, N. (2015). Cyber Attacks, Prevention, and Countermeasures. In *Counterterrorism and Cybersecurity* (pp. 249–286). Springer, Cham.

[40] Chatterjee, D. (2021). *Cybersecurity Readiness: A Holistic and High-performance Approach*. SAGE Publications.

[41] Aissa, S. B., & Goaied, M. (2016). Determinants of Tunisian Hotel Profitability: The Role of Managerial Efficiency. *Tourism Management, 52*, 478–487.

[42] Mwiki, H., Dargahi, T., Dehghantanha, A., & Choo, K. K. R. (2019). Analysis and Triage of Advanced Hacking Groups Targeting Western Countries Critical National Infrastructure: APT28, RED October, and Regin. In *Critical Infrastructure Security and Resilience* (pp. 221–244). Springer, Cham.

[43] Aïmeur, E., & Schönfeld, D. (2011, July). The Ultimate Invasion of Privacy: Identity Theft. In *2011 Ninth Annual International Conference on Privacy, Security and Trust* (pp. 24–31). IEEE.

293

[44] Chen, R. J. (2015). From Sustainability to Customer Loyalty: A Case of Full Service Hotels' Guests. *Journal of Retailing and Consumer Services, 22*, 261–265.

[45] Lamba, A. (2018). Protecting 'Cybersecurity&Resiliency'of Nation's Critical Infrastructure–Energy, Oil & Gas. *International Journal of Current Research, 10*, 76865–76876.

[46] Pantoulas, E. (2022). *Description, Analysis and Implementation of a Web Application Firewall (WAF). Creation of Attack Scenarios and Threats Prevention* (Master's Thesis, Πανεπιστήμιο Πειραιώς).

[47] Modi, B., Chourasia, U., & Pandey, R. (2022, March). Design and Implementation of RESTFUL API based Model for Vulnerability Detection and Mitigation. In *IOP Conference Series: Materials Science and Engineering* (Vol. 1228, No. 1, p. 012010). IOP Publishing.

[48] Latif, D. V., Arsalan, S., Hadian, N., Octavia, E., Tresnawati, R., & Mudzakar, M. K. (2022). The Adoption of Self Ordering Machine Technology by Customers and the Influence on Customer Satisfaction. *Central Asia And The Caucasus, 23*(1).

[49] Sahu, A. K., & Gutub, A. (2022). Improving Grayscale Steganography to Protect Personal Information Disclosure Within Hotel Services. *Multimedia Tools and Applications*, 1–21.

[50] Butler, J.: Not Just Heads In Beds – Cybersecurity for Hotel Owners. https:// www.hospitalitynet.org/ opinion/4073687.html (2016), Last visited on December 26, 2019.

Automation and Computation – Vats et al. (Eds)
© 2023 the Author(s), ISBN 978-1-032-36723-1

A review on machine learning approach for predicting cryptocurrency prices

Shweta Singh
Department of MCA, Greater Noida Institute of Technology, Greater Noida, India

Deepak Kumar Verma
Department of Computer Engineering, Marwadi University, Rajkot, Guajrat, India

Kuldeep Malik
Department of CSE, ITS Engineering College, Greater Noida, India

ABSTRACT: Cryptocurrency has become a global phenomenon in the financial sector, and it is now one of the most widely traded financial assets on the planet. Cryptocurrency is regarded as a confusing topic in finance due to its significant volatility. It is not just one of the most sophisticated and arcane fields among financial products. The key issue in this development is the fast pace of Cryptocurrency swings. The authors have proposed a number of machine learning strategies for resolving the above mentioned issue. The proposed technique would be a great choice to anticipate and forecast Cryptocurrency prices using the index and constituents. The comparison of the proposed mechanism with the existing methods is also accomplished. In this paper, two models have been proposed named as Light Gradient-boosting Machine (LGBM) and Auto-regressive Integrated Moving Average (ARIMA). It is witnessed that the ARIMA model gives more accurate results while forecasting values, whereas LGBM used lag values and hyper parameters based on closing variables with 7-fold cross validation technique to train the model.

1 INTRODUCTION

Cryptocurrency gives people more reasons to become accustomed to the convenience of online shopping, as it reduces the risk of fraud for both merchants, sellers, and buyers and alleviates consumers' concerns about the security of their personal information. It is also expected that Cryptocurrency will make it easier to buy online more efficiently than fiat currency as it does not have to comply with international laws or comply with trading restrictions. Crypto currencies challenge the standard financial systems in many sectors. One of these areas is the transfer of funds. Digital currencies such as bit coin are as freely exchangeable as it could be. In addition, their fast-moving time and low cost of purchasing make them ideal resources for cross-border transfers. Not surprisingly, a ripple-like currency takes over the retail trade.

Cryptocurrency has become a global phenomenon in the financial sector, and it is now one of the most widely traded financial assets on the planet. Cryptocurrency is regarded as a confusing topic in finance due to its significant volatility. It is not just one of the most sophisticated and arcane fields among financial products. The key issue in this development is the fast pace of Cryptocurrency swings. The authors have proposed a number of machine learning strategies for resolving the above mentioned issue. The proposed technique would be a great choice to anticipate and forecast Cryptocurrency prices using the index and constituents. The comparison of the proposed mechanism with the existing methods is also accomplished. In this paper, two models have been proposed named as Light Gradient-boosting Machine (LGBM) and Autoregressive Integrated Moving Average (ARIMA). It is

DOI: 10.1201/9781003333500-33

witnessed that the ARIMA model gives more accurate results while forecasting values, whereas LGBM used lag values and hyper parameters based on closing variables with 7-fold cross validation technique to train the model.

2 LITEARATURE REVIEW

An optimized svm (support vector machine) based on the pso (particle swarm optimization) has been introduced for predicting the future prices of crypto currencies. examiners and financial backers as a rule consolidate essential and specialized examination prior to settling on the best cost to execute an exchange. at times, calculations are utilized to execute exchanges. foreseeing results utilizes a straightforward svm calculation that is not exactly encouraging. particle swarm optimization (pso) [8]. in order to forecast the price of bitcoin, the authors proposed a stochastic neural network model. the suggested model is based on random walk theory, which is frequently employed to forecast stock values in financial markets. to replicate market volatility, the proposed model introduces randomness at each layer in the activation of the observable properties of the neural network [11].

In the proposed scheme, the price of crypto currencies is predicted using a stochastic neural network model. The user uses the strategy to identify the pattern of the market's reaction to new information. The findings concluded that the proposed theory as being not only accurate but also very useful in predicting market volatility. Nearly all models of stochastic neural networks performed better than their deterministic counterparts [1]. The prediction accuracy of the scheme does not fluctuate much over the various prediction ranges [7]. However, to generate models with better prediction accuracy at shorter time intervals it is possible due to large sample size of observations at the minute level, assuming more complex models or more features with higher return frequencies.

In general, the prescient model in this paper shows exact outcomes that are near the genuine digital currency cost [12]. The significance of these models is that these can have a critical financial effect by assisting financial backers and merchants with restricting the trading of digital forms of money. Based on the machine learning model, the authors have provided evidence that the proposed model can execute successful intraday transactions even after considering transaction costs [4]. Appropriate handling of transaction costs and slippage based on order book data is also desirable to increase the reliability of transaction results. Alessandretti et al. [2] has proposed a methodology for generating the anomalous returns using the hypothesis testing. It has used machine learning and artificial intelligence schemes for the better trading growth in the market. The authors have analyzed the data on daily basis for crypto currencies of the interval November 2015 and April 2018. It is clear from the results that the proposed scheme outpaces ordinary standards.

Chen and Sun [3] discussed about the popularity of bitcoin as the greater investment in the form of Cryptocurrency. A good prediction is needed for the best investment choice as it has an extremely instable landscape. The authors have concentrated on the practicability of heterogeneous models to find the samples with distinct structures and dimensionality. Gautam et al. [6] inspected the various categories of Crypto currencies along with their axioms and solutions with respect to the growth of the market. The authors have given a detailed study on the Cryptocurrency market prediction in the era of digitization. Hitam et al. [9] provided a deep study in the financial industry to improve the applicability of support vector machine for predicting the future of Cryptocurrency. Social media platforms are playing a vital role in the advancement of support vector machine research. Jang and Lee [10] explored the application of Bitcoin in the arenas of computer science, cryptography and economics due to its integral landscape of joining monetary units and encryption techniques. A vital research has done by Kurbucz [13] on the Bitcoin transactions in the past years. He has defined the prediction influence of the network for the bitcoin transactions for saving the future.

Liew et al. [14] analyzed the facts and findings on the daily return generating procedure.

They found that the short-term predictableness is precise questionable, which recommends that near-term Cryptocurrency marketplaces are semi-strong form effectual and hence, day trading crypto currencies may be right challengeable [15]. Mor *et al.* [16] has proposed an alternative blockchain for the outmoded remittance methodologies, which is an effectual, open and reliable technique for international payment system. Verma *et al.* [17] provided a solution to electronic health record security in the covid-19 era based on Blockchain. The advent of blockchain expertise fetches a new-fangled notion to decipher the abovementioned security issues due to its features of immutability, verifiability, and decentralization. Using a machine learning approach, we are able to create a short term forecasting model for bitcoin prices. The modified Binary Autoregressive Tree (BART) model is based on the regression tree and time series data standard model. BART is a hybrid of the standard classification and regression tree (C & RT) algorithms with the ARIMA auto regressive model. The user developed short-term projections (5–30 days) of the three most capitalized crypto currencies using the BART model (Bitcoin, Ethereum, Ripple) [5]. In forecasting bitcoin time series, the suggested technique was found to be more accurate than the ARIMA ARFIMA model, both in the period of sluggish growth (falling) and the period of transition dynamics (trend change). Regardless of whether the target data was obtained before, during, or after the recession, the BART approach suggested evaluating and predicting bitcoin time series is compared to standard time series methods. It has been shown to be highly efficient in generating predictive estimates.

3 PROPOSED MODEL

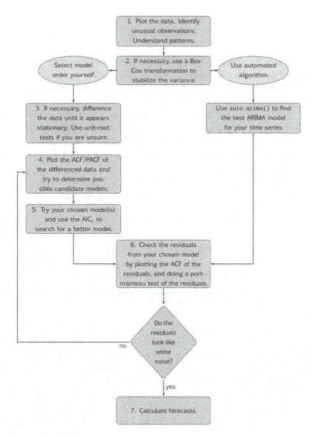

Figure 1.   Proposed model architecture.

The authors have proposed two models to forecast the crypto currencies. The first model is based on ARIMA (Autoregressive Integrated Moving Average). In which the user analyses the moving averages of crypto currencies based on the various window sizes (5, 10, 15, 30, 60) and made simple moving average-based anomaly detection method. Then, using an exponential smoothing approach for univariate time-series forecasting, which may be extended to data with a systematic trend or seasonal component, was applied.

4 CONCLUSION

The proposed technique focuses on the improvement of the existing techniques for predicting the price of Cryptocurrency. The proposed model is efficient for solving real-world problems related to finance. The authors have improved the existing schemes by using ARIMA and LGBM models. Any user can deploy this model for forecasting crypto currencies using the techniques such as 7-fold cross validation, dicky fuller test, exponential smoothing. The prime concern of the proposed scheme is to predict the Cryptocurrency prices over a given period of time series data as the user knows crypto currencies are very volatile so it is necessary to make a Machine Learning Model for predicting their prices. An investor or an assurer can use this model to find the strategies for market prediction. So that he/she can make a profit in their investment and having minimal chances for any loss in their investment. This research may be further extended for the prediction of health edges and banking sector.

REFERENCES

[1]   Akyildirim, E., Goncu, A., & Sensoy, A. (2021). Prediction of Cryptocurrency Returns Using Machine Learning. *Annals of Operations Research*, *297*(1), 3–36.
[2]   Alessandretti, L., ElBahrawy, A., Aiello, L. M., & Baronchelli, A. (2018). Anticipating Cryptocurrency Prices Using Machine Learning. *Complexity*, *2018*.
[3]   Chen, Z., Li, C., & Sun, W. (2020). Bitcoin Price Prediction Using Machine Learning: An Approach to Sample Dimension Engineering. *Journal of Computational and Applied Mathematics*, *365*, 112395.
[4]   Chowdhury, R., Rahman, M. A., Rahman, M. S., & Mahdy, M. R. C. (2020). An Approach to Predict and Forecast the Price of Constituents and Index of Cryptocurrency Using Machine Learning. *Physica A: Statistical Mechanics and its Applications*, *551*, 124569.
[5]   Derbentsev, V., Datsenko, N., Stepanenko, O., & Bezkorovainyi, V. (2019). Forecasting Cryptocurrency Prices Time Series Using Machine Learning Approach. In *SHS Web of Conferences* (Vol. 65, p. 02001). EDP Sciences.
[6]   Gautam, K., Sharma, N., & Kumar, P. (2020, June). Empirical Analysis of Current Cryptocurrencies in Different Aspects. In *2020 8th International Conference on Reliability, Infocom Technologies and Optimization (Trends and Future Directions)(ICRITO)* (pp. 344–348). IEEE.
[7]   Hamayel, M. J., & Owda, A. Y. (2021). A Novel Cryptocurrency Price Prediction Model Using GRU, LSTM and bi-LSTM Machine Learning Algorithms. *AI*, *2*(4), 477–496.
[8]   Hitam, N. A., Ismail, A. R., & Saeed, F. (2019). An Optimized Support Vector Machine (SVM) based on Particle Swarm Optimization (PSO) for Cryptocurrency Forecasting. *Procedia Computer Science*, *163*, 427–433.
[9]   Hitam, N. A., Ismail, A. R., Samsudin, R., & Alkhammash, E. H. (2022). The Effect of Kernel Functions on Cryptocurrency Prediction Using Support Vector Machines. In *International Conference of Reliable Information and Communication Technology* (pp. 319–332). Springer, Cham.
[10] Jang, H., & Lee, J. (2017). An Empirical Study on Modeling and Prediction of Bitcoin Prices with Bayesian Neural Networks based on Blockchain Information. *Ieee Access*, *6*, 5427–5437.
[11] Jay, P., Kalariya, V., Parmar, P., Tanwar, S., Kumar, N., & Alazab, M. (2020). Stochastic Neural Networks for Cryptocurrency Price Prediction. *IEEE Access*, *8*, 82804–82818.
[12] Kumar, A. (2021). Short-Term Prediction of Crypto-Currencies Using Machine Learning. *Available at SSRN 3890338*.

[13] Kurbucz, M. T. (2019). Predicting the Price of Bitcoin by the Most Frequent Edges of its Transaction Network. *Economics Letters*, *184*, 108655.

[14] Liew, J., Li, R. Z., Budavári, T., & Sharma, A. (2019). Cryptocurrency Investing Examined. *The Journal of The British Blockchain Association*, 8720.

[15] Mittal, A., Dhiman, V., Singh, A., & Prakash, C. (2019, August). Short-term Bitcoin Price Fluctuation Prediction Using Social Media and Web Search Data. In *2019 Twelfth International Conference on Contemporary Computing (IC3)* (pp. 1–6). IEEE.

[16] Mor, P., Tyagi, R. K., Jain, C., & Verma, D. K. (2021, July). A Systematic Review and Analysis of Blockchain Technology for Corporate Remittance and Settlement Process. In *2021 Fourth International Conference on Computational Intelligence and Communication Technologies (CCICT)* (pp. 121–128). IEEE.

[17] Verma, D. K., Tyagi, R. K., & Chakraverti, A. K. (2022). Secure Data Sharing of Electronic Health Record (EHR) on the Cloud Using Blockchain in Covid-19 Scenario. In *Proceedings of Trends in Electronics and Health Informatics* (pp. 165–175). Springer, Singapore.

Automation and Computation – Vats et al. (Eds)
© 2023 the Author(s), ISBN 978-1-032-36723-1

Communication structure for Vehicular Internet of Things (VIoTs) and review for vehicular networks

Rakesh Kumar & Sunil K. Singh
CSE Department, CCET, Punjab University, Chandigarh, India

D.K. Lobiyal
School of Computer & System Sciences, Jawaharlal Nehru University, New Delhi, India

ABSTRACT: IoT-enabled intelligent vehicles are increasing on the roadside traffic network. The demand for secure, effective, and privacy of data, including driving conditions, is the primary motivation of the smart ITS for the Vehicular internet of things. We have introduced a communication structure for VIoTs, which provides a framework for the V2X (vehicle to anything) communication. A gateway selection algorithm performs communication between vehicles and infrastructure [30] on LTE/5 G-based cellular networks. Gateway selection-based cellular network is making power to the roadside unit, and the cellular network is working as a roadside unit. They have high computation power by which operations are effectively performed by encryption/decryption, group management, etc. A review of various studies summarizes various characteristics and challenges in the IoT environment and VANETs, and these help us design the VIoT structure.

Keywords: VIoT, Communication, VANET, ITS, IoT, Group Key management

1 INTRODUCTION

To make intelligent motor vehicles, the intelligent transportation system (ITS) [11] is providing connectivity among motor vehicles, and Intelligent Transportation Systems (ITS) is making revolutionary changes in the IoV, VANETs as well as VIoT by connecting them with the Internet of Things (IoT) enables vehicles to a broader reach. Further, a key motivation for ITS is to expand road protection, the privacy of data, and driving situations. This critical driving information is shared with the *Vehicular Internet of Things (VIoTs)* [33]. The Vehicular Internet of Things (VIoTs) is a wireless sensor network consisting of many sensors based on WSNs. The transmission channel established by VIoTs-based wireless sensor networks is known as VANETs. *Vehicular Ad-hoc Networks (VANET) are comprised of two types of transmission* one is *vehicle-to-vehicle (V2V)*, and another is *vehicle-to-infrastructure (V2I)* transmission [12]. In vehicle-to-vehicle (V2V) transmission, each vehicle sends out end-to-end vehicles to conversation information, and in vehicle-to-infrastructure (V2I) transmission, each vehicle transmission with *roadside units* (RSUs). V2V *and* V2I are small-range transmissions, and V2X transmission is long-range communication based on technology such as LTE [25].

The unique characteristic of VANET is High portability and unpredictability; further, the network is exposed to several kinds of internal and external attacks due to network vulnerability [13]. The VIoTs design suffers from these attacks and has three main concerns, i.e., security, privacy, and trust.

Various research [2, 4, 5, 21. 22] have proposed different techniques to guarantee security, conserve *privacy*, and create *belief direction* for *VANETs*. Further, approaches such as group

DOI: 10.1201/9781003333500-34

access authentication include a framework for VANET, routing protocol using beacon message, RCoM scheme for road conditions, source authentication, cloud model, master token encryption, etc. These multiple approaches have been discussed by many researchers in the field of VIoTs, and some of the methods are listed below-

Approaches to resolving time constraints in which an effective routing protocol is required to transmit the alerting *transmissions* approach to determine the group's high dynamicity and scalability for group *transmission*. Techniques that improve beaconing messages among vehicles for navigation and alerting of vehicles. *These methods create the requirement for an efficient VIoT network structure for communication. The proposed communication network structure is discussed in section 4.2.*

The general architecture of vehicular Internet of Things is given in Figure 1 – which shows three types of transmission -

Inter-vehicle transmission is called vehicle-to-vehicle transmission and communication; it uses the IEEE802.11p standard.

Vehicle to roadside transmission – vehicle transmission to roadside infrastructure is called a vehicle-to-roadside transmission and uses the IEEE802.11p standard.

Inter roadside transmission – The transmission from the roadside network to other road-side networks is called inter-roadside communication.

This architecture has three main components which show a chief role in the VIoT, details of which are given below: -

Roadside Unit (RSU): These are devices positioned along the road at dedicated locations such as crossings, intersection points, and parking lots. The key features of roadside units are providing coverage for the communication range of the VIoT network through a message to other *OBUs and RSUs.* Therefore, the running security demands of the network such as traffic situation (congestion or not) reporting, accident alert, and internet connectivity to *OBUs.*

Onboard Unit (OBU): In vehicles, different types of sensors, resource command processors (RPC), network devices, sets of radars, and storage make a transceiver unit that is called the onboard unit (OBU), with which every vehicle is enabled. The global *positioning system (GPS)* is a sensor that collects the signals and sends them to the OBUs, and Radar finds the hindrances and confirms the neighbor's position.

Regional Trusted Authority (RTA): A regional trusted Authority (RTA) [17] is accountable for producing cryptographic essential key parameters for *RSUs* and the vehicles in the surrounding regions. These key parameters are delivered over secure channels to them. An RTA also manages the list of vehicles having their participation over and periodically updating the list of vehicles.

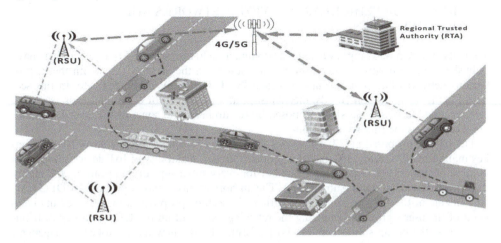

Figure 1.   Architecture of Vehicular Internet of Things (VIoT) [14].

1.1 Background

Day by day, the demand for IoT-enabled devices (Smart devices) is increasing worldwide. Similarly, the need for IoT-enabled vehicles also increases, which is playing a vital role in the VIoTs. The demand for devices enabled by IoT will be 75.44 billion in 2025 [15]. Further, developing a 5G network will provide massive device connectivity, increasing the demand for VIoT. Therefore, the massive vehicular connectivity increase will pose routing and security challenges for VIoTs. Consequently, it is a promising research area that can be explored to look for a timely solution for the congestion, routing, and securities issues in VIoT. There is a wide range of applications of VIoTs, and some of the applications are discussed in the following section.

1.2 Applications of Vehicular Internet of Things (VIoT)

The vehicular Internet of things (VIoTs) Network supports a wide range of applications. Some applications are given in the following diagram-

Platooning- Platooning is a systematic transportation technique in which all vehicles follow the head vehicle signals during the journey. Group motion is controlled by a group motion controller that uses ITS-based vehicle sensors and external data to control the behavior of platooning vehicles and drivers. Four operations [27], join, merge, leave, and split, are performed during the platooning.

Traffic information systems use VIoTs transmission to provide real-time traffic information and report to the vehicles through a satellite-based navigation system. Based on real-time data, users can decide which root will be suitable for the destination. The traffic information system saves energy, time, and cost of distance or fuel.

Road Transportation Emergency Services – The road transportation system is the necessary entity that plays a significant role during the emergency services, and the VIoTs communications, and VIoTs networks, are used to reduce delay in the dissemination of road protection alerts and position data to hurry up emergency delivery action to save the life of those injured on the roadside accident.

On-The-Road Services – It is also envisioned that the future transportation highway would be "information-driven" or "wirelessly enabled." VIoTs can help advertise facilities to the driver and even send warnings of any alert going on at that instant.

Electronic Emergency brake lights allow drivers to respond to vehicles slowing down even though they may be clouded.

2 A REVIEW OF VEHICULAR IOT (VIOT) NETWORKS AND COMMUNICATION

Vehicular Internet of Things (VIoTs) is a prominent field of research. Many researchers have published vehicle-to-anything communication studies in the last few years with the help of various networks like *vehicular ad-hoc networks (VANET)*, WBAN, IoT, etc. In this scenario, many techniques, methods, mechanisms, and protocols for communication, security, privacy, and trustworthiness are proposed by researchers, and some of the essential studies are discussed in the following section-

The scheme proposed in [22] overcomes the challenges such as access control-based group key management (GKM), IoT-based centralized model, scalability of IoT devices, improved number of subscribers, dependent group key for every subgroup communication, and highly dynamic behavior of users (subscribers). The authors in this research proposed a DLGKM-AC scheme in the IoT environment. Further, the author proposed improved key management of the user's group, minimizing the rekeying overhead on the KDC and key distribution across the group subscribers (users) performed by the new master token management protocol. The proposed scheme improved transmission, computation, and *storage overhead* during group joining and leaving the event.

The authors in [26] proposed a scheme based on GKM's two-tier architecture called *GROUPIT*. The proposed scheme overcomes the challenges in the group key management, such as the massive number of *IoT devices*, the dynamicity of the user membership, and the dynamical change in the IoT devices. The proposed architecture of GKM has effectively handled devices by assigning them to one of many predefined groups, and key management improves efficiency within each group. The authors also provide theoretical and implementation proof based on Alljoyn, which is an open-source IoT framework where the feasibility of GROUPIT is demonstrated.

The attribute-Based Signature (ABS) scheme is provided for privacy preservation for a vehicle in [1] without considering pseudonym/private key overhead. This scheme carries the vehicle revocation and authentication by the accountable authority. The scheme's performance is minimized in ante-computation, server-attached computation, code escalation, and cryptographic equipment.

Contributory group key agreement, group joining, group leaving, group merging, and group participation are all employed to maintain the group dynamic, according to the Author [2]. Light-Weight Group, Access Authentication technique, is described by the author based on Message Authentication Code (MAC).

Lightweight Group, Access Authentication protocol prevents DoS attacks, effective and secure transmission against eavesdroppers, and several other attacks specific to group settings. The efficiency of the *SEGM-based* protocol is minimized in terms of access authentication overheads, transmission overheads, bandwidth consumption, and computation complexity.

In [3], the author proposed Fog computing to minimize the problem of network congestion and delay in cloud computing. The author minimized latency, response time, and virtual machine cost with the help of the Fog-enhanced vehicular services (FEVS) algorithm. Their scheme also increased resource utilization by up to 35–40% concerning cloud models.

In [4], the author proposed a scheme that reduced the overhead of Beacon-based routing at the road segment of the city environment in Vehicular Ad-Hoc Networks. The key concern of the scheme was to reduce the resulting overhead with the help of two directional antennas in place of only one from the periodic exchange of Beacon messages while preserving the same network awareness.

In this paper [5], the author discussed the requirement of real-time road condition monitoring for the authority to maintain security and privacy-related issues. In this situation, the author proposed a cloud-based road condition monitoring (RCoM) scheme, which solved the security and privacy challenges of the RCoM scheme. The encrypted information on road conditions is sent to authority and Cloud servers. A cloud server and Authority should validate this reported information, i.e., to check whether legitimate vehicles are reported.

The authors in [6] proposed a blockchain-based anonymous reputation system (BARS) to split the likability between real identities and public keys to maintain privacy. With the support of a reputation management system, this approach also held the reliability of broadcast messages and effectively protected vehicle privacy. Furthermore, this technique reduces the computational overhead of the public key by up to 10%.

In [7], the author discussed some issues in the VANET, such as hardware problems, high vehicle density, increased mobility of vehicles, high computation overhead, and limited bandwidth. VANET cannot receive group certificates from collapsing roadside units due to these difficulties. Therefore, the author provided group certification solutions using the Vehicular Backbone Network (VBN) structure and *Mobile Backbone Network (MBN)*. Vehicles can serve as both a router and a terminal node in VANETs within VBNs.

To improve security and privacy in VANET, the authors in [8] presented an Identity-Based Batch verification approach. When the receiver has to confirm huge messages, the underlying notion behind batch verification of several messages is more effective than one-by-one verification. For the proposed IBV scheme in VANET, the author improved the average message delay and message loss rate.

In [9], the author discussed a -group key agreement technique based on the Chinese remainder theorem (CRT). The group key must be updated when a vehicle joins or departs from a group. Compared to other existing methods, the authors proposed minimizing the computation cost and transmission overheads in this approach.

Compared to current systems, the research [10] used a link selection algorithm that minimized latency and maximized throughput and packet delivery ratio. The total delay in optimum resource allocation at the cellular eNodeB is minimized by selecting the most suitable receiver vehicle to find a V2V link and allocating a proper channel. The link selection algorithm addresses the problem known as the NP-hard problem, which is identical to the maximum weighted independent set problem (MWIS-AW) with associated weights.

In [24], the author proposed clustering approaches for VANET to increase routing protocol reliability and scalability across urban regions. Clustering algorithms are critical in the dynamic environment of the nodes (vehicles) for V2V transmission, where all nodes move at the same velocity on the road. When a node (vehicle) joins and leaves a cluster, clustering techniques provide an efficient and secure channel by maintaining backward and forward secrecy. Clustering techniques can improve the routing scalability and dependability of VANET. These clusters can be used for accident or traffic jam notification, information dissemination, warning alerts, entertainment, and other applications. This can be seen as a benefit of cluster routing.

In the research study [25], the authors covered system-level analysis for ITS-G5 and LTE-V2X communications. They discussed short-range communications and long-range communications. Short-range communication is further separated into V2V transmission and V2I communication. V2V communication occurs between vehicles, while V2I communication occurs between vehicles and infrastructure. Long-range transmission, in this case, is LTE-based technologies, such as (V2X) connectivity. V2X transmission is communication between vehicles and anything, and PC5 mode 4 is the direct interface for V2V and V2I communication in C-V2X (Cellular-V2X) communication. When comparing ITS-G5 with PC5-mode4 does not include latency in the design. In the following Table 1, we have summarized the research work of the literature review on the bases of the key focus area, research gap, and environment, including publications year as:

Table 1. Summary of literature review.

References	Year	Key Focus Area	Research Gap	Environment
[33]	2022	• Group Key management for multiple processes • Master key encryption Approach • Reduced rekeying at the time of joining & leaving the group • Reduced overhead	• General applications • Scope for lightweight approaches	VIoT
[34]	2022	• Batch-based GKM for authentication • LSTM Neural network approach • Average waiting time	• Waiting time can improve through various approaches • Computation overhead • Comparative study	VANET

(continued)

Table 1. Continued

References	Year	Key Focus Area	Research Gap	Environment
[21]	2020	• Routing • throughput, • Packet drop ratio • End-to-end delay • Trust	• The communication model is not discussed for VIoT • Emphasizing security attacks	VIoT
[22]	2020	• Hierarchical Architecture • Master token encryption algorithm • Security • Mobility	• The communication model for the IoT environment only • Limited security attack.	IoT
[23]	2019	• Authentication, • Security attacks like Replay Attack-DoS • Impersonation attack • session key agreement	• The efficiency of the scheme is improvable without compromising security requirements	IoT
[1]	2019	• Attribute-based framework • message authentication • integrity and protect vehicle privacy	• pre-computation, • server-aided computation • code optimization and cryptographic hardware	VANET
[2]	2019	• Group access authentication • a framework for VANET	• access authentication overheads, • communication overheads, • bandwidth consumption	VANET
[4]	2019	• Beacon-based routing • Generated overhead • Use of Antennas	• Beacon-based routing • Overhead is improvable	VANET
[3]	2018	• Fog Commuting • cloud computing • congestion and delay	• the resource utilization • cloud models	VANET
[5]	2018	• Road condition monitoring (RCoM) mechanism • Source Authentication • Cloud model	• Communication channel for services • Authentication	VANET
[6]	2018	• Blockchain model • Trust • privacy	• Public key Computation	VANET
[7]	2018	• Vehicular Backbone Network (VBN) structure • Mobile Backbone Network (MBN)	• high vehicle density • high mobility of vehicles • computation overhead • limited bandwidth	VANET
[9]	2018	• Group key agreement mechanism • Chinese remainder theorem	• computation cost • communication overheads	VANET
[8]	2017	• Identity-based batch verification • security and privacy	• average message delay • Message loss rate	VANET

2.1 *Comparative study for group key management*

From the literature review, we have summarized distributed key management mechanism in Table 2 as follows-

Table 2. Distributed key mechanism in VANET.

Parameters	[2]	[31]	[32]
Group Settings	Yes	Yes	No
Group Joining	Yes	Yes	Yes
Group Leaving	Yes	Yes	Yes
Group Combine	Yes	No	No
Group split	Yes	No	No
RSU Assistance	Limited	Yes	Yes
Privacy	Yes	Yes	Yes

Comparative studies of distributed key mechanism in Vehicular Ad-hoc Networks –
We have identified the scope of the Vehicular Internet of Things (VIoT) for developing a communication network structure. The structure [2,31,32] efficiently performs the tasks mentioned in Table 2 in the vehicular network. Still, the paper [2] utilizes limited assistance of RSU during communication, and the paper [31,32] is not performing group-related tasks in Table 2, such as group combination and group split.

2.2 *Problem analysis*

With the growing applicability of VANETs, the increasing volume of vehicles due to the demands from people, and technological development in the field of IoT, the area of VIoT is becoming a new sub-domain of VANET research. Therefore, it is opening new promising avenues of research for academia, industry, and scientists. The massive number of connected vehicles in a given area may increase collision and require better contention mechanisms. Further, better routing protocols will be necessary for the quick and timely delivery of alerts and warnings. Additionally, the privacy of the drivers and the security of resources and information become a challenging issues. Further, the delivery of infotainment content will also require algorithms that can efficiently utilize bandwidth, the battery power of sensors, etc. We propose a communication structure to maintain the rekeying and bandwidth consumption and the complexity of the VIoTs networks due to overheads.

3 NETWORK STRUCTURE FOR VEHICULAR INTERNET OF THINGS (VIOT) COMMUNICATION

3.1 *Networks structure requirements for VIoTs scheme*

We have studied many research papers to identify the parameters and framework for VIoT. These parameters play a key role in designing the network structure for VIoTs schemes. In this study, we have considered parameters to create a news network for VIoTs, and they help us to make the design requirements for it. The design requirements for developing a new network structure for VIoTs are given below:

- Information should be trustworthy of real-time traffic and be provided to the end-user. This information makes an efficient transportation mechanism scheme.
- The transportation mechanism scheme should be fast in making attractive conclusions.

- Priority should be highest for the emergency vehicles at the crossings to save lives and assets.
- The transportation mechanism scheme should be sensing roadside mishaps.
- The capability of the intelligent transportation mechanism should deliver safety and privacy.

3.2 *Network structure*

The vehicular Internet of Things communication structure, shown in Figure 2, consists of some communication entities. These communication entities play a vital role in vehicular internet of things communication. Some communication entities are listed below-

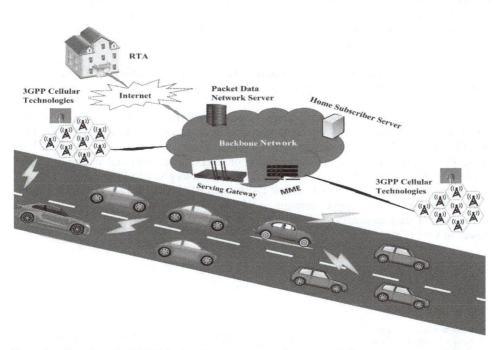

Figure 2.   Component of OBU for intelligent transportation system [31].

3GPP-Cellular Technologies- 3rd Generation Partnership Project in cellular technologies is the long-term evolution (LTE, LTE-V) networks given rise to the new era of mobile networks, and LTE is the fastest deployed mobile network. Long-Term Evolution-Vehicle (LTE-V) plays a vital role in vehicular communication and making effective communication between vehicles (V2V) and between vehicles and roadside units (V2I).

In other words, vehicles are using LTE-V technology for communication. This technology is inbuilt into the cellular unit to connect the 3GPP-evolved universal mobile tele-communication system terrestrial radio access network (E-UTRAN) with the help of eNodeB. LTE-V technology is divided into two units. One for the centralized networks is LTE-V-Cell, and another unit for decentralized networks is LTE-V-Direct. LTE-V cellular technologies are the extension of the existing technologies and design. The previous cellular technologies and innovations are mainly used for traditional internet services. Further, LTE has introduced Node-to-Node (N2N) communication, also known as V2V communication, in vehicular networks.

Backbone Network- Backbone network is the central unit of VIoTs Communication networks, which consists of several gateways for mobility management, home subscribers' management, service management, and data management. Every gateway plays an influential role in the VIoTs communication network; one such case is *MME* [28], which is the mobility management entity that is responsible for the exact location of the User Entity' (UEs) in the VIoT networks and manages the mobility-related messages between UEs and server network, *Serving Gateway* is responsible for storing and management of vehicles for gateway and handover, *Home subscriber's server* is responsible for authentication of vehicles for mutual authorization when vehicles are connected with the backbone network. Then MME sends the request to the HSS for the authentication of vehicles. After that vehicle is mutually authenticated, the *Packet data network server* works as a gateway server for *a* central unit of the network in which UEs are connected with the packet data network (PDN) and other words; this is responsible for interfacing vehicles from the backbone network to Regional Trusted Authority (RTA) through public network Internet.

Regional Trusted Authority (RTA)- The registration process of vehicle OBUs, RSUs, and vehicle users is done by RTA. Every region of the state has various RTAs to register vehicle entities.

OBU Component for Intelligent Transportation System- Every vehicle of VIoT is OBU-enabled. OBU consists of various components to interact with RTA with the help of the WAVE interface (RSU) for communication. OBU consists of six main components as Rule Base, Fuzzy Interference Engine, Data collection agent, spatiotemporal reasoning Agent, decision-making agent, and encryption/decryption agent, as given in Figure 3.

Furthermore, OBU [31] interacts with RTA through RSU with the help of vehicle sensors, temporal proof devices, DSRC communication medium, smart card devices, fingerprint devices, event data recorders, and human-machine interfaces to perform decision-making based on vehicle movement.

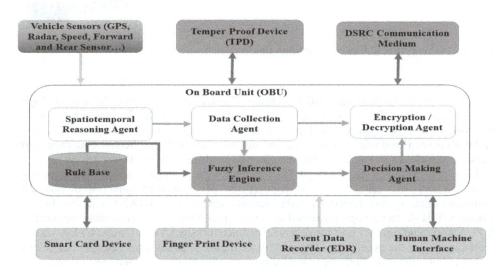

Figure 3.   Network structure for vehicular internet of things (VIoTs) communication.

4 CONCLUSIONS

In this paper, we have proposed a communication network structure for VIoT, leading to communication for the network. In this scenario, connectivity is the primary concern in the mobile vehicular network for secure transmission. Intelligent Transportation System (ITS) has adopted IEEE standards [29] IEEE802.11p and IEEE1609 for connectivity and WAVE architecture, respectively. IEEE 802.11p Standard provides traditional connectivity between systems based on geographical areas like LAN and MAN networks. WAVE (Wireless Access in Vehicular Environments) architecture provides an interface for secure communication in mobile vehicular Networks and provides a secure interface for V2V and V2X communication based on IEEE 1609 standards. In other words, IEEE802.11p supports the WAVE interface to perform operations effectively. The proposed architecture is effective and secure for vehicular communication networks such as V2V and V2X.

REFERENCES

[1] Cui, H., Deng, R. H., & Wang, G. (2019). An Attribute-based Framework for Secure Communications in Vehicular ad Hoc Networks. *IEEE/ACM Transactions on Networking*, 27(2), 721–733.

[2] Lai, C., Zheng, D., Zhao, Q., & Jiang, X. (2018). SEGM: A Secure Group Management Framework in Integrated VANET-cellular Networks. *Vehicular Communications*, 11, 33–45.

[3] Sutagundar, A. V., Attar, A. H., & Hatti, D. I. (2019). Resource Allocation for Fog-enhanced Vehicular Services. *Wireless Personal Communications*, 104(4), 1473–1491.

[4] Zahedi, K., Zahedi, Y., & Ismail, A. S. (2019). Using Two Antennas to Reduce the Generated Overhead of Beacon-based Protocols in VANET. *Wireless Personal Communications*, 104(4), 1343–1354.

[5] Wang, Y., Ding, Y., Wu, Q., Wei, Y., Qin, B., & Wang, H. (2018). Privacy-preserving Cloud-based Road Condition Monitoring with Source Authentication in VANETs. *IEEE Transactions on Information Forensics and Security*, 14(7), 1779–1790.

[6] Lu, Z., Wang, Q., Qu, G., & Liu, Z. (2018, August). BARS: A Blockchain-based Anonymous Reputation System for Trust management in VANETs. In *2018 17th IEEE International Conference On Trust, Security, And Privacy In Computing And Communications/12th IEEE International Conference On Big Data Science And Engineering (TrustCom/BigDataSE)* (pp. 98–103). IEEE.

[7] Yue, X., Chen, B., Wang, X., Duan, Y., Gao, M., & He, Y. (2018). An Efficient and Secure Anonymous Authentication Scheme for VANETs based on the Framework of Group Signatures. *IEEE Access*, 6, 62584–62600.

[8] Tzeng, S. F., Horng, S. J., Li, T., Wang, X., Huang, P. H., & Khan, M. K. (2015). Enhancing Security and Privacy for Identity-based Batch Verification Scheme in VANETs. *IEEE Transactions on Vehicular Technology*, 66(4), 3235–3248.

[9] Cui, J., Tao, X., Zhang, J., Xu, Y., & Zhong, H. (2018). HCPA-GKA: A Hash Function-based Conditional Privacy-preserving Authentication and Group-key Agreement Scheme for VANETs. *Vehicular Communications*, 14, 15–25.

[10] Abbas, F., Fan, P., & Khan, Z. (2018). A Novel Low-latency V2V Resource Allocation Scheme based on Cellular V2X Communications. *IEEE Transactions on Intelligent Transportation Systems*, 20(6), 2185–2197.

[11] Liu, N. (2011). Internet of Vehicles: Your next Connection. *Huawei WinWin*, 11, 23–28.

[12] Dua, A., Kumar, N., & Bawa, S. (2014). A Systematic Review of Routing Protocols for Vehicular Ad Hoc Networks. *Vehicular Communications*, 1(1), 33–52.

[13] Engoulou, R. G., Bellaïche, M., Pierre, S., & Quintero, A. (2014). VANET Security Surveys. *Computer Communications*, 44, 1–13.

[14] Verma, S., & Mittal, S. (2018). Implementation and Analysis of Stability Improvement in VANET using Different Scenarios. *International Journal of Engineering & Technology*, 7(1.2), 151–154.

[15] https://www.statista.com/statistics/471264/iot-number-of-connected-devices-worldwide/Dec 2020.

[16] Jain, B., Brar, G., Malhotra, J., Rani, S., & Ahmed, S. H. (2018). A Cross Layer Protocol for Traffic Management in Social Internet of Vehicles. *Future Generation computer systems*, 82, 707–714.

[17] Rajadurai, R., & Jayalakshmi, N. (2013). Vehicular Network: Properties, Structure, Challenges, Attacks, Solutions for Improving Scalability and Security.

[18] Mughal, M. A., Shi, P., Ullah, A., Mahmood, K., Abid, M., & Luo, X. (2019). Logical Tree-based Secure Rekeying Management for Smart Devices Groups in IoT-Enabled WSN. *IEEE Access, 7*, 76699–76711.

[19] Mawlood Hussein, S., López Ramos, J. A., & Álvarez Bermejo, J. A. (2020). Distributed Key Management to Secure IoT Wireless Sensor Networks in Smart-agro. *Sensors, 20*(8), 2242.

[20] Kouicem, D. E., Bouabdallah, A., & Lakhlef, H. (2018). Internet of things Security: A Top-down Survey. *Computer Networks, 141*, 199–221.

[21] Sohail, M., Ali, R., Kashif, M., Ali, S., Mehta, S., Zikria, Y. B., & Yu, H. (2020). Trustwalker: An Efficient Trust Assessment in Vehicular Internet of Things (viot) with Security Consideration. *Sensors, 20*(14), 3945.

[22] Dammak, M., Senouci, S. M., Messous, M. A., Elhdhili, M. H., & Gransart, C. (2020). Decentralized Lightweight Group Key Management for Dynamic Access Control in IoT Environments. *IEEE Transactions on Network and Service Management, 17*(3), 1742–1757.

[23] Guerrero-Ibáñez, J., Zeadally, S., & Contreras-Castillo, J. (2018). Sensor Technologies for Intelligent Transportation Systems. *Sensors, 18*(4), 1212.

[24] Cooper, C., Franklin, D., Ros, M., Safaei, F., & Abolhasan, M. (2016). A Comparative Survey of VANET Clustering Techniques. *IEEE Communications Surveys & Tutorials, 19*(1), 657–681.

[25] Cooper, C., Franklin, D., Ros, M., Safaei, F., & Abolhasan, M. (2016). A Comparative Survey of VANET Clustering Techniques. *IEEE Communications Surveys & Tutorials, 19*(1), 657–681.

[26] Kung, Y. H., & Hsiao, H. C. (2018). GroupIt: Lightweight Group Key Management for Dynamic IoT Environments. *IEEE Internet of Things Journal, 5*(6), 5155–5165.

[27] Kulla, E., Jiang, N., Spaho, E., & Nishihara, N. (2018, July). A Survey on Platooning Techniques in VANETs. In *Conference on Complex, Intelligent, and Software Intensive Systems* (pp. 650–659). Springer, Cham.

[28] Alsaeedy, A. A., & Chong, E. K. (2020). A Review of Mobility Management Entity in LTE Networks: Power Consumption and Signaling Overhead. *International Journal of Network Management, 30*(1), e2088.

[29] https://www.standards.its.dot.gov/Factsheets/Factsheet/80.

[30] El Mouna Zhioua, G., Tabbane, N., Labiod, H., & Tabbane, S. (2014). A Fuzzy Multi-metric QoS-balancing Gateway Selection Algorithm in a Clustered VANET to LTE Advanced Hybrid Cellular Network. *IEEE Transactions on Vehicular Technology, 64*(2), 804–817.

[31] Vijayakumar, P., Azees, M., Kannan, A., & Deborah, L. J. (2015). Dual Authentication and Key Management Techniques for Secure Data Transmission in Vehicular ad Hoc Networks. *IEEE Transactions on Intelligent Transportation Systems, 17*(4), 1015–1028.

[32] Lu, R., Lin, X., Liang, X., & Shen, X. (2011). A Dynamic Privacy-preserving Key Management Scheme for Location-based Services in VANETs. *IEEE Transactions on Intelligent Transportation Systems, 13*(1), 127–139.

[33] Kumar, R., Singh, S. K., Lobiyal, D. K., Chui, K. T., Santaniello, D., & Rafsanjani, M. K. (2022). A Novel Decentralized Group Key Management Scheme for Cloud-Based Vehicular IoT Networks. *International Journal of Cloud Applications and Computing (IJCAC), 12*(1), 1–34.

[34] Shen, X., Huang, C., Pu, W., & Wang, D. (2022). A Lightweight Authentication with Dynamic Batch-Based Group Key Management Using LSTM in VANET. *Security and Communication Networks, 2022.*

Automation and Computation – Vats et al. (Eds)
© 2023 the Author(s), ISBN 978-1-032-36723-1

Optimization-based watermarking scheme for security

Abhay Singh Rawat, Pankaj Kumar, Sumeshwar Singh & Resham Taluja
Department of Computer Science and Engineering, Graphic Era Hill University, Dehradun, India

Rajeev Gupta
Department of Electronics and Communication Engineering, Graphic Era Hill University, Dehradun, India

ABSTRACT: In the current day, we transfer sequential data online from one person to another, or we can say, from one source to another source, for this transfer of data privacy or security, we may use the optimization-based watermarking system. To stop sharing Sequential data without the owner's consent or, more accurately, to prevent the unlawful sharing of personal data, is a further technique to apply the optimization-based watermarking scheme. We provide an innovative optimization-based watermarking approach for exchanging sequential data to achieve this feature. The best embedding strength factor, which helps preserve a desirable connection between resilience and quality, is initially computed using firefly optimization in the algorithm. Next, a spatial and transformed domain scheme combination is used. To determine the ideal embedding strength factor, which is important for preserving a beneficial link between resilience and quality, the technique initially employs firefly optimization. The appropriate strength factor is then used to combine spatial and transformed domain techniques to hide the various markings on the several carrier media channels. Before being hidden by the media channel, picture marking is encrypted to make it safer. The influence of various outcomes on the Kodak and USI-SIPI datasets was extensively investigated through tests, and the results showed that the recommended approach has increased resilience, great quality, and high capacity. Superior To other known systems withed watermark capacity, the suggested approach achieves robustness of up to 11%.

1   INTRODUCTION

Stepwise data consists of genomic data and other organized data, time-series data, such as speech, stock market, location trends, and geographic data. Data is utilized for several purposes, and people trade different types of data to obtain tailored services from internet service providers (SPs). The data gathering and processing activities carried out by these SPs may reveal personally identifiable data about people. As a result, the way these SPs use the data they have collected compromises people's privacy, and thus calls for giving individuals choice over how the SPs collect and use their data. A person wants to make sure that their data is appropriate for a certain purpose before providing it to an SP. won't be by anybody else outside. When SP shares a person's personal information with third parties in other ways, it is a privacy breach (e.g. For instance, for monetary gain). It is important to stop SPS from participating in such illegal sharing. the development of technological tools to hold them accountable for such unlawful sharing Situations is a common device for this type of watermark. Before sharing their data with each SP, someone might add a specific watermark, and if their information is shared with others without their permission, they can link the sharing without permission. Watermarking is one common strategy for addressing liability issues, particularly with multimedia data [1]. Using Watermarks are applied to multimedia data by changing certain pixel values due to the high amount of data redundancy and the fact that the human eye cannot identify minute variations in pixel values. Applying watermarking to sequential data, such as genetic or geographic patterns, is challenging. In order to inject a watermark into sequential data, the watermark must update the original data, which reduces the amount of service that the SPs can provide. As a result, watermarking sequential data while retaining data usability presents unique challenges. The longevity (or identity) of the watermark is an additional difficulty for watermarking sequential data. If the SP detects the data points with the associated watermark and eliminates (as well as interferes with) the watermark prior to the unlawful distribution, the SP cannot be identified as the source of the data breach. Additionally, rather than disclosing all of the individual's information, an SP may change or only partially publish the data (to damage the watermark). As a result, it becomes more difficult to determine the leak's source. An SP may make use of various types of auxiliary data to identify (and subsequently alter) the data's watermark. We provide a special watermarking method for distributing sequential data to solve these robustness and utility issues solve these robustness and utility issues, we provide a special watermarking method for distributing sequential data. We initially presume that the data are unconnected and make suggestions. Which data points should be watermarked is determined by an algorithm that solves a – anti-optimal control problem. This approach was developed to be impervious to unreliable SPs' cooperation. Then, we go through how the suggested method handles correlated data. As a result, we reduce the possibility of correlation assaults from bad SPs who are aware of the pairwise correlations between data points. On a genomic dataset, we assess the usefulness and security (robustness) of the suggested approach. When sharing, digital watermarking has become very popular. Getting information through the Internet has become commonplace. Exchange of files You never know who could use them without your permission online. consent. Your files should never be used in commerce without your permission. Either put them online in the worst quality possible or don't distribute anything valuable at all. It is not a practical solution to the issue. Is the possibility of unauthorized usage a concern? You should thus look for more. measures for copyright protection that are successful, such as digital watermarking understand watermarks better we classified the Watermarking based on various Parameters. These watermarks need some unique data In this research, we looked at the different kinds of watermarking to find the embedded data in the wat. approaches based on several criteria, such as human consciousness, robustness, etc. based on the human As a result, we may categorize watermarking into two categories: Watermarking that is visible and invisible. Considering the user approval for the watermark's detection, There are two types of watermarking: public and private.signal.watermarking using remarks. Depending on a variety of applications, Three categories of watermarking exist fragile-Fragile Watermarking, Strong Watermarking. Future efforts will be made to create some novel watermarking methods that are reliable and usable and used in a variety of practical applications. The information model, model scheme, and security model are all described in this section. Table 1 lists the symbols and notations that are often used.

Table 1. Symbols and notations that are often used.

$X_1, \cdots, X_L'$	A collection of ordered data points
$d_1, \cdots, dm$	A data point's possible values (states)
I_i	The index collection of data points exchanged with the SP_i
D_{li}	a collection of data points in I_i W_{li}
W_{li}	Ii data point set after watermarking
Z_{li}	W_{li} data points with watermarks

2   WATERMARKING METHODS

In digital watermarking, we take data (video, image, audio, etc.) and hide the massage with it so that outsider can not make changes in the file. The information model, model scheme, and security model are all described in this section. Table 1 lists the symbols and notations that are often used in this work. Let's take a case where we have an image, we use watermarking to hide a message, they secured it with a secret key and transferred it to the receiver who opens the watermarking to read the message. Digital watermarking is the process of concealing an indication of a digital signal contained inside the signal itself, such as a picture, music, or video [2]. The majority of the time, for copy protection and copyright, digital watermarks are used are also utilized [1,3–5] which has applications ranging from broadcast monitoring [6] to other scenarios both content authentication and transaction tracking [7].

The use of digital watermarking watermarks for copy protection and preventing copying of multimedia information. The high level of redundancy in multimedia watermarking techniques [8] is advantageous. A fingerprint is an imprint made by friction ridges on a human finger, according to Boneh and Shaw. Obtaining partial fingerprints from a crime scene is a crucial forensic science technique. Fingerprints left on metal or glass surfaces by moisture and oil on the finger. By carefully applying ink or other substances from the peaks of friction ridges on the skin to a smooth surface, such as paper, one can obtain impressions of entire fingerprints. Though fingerprint cards also typically record portions of the lower joint areas of the fingers as a watermark, fingerprint records typically contain impressions from the pad on the last joint of fingers and thumbs. They do not believe that sophisticated assaults on the data are the method of watermarking. Since there are redundant positions, Non-media data are often not as high, making it more difficult to watermark such data. The methods of watermarking suggested for text and other non-media [9]. Several articles have suggested watermarking techniques for sequential data, including time-series data and spatiotemporal data. A method for concealing confidential material, like the "ssn" or birthdate, in electrocardiograms was proposed by Kozat *et al.* (ECG) without altering crucial ECG properties to verify the ownership of data [10]. A low-complexity watermarking approach was also devised by Panah et al to tamper-proof ECG signals at the sensory nodes. Due to the limited data redundancy, watermarking spatiotemporal data is especially difficult. Watermarking track records that include the orbits of many objects has been the main focus of research in this area . We do, however, consider the situation in which someone wishes to share their specific data with a number of SPs after watermarking it. A generic watermarking (fingerprinting) method that is resistant to collusion was put out by Boneh and Shaw. Their plan creates fingerprints in a way that prevents any group of attackers from locating one. Although there are still some practical issues with this plan. The fingerprint length may initially be fairly long in order to give resilience to collusion, which lowers the value of the data. Additionally, the system does not take into account sophisticated assaults against the watermarking algorithms, such as those that use auxiliary data or alter the watermark, nor does it take into account data correlations. We talk about these shortcomings in our suggested plan.

2.1   *Information model*

Sequential data is composed of arranged data components $X_1,..., X_L$, where is the length of the information. The amount of information pieces xi can vary based on the kind of data and can come from the set [d1," "dm."]. Examples include accurate latitude and longitude coordinate pairs for GPS data, place semantics for check-in data (such as a coffee shop or restaurant), and the relevance of polymorphism or point mutations in genomics data. which, when paired, may be correlated. Linkage disequilibrium is the term for such pairwise correlations, which are not always between successive data points. The state of each data point may have an impact on the correlation value, and there is a frequent asymmetry in the

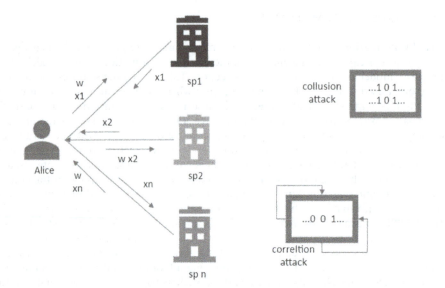

Figure 1.   Overview of the system and threat models.

correlation between the data points. Furthermore, it has been demonstrated that the human genome can display higher-order correlations. For the sake of clarity, we design our solution first for sequential data that is not connected before extending it to correlated data.

We consider a system in which Alice is the data owner and there are several service providers (SPs). A medical establishment, a biological researcher, or an immediate service provider might be the SP for genetic information. Any position service provider can act as the SP for GPS data. For the sake of clarity, we discuss several numbers of algorithmic elements in the description of the recommended scheme using raw code; nevertheless, the suggested scheme may be expanded to include nonbinary data. To analyze the proposed strategy, we focus on genetic variations in omics with values spanning from 0 to 1, 2. Alice grants the SPs access to some of her data in return for some services. The SP for genetic data might be a pharmaceutical company, a genomic researcher, or a basic service provider. The SP for location data might be any company that offers location-based services. The suggested technique may be changed to work with non-binary data, however for the sake of clarity, we describe many of the algorithmic elements in the suggested scheme using binary data. In reality, To assess the recommended method, we concentrate on gene mutation in DNA sequencing with values ranging from 0 to 1, 2. Alice gives the SPs access On the one hand, Alice wants to make sure that when she provides information to an SP, her data won't be shared with anybody else. by the pertinent SP as well as other parties. In case there is any more unauthorized sharing, she asks for information on the SP who is in responsible of this leak As a consequence, Alice adds a different watermark each time she exchanges her data with another SP. An SP, on the other hand, is permitted to disclose Alice's information to third parties without Alice's permission. The SP would really like to locate this watermark in the material and erase it so that it is no longer visible. Only a fraction of Alice's data can be shared with a 3rd party via an SP, not the complete collection. A number of techniques, including as removal, collusion, masking, insertion/deletion, and reordering assaults, have been developed to apply a watermark to an image or a piece of text [12,21]. By using previously established attacks on sequential data and developing novel attacks for sequential data, including the correlation attack, we take into consideration the following attacks on the proposed watermarking system. Uncorrelated data is subjected to a single SP assault. Assume Alice gives an SP access to her (uncorrelated) w-length sequential data as well as a w-length watermark. Because the data are uncorrelated and each data point is independent of the others, the SP calculates the probability that a data point is watermarked as w/' for each data point. Instead of attempting to identify and damage the watermark, the possibly harmful SP may alter the data.

2.2   *Possible solutions*

Here, we first give a general overview of the suggested protocol before going into further specifics on the suggested watermarking technique. The following procedure is used when Alice wishes to exchange her info with an SPi. The requested indices of Alice's data are sent by the SP I indicated by Ii. Alice produces $D_{Ii} = U_{I \ E \ Ii} \ x_i$. Given that she has previously shared her data, Alice believes that the data points are watermarked. Our suggested watermarking algorithm, which is used for this phase, is covered in depth in this section. Alice creates the watermarked data Wii by adding a watermark to the data points in D_{Ii}. Alice keeps track of the SP's ID and Z_{Ii} (the SP's watermark pattern). Wii is sent to SP I by Alice. The suggested watermarking technique explains how to choose which data points in sequential data should be watermarked. In order to protect the watermark against the threats covered in Section 3.3. A watermark has been applied to a data point. It merely changes the status of this data item. If the change is binary, it moves from 0 to 1 or vice versa. If every value point in the collection "d1," "dm," and the alteration is from the current condition to a different state j. Section 4.1 includes ensuring that there is no relationship between the data, data point xi changed into a state d consistently at random. In order to protect the watermark against the assaults covered in Section 3.3, the proposed watermarking method specifies how to choose which data points within the sequential data should be watermarked. By altering the state of a data point, we may add a watermark to it. Consider the case when data is. In Section 4.2, it is decided that the new state d*j of a data point x_i will decrease the likelihood of a correlation attack due to the correlation in the data. Following a detailed explanation of the solution for time series data without correlates (where data points are independent of one another), steps are provided for expanding it to correlated sequential data. We initially give the following notations to aid in the discussion before getting into the specifics of the recommended plan of action. The quantity of watermarked data points (n H I) The entire collection of data is divided among H SPs I times. Watermarked sample values are provided as y h l when the full dataset is transferred h times while they are not watermarked following $(h + 1)^{-th}$ sharing. When the entire set of information is shared h times, the number of watermarked points I times that will be watermarked in the $(h + 1)^{-th}$ sharing is y h l. To reduce the likelihood of a collide attack when Alice exchanges her data with a new SP, the watermark places in the information are first specified for the new request by watermarking distinct patterns in previously shared material. These definitions show that n H I = y H I + y H I, which means that of the n H I information points that are watermarked I time is the time after h gradients, y H I will be shared in the $(h - 1)^{-th}$ sharing while the remaining y H I will not. As a consequence, the recommended method determines the values of y h I & y H I to reduce the likelihood of a collision assault. Since the measured values are uncorrelated, after computing these values, any y H I of both the n H I watermarked pieces of information can be chosen to contain the watermark.

2.3   *Experimental results*

We looked at the proposed watermarking approach from a variety of angles. We focused on its security (robustness) against watermark inference and alteration. Run collision and correlation attacks to test your resistance to watermark inference. We presumptively assume that Alice and the SPs share an identical data piece in each collusion attack scenario. This presumption gives the malicious SPs the most information possible. Malicious SPs may

315

launch a collusion attack if various sets of data points are shared with them, which they can do by intersecting these data points. The performance of the data owner's (watermark) detection is used to measure robustness against watermark change under various assaults. The outcomes also contain an assessment of the reduction in data usefulness caused by the insertion of watermarks. We performed each experiment 1000 times and then reported the average results. We write r = w/' to represent the percentage of data that is watermarked. The watermark proportion r likewise depicts the lost of utility in the data transfer caused by the additional watermark.

3 FURTHER POSSIBILITIES AND CONCLUSIONS

The use of multimedia technology is rising, and providing permitted data and protecting secret information from illegal use is a challenging and complex procedure. Watermarking allows only authorized users to access the data. Digital watermarking [11] is a frequently used approach for digital data security.Smart healthcare systems have grown in popularity in recent years due to the simplicity of exchanging e-patient information via an open network. Many researchers have been interested in the challenge of preserving the security of this information. Thus, for CT scan pictures of COVID-19 patients [12], robust and dual watermarking based on redundant discrete wavelet transform (RDWT), Hessenberg Decomposition (HD), and randomized singular value decomposition (RSVD) are proposed. The fundamentals of Quantum Information Processing (QIP) [13] are presented below, along with a thorough examination of quantum image-based data concealing strategies. QIP is more efficient and secure than traditional information processing. There is also a comparison of the Flexible Representation of Quantum Images (FRQI) and Novel Enhanced Quantum Representation (NEQR) models.

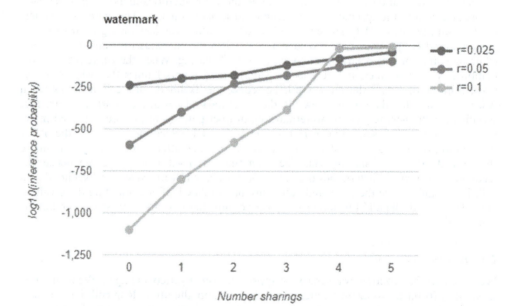

Figure 2.   The likelihood that all watermarked spots in a collusion attack will be recognized when a group of malicious SPs collaborates. The percentage of watermarked data is represented by r.

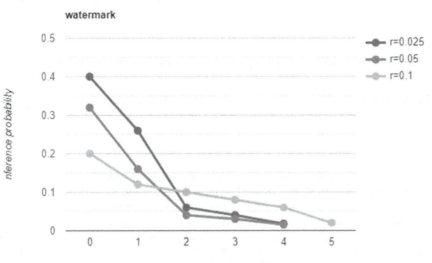

Figure 3. When six malicious SPs are collaborating, it is probable to detect different percentages of the watermarked spots in the collusion attack. This percentage is denoted by the letter r.

REFERENCES

[1] Memon, N. and Wong, P.W., 2001. A Buyer-seller Watermarking Protocol. *IEEE Transactions on Image Processing*, *10*(4), pp.643–649.

[2] Cox, I.J., Miller, M.L., Bloom, J.A. and Honsinger, C., 2002. *Digital Watermarking* (Vol. 53). San Francisco: Morgan Kaufmann.

[3] Bloom, J.A., Cox, I.J., Kalker, T., Linnartz, J.P., Miller, M.L. and Traw, C.B.S., 1999. Copy Protection for DVD Video. *Proceedings of the IEEE*, *87*(7), pp.1267–1276.

[4] Chung, T.Y., Hong, M.S., Oh, Y.N., Shin, D.H. and Park, S.H., 1998. Digital Watermarking for Copyright Protection of MPEG2 Compressed Video. *IEEE Transactions on Consumer Electronics*, *44* (3), pp.895–901.

[5] Maes, M., Kalker, T., Linnartz, J.P., Talstra, J., Depovere, F.G. and Haitsma, J., 2000. Digital Watermarking for DVD Video Copy Protection. *IEEE Signal Processing Magazine*, *17*(5), pp.47–57.

[6] Liu, L. and Li, X., 2010, May. Watermarking Protocol for Broadcast Monitoring. In *2010 International Conference on E-Business and E-Government* (pp. 1634–1637). IEEE.

[7] Emmanuel, S., Vinod, A.P., Rajan, D. and Heng, C.K., 2007, February. An Authentication Watermarking Scheme with Transaction Tracking Enabled. In *2007 Inaugural IEEE-IES Digital EcoSystems and Technologies Conference* (pp. 481–486). IEEE.

[8] Lee, S.J. and Jung, S.H., 2001, June. A survey of watermarking techniques applied to multimedia. In *ISIE 2001. 2001 IEEE International Symposium on Industrial Electronics Proceedings (Cat. No. 01TH8570)* (Vol. 1, pp. 272–277). IEEE.

[9] Kamaruddin, N.S., Kamsin, A., Por, L.Y. and Rahman, H., 2018. A review of text watermarking: theory, methods, and applications. *IEEE Access*, *6*, pp.8011–8028.

[10] Kozat, S.S., Vlachos, M., Lucchese, C., Van Herle, H. and Yu, P.S., 2009. Embedding and retrieving private metadata in electrocardiograms. *Journal of medical systems*, *33*(4), pp.241–259.

[11] Mohanarathinam, A., Kamalraj, S., Prasanna Venkatesan, G.K.D., Ravi, R.V. and Manikandababu, C.S., 2020. Digital watermarking techniques for image security: a review. *Journal of Ambient Intelligence and Humanized Computing*, *11*(8), pp.3221–3229.

[12] Anand, A. and Singh, A.K., 2022. Dual Watermarking for Security of COVID-19 Patient Record. *IEEE Transactions on Dependable and Secure Computing*.

[13] Cox, I.J., Miller, M.L., Bloom, J.A. and Honsinger, C., 2002. *Digital watermarking* (Vol. 53). San Francisco: Morgan Kaufmann.

Automation and Computation – Vats et al. (Eds)
© *2023 the Author(s), ISBN 978-1-032-36723-1*

Video watermarking techniques for security: An overview

Abhishek Negi, Pankaj Kumar, R. Gowri & Dibyahash Bordoloi
Department of Computer Science and Engineering, Graphic Era Hill University, Dehradun, India

ABSTRACT: The amount of data sent online and the widespread use of digital technologies have both increased significantly. The growth in demand for copyrighted material of digital material has undoubtedly contributed to the interest in digital watermarking during the past ten years. Video watermarking is increasingly being used for copy protection, broadcasting tracking, video authentication, fingerprinting, and other purposes. Security, capacity, and resilience are the three main components of information concealing. The amount of data that can be disguised is the subject of capacity. Robustness, which describes the resistance to alteration of the cover material until hidden information is lost, relates to the capability of anybody discovering the information. Algorithms for video watermarking often favour robustness. A reliable algorithm does not use eliminating the watermark without exactly altering the cover's content is possible. The criteria needed to develop a trustworthy watermarked movie for a useful application are discussed in this study. We explore a variety of approaches and present a widely utilised essential strategy. This research aims to concentrate on the different video watermarking technology fields. The notion of algorithm robustness is highlighted in the best portion of the studied solutions based on video watermarking. The discipline of video watermarking is fresh and expanding quickly in the multimedia sector. The goal of this study is to emphasis on the various video steganographic technique domains. Keywords: Video Watermarking, Robust Techniques, Authentication, DCT, DWT, PSNR, & MSE.

1   INTRODUCTION

The practise of integrating online coding into digital multimedia is known as digital video watermarking (images, audio and video sequence). A serial number, randomly sequence, ownership identifiers, trademark messages, signals, transactional dates, information about the authors of the work, bi-level or gray-level pictures, text, or other digital data formats can all be included as embedded information or watermarks. Watermark production, watermark embedding, or watermark extraction or detecting should all be included in a full digital watermarking system [2]. Information concerning provenance, ownership, copy control, etc. is contained in a watermark. Multimedia files include this information encoded in them. According to the specifications, the watermark is implanted and removed. Typically, a watermark is used to indicate the creator or authorised person, provide usage restrictions, confirm the validity of control the use and dissemination of a data, or assure its integrity. This ability to embed information more reliably and redundantly due to extra data shown with watermarking different from image watermarking. A digital video is an image sequence or collection. Payload refers to the volume of data that may be included in video sequence.

DOI: 10.1201/9781003333500-36

2   VIDEO WATERMARKING OVERVIEW

Different kinds of video steganographic techniques can be classified into different categories. These watermarks can be utilized in either the frequency domain or the spatial domain. The following figure illustrates many watermark kinds.

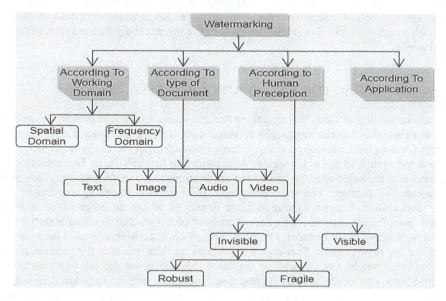

Figure 1.   Different types of watermarking.

Mainly there are categories of watermarking that may be made based on the kind of document like Text Watermarking, Audio Watermarking, and Video Watermarking, and these may be categorised into three categories based on how people perceive it. There are three types of watermarks: visible, invisible, and invisible-fragile dual.

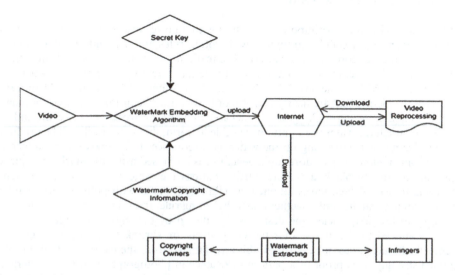

Figure 2.   Video watermarking outlines for security.

3 ROBUSTNESS

The ability to resist any form of attack is robustness. Attacks can be direct and indirect. One example of a passive assault is the quality reduction brought on by noise signals. And a common active assault includes erasing the watermark besides deleting some few frames. In studies, compressed video containing a watermark is first decompressed, then it undergoes various assaults before being compressed once more. Recompression attack is therefore vulnerable to other attacks. Following, we go over the suggested algorithm's robustness from several angles.

4 WATERMARKING FOR VIDEOS: APPLICATIONS

Applications for digital video watermarking are numerous. Copyright protection: In digital video distribution networks, copyright of video data is a crucial issue. There are several methods for copyright protection using video watermarking. One of the methods involves adding a watermark to the video signal. Authentication through video The authenticity of the photos and the video must be checked in applications that use, for example, surveillance camera-captured footage. The most often used rules for watermarking are fragile, semi-fragile, and robust. Fragile watermarks are destroyed when the cover video is slightly changed. Semi-fragile watermarking may withstand processes that preserve material and is sensitive to changes in the content. Duplicate protection is a commonly used usage in video embedding for copy control. This uses a watermark to show whether the video material is protected by copyright. Only a severely degraded video sequence will erase this watermark. The clip is uniquely identifiable by its resulting fingerprint, which is produced by software that detects, extracts, and then compresses a video's distinctive elements. This approach is known as fingerprinting. Key frame analysis, colour changes, motion changes, and other aspects of a video sequence are some of the characteristics that go into video fingerprinting analysis. With this method, watermarks are included into the video like fingerprints. The fingerprints in the movie are extracted using several fingerprinting techniques. The video material is then evaluated and identified by comparing the retrieved fingerprints.

5 FREDRICH'S ALGORITHM

The technique relies on superimposing a pattern with a power distribution that is mostly in the low frequencies. A cellular automata with voting rules and a pseudo-random number generator are used to construct the pattern. The method's resilience has been demonstrated by Fredrich 6J. The technique also addresses a potential flaw in [2]'s approach, which may be circumvented by using the possibility that some of the watermark pattern would appear in portions of the picture that are nearly uniform or have a nearly constant brightness gradient. Friedrich employs a watermarking technique based on patterns overlaying to get around this flaw. Even though the watermark pattern is visible in tiorm, the pattern will still be produced in a sensitive manner depending on the watermark sequence. It is impossible to launch an attack in some regions. A pseudorandom generator is initialised using the mark bit sequence to produce a random black-and-white starting pattern the same size as the picture. Until converging to a fixed location is attained, a celhilar automata with polling rides is imple-mented. Random patches link together under the voting rule. In order to shift the majority of the energy to low frequencies, the design is the filtered by the smoothing filter. The final pattern is applied to the image after the Gray levels are shrunk to a narrow range. The watermarked image does not clearly exhibit any damage from the overlay patten, but it is ingrained strongly. It is proof of a pattern's existence in photographs that have undergone filtering, JPEG compression with a quality factor as low as 5%, cropping, resampling, blur, down-sampling, or noise addition. Regarding the collusion assault, the watermark also seems to be resistant (averaging several watermarked images to remove the watermark). The requirement for the original, unwatermarked image during the recovery procedure is the first major drawback. Ibis is not appropriate for videos since we would have to show the entire video to establish the watermark. Our solution addresses this flaw by retrieving an identifier without the origin using simple statistical techniques. The watermarking process only embeds one of the 73 patterns formed by a cellular automata with voting rules and a pseudo-random number generator, which is the second major drawback. No in-depth information on the writer or producer is integrated. Only true or false are returned by the retrieval procedure if the match was successfully located. In order to include code words for specific information like author names or addresses, we want to extend the method.

6 PROCESSES FOR WATERMARKING

Watermarking may be done in two different ways: in the spatial domain and in the frequency domain. Each method has benefits and drawbacks of its own. A typical photo cropping technique may remove the watermark, which is a fundamental drawback of spatial domain watermarking. Additionally to image pixels watermarking, frequency domain methods have been suggested. Multimedia watermarking uses the wideband communication technology as well. Most watermarks in earlier works are symbols or random numbers made up of a series of bits, and they can only be "detected" by using the "detection theory" Review of Video Postprocessing Techniques We may easily add a watermark on a host picture in the image space by adjusting the grey levels. merely a few pixel in the host picture, but computer analysis may immediately spot the injected data. within the frequency range. A modified signal's coefficients can have a watermark added to them. **Discrete Fourier Transform (DFT), Discrete Cosine Transform (DCT), and Discrete Wavelet Transform (DWT)** techniques are utilised in frequency domain watermarking. In the DCT method, the host signal's transform domain contains the watermark. In order to insert the watermark without affecting the image's appearance, the medium frequency coefficients are modified. Its **Discrete Wavelet Transform (DWT)**, which has strong spatial localisation and multi-resolution capabilities, is frequently used in picture watermarking applications. Typically, the frequencies lower sub-bands are where the majority of the visual energy is concentrated, thus incorporation Watermarks in the lower frequencies sub-bands may drastically worsen the image while significantly increasing resilience. However, if we embed too much information in the frequency domain, the host image's quality would be drastically degraded. In other words, the watermark's size should be less than that of the host picture.

7 DOMAIN SPATIAL

Watermarking in the spatial domain is often done in the brightness and colour components and is evaluated as a cheap and simple complex approach. There are, however, a few significant restrictions. With simply geographical analytic methods, watermark optimization is difficult. The following are some different spatial domain watermarking The following are some different spatial domain watermarking techniques: The least significant bits are used to incorporate the watermark in the LSB approach, which is straightforward and easy to understand. High capacity is provided by this technology. This method resists cropping but is vulnerable to noise addition, lossy compression, and LSB resetting:

a) This method resists cropping but is vulnerable to noise addition, lossy compression, and LSB resetting The LSB approach can increase security and make it impossible for someone else to find the watermark.

b) Techniques based on correlation Using a correlation-based approach, a randomly produced noise is added to the cover media pixels' brightness. The resilience can be enhanced by the watermark strength. When the correlation rises beyond the threshold, the watermark is considered to have been spotted. To recreate the fake random noise during detection, the key is required as side information. The appropriate seed needs to be accessible in order to get a high correlation.

8 DWT-DCT ALGORITHM HYBRID

The DWT and DCT algorithms are usually applied with each other in watermarking as a result of the development of watermarking algorithms. DWT is often applied to the full picture and permits image breakdown into distinct subbands for the purpose of displaying representations with multiple resolutions.For instance, choosing the LL subband can result in a strong feature, while choosing the HH sub-band can maintain the finer details. DCT, on the other hand, only processes square sized portion ofthe image rather than the whole thing This makes it possible to employ DCT for compression with a high compression ratio. To achieve resilience for watermarking, the watermarks might be placed onto the DC coefficient. In the meanwhile, watermarks can be incorporated onto either the high- or low-frequency AC coefficients in order to save the specific information. The jointed DCT-DWT technique, which is often used in picture watermarking , fully exploits their strengths. There are, however, currently relatively few hybrid DCT-DWT watermarking algorithms available for digital videos.

9 WATERMARKING USING WAVELETS

Use of a multi-resolution data combining approach is employed to convert to the discrete wavelet domain. The picture and watermark are transformed. A watermark is introduced to the host picture at each wavelet based level for embedding purposes. A wavelet decomposition average is calculated at the moment of detection. It has sufficient strength to withstand compression, noise reduction, and filtering processes.

10 VIDEO WATERMARKING ATTACKS

Frame average, statistics, frame skipping, lossy compression, cropping, and different signal processing and geometric techniques are examples of video watermarking attacks. There are two types of assaults on video watermarking: purposeful attacks and inadvertent attacks.

a) Intentional attacks: Intentional attacks against watermarks include noise addition, contrast and colour augmentation, and filtering of single frames. attacks using statistics, such as collision and average assaults.
b) Unintentional attacks: These can result from degradations that can occur during lossy copying, from compressing of the video while reencoding, or from changes in frame rate and resolution.

11 ALGORITHMS FOR VIDEO WATERMARKING IN THE ENCODING PROCESS

The quantization discrete cosine transform (DCT) components, predictive modes, motion vectors, and other redundant video spaces are often modified by the video watermarking method during the encoding phase to achieve watermark embedding. The benefits of this kind of algorithms include I the ability to directly combine with corresponding points coding standards and (ii) the ability to embed and extract the watermark in real time through into the adjustment of the encoder, which is simple, efficient, and has little impact on the data rate of video streams. Its drawbacks include I the requirement to adjust the encoder and (ii) the watermark's embedding capability is impacted by video coding settings. and decoder, which to some extent restricts the use of particular watermar king methods.

12 ALGORITHMS FOR VIDEO WATERMARKING FOLLOWING COMPRESSION

After compression, this video watermarking algorithm looks for blank space in the reduced bit stream and inserts watermark data there. These algorithms' merits include I their great efficiency and independence from the associated codec; (ii) their little computational redundancy and high fidelity. Its drawbacks include I a relatively tiny amount of redundant space is available for watermark embedding, which limits capacity, and (ii) the algorithm's resilience is inadequate. In conclusion, every algorithm has pros and cons of its own. In actual use, many watermark kinds are chosen in accordance with various circumstances and requirements.

13 CONCLUSIONS

This overview of the literature examines several video watermarking approaches that aca-demicians and businesspeople have developed in a variety of spatial and transformation domains, including issues like robustness, imperceptibility, and payload capacity. To make reading easier for readers at the conclusion of each group, a concise overview has been provided for each created group of video watermarking procedures. There is little doubt that new algorithms will emerge and they may combine existing technology. However, there are still many problems to be solved, like the "video watermarking in real-time environment," "ambiguity of watermarks," "collusion attack," "elapsed time to embed the watermark," "video specific assaults," and "applying more than single attack at a time." As previously indicated, this review described the majority of those studies. where original uncompressed video is taken into account. The remainder of video watermarking, wherein watermark is placed in encrypted video or even during encoding, will be discussed very shortly.

REFERENCES

[1] Mohanty, S.P. and Kougianos, E., 2011. Real-time Perceptual Watermarking Architectures for Video Broadcasting. *Journal of Systems and Software*, *84*(5), pp.724–738.
[2] Cross, D. and Mobasseri, B.G., 2002, September. Watermarking for Self-authentication of Compressed Video. In *Proceedings. International Conference on Image Processing* (Vol. 2, pp. II–II). IEEE.
[3] Sun, Q., He, D. and Tian, Q., 2006. A Secure and Robust Authentication Scheme for Video Transcoding. *IEEE Transactions on Circuits and Systems for Video Technology*, *16*(10), pp.1232–1244.
[4] Xu, D., Wang, R. and Wang, J., 2011. A *novel watermarking scheme* for H. 264/AVC video Authentication. *Signal Processing: Image Communication*, *26*(6), pp.267–279.
[5] Agarwal, H. and Husain, F., 2021. Development of Payload Capacity Enhanced Robust Video Watermarking Scheme based on Symmetry of Circle using Lifting Wavelet Transform and SURF. *Journal of Information Security and Applications*, *59*, p.102846.
[6] Sun, J., Jiang, X., Liu, J., Zhang, F. and Li, C., 2020. An Anti-recompression Video Watermarking Algorithm in Bitstream Domain. *Tsinghua Science and Technology*, *26*(2), pp.154–162.

[7] Sun, X.C., Lu, Z.M., Wang, Z. and Liu, Y.L., 2021. A Geometrically Robust Multi-bit Video Watermarking Algorithm based on 2-D DFT. *Multimedia Tools and Applications, 80*(9), pp.13491–13511.

[8] Sahu, A.K., 2022. A Logistic Map based Blind and Fragile Watermarking for Tamper Detection and Localization in Images. *Journal of Ambient Intelligence and Humanized Computing, 13*(8), pp.3869–3881.

[9] Munir, R., 2019, October. A Secure Fragile Video Watermarking Algorithm for Content Authentication Based on Arnold Cat Map. In *2019 4th International Conference on Information Technology (InCIT)* (pp. 32–37). IEEE.

Automation and Computation – Vats et al. (Eds)
© 2023 the Author(s), ISBN 978-1-032-36723-1

Emerging IoT applications related to industry 4.0

Raj Kumar
Department of Computer Application, Uttaranchal University, Dehradun, Uttarakhand, India

Dev Baloni & Prateek Joshi
Uttaranchal Institute of Technology, Uttaranchal University, Dehradun Uttarakhand, India

Mukesh Kumar
Department of Computer Science & Engineering, Graphic Era Hill University, Dehradun Uttarakhand, India

ABSTRACT: Today, more companies and organisations than ever before are utilising the advantages of IoT. Machine learning, artificial intelligence, immediate critique, remote checking, and tasks are already in use today and are not going away any time soon. They are also the future. Due to the risky adoption and usage of IoT, businesses who join the IoT insurgency early have some great possibilities. Partnerships that learn how to adapt and immerse themselves in the benefits of IoT may profit clearly. Here are only a few reasons why IOT will eventually have an impact on society rather than remaining a straightforward pattern. The consumer market shouldn't be overlooked, despite the fact that 57% of IOT spending is now made by businesses. Consumer IOT will become the third-largest industry for IoT investment by the end of 2020, with more than 66% of all customers likely to purchase IOT devices for their homes. Within a few years, almost every organisation will be utilising the Internet of Things in some form. Manufacturing, retail, transportation, government, and healthcare are the keys IoT sectors continue to develop state-of-the-art IoT applications and solutions for their daily operations. IoT applications are expected to provide a return on investment of 30% in the future, including cost reserve funds and efficiency improvements, according to 88% of current IoT company users who consider the technology to be "important to their business accomplishment."

1 INTRODUCTION

The IoT technology broadens the range of actuators, sensors, and other linked services that can be handled by people and computers [1]. Due to the Internet's quick development, several connecting techniques have been created. The Internet of Things (IoT) is one technique for transferring current Internet communication to a Machine-to-Machine (M2M) foundation [2]. Offering plug-and-play solutions to end users is the ultimate objective of the Internet of Things [3]. The Internet of Things is revolutionising how we view the world (IoT). A network of interconnected devices is part of IoT [4]. A cloud-centric vision for the mainstream adoption of the Internet of Things is presented in this article. The key auxiliary technologies are discussed [5].

The global internet of things market, according to Fortune Business Insights, was estimated to be worth USD 250.72 billion in 2019 and is expected to reach USD 1,463.19 billion by 2027, growing at a CAGR of 24.9% over the forecast period.

DOI: 10.1201/9781003333500-37

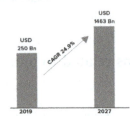

Figure 1.   Market size of global internet of things.

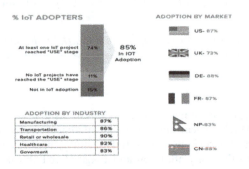

Figure 2.   Worldwide adoption of IoT.

By 2020, it is expected that the market for Internet of Things technology in the US would be worth $45.7 billion. The target market size for China, the second-largest economy in the world, is US$169.1 billion, and growth is anticipated to occur at a CAGR of 29.9% from 2020 to 2027. All spheres of society and all continents are anticipating the mainstream deployment of the Internet of Things. In a research by Microsoft of workplace IoT decision-makers, it was discovered that 85% of them said they were working on one or more IoT projects at any one time, including ones for learning, proof-of-concept, buying, and using. The core sectors of manufacturing, retail/wholesale, transportation, government, and healthcare all have adoption rates that are comparable to those of the other examined countries, as do the US, UK, Germany, France, China, and Japan. By the end of 2021, 94% of enterprises will use the Internet of Things (IoT) in some capacity. 88 percent of IoT business users believe that their company's success depends on the technology.

By 2020, it was predicted that global spending on IoT technology would total $749 billion. Despite rising during the previous year, the global coronavirus pandemic will cause spending to grow at a slower rate in 2020 than anticipated. By 2023, it was estimated that 1.1 trillion US dollars will be spent worldwide on IoT. Globally, the IoT market was dominated by the Asia Pacific region. In terms of significance, Africa was ranked below the Middle East, Europe, and North America [6].

2   AN OUTLINE OF IOT

2.1   *Architecture*

Hardware, networks, and cloud technologies make up the IoT architecture, which enables connections between IoT devices. A basic Internet of Things architecture is composed of these three levels:

- Perception Layer: includes the sensors, gadgets, and other devices
- Layer of the Network: ensures the connectivity between devices
- Layer of Application: the layer the user interacts with

Figure 3.   Layers of IoT architecture.

In order to help IoT devices, these layers gather and process data. Through the incorporation of data transformation into useable information, this architecture expands the OSI model. These insights give businesses the ability to act right now thanks to automation, machine learning, and artificial intelligence.

An IoT cannot operate until the network's objects are interconnected. The IoT's functionality must be guaranteed by the design of the system connecting the digital and physical worlds. The design of IoT architecture considers a number of variables, including networking, communication, procedures, and other features. Consider the devices' usability, scalability, and extensibility while creating an IoT architecture. Since things may change and require in-the-moment interaction, IoT architecture should be flexible enough to enable devices to interact with one another in a variety of ways [7]. Additionally, IoT should be diverse and decentralised.

With a single application, managing IoT gateways and other linked devices is feasible (Innovation Center IoT Application, or ICIoT). The ICIoT continuously assesses all the data at its disposal with the aim of quickly producing the ideal operating room conditions. Due to the ability to easily add additional IoT gateways or devices by simply upgrading the ICIoT application, the architecture is also incredibly adaptable. Future telemedicine scenarios will benefit from SDN because it makes it possible to dynamically allocate network resources for low latency and high data rates, including remote procedures with 4K video transmission [8].

2.2   Industry 4.0 and IoT

Industry 4.0 marks the official start of the Fourth Industrial Revolution. The Internet of Things (IoT), cloud computing, and cyberphysical systems are some of the "Business 4.0" industrial automation technology trend's enabling technologies (CPS). The phrase "industry 4.0" refers, according to GTAI (2014), to the technical transition from embedded systems to cyber-physical systems. For the operation and usefulness of many existing industrial systems inside the complex industrial ecosystems of the future, known as Industry 4.0, IoT is anticipated to offer promising disruptive solutions [9]. IoT is reportedly entering production while simultaneously revolutionising current manufacturing systems, and as a result, it is viewed as a crucial enabler for the next generation of sophisticated manufacturing, according to GTAI (2014).

2.3   IoT's Large scale connectivity

Thanks to the addition of the NB-IoT standard (narrowband IoT) into the 3GPP roadmap in June 2016, we will soon be able to link more than 50,000 devices per base station cell, each linked at a very low data rate [10]. A huge number of sensors will be made possible by the ability to link and send data in unpredictably spaced-out small packets. It was obvious that the NB-IoT standard's architecture had to abide by a pricing corridor that supported "terminals" in the US$/EUR 1 price range given the substantial number of linked sensors [11].

Let's talk about the IOT as it is currently evolving now that you are aware of the IOT future Figures and facts. In 2021, the development of IOT new technologies will be influenced by the development of 5G technology, the most recent developments in AI and blockchain technology, and the ongoing pandemic problem. IOT will undoubtedly improve in intelligence, security, and dependability with time, though. To understand how:

3 THE EMERGING INDUSTRY WISE IOT APPLICATIONS

3.1 *IoT security*

Due to the high degree of connectedness, one of the main issues to far has been security. Only in the first half of 2019, a security company discovered over 100 million assaults on IOT endpoints, underscoring the ongoing risk to unprotected connected devices. As a result, security is an emerging IOT trend, and several businesses worldwide are creating IOT security solutions utilising various technologies. At each layer of the IoT architecture, security is the aspect that researchers are concentrating on the most. When an IoT system is built with security in mind, it will be able to implement it successfully. Therefore, security considerations must be made while developing the application for managing and maintaining IoT networks [12].

3.2 *IoT fuelled with 5G technology*

The basis for realising IOT's full potential and redefining technological advancement is 5G technology, which is more than just a new generation of wireless technology. Without a question, consistent connectivity will lead to better-performing IOT devices, making it one of the most significant developing technologies for the internet of things in 2021. The benefits of 5G include reduced latency, real-time data processing, network slicing, broad coverage and faster transmission speeds. Finding the distinguishing characteristics was a challenge as the definition of 5G developed about 6 years ago. It is obvious that the pursuit of higher data rates would continue, necessitating a 10x increase in data rates every five years [13,14].

3.3 *Blockchain*

The expanding application of block chain technology is one of the most recent IOT innovations.

The use of block chains in IOT devices may ensure data security. IOT applications are a great fit for block chain since they are distributed by nature, allow for efficient communication between different network nodes, and provide safe data storage.

3.4 *AI software with IoT capabilities*

Artificial intelligence and the internet of things are two very dissimilar technologies that can work together to provide useful business solutions. In order to produce good and efficient results, AI systems currently require a relatively little amount of data. IOT development services for businesses using these two technologies may aid in automating a number of operations, lowering downtime, lowering operating costs, boosting productivity, and enabling predictive maintenance.

3.5 *Digital twins*

The technical concept of the "digital twin" gained widespread in 2020. Digital twins are virtual replicas that serve as real-time digital substitutes for physical objects or processes. A digital twin has several uses, including controlling asset performance and usage, monitoring, diagnosing, and optimising asset performance. It is predicted that the market for

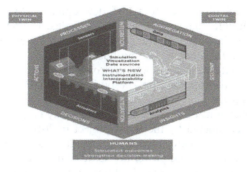

Figure 4.   Digital twin model.

digital twins would expand at a compound annual growth rate of 8.3%, from US$3.8 billion in 2019 to US$35.8 billion in 2025.

3.6   *Voice activated IoT devices*

With the help of Google Assistant, Siri, and Amazon Echo, voice-based user interfaces have improved. We will soon be able to change settings, give orders, and receive replies from our smart devices across a range of industries by using voice interactions. Banks and Fitch was one of the first businesses to employ voice recognition and speech recognition technology. Speech biometry is a significant advancement in voice recognition technology. Organizations can digitise a person's speech by analysing a range of distinguishing traits including tone, pitch, intensity, dynamics, dominating frequencies, etc. This procedure will undoubtedly be safer than the ones now in use, according to businesses.

3.7   *Smart cities*

Smart cities cannot be ignored while analysing the evolution of IOT technology. Several governmental organisations have started IOT technology programmes during the past five years that will affect whole cities. The government would be able to adopt a variety of intelligent solutions for a number of challenges, such as traffic congestion, citizen safety, energy consumption, sustainable growth, etc. with the help of enormous volumes of data.

For instance, Singapore uses linked sensor data to enhance transportation, urban planning and public safety. Singapore then analyses this data and disseminates it.

3.8   *Edge computing*

Edge computing is predicted to increase dramatically by the end of 2021 as a result of remote work and COVID-19. This innovation leads to the development of new business models. Forrester asserts that the development of edge solutions with cloud-like features by big retailers like HPE, Dell, and IBM may enable end-of-life marketers to keep and hold a significant portion of the public market of cloud.Edge computing helps companies to service customers in new areas while maintaining flexibility and control over an increasing workforce.

3.9   *Traffic management*

You can't bear the notion of leaving for work the first thing in the morning. Traffic and gridlock are big annoyances that force people to spend more time and energy on the road during the day in any city in the world. Since lengthy commutes have a detrimental effect on economic activity, traffic congestion is a serious issue in urban places all over the world. A vehicle monitoring system has been created in the present to collect data on vehicle movement in real-time and check its performance [15]. IOT technology is suggested to be suited to handle this difficulty by current IOT developments. Many businesses today are offering plans and solutions that aim to use IOT-installed technology in cars and traffic systems to develop more intelligent traffic networks, which are expected to reduce needless traffic and congestion. IOT is being used by business owners, entrepreneurs, and governments to boost productivity, save costs, and improve user experience. The main sectors are significantly impacted by the top IOT apps, despite the fact that many of them are still in the early phases of development.

The situational applications of the internet of things may be classified into 14 categories, according to a survey conducted by the IoT-I project in 2010 [16]. Transportation, smart homes and cities, fashion, healthcare, agriculture, emergency situations, culture and tourism, supply chains, and environment and power are some of these areas. The results of the poll, which were based on 270 responses from 31 different nations, indicated that the most fascinating applications were those for smart homes, smart cities, transportation, and healthcare. The applications of the IoT in the areas of transportation, healthcare, smart homes or cities, and social and personal connections will be briefly covered in this talk.

3.10 *Healthcare*

The present turn of events has resulted in a huge explosion of IOT innovation in the healthcare sector. The Covid-19 issue has significantly raised the demand for IOT, enabling real-time patient health monitoring and alerting via telemedicine, digital diagnostics, remote monitoring, etc. IOT has enhanced patient participation and satisfaction with doctors by enabling easier, more efficient interactions and processes. Humanity will greatly benefit from the emerging field of leveraging IoT for diagnosis and control of infectious diseases. In order to identify emerging public health risks, for instance, "disease monitoring programmes like HealthMap and EpiCaster at Virginia Tech Network Dynamics and Simulation Laboratory are combining demographic data with IoT data,land-use information, GIS data, social media streams, and many other sources." [17,18].

4 THE FUTURE OF IOT

The Internet of Things (IoT) can fundamentally change how people connect on the Internet today and present a number of beneficial possibilities for research and development in the real world [19]. Future technology will be driven by IoT innovation, and many inventive and imaginative items will be created. Remote monitoring and intervention are made possible by IoT. It involves being aware of events occurring where one is not. Bill Gates' book The Road Ahead [20] has several concepts for applications. Simple applications include things like receiving bus arrival notices. A higher perceived quality results from this. This concept is comparable to that of ambient intelligence [21]. The key distinction with IoT is that we measure things not just locally but also remotely. In the past, we referred to it as remote telemetry [22].

- The IoT market will be valued $1111.3 billion by 2026.
  IOT is expanding exponentially. Well-known businesses like Facebook, Google, Microsoft, Apple, Dell and Cisco, are among those making big investments in IOT applications, according to Cision and PR Newswire. The IOT industry is expanding, with a startling compound annual growth rate of 24.7%. These organisations' growing reliance on cutting-edge technology is fueling the global IOT business. Due to consumer demand for technologies like artificial intelligence and cloud computing, which have numerous uses for manufacturing, security, retail, government, healthcare, and financial institutions, to mention a few significant areas, the IOT business is still expanding year after year.
- In 2019, there were 26.66 billion IOT devices in use.
  Every year, more people use smart devices to access the Internet of Things; in 2019, there will be approximately 26 billion active gadgets. The average number of connected devices

per person in 2020 is predicted to be 6.58. Disruptive Asia claims that although businesses currently make up 57% of all IOT investments, the consumer market should not be ignored. By the end of 2020, it's projected that more than two-thirds of consumers will have purchased IOT devices for their homes, with consumer IOT ranking third in terms of spending on Iot.

- By 2020, the market for "Smart Cities" will be worth more than $400 billion. "Smart Cities," or urban areas where IOT sensors are used to gather data and offer insights for improved management, have already developed into a significant sector. The majority of IOT projects in 2018 were "smart city" efforts due to the use of IOT-driven applications in cities including London, Rio de Janeiro, San Francisco, and Copenhagen. Smart metres are planned to be used in these cities indefinitely, with an increase in investment of more than 1.1 billion predicted by 2022.

- By the end of 2021, 94% of enterprises will be utilising IOT.
  According to a recent prediction from Microsoft, almost all businesses will be utilising IOT in some way by the end of the year. The primary IOT industries—manufacturing, retail, transportation, government, and healthcare—continue to incorporate new IOT applications and solutions into their routine business operations. Eighty eight percent of IOT company adopters say IOT is "essential to their business performance" and forecast a 30% ROI from IOT applications in two years, including cost savings and efficiency.

- IOT will cost $1.1 trillion in 2023. The expected value of global IOT spending is above 1.1 trillion dollars and is increasing, according to Statista. The global economy is predicted to save over 5.6 trillion dollars by 2050 thanks to autonomous vehicles, while food prices are predicted to drop by nearly half thanks to IOT in agriculture. Despite the huge amount, IOT has also proven its capacity to save us money. The IOT will help reduce traffic accidents and the amount of time that cars are stalled in traffic or hunting for parking by being included into 75% of new vehicles by the end of 2020.

4.1 *Why Iot?*

Businesses are growing more and more concerned about security. In fact, the majority of European IoT users who were asked why they were embracing the technology said it was to increase security [6]. In fact, worldwide spending on IT security reached over 120 billion dollars in 2019. Other common reasons for adopting IoT included cost savings and increased productivity.

5 CONCLUSION

Global adoption of IOT solutions and IOT future trends will be significant in the near future. Blockchain, artificial intelligence, 5G, cloud computing, and other significant technologies will be crucial to the growth of the internet of things and global connectivity. We came to the conclusion that because IOT is affecting every aspect of our lives, its value rests not just in data collection but also in developing business cases that make use of the knowledge gained through data analysis to produce value.

REFERENCES

[1] Jyoti Neeli, Shamshekhar Patil "Insight to Security Paradigm, Research Trend & Statistics in Internet of Things(IOT), *Global Transitions Proceedings,*" Volume 2, Issue 1, 2021, pp. 84–90

[2] Kraijak S. and Tuwanut P., "A Survey on IOT Architectures, Protocols, Applications, Security, Privacy, Real-world Implementation and Future Trends," *11th International Conference on Wireless Communications, Networking and Mobile Computing (WiCOM2015),* 2015, pp. 1–6

[3]   Shafique K., Khawaja B. A., Sabir F., Qazi S. and Mustaqim M. "Internet of Things (IOT) for Next-Generation Smart Systems: A Review of Current Challenges, Future Trends and Prospects for Emerging 5G-IOT Scenarios," In *IEEE Access*, vol. 8, pp. 23022–23040, 2020

[4]   Bang Nguyen & Lyndon Simkin (2017) The Internet of Things (IOT) and Marketing: the State of Play, Future Trends and the Implications for Marketing, Journal of Marketing Management, 33:1–2, pp.1–6

[5]   Jayavardhana Gubbi, Rajkumar Buyya, Slaven Marusic, Marimuthu Palaniswami, Internet of Things (IOT): *A Vision, Architectural Elements, and Future Directions,Future Generation Computer Systems*, Volume 29, Issue 7, 2013, pp. 1645–1660https://appinventiv.com/blog/emerging-IOT-technologies-in-coming-years/

[6]   Internet of Things Spending Worldwide 2023 | *Statista*

[7]   Gokhale, P., Bhat, O., & Bhat, S. (2018). Introduction to IOT. International Advanced Research Journal in Science, Engineering and Technology, 5(1), 41–44.

[8]   Miladinovic I. and Schefer-Wenzl S., "NFV Enabled IoT Architecture for an Operating Room Environment," 2018 *IEEE 4th World Forum on Internet of Things* (WF-IoT), 2018, pp. 98–102, doi:10.1109/WF-IoT.2018.8355128.

[9]   Li Da Xu, Eric L. Xu & Ling Li (2018) Industry 4.0: State of the Art and Future Trends, International Journal of Production Research, 56:8, 2941–2962, DOI: 10.1080/00207543.2018.1444806

[10]   GSMA: Mobile IoT: *Low Power Wide Area Connectivity – Industry Paper, White Paper by GSMA*, pp.1–20.

[11]   Fettweis, G. P. (2016, September). 5G and the Future of IoT. In ESSCIRC Conference 2016: 42nd European Solid-State Circuits Conference (pp. 21–24). *IEEE*.

[12]   Yugha, R., & Chithra, S. (2020). A Survey on Technologies and Security Protocols: Reference for Future Generation IoT. Journal of Network and Computer Applications, 169, 102763.

[13]   Fettweis G., *"A 5G Wireless Communications Vision," in Micro wave Journal*, December 2012.

[14]   Fettweis G. and Alamouti S., "5G: Personal Mobile Internet beyond What Cellular Did to Telephony," IEEE Communication s Magazine, Vol. 52, February 2014, pp. 140–145.

[15]   Wang, Shulong, Yibin Hou, Fang Gao, and Xinrong Ji. "A Novel IoT Access Architecture for Vehicle Monitoring System." In Internet of Things (WF-IoT), 2016 IEEE 3rd World Forum on, pp. 639–642. *IEEE*, 2016.

[16]   Vermesan, O., Friess, P. and Furness, A. (2012) The Internet of Things 2012. *By New Horizons.*

[17]   Shah, H. (2017). *How IoT Can Help Detect and Control Infectious Disease Outbreaks in Real-time.* https://www.idigitalhealth.com/news/how-internet-of-things-helps-detectand-control-infectious-disease-outbreaks-in-realtime.

[18]   Lin, Q., Zhao, Q. (2021). IoT Applications in Healthcare. In: García Márquez, F.P., Lev, B. (eds) Internet of Things. *International Series in Operations Research & Management Science*, vol 305. Springer, Cham.https://doi.org/10.1007/978-3-030-70478-0_7

[19]   Kraijak S. and Tuwanut P., "A Survey on IoT Architectures, Protocols, Applications, Security, Privacy, Real-world Implementation and Future Trends," *11th International Conference on Wireless Communications, Networking and Mobile Computing (WiCOM 2015)*, 2015, pp. 1–6, doi:10.1049/cp.2015.0714; Lin, Q., & Zhao, Q. (2021). *IoT Applications in Healthcare. In Internet of Things* (pp. 115–133). Springer, Cham.

[20]   Gates B. et al., The Road Ahead, 1995.

[21]   Ramos C., Augusto J. C., and Shapiro D., *"Ambient Intelligence — the Next Step for Artificial Intelligence,"* IEEE Intelligent Systems, 23.2, 2008, pp. 15–18.

[22]   Ploennigs J., Cohn J. and Stanford-Clark A., "The Future of IoT," In *IEEE Internet of Things Magazine*, vol. 1, no. 1, pp. 28–33, SEPTEMBER 2018, doi:10.1109/IOTM.2018.1700021.

Automation and Computation – Vats et al. (Eds)
© 2023 the Author(s), ISBN 978-1-032-36723-1

Introduction of proprietary software to students for an effective online learning

Saksham Mittal, Amit Kumar Mishra, Anmol Kundlia & Deepak Upadhyay
Department of CSE, Graphic Era Hill University, Dehradun, India

Neha Tripathi
Department of CSE, Graphic Era Deemed to be University, Dehradun, India

ABSTRACT: In these pandemic times due to COVID-19, there has been a huge shift experienced in the education sector in the last two and half years. The offline mode of learning has been shifted to online mode of learning. Physical classrooms have been converted to virtual classrooms. Therefore, our paper targets to explain the benefits of the online mode of learning and what tools, techniques, and software we can use to provide a smooth and great learning experience to students. In this paper, first we have discussed about the traditional (offline) mode of learning, followed by the discussion about shift to online mode of learning due to COVID-19 pandemic. After that, we have discussed about tools and technologies required for effective online learning. At last, we have discussed about the benefits of online learning.

Keywords: Online Mode, Offline Mode, MS Teams, Zoom, Moodle, Hackerearth, Hackerrank

1 INTRODUCTION

Education plays a very important role in one's life to have a better career and overall personality development [1]. Education is a key to success. To gain knowledge and education, we all go to schools and universities. The trend of going to schools and universities and attending the classes and lectures there (in the offline mode) has been followed for many centuries.

But, for the last two and half years, due to COVID-19 pandemic, this trend has been affected very badly. It is experienced that there is a very huge shift from offline mode of learning to online mode of learning. We would have never thought of such a sudden and big shift. The pandemic has not only impacted the education or learning, but also impacted the life of working professionals. At many sectors, work from office has been shifted to work from home.

As change is the nature of the life, so we need to adapt to every change and situation. Shifting from offline to online mode of learning is a challenging task, but we must thank the IT Sector which has made it possible by providing various software tools, and technologies to us.

One of the major challenges in this shift is how to gather all the students and people at one place in the online mode of learning [2]. But thanks to applications like Microsoft Teams, ZOOM, etc., which has made it possible.

Another challenge is how to provide lecture material such as e-books, assignments, notes, etc. to the students for smooth learning. We are fortunate that we have web applications like MOODLE, etc. which has helped in achieving the above task.

Another challenge is how to educate students, faculty members and people about such tools and technologies and how to make an effective use of it. As in this digital and smart era, knowledge about such tools, technologies and proprietary software is a necessity. So, we can prepare the content related to usage of such applications and software and make it

DOI: 10.1201/9781003333500-38

available to people at some common platform where everyone can access. Another thing we can do is educate the students about it during their Induction Program in the university (program held in colleges or universities to welcome new students) because it will be useful for them in the rest of their life. In this paper, later we have discussed about such technologies and software in detail.

Below we have discussed about traditional or offline mode of learning followed by shift from offline to online mode of learning. After that we have discussed about benefits of online mode of study or learning. At last, discussed about various challenges in the proposed model and can be overcome using combination of various technologies such as Cloud Computing [11], Machine Learning [12] and Artificial Intelligence [13], IoT [14], etc. which can help in establishing strong and smart education infrastructure system.

2 TRADITIONAL OR OFFLINE MODE OF LEARNING & TEACHING

It has been a tradition for centuries for people to attend schools and universities and listen to the lectures there. This mode of learning is known as an offline mode of learning in the present times.

In the offline mode, teachers or faculty members provide lectures to students in the physical classrooms. They use traditional tools and methods of teaching like whiteboard/ blackboard, duster, marker, chalks, projectors in smart classrooms, etc. Students are required to attend their lectures or sessions on time since there are no pre-recorded videos or notes that are easily available to the students. Hence, students are required to follow a predetermined and strict schedule as set by their educational institute [3].

The communication and collaboration between students and teachers happen face-to-face. Students have regular live interaction with the teachers. Moreover, offline education also allows teachers to monitor the responses and behavior of their students and accordingly address them as and when required.

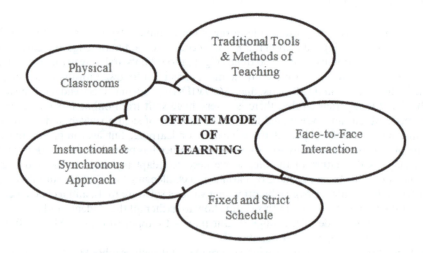

Figure 1.   Characteristics of offline mode of learning [3].

3 SHIFT TO ONLINE MODE OF LEARNING

An unprecedented situation overcame the regular lives when COVID-19 pandemic hit the world. The most affected infrastructure were the schools and colleges. Due to the large

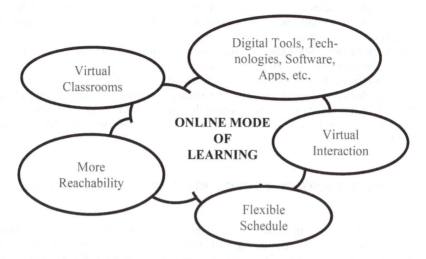

Figure 2.   Characteristics of online mode of learning [3].

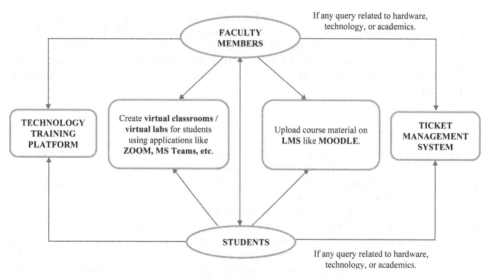

Figure 3.   Proposed model.

number of students sitting close together in class and fear of the virus spreading among them, it was decided to shut down schools, colleges and universities and use technology to propagate learning to the young masses. This shift required a lot of changes in teaching methodologies and technologies (teaching aids) used. Teachers had to educate themselves to use better technology and software that could effectively convey their classes. Students had to adjust themselves to the changing scenario wherein they could attend classes from the comfort of their homes.

This resulted in an increase in the online presence of every person. Quick advancements were made in the software industry to accommodate the increasing use of technologies and a very high increase in the number of users.

Attempts were made to shift learning resources to online. The most important requirement that has risen due to this scenario is to educate the students and teachers to use these technologies to effectively learn and impart knowledge. Our paper targets this standpoint and aims to implement a system of such proprietary software to make a comprehensive system that can accommodate all these demands using a framework that can provide a common portal for both learning the subject and learning how to use technology effectively [4].

4   TOOLS AND TECHNOLOGIES REQUIRED FOR EFFECTIVE ONLINE LEARNING

There are many tools and technologies available in the market that can facilitate the exchange between the students and the Schools/Universities. This system is most beneficial to the students of Software Engineering. Below are some of the software that form the foundation of this system:

4.1   *Microsoft-teams for virtual classrooms*

Use of collaboration software like MS-Teams has proven to be an effective way to interact with students and hold online live interaction lectures. MS-Teams can also be used to hold quizzes and give assignments to the students. An array of different apps that integrate with Teams helps to give an organization-like experience to the students which will be highly beneficial to them in the future when they themselves join corporate environment [6].

4.2   *Moodle (LMS)*

"Moodle is a free and open-source learning management system (LMS) written in PHP and distributed under the GNU General Public License. Developed on pedagogical principles, Moodle is used for blended learning, distance education, flipped classroom and other e-learning projects in schools, universities, workplaces, and other sectors." [5]

The most important proprietary software needed in the framework is a Learning Management System (LMS). It can be used for providing course content on a common platform, conducting live lectures, monitoring students' attendance, and conducting online proctored examinations. The preferred platform used in almost all the institutions is Moodle. It is an open-source software, thus does not require much in cost while setting up. It is ideal for scenarios where small institutions which are tight on budget want to set up their own infrastructure on online learning.

4.3   *Ticketing system*

As there are many users in this scenario, there are also a lot of queries and technical help required to help when a user faces technical difficulties. A ticketing system, as used in the corporate scenario, is the most appropriate tool that can be applied in this situation. Users can raise tickets which will then be assigned to the required person and then view the ticket status until it gets resolved. We target to develop a customized ticketing system which can cater to the specific needs of a particular institution.

4.4 *Technology training platform*

It is required to provide training and hands-on experience on various tools and technologies to the students and faculty members for the smooth delivery of lectures. To do so, the content related to usage of such applications and software is made available to people at some common platform where everyone can access. Moreover, we can invite the IT experts from the industry for virtual guests' lectures on usage of such tools and technologies.

4.5 *Virtual labs*

Platforms like Hackerearth [7], Hackerrank [8], etc. can be used to conduct the coding labs virtually.

5 BENEFITS OF ONLINE LEARNING

Following are the benefits of online mode of learning:

- *Flexible*
  Students can access the lectures and the course content from anywhere or anytime. They can attend their classes virtually from their places. There is no such restriction of the location [9].
- *Time management and more scope of learning*
  Besides academic curriculum, students can manage and utilize their time in pursuing their hobbies. As various courses are available online, they can invest their time to learn new things which will help them in their career and personality development.
- *More reachability*
  As everything is available online, people who live in remote areas (where there is no school or university) can get an opportunity to get an education from experienced professional educators. Classroom learning with experienced educators can now be extended to remote areas through virtual or online mediums.
- *Common place for learning*
  All the courses based on academic curriculum are made available at one place, so learning becomes easier and more effective [10].
- *Gaining knowledge about new Tools, Technologies, Software, etc. which are in great demand in the Industry or Corporate Sector.*
  Before moving into the industries after the placements, the students get knowledge and hands-on experience of such tools and technologies which are in great demand in the industries. Moreover, in this digital era, non-technical people also get experience of using such new technologies.

6 CONCLUSION

Online mode of learning seems to have a great future in the upcoming years. But every new thing comes with some challenges. Though we have discussed about many applications and benefits of online studies but there are some challenges too in implementing this model successfully. Some of the major challenges are security (as everything is available online over internet), smooth conduction of such labs which require hardware for hands-on or practice, to educate everyone about new tools and technologies, etc. These can be overcome using combination of various technologies such as Cloud Computing, Machine Learning and Artificial Intelligence, IoT, etc. which can help in establishing strong and smart education infrastructure system.

REFERENCES

[1]   Al-Shuaibi, Abdulghani. (2014). The Importance of Education., *ResearchGate*

[2]   Online versus Offline Mode of Education Is India Ready to Meet the Challenges of Online Education in Lockdown? *ResearchGate*, Naman Wadhwa1 (Student, B.Tech Biotechnology), Sunita Khatak (Assistant Professor), Department of Biotechnology, University Institute of Engineering & Technology, Kurukshetra University, Kurukshetra-136119, Haryana, India Poonam2 (Assistant Professor), Department of Applied Psychology, Guru Jambheshwar University of Science & Technology, Hisar-125001, Haryana, India

[3]   https://leverageedu.com/blog/online-classes-vs-offline-classes/

[4]   Singh, S., Rylander, D. H., & Mims, T. C. (2012). Efficiency of online vs. offline learning: A comparison of inputs and outcomes. International Journal of Business, Humanities and Technology, 2(1), 93–98.b

[5]   Athaya, Hisyam, et al. "Moodle Implementation for E-Learning: A Systematic Review." *6th International Conference on Sustainable Information Engineering and Technology* 2021. 2021.

[6]   *A Critical Study on the Efficiency of Microsoft Teams in Online Education*, TY – CHAP, AU – A.K., Saranya, PY – 2020/11/07, SP – 310,EP – 323, SN – 978-93-89515-28-2

[7]   Ghazouli, Kiana. *"The Impact of Time Constraints on HackerRank Assessments."* (2018).

[8]   Sreeram, N., V. Uday Kumar, and L. Sridhara Rao. "A Survey Paper on Modern Online Cloud-Based Programming Platforms." International Journal of Engineering & Technology 7.2.32 (2018): 352–354.

[9]   Anggrawan A., Yassi A. H., Satria C., Arafah B. and Makka H. M., "Comparison of Online Learning Versus Face to Face Learning in English Grammar Learning," 2019 *5th International Conference on Computing Engineering and Design (ICCED)*, 2019, pp. 1–4, doi:10.1109/ICCED46541.2019.9161121.

[10]   Robinson P. E. and Carroll J., "An Online Learning Platform for Teaching, Learning, and Assessment of Programming," 2017 *IEEE Global Engineering Education Conference (EDUCON)*, 2017, pp. 547–556, doi:10.1109/EDUCON.2017.7942900.

[11]   Mittal, Saksham, et al. "Integration of Big Data and Cloud Computing for Future Internet of Things." Communication and Computing Systems. CRC Press, 2019. 9–17.

[12]   Pundir, S., Obaidat, M. S., Wazid, M., Das, A. K., Singh, D. P., & Rodrigues, J. J. (2021). MADP-IIME: Malware Attack Detection Protocol in IoT-Enabled Industrial Multimedia Environment using Machine Learning Approach. Multimedia Systems, 1–13.

[13]   Tiwari, P., Upadhyay, D., Pant, B., & Mohd, N. (2022). Multiclass Classification of Disease Using CNN and SVM of Medical Imaging. In International Conference on Advances in Computing and Data Sciences (pp. 88–99). Springer, Cham.

[14]   Singh, Prabhdeep, Devesh Pratap Singh, and Dibyahash Bordoloi. "Internet Of Things And Artificial Intelligence For Sustainable Development: New Opportunities And Risks In Technology And Society." Ilkogretim Online 20.1 (2021): 7284–7297.

Automation and Computation – Vats et al. (Eds)
© 2023 the Author(s), ISBN 978-1-032-36723-1

Audio emotion recognition using machine learning

Aparna Saini, Sujata Kukreti, Aditya Verma & Ankit Tomar
Auriss Technologies, Graphic Era Hill University, Graphic Era Deemed to be University, Dehradun, Uttarakhand, India

ABSTRACT: Over the last decade, the study of emotion recognition has attracted a lot of attention in the area of human-computer interaction. Current recognition accuracy can be improved, but more research into the fundamental temporal link between speech waveforms is needed. A method for speech recognition is proposed that takes advantage of difference in emotional saturation between time frames (RNN) using combination of speech features with attention-based Long Short-term Memory (LSTM) recurrent neural networks. In place of standard statistical features, frame level speech features were derived from the waveform to retain the original speech's temporal relations through sequence of frames. Two LSTM enhancement algorithms based on the attention mechanism have been presented to distinguish emotional saturation in distinct frames. An Emotion Recognition in Conversion system that is capable of recognizing face emotion in real-time was one of the systems that was proposed.

1 INTRODUCTION

When it comes to human-computer interaction, emotion recognition has been a major topic. The process of identifying human facial expressions, as well as verbal and nonverbal cues like tone of voice, gestures, etc., is known as emotion recognition [25]. The current recognition accuracy can be increased, but more research into the basic temporal relationship between speech waveforms is needed. A novel approach to speech recognition, Frame-level speech parameters in combination with attention-based Long Short-Term Memory (LSTM) recurrent neural networks can be used. This combination is used to take advantage of the difference in emotional saturation that occurs between different time frames (RNN). The waveform was used instead of typical statistical features to construct frame level speech features that preserve the original speech's temporal relationships across frames. It has been shown that two attention-based LSTM augmentation algorithms can discriminate emotional saturation in different frames.

1. This is accomplished via the algorithm's modification of the conventional LSTM forgetting gate calculation approach, which, in turn, ensures that recognition accuracy is maintained despite the reduction in computational time complexity.
2. For final output of LSTM, instead of employing the most recent iteration of the standard method, time and feature dimensions are applied to get superior outcomes.

2 EMOTION RECOGNITION

The act of recognizing human feelings and emotions is referred to as emotion recognition. Research in this area is still in its infancy, it focuses on the use of technology to assist people

DOI: 10.1201/9781003333500-39

with emotion recognition. When applied to a given setting, the technology functions most effectively when it combines numerous modes of operation. Until now, the majority of R&D efforts have been focused on developing automated methods for recognizing facial expression from video, spoken gestures from audio, written expressions from text, and physiology as observed by wearables.

Emotion recognition in conversation, or ERC, has number of applications in a variety of fields, including recommendation systems, human-machine interactions, and medical care. To the exclusion of all else, conversation is a dynamic process of contact in which people routinely communicate emotional messages by relying on context and knowledge that they intuitively possess. Figure 1 depicts the steps involved in running an ERC system. Voice activity detection, audio feature extraction, and sound classification are all part of the process. Finally, people are acknowledging their emotions.

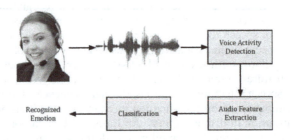

Figure 1.   Overview of an ERC system [20].

3   LITERATURE REVIEW

In the last few years, many efforts have been made in developing an ERC system, but less work is done. A literature survey is done on the systems that were developed for emotion recognition in conversation.

Caroline Etienne et.al. (2018) [8] designed a neural network which helps in recognizing emotions in speech with the use of IEMOCAP dataset. They used the architecture of neural network that involves two convolutional layers which helps in spectrograms' high-level and recurrent feature extraction, for accumulating long term dependencies. They scrutinized the technique of data augmentation and batch normalization of recurrent layers. 64.50% weighted accuracy and 61.70% un-weighted accuracy was achieved.

Yue Xie *et al.* [16] proposed a speech recognition approach that combines frame-level speech features with attention-based Long Short-Term Memory (LSTM) RNNs. Based on the attention technique, the author also proposed two LSTM improvement strategies. Initially, the algorithm reduces time complexity by removing the gate calculation method from traditional LSTM. To improve the results of the LSTM, an attention technique is applied to both feature and time aspects.

Jingwen Hu *et al.*(2021) [3] proposed a multi-model fused graph Convolutional network-based model (MMGCN). The MMGCN stores speaker information to model inter- and intra-speaker reliance. The author used two standard datasets IEMOCAP and MELD to evaluate the proposed model. The MMGCN outperforms other SOTA methods.

Devamanyu Hazarika *et al.* (2019) [17] proposed a method TL-ERC. They trained a neural dialogue generation model before transferring parameters to begin their target emotion classifier. They also assimilate parameter transfer from the reiterative module that model inter-sentence conditions in the whole conversation. They perform many experiments with multiple datasets based on this idea. TL-ERC attains better validation performances in fewer epochs.

Zheng Lian *et al.* (2021) introduced a multimodal learning substructure for identifying conversational emotion [18]. The main objective of this framework is to model intra-modal and cross-modal interaction among multi-modal features using a transformer-based structure. They assessed the proposed method's performance using two datasets, IEMOCAP and MELD. On a weighted average F1, their method outperforms existing strategies by 62%.

Zixuan Peng *et al.* (2021) [1] proposed a system using multi-scale convolution neural network (MSCNN) to acquire audio as well as text hidden patterns. They used a statistical pooling unit (SPU) for feature extraction of each module. To improve performance, an attention module can be added on top of the MSCNN-SPU (audio) and MSCNN (text). They performed different tests on the proposed model. The proposed model outperforms on IEMOCAP dataset with 4 emotion classes which are angry, happy, sad, and neural. The proposed framework outperforms in both weighted accuracy and un-weighted accuracy with an enhancement of 5.00 % and 5.20% respectively.

Jingwen Hu *et al.* (2021) MMGCN's efficiancy, which beats other state-of-the-art under the multimodal discussion, state-of-the-art methods outperformed state-of the-art methods by a wide margin setting. They planned a new mode MMGCN (multimodal fused graph convolutional network) which is used to examine both multimodal and long-distance contextual data.

Soujanya Poria *et al.* (2019) continuous emotion detection with fine-grained speaker-specificity could be useful for tracking emotions during long monologues. Chitchat conversation can be significantly enhanced, as well as some aspects of task-oriented dialogue, by an efficient emotion-shift recognition model and context encoder.

4 AUDIO FEATURE EXTRACTION

The process of extracting features from an audio signal and identifying those components of the signal that are good for identifying linguistic content is an important step in any automatic speech recognition system. This step involves removing all the other information that the signal carries, such as background noise, emotion etc.

Mel Frequency Cepstral Coefficients (MFCCS) is a characteristic that is utilized extensively in computerized speech analysis and speaker recognition. It was invented by Davis and Mermelstein in the 1980s, and they have remained the most cutting-edge technology available ever since. It was discovered that extracting features from the audio signal and using those features as inputs to the model significantly improved the performance of the base model over directly considering raw audio signal as inputs.

MFCC algorithm- In the process of computing the MFCC, the windowing of the speech signal comes first, followed by the framing of the speech signal. Because the amplitude of the high frequency formants changes less than that of the low frequency formants, the high frequencies are accentuated to achieve an amplitude that is comparable across all formants. The Fast Fourier Transform (FFT) is used to calculate the power spectrum of each frame after the windowing procedure. Following that, filter bank is processed using the mel-scale on the power spectrum. To calculate the MFCC coefficients, first we need to convert power spectrum to log domain, and then apply DCT to the speech signal.

$$mel(f) = 2595 * \log 10 \left(1 + \frac{f}{700}\right) \tag{1}$$

where mel(f) is the frequency (mels) and f is the frequency (Hz).

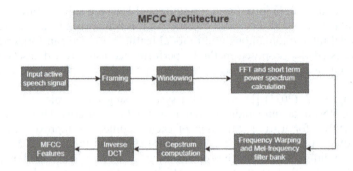

Figure 2. MFCC algorithm architecture.

The equation mentioned below is used to calculate MFCCs in which mel cepstrum coefficients is determined by k and output of filter bank by $\hat{S}k$ and $\hat{C}n$ is the final mfcc coefficients

$$\hat{C}n = \sum kn = 1(\log \hat{S}k)\cos\left[n\left(k - 1/2\right)\pi/k\right] \tag{2}$$

5 CLASSIFICATION

The phase of emotion classification occurs after the phase of feature extraction. LSTM networks are utilized for the purpose of emotion classification in the system that has been developed. A biological neural network serves as the model for an artificial neural network, which is a multi-tiered structure made up of connected neurons. It is not a single algorithm but rather a mix of algorithms that gives us the ability to perform sophisticated actions on data.

The phrase "long short-term memory networks" refers to a technique that is utilized in the field of "deep learning". There are numerous types of recurrent neural networks (RNNs), and especially shine when it comes to solving problems involving sequence prediction. LSTM is equipped with feedback connections, which means that it can process the full sequence of data, except for single data points like images. This has several applications including speech recognition and automatic translation, among others. The recurrent neural network (RNN) known as the LSTM is a subtype that excels at solving a wide range of challenging issues. An important part of an LSTM model is a memory cell that is referred to as a "cell state," and it is responsible for keeping its state consistent throughout the course of time.

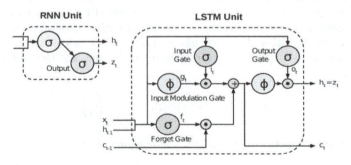

Figure 3. Structure of RNN and LSTM [22].

6 PROPOSED CNN ARCHITECTURE

There is no one-size-fits-all answer when it comes to recognizing emotions. Soon, there will be a slew of new solutions that will be more efficient, accurate, and user-friendly. Rather than unimodality dominating the field of emotion recognition, previous and current studies demonstrate that multimodality is now the norm. According to the most recent studies, the best results are obtained when EEG and audio-visual signals are combined. We believe that the best technique for dealing with multimodality is LSTM-RNN. When it comes to EEG and audio-visual signals, LSTM-RNN can be used to recognize emotions. We must, however, work to develop the model so that it can be trained using both EEG and audiovisual data at the same time, and so that the model can still generate the desired result even if one form of data is not accessible. So, the training will be divided into two parts: training for the data and training for understanding the data's connections. A five-layer architecture and an optimizer, such as Adam, were utilized in the development of the suggested sign language recognition system which includes two LSTM layers, two Dense layers, and one Dropout layer, respectively. The convolution neural networks proposed here are trained using the dataset that was described earlier. Figure 4 depicts the architecture of the suggested CNN model's internal workings.

```
Model: "sequential"
_____
Layer (type)                 Output Shape              Param #
=================================================================
lstm (LSTM)                  (None, None, 128)         72704
_____
lstm_1 (LSTM)                (None, 64)                49408
_____
dense (Dense)                (None, 64)                4160
_____
dropout (Dropout)            (None, 64)                0
_____
dense_1 (Dense)              (None, 6)                 390
=================================================================
Total params: 126,662
Trainable params: 126,662
Non-trainable params: 0
_____
```

Figure 4. Proposed CNN architecture.

7 RESULTS

When we begin the process of putting our model into action, the first thing we do is comparing all the feelings that are present in our dataset. The feelings included repugnance, happiness, sadness, neutrality, fear, and anger. The count of emotions can be observed in the graph that is presented below in Figure 5.

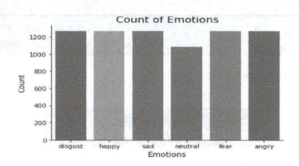

Figure 5. A bar graph representing count of emotions present in the dataset.

After showing a bar graph representing the total number of emotions, we then plot the wave plot and spectrogram of specific emotion, such as anger.

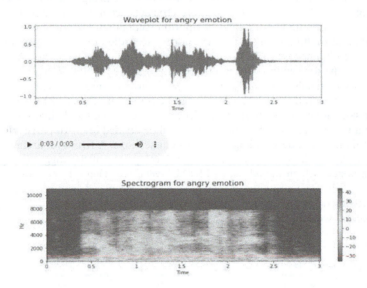

Figure 6. The wave plot and spectrogram of angry emotion.

The accuracy is dependent on a wide variety of parameters and varies as a direct result of these elements at each stage of the process. The training set is used to train the model for a total of 30 iterations and 168 steps. Following the conclusion of each period, both the training and the testing accuracy are assessed. This system attained an accuracy in training of 65.88 percent (Figure 7) and an accuracy in testing of 60.53 percent (Figure 8).

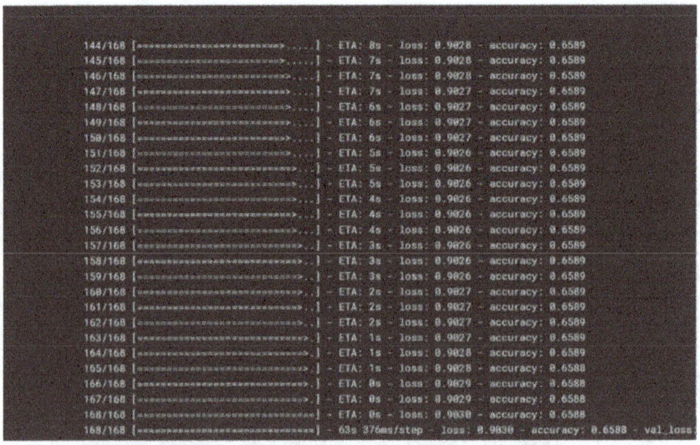

Figure 7. Training of proposed model.

Figure 8. Testing accuracy of a model.

8 CONCLUSION

We used the CREMA-D dataset to train a neural network that can identify emotional states in speech. We employed a combined CNN-LSTM architecture, which exploits the ability of convolutional layers to extract high-level representations from raw inputs, in accordance with current developments in speech analysis. As it turns out, the parameters of convolutional and LSTM layers are trained at different rates. By changing update rule settings per layer, we tried to take advantage of this discovery but were not able to draw any conclusions. This is an interesting direction to pursue, and more extensive testing may yield a better outcome. To achieve the desired result, the process was followed step-by-step. Keras, TensorFlow, and retraining principles were utilized to follow the framework and identify emotion expression patterns more effectively, as well as the deep learning RNN algorithm to identify emotion expression patterns more accurately. The real-world image could be analyzed using these techniques to determine the emotions and types of emotions present within it. As a result, decision tree approaches were devised to help determine which emotion percentage were high and low. Now, the high proportion of emotions receives the most exact emotions that are conceivable. There is a direct correlation between a low percentage of emotions and a low chance of surviving. As a result of this discovery, it is now able to discern emotions with accuracy. Automates, on the other hand, are better at identifying emotional states. They can then react appropriately and help to avoid a recurrence. In addition, this machine can be used to substitute a human being.

REFERENCES

[1] Peng Z., Lu Y., Pan S., and Liu Y., "Efficient Speech Emotion Recognition Using Multi- Scale CNN and Attention," ICASSP 2021 – 2021 *IEEE International Conference on Acoustics, Speech, and Signal Processing (ICASSP)*, 2021, pp. 3020–3024.

[2] Yoon S., Byun S., and Jung K., "Multimodal Speech Emotion Recognition Using Audio and Text", In: 2018 IEEE Spoken Language Technology Workshop, SLT 2018, Athens, Greece, 18–21 December 2018, pp. 112–118. IEEE (2018).

[3] Jingwen Hu, Yuchen Liu, Jinming Zhao, and Qin Jin, *"MMGCN: Multimodal Fusion via Deep Graph Convolution Network for Emotion Recognition in Conversation"*, arXiv preprint arXiv:2107.06779 (2021).

[4] Tzirakis P., Trigeorgis G., Nicolaou M. A., Schuller B. W., and Zafeiriou S., "End-to-End Multimodal Emotion Recognition Using Deep Neural Networks," In *IEEE Journal of Selected Topics in Signal Processing*, vol. 11, no. 8, pp. 1301–1309, Dec. 2017.

[5] Poria S., Majumder N., Mihalcea R. and Hovy E., "Emotion Recognition in Conversation: Research Challenges, Datasets, and Recent Advances," In *IEEE Access*, vol. 7, pp. 100943–100953, 2019.

[6] Deepanway Ghosal, Navonil Majumder, Soujanya Poria, Niyati Chhaya, and Alexander Gelbukh, "Dialoguegcn: *A Graph Convolutional Neural Network for Emotion Recognition in Conversation.*" arXiv preprint arXiv:1908.11540 (2019).

[7] Soujanya Poria, Devamanyu Hazarika, Navonil Majumder, Gautam Naik, Erik Cambria, and Rada Mihalcea, "Meld: *A Multimodal Multi-party Dataset for Emotion Recognition in Conversations.*" arXiv preprint arXiv:1810.02508 (2018).

[8] Caroline Etienne, Guillaume Fidanza, Andrei Petrovskii, Laurence Devillers, and Benoit Schmauch, "CNN+LSTM Architecture for Speech Emotion Recognition with Data Augmentation". 21–25.

[9] Wang J., Xue M., Culhane R., Diao E., Ding J. and Tarokh V., "Speech Emotion Recognition with Dual-Sequence LSTM Architecture," *ICASSP 2020 – 2020 IEEE International Conference on Acoustics, Speech, and Signal Processing (ICASSP)*, 2020, pp. 6474–6478.

[10] Xing S., Mai S., and Hu H., "Adapted Dynamic Memory Network for Emotion Recognition in Conversation", *IEEE Transactions on Affective Computing*, doi: 10.1109/TAFFC.2020.3005660.

[11] Ren M., Huang X., Shi X., and Nie W., "Interactive Multimodal Attention Network for Emotion Recognition in Conversation," in *IEEE Signal Processing Letters*, vol. 28, pp. 1046–1050, 2021.

[12]   Cowie R., Douglas-Cowie E., Tsapatsoulis N., Votsis G., Kollias S., Fellenz W., and Taylor J.G., "Emotion Recognition in Human-computer Interaction," In *IEEE Signal Processing Magazine*, vol. 18, no. 1, pp. 32–80, Jan 2001.

[13]   Shashidhar G. Koolagudi and K. Sreenivasa Rao, "Emotion Recognition from Speech: A review", *Int J Speech Technol* 15, 99–117 (2012).

[14]   Björn Schuller, Gerhard Rigoll, and Manfred Lang, "Hidden Markov Model-based Speech Emotion Recognition", ICASSP, *IEEE International Conference on Acoustics*, Speech, and Signal Processing – Proceedings 2:401–404, 2003. 10.1109/ICME.2003.1220939.

[15]   Alice Baird, Shahin Amiriparian, Manuel Milling, and Björn W. Schuller. "Emotion Recognition in Public Speaking Scenarios Utilisinganlstm-rnn Approach with Attention." In 2021 *IEEE Spoken Language Technology Workshop (SLT)*, pp. 397–402. IEEE, 2021.

[16]   Yue Xie, Ruiyu Liang, Zhenlin Liang, Chengwei Huang, Cairong Zou, and Björn Schuller, "Speech Emotion Classification Using Attention-based LSTM." IEEE/ACM Transactions on Audio, Speech, and Language Processing 27, no. 11 (2019): 1675–1685.

[17]   Devamanyu Hazarika, Soujanya Poria, Roger Zimmermann, and Rada Mihalcea, "*Emotion Recognition in Conversations with Transfer Learning from Generative Conversation modeling.*" arXiv preprint arXiv:1910.04980 (2019).

[18]   Lian, Zheng, Bin Liu, and Jianhua Tao, "CTNet: Conversational Transformer Network for Emotion Recognition" IEEE/ACM Transactions on Audio, Speech, and Language Processing 29 (2021): 985–1000.

[19]   Kwong J. C. T., Garcia F. C. C., Abu P. A. R. and Reyes R. S. J., "Emotion Recognition via Facial Expression: Utilization of Numerous Feature Descriptors in Different Machine Learning Algorithms," TENCON 2018 – 2018 *IEEE Region* 10 Conference, 2018, pp. 2045–2049.

[20]   Chew L. W., Seng K. P., Ang L. -M., Ramakonar V. and Gnanasegaran A., "Audio-Emotion Recognition System Using Parallel Classifiers and Audio Feature Analyzer," 2011 *Third International Conference on Computational Intelligence, Modelling & Simulation*, 2011, pp. 210–215.

[21]   Ghazi Al-Naymat, Mouhammd Alkasassbeh, Nosaiba Abu-Samhadanh, Sherif Sakr, "Classification of VoIP and Non-VoIP Traffic using Machine Learning Approaches" *Journal of Theoretical and Applied Information Technology*, 2016, 3192.

[22]   Aliaa Rassem, Mohammed El-Beltagy, and Mohamed Saleh, "Cross-Country Skiing Gears Classification using Deep Learning", 2017.

[23]   Shweta Singhal and Rajesh Dubey, "Automatic Speech Recognition for Connected Words using DTW/ HMM for English/ Hindi languages", 2015, 199–203.10.1109.

[24]   Muhammed Talo, "*Convolutional Neural Networks for Multi-class Histopathology Image Classification.*" arXiv preprint arXiv:1903.10035 (2019).

[25]   Tomar A., Gera S., Kumar S. and Pant B., "*HHFER: A Hybrid Framework for Human Facial Expression Recognition,*" 2021 International Conference on Data Analytics for Business and Industry (ICDABI), 2021, pp. 537–541, doi:10.1109/ICDABI53623.2021.9655808.

Automation and Computation – Vats et al. (Eds)
© 2023 the Author(s), ISBN 978-1-032-36723-1

Banana leaf disease classification using deep learning and various boosting techniques

Prateek Srivastava & Saumitra Chattopadhyay
Department of Computer Science and Engineering, Graphic Era Hill University, Dehradun, Uttarakhand, India

ABSTRACT: The banana is indeed a common fruit because it is inexpensive and full of nutrients that can be eaten raw or cooked. This grows in all tropical regions and is important to the financial systems of several underdeveloped nations. As a result, many farmers rely on it for income. India leads the world in banana production and ranks third in terms of land area used. Diseases that impact crop production seem to be common and result in significant monetary losses for ranchers. Farmers could save plants from diseases if they were able to identify them in their beginning phases, resulting in better quality and quantity. To address this issue, various researchers have proposed different cutting-edge systems that use computer vision, deep learning, and machine learning strategies to instantly classify the diseases. However, accuracy is indeed an issue that needs to be addressed. This paper focuses on the development of a deep learning model with the inclusion of various boosting techniques to make a more accurate system for the classification of banana leaf diseases. The dataset was obtained from the Kaggle dataset repository. The proposed deep learning model is trained with various hyperparameter variations such as learning rate, decay, regularizers, dropouts, normalizations, and so on. Finally, it has been discovered that this model performs best with the XGBoost algorithm, which has 96.25% accuracy. The envisaged model's high accuracy makes it entirely implementable.

1 INTRODUCTION

After mango, banana is India's second biggest fruit crop. Around 120 nations grow bananas and plantains. Bananas are indeed the world's fourth most important food crop after rice, wheat, and maize in terms of revenues value of production (FAO 2002). The statistic presented in (Statista 2022a) depicts the top banana yielding countries in the world in 2020. India led the world in banana production that year, with 31.504 million metric tonnes. China came in second place among these countries, producing approximately 11.51 million metric tonnes of bananas. Since agriculture employs 70% of the Indian population and accounts for 23% of GDP, it engaged 59% of the nation's entire workforce in 2016 (Economy of India 2022). In the financial year 2020, bananas significantly contributed 354 billion Indian rupees to the Indian economy. This has been the greatest estimated value registered in recent years (Statista 2022b). This represents the importance of banana farming. However, various diseases are there, which are major challenges that affect the productions and hence cause a huge loss to the economy and to our farmers too. A few major diseases like Pestaliopsis leaf blight, Sigatoka, Cordana have all been reported (Mahmud 2022). Machine learning becomes a technique in which researcher considers features and then train and predict models using the feature set. However, due to the large volume of datasets, it is frequently hard to ascertain the best features. Discovering features manual process is thus a tough and frequently inaccurate task (Thongsuwan *et al.* 2021). As a result, the majority of the moment, research depends on automatic feature extraction, that could be discovered using CNN. With the advent of computer vision and deep neural networks, it became possible to predict diseases by learning textual features of leaves. Convolution neural networks, that are comprised of various neurons, kernels, pooling layers, and activation functions, can be employed for one such intent. CNNs are employed to retrieve the best features and develop a feature map for disease classes. CNNs are a subtype of feed forward neural networks. It is commonly used in machine vision. Figure 1 depicts the CNN architecture.

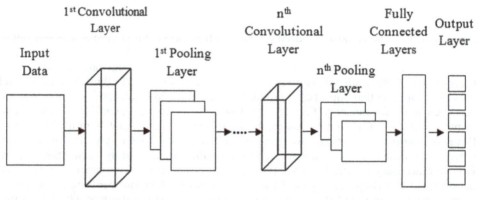

Figure 1.   Architecture of CNN.

The output of CNN is converted into long vectors using flattering layers and then fed to classifiers like Neural networks, SVM, Decision tree, random forest etc. There are various boosting algorithms are given for better classification accuracy. The XGBoost, LightGBM, and CatBoost are considered popular gradient boosting techniques for a variety of applications (Chen & Guestrin 2016).

1.1   *XGBoost*

XGBoost is a gradient-boosted decision tree deployment. This technique generates decision trees in a sequential fashion. The weights are very important in XGBoost. All of the independent variables are given weights, that are then supplied into the decision tree, that forecast outcomes. The weight of variables that the tree projected incorrectly is elevated, and these factors are then supplied into another decision tree. Individual classifiers are then put together to create a much more potent and accurate model.

Since trees are employed to make decisions, they can sometimes lead to incredibly difficult states. XGBoost uses both Lasso and Ridge Regression regularisation to avoid being an overly complicated model. To split the branch, XGBoost utilises the max_depth configuration defined in the halting criteria and begins pruning trees (Chen & Guestrin 2016). However, if the feature dimension is huge and the data volume is massive, the performance and scalability remain inadequate (Guolin *et al.* 2017).

1.2   *LightGBM*

It is based on gradient boost decision trees that improve model performance while lowering memory consumption and the risk of overfitting. It employs two innovative methods:

Gradient-based One Side Sampling and Exclusive Feature Bundling (EFB), that overcome the shortcomings of the histogram-based method used in every GBDT (Gradient Boosting Decision Tree) architectures. The features of the LightGBM Method are formed by the techniques of GOSS and EFB. They work together to ensure that the framework works efficiently and give it a competitive advantage. There are various parameters like max_-depth, categorical_feature, bagging_fraction, num_iterations, num_leaves, max_bin, min_data_in_bin, task, and feature_fraction that can be tuned to find better results (Guolin *et al.* 2017).

1.3 *CatBoost*

Yandex's CatBoost is an open-source machine learning technique (Prokhorenkova *et al.* 2018). It blends seamlessly with deep learning architectures such as Google's TensorFlow and Apple's Core ML. It produces cutting-edge outcomes without requires substantial data training and offers strong exceptional support for more descriptive data formats. The title "CatBoost" is derived from the words "Category" and "Boosting." It performs effectively with a variety of data types, such as audio, text, image, and historical data. CatBoost produces cutting-edge findings, requiring no manual pre-processing to generate categorical data to numbers. It minimises necessity comprehensive hyper-parameter tuning and decrease the probability of overfitting, resulting in more generalised frameworks.

2   DATASET AND PREPROCESSING

This article utilizes pictures of banana leaves from the Kaggle dataset for 4 medical conditions: Pestaliopsis leaf blight, Sigatoka, Cordana, and healthy, which have 173, 473, 162, and 129 pictures, respectively (Mahmud 2022). A data augmentation strategy with key variables is used to produce images in order to expand the dataset. It contributes to increasing the data quality and the number of images as well. Figure 2 depicts the 4 image classes under consideration here. After normalizing the image data, the model is trained and analysed. Each image is normalized in order to scale the data to an appropriate range for the model.

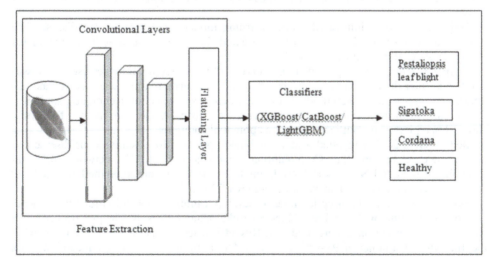

Figure 2.   Proposed model blended with CNN and Boosting techniques.

3 LITERATURE REVIEW

(Jogekar *et al.* 2020) have presented an intense survey considering various techniques to identify diseases in various fruits like apple, banana, maize, soyabean, blueberry etc. Finally authors have presented a table showing performance value for corresponding fruit a technique used.

(Criollo *et al.* 2020) presented CNN models to identify banana leaf diseases in three ways: first without regularisation, then with Dropout, and finally with weight regularisation. This paper concludes that model without regularization works better than a model with regularization.

(Tuazon *et al.* 2021) have proposed a portable model that can detect Sigatoka spot disease on a banana leaf in order to assist farmers in accurately identifying infected spot on banana leaves. The model's outcome depicts the classification of the leaf as well as the leaf area. The investigators used a USB camera to record banana leaf images, which were then stitched, equalised, segmented, feature extracted, and classified using Support Vector Machine. All in all, the suggested scheme classified with 90% accuracy.

The model suggested by (Aruraj *et al.* 2019) comprises of two major phases: (i) retrieval of texture characteristics of images and (ii) identification of infected and healthy banana plants. Support vector machine and K-nearest neighbour classifiers were fed the extracted features. The Plant Village dataset is taken into account here. Using SVM, the suggested model achieved accuracy of 89.1% and 90.9% for two test cases.

(Chaudhari & Patil 2020) have described a model for classifying banana leaf diseases using colour, shape, and texture features. For the classification, Support Vector Machine (SVM) has been applied. The envisaged model demonstrated an accuracy of 85% in identifying diseases.

(Devi *et al.* 2019) suggested a model that uses image processing and IoT to process plant pictures and retrieve texture characteristics. The Random Forest Classification (RFC) technique is used to classify the GLCM attributes. Many times, changes in the environment cause plant diseases. As a result, many sensors based on the RaspBerry Pi3 have been deployed. The proposed model has a classification accuracy of 99%.

(Nandhini & Kumar 2022) presented an improved DenseNet-121 structure to improve accuracy rate for various plant images identification. This model has considered fourteen leaf classes and classified images with 98.7% accuracy.

(Srivastava 2022) illustrated an efficient CNN for identification of various diseases in corn leaves.

(Gajjar *et al.* 2022) illustrated a deep learning model for detecting crop disease. This model is used on specific hardware to recognise and classify plant diseases. This model has a disease classification accuracy of 96.88%.

(Bhagat & Kumar 2022) provided an overview of the different strategies used to classify plants and their diseases. It gives an overview of a few current approaches, as well as their limitations, pre-processing methods, feature extraction and selection techniques, sources of data used, classification methods, and performance metrics. Aside from these, a few obstacles and open issues are outlined, as well as potential solutions.

(Bedi & Gole 2021) suggested a new combined approach for instantaneous disease prediction depending on the Convolutional Autoencoder (CAE) network and the Convolutional Neural Network (CNN). Using leaf pictures, the proposed method can detect disease in peach crops and achieves an accuracy of 99.35%.

(Sanga *et al.* 2020) used deep learning to develop a mobile application for timely identification of Fusarium wilt race 1 and black Sigatoka banana illnesses. The pre- trained CNN Resnet152 and Inceptionv3 were used. The Resnet 152 accomplished 99.2% accuracy, while the Inceptionv3 attained a 95.41% accuracy. The authors have selected Inceptionv3 for implementation using Android mobile phones because it has low computational requirements than Resnet152.

4 THE PROPOSED MODEL

The proposed model is mentioned in Figure 2. This work makes use of the power of convolutional neural networks for automatic image feature extraction.

To find the important features, neural networks use kernels that slide over pixels in images. Many parameters, such as stride and padding, can be used to control kernel movement. The choice of convolution kernel does indeed have a significant impact on convolution; on the one side, using a tiny kernel increases the number of convolution process times required; on the other, using a large kernel makes it hard to precisely acquire the features of the input data.

A feature map is obtained after applying various convolutional layers, activation functions, and down sampling layers. By flattening the layer, this feature map can be converted into a single long vector. This vector will now be used as an input for various classifiers. For classification, various boosting techniques were used. A few popular boosting algorithms, such as XGBoost, CatBoost, and LightGBM, have been considered for classification in this article.

5 THE PROPOSED CONVOLUTIONAL NETWORK

Here, four convolutional layers are considered. To add non-linearity the ReLU activation function is applied to each layer. The Maxpool2D class is used to downsample the inputs along height and width. The image shape (160, 160, 3) is considered here. To avoid overfitting, various combinations of dropout are applied to each layer. The first layer is comprised of 512 neurons and 128 kernels of (6, 6) size. A batch normalisation layer and a 25% dropout are used. The second layer is composed of 256 neurons and 64 kernels of size (5, 5). A dropout rate of 40% is applied. The third layer is comprised of 128 neurons and 32 kernels of size (4, 4). A dropout of 30% is applied. The fourth layer is made up of 64 neurons and 32 kernels (3, 3). A dropout of 20% is applied. Finally, the feature map produced by the final convolutional layer will be passed to a flattening layer. This layer will produce a single, long vector of feature values that will be input to classifiers for banana leaf disease classification.

6 EXPERIMENTAL SETUP

All of the convolutional networks discussed in the earlier section were devised in a Jupyter notebook using Google Colab. The default GPUs at Google Colab were used for model training and validation. The shape of the input images is taken into account (160, 160, 3). Data augmentation is applied to expand the number of images with different variations. Early Stopping (with patience value = 4) and Model Check Point (with monitor = validation loss and mode = minimum) callback methods were used to control overfitting.

To achieve the best results, each model is trained separately and evaluated numerous times. ImageDataGenerator is very crucial during model training and validation. It generates a set of augmented images (here, batch = 50) and provides for training and validation for the each iteration. For each iteration, the same batch will be formed until the entire training dataset has been exhausted. In this case, 75% of the data is used for training, while the remaining 25% is taken for validation. For testing purposes, 40 images of each class have been considered. During model compilation, the Adam optimizer (learning rate = 1e-03 and decay = learning rate / 1000) is used.

7 RESULT AND DISCUSSION

The various boosting techniques like XGBoost, LightGBM and CatBoost have been applied with convolutional neural networks. The classification report and confusion matrix for proposed model with the combination of CNN and XGBoost are given in Table 1 and Figure 3, respectively.

Table 1 shows the proposed CNN's average accuracy with XGBoost, which is 96.25%. According to the confusion matrix shown in Figure 3, all 40 healthy leaves are correctly classified.

Table 1. Classification report for XGBoost using proposed convolutional neural network.

	Precision	Recall	F1-Score	Support
Pestaliopsis leaf blight	1.00	0.95	0.97	40
Sigatoka	0.93	0.95	0.94	40
Cordana	1.00	0.95	0.97	40
Healthy	0.93	1.00	0.96	40
Average	0.96	0.96	0.96	40
Accuracy	0.9625			

Confusion Matrix		Actual (True) values			
		Pestaliopsis leaf blight	Sigatoka	Cordana	Pestaliopsis leaf blight
Predicted Values	Pestaliopsis leaf blight	38	1	0	1
	Sigatoka	0	38	0	2
	Cordana	0	2	38	0
	Healthy	0	0	0	40

Figure 3. Confusion matrix by XGBoost model using proposed convolutional neural network.

The classification report and confusion matrix for LightGBM are given in Table 2 and Figure 4, respectively.

Table 2. Classification report for LightGBM using proposed fully connected neural network.

	Precision	Recall	F1-Score	Support
Pestaliopsis leaf blight	0.97	0.85	0.91	40
Sigatoka	0.90	0.90	0.90	40
Cordana	0.82	0.93	0.87	40
Healthy	0.90	0.90	0.90	40
Average	0.90	0.89	0.89	
Accuracy	0.8938			

Confusion Matrix		Actual (True) values			
		Pestaliopsis leaf blight	Sigatoka	Cordana	Healthy
Predicted Values	Pestaliopsis leaf blight	34	0	4	2
	Sigatoka	0	36	2	2
	Cordana	0	3	37	0
	Healthy	1	1	2	36

Figure 4. Confusion matrix for LightGBM model using proposed convolutional neural network.

Table 2 shows the proposed CNN's average accuracy with LightGBM, which is 89.38%. According to the confusion matrix shown in Figure 4, out of 40 healthy leaves 36 were classified correctly.

The classification report and confusion matrix for CatBoost are given in Table 3 and Figure 5, respectively.

Table 3 shows that the proposed CNN's average accuracy with CatBoost is 93.13%. The confusion matrix in Figure 5 shows that 39 of 40 healthy leaves were correctly classified.

Table 3. Classification report for CatBoost using proposed fully connected neural network.

	Precision	Recall	F1-Score	Support
Pestaliopsis leaf blight	1.00	0.93	0.96	40
Sigatoka	0.95	0.88	0.91	40
Cordana	0.90	0.95	0.93	40
Healthy	0.89	0.98	0.93	40
Average	0.93	0.93	0.93	
Accuracy	0.9313			

Confusion Matrix		Actual (True) values			
		Pestaliopsis leaf blight	Sigatoka	Cordana	Healthy
Predicted Values	Pestaliopsis leaf blight	37	0	0	3
	Sigatoka	0	35	3	2
	Cordana	0	2	38	0
	Healthy	0	0	1	39

Figure 5. Confusion matrix for CatBoost model using proposed fully connected neural network.

8 CONCLUSIONS

This work focuses on the identification of banana leaf diseases with the help of convolutional neural networks and boosting techniques. The CNN has proven its significance for automatic feature extraction, and boosting techniques are also well proven for accurate prediction. In this paper, both techniques are blended. The experimental result shows that the proposed CNN works better with XGBoost in comparison to LightGBM or CatBoost, with an average accuracy of 96.25%, 89.38%, and 93.13%, respectively. It may be possible to improve the accuracy of this model by using pretrained deep neural networks as a feature extractor. The inclusion of more sample images may also increase the performance of the model. The proposed model's high accuracy makes it perfectly implementable.

CONFLICT OF INTEREST STATEMENT

The author declares that there is no conflict of interest.

REFERENCES

Aruraj, A., Alex, A., Subathra, M. S. P., Sairamya, N. J., George, S. T., & Ewards, S. V. 2019. Detection and Classification of Diseases of Banana Plant using Local Binary Pattern and Support Vector Machine. In 2019 2nd *International Conference on Signal Processing and Communication (ICSPC)* (pp. 231–235). IEEE.

Bedi, P., & Gole, P. 2021. Plant Disease Detection using Hybrid Model Based on Convolutional Autoencoder and Convolutional Neural Network. *Artificial Intelligence in Agriculture*, 5, 90–101.

Bhagat, M., & Kumar, D. 2022. A Comprehensive Survey on Leaf Disease Identification & Classification. *Multimedia Tools and Applications*, 1–29.

Chaudhari, V., & Patil, M. 2020. Banana Leaf Disease Detection using K-means Clustering and Feature Extraction Techniques. *In 2020 International Conference on Advances in Computing, Communication & Materials (ICACCM)* (pp. 126–130). IEEE.

Chen, T., and Guestrin, C. 2016. Xgboost: A Scalable Tree Boosting System. *In Proceedings of the 22nd ACM Sigkdd International Conference on Knowledge Discovery and Data Mining* (pp. 785–794).

Criollo, A., Mendoza, M., Saavedra, E., & Vargas, G. 2020. Design and Evaluation of a Convolutional Neural Network for Banana Leaf Diseases Classification. In 2020 *IEEE Engineering International Research Conference (EIRCON)* (pp. 1–4). IEEE.

Devi, R. D., Nandhini, S. A., Hemalatha, R., & Radha, S. 2019. IoT Enabled Efficient Detection and Classification of Plant Diseases for Agricultural Applications. In 2019 *International Conference on Wireless Communications Signal Processing and Networking (WiSPNET)* (pp. 447–451). IEEE.

Economy of India. Wikipedia. 2022 https://en.wikipedia.org/wiki/Economy_of_India.WTTCBenchmark-146 (Retrieved on November 30, 2022).

Food and Agriculture Organization of the United Nation. 2002. https://www.fao.org/3/y5102e/y5102e03.htm (Retrieved on November 22, 2022).

Gajjar, R., Gajjar, N., Thakor, V. J., Patel, N. P., & Ruparelia, S. 2022. Real-time Detection and Identification of Plant Leaf Diseases Using Convolutional Neural Networks on an Embedded Platform. *The Visual Computer*, 38(8), 2923–2938.

Guolin Ke, Qi Meng, Thomas Finley, Taifeng Wang, Wei Chen, Weidong Ma, Qiwei Ye, and Tie-Yan Liu. 2017. LightGBM: A Highly Efficient Gradient Boosting Decision Tree. In *Proceedings of the 31st International Conference on Neural Information Processing Systems* (NIPS'17). Curran Associates Inc., Red Hook, NY, USA, 3149–3157

Jogekar, R., & Tiwari, N. 2020. Summary of Leaf-based Plant Disease Detection Systems: A Compilation of Systematic Study Findings to Classify the Leaf Disease Classification Schemes. In 2020 *Fourth World Conference on Smart Trends in Systems, Security and Sustainability (WorldS4)* (pp. 745–750). IEEE.

Mahmud K. A., Kaggle, Banana Leaf Dataset. 2021 https://www.kaggle.com/datasets/kaiesalmahmud/banana-leaf-dataset (Retrieved on December 05, 2022).

Nandhini, S., & Ashok kumar, K. 2022. An Automatic Plant Leaf Disease Identification using DenseNet-121 Architecture with a Mutation-based Henry Gas Solubility Optimization Algorithm. *Neural Computing and Applications*, 34(7), 5513–5534.

*National Horticulture Board.*India. https://nhb.gov.in/report_files/banana/BANANA.htm (Retrieved on December 02, 2022).

Prokhorenkova, L., Gusev, G., Vorobev, A., Dorogush, A. V., & Gulin, A. 2018. CatBoost: Unbiased Boosting with Categorical Features. *Advances in Neural Information Processing Systems*, 31.

Srivastava P. 2022. Corn Leaf Disease Identification with Improved Accuracy. Workshop on Advances in Computation Intelligence, its Concepts & Applications at ISIC 2022, May 17–19, Savannah, United States.

Statista. 2022a. https://www.statista.com/statistics/811243/leading-banana-producing-countries/ (Retrieved on November 25, 2022).

Statista. 2022b. https://www.statista.com/statistics/1080404/india-economic-contribution-of-bananas/ (Retrieved on December 01, 2022).

Thongsuwan S., Jaiyen S., Padcharoen A., and Agarwal P. 2021. ConvXGB: A New Deep Learning Model for Classification Problems based on CNN and XGBoost, *Nuclear Engineering and Technology*, Vol. 53, Issue 2, Pages 522–531, ISSN 1738–5733.

Tuazon, G. L. H., Duran, H. M., & Villaverde, J. F. 2021. Portable Sigatoka Spot Disease Identifier on Banana Leaves Using Support Vector Machine. In 2021 *IEEE 13th International Conference on Humanoid, Nanotechnology, Information Technology, Communication and Control, Environment, and Management (HNICEM)* (pp. 1–6). IEEE.

Automation and Computation – Vats et al. (Eds)
© *2023 the Author(s), ISBN 978-1-032-36723-1*

Internet traffic load during Covid-19 pandemic

Manisha Aeri
Assistant Professor, Department of Computer Science & Engineering, Graphic Era Hill University, Dehradun, India

Shiv Ashish Dhondiyal
Assistant Professor, Department of Computer Science & Engineering, Graphic Era Deemed to be University, Dehradun, Uttarakhand, India

Sujata Negi Thakur
Assistant Professor, Department of Computer Science & Engineering, Graphic Era Hill University, Haldwani, Uttarakhand, India

Manika Manwal & Sonali Gupta
Assistant Professor, Department of Computer Science & Engineering, Graphic Era Hill University, Dehradun, Uttarakhand, India

ABSTRACT: Due to huge streaming content and large amount of work throughout all metropolitan cities, including tier two cities, Internet peak traffic increased by 40% in India, as did downloads and uploads per user. The largest wired Internet service provider in India, ACT Fibernet, conducted research titled as "State of Internet Traffic Trend" found that the average monthly uploads increased by 37% while average monthly downloads rose by 66% per subscriber. This study provides a brief overview analysis of internet traffic load and increase in various online content during the Covid 19 pandemic, as well as average data consumption per day. A survey clearly reveals that internet traffic load during pandemic was high.

Keywords: Internet traffic, Covid-19, downstream load, upstream load, database management.

1   INTRODUCTION

A once-in-a-generation worldwide phenomenon, the COVID-19 pandemic has transformed billions of people's lives while also destabilising the interdependent global economy. What began as a localised health concern in Asia at the ending of year 2019 quickly expanded to become a worldwide issue when the first cases began to appear on other continents at the beginning of 2020. In attempt to halt the spread of COVID-19, many nations throughout the world imposed harsh restrictions on economic and social activity. In March 2020, the World Health Organization proclaimed the disease to be pandemic. These regulations have a significant impact on the behaviour of a sizable portion of the global population, which increasingly relies on home Internet access for education, socialization, work, & leisure. [1]

In this study, we examined data from Internet traffic during a two-year period, including the initial pandemic year, to investigate how the COVID-19 epidemic has affected internet traffic. More precisely, we describe the changes in demand for individual applications that

DOI: 10.1201/9781003333500-41

quickly rose to prominence as well as the adjustments in general traffic patterns. Throughout the procedure, we attempt to comprehend whether Internet traffic has a "new normal" and to observe how the Internet has responded to these unexpected circumstances. [2] After compiling our investigation for 2020 spring wave from Feb 2020 - June 2020, we expanded the research work for the autumn 2020 wave i.e., Sept 2020 - Feb 2021. To do this, we gather and analyse network traffic data from a variety of sources, including a major European ISP, 3 IXP in Europe and US, a phone operator, and a major European academic network.

Our key findings are as follows:

- Variations in demand cause changes in traffic volume, resulting in a 15 to 20 percent reduction in traffic. surge for the ISP/IXPs in our analysis during the fall 2020 shutdown. After the economy reopened in summer 2020, one IXP saw an increase of roughly 20%, but only 6% at ISP Tier 1. The influence of the fall 2020 wave was also felt, with the annual traffic rise in 2020 being larger than in a usual year. The reported increase in traffic occurs largely during non-traditional peak hours.
- During the spring 2020 lock down, daily traffic patterns shifted to weekends like patterns. Traffic connected to remote working apps, such as video conferencing apps and virtual private network connection apps increased by greater than 200%. Virtual Private Network traffic appears to be at high during the autumn 2020 wave also.
- Traffic alteration fluctuate among networks. For e.g., after spring 2020 lockdown, there was a considerable decline approx. 55 percent in traffic volume in Redi Madrid campus network on working days because most individuals were not in campus, but a hype during 2020 lockdown. Traffic at ISP and IXP fluctuates as well, based on the enforced lockout policy and customer profiles.

Prior to the closure, a swarm of commuters clogged the roadways during rush hour. However, as more people worked from home, the Internet became congested.

The average cell phone data speed has dropped by more than 20% in recent years. The speed of fixed broadband has decreased by 8%. The report is based on data from Ookla, a global provider of Internet speed information. [3]

It examined data from the week ending April 5. This was compared to the four-week average speed before the lockdown.

It revealed that the average mobile download speed was down 20.9 percent for the week ending April 5. The upload speed had decreased by 16.3 percent. On the fixed segment, download speed was down 8%, while upload speed was down 9.9%.

According to a new survey by Bobble AI, which measures mobile app usage, individuals are increasingly turning to the Internet to meet business, social, and leisure demands. [4, 5]

The use of platforms like Hot star, Netflix and Amazon Prime has increased by 82.6 percent. People are spending 46.3 percent more time on social media applications like WhatsApp Facebook, Twitter and Instagram, Education and Fitness were also in high demand as people seek to achieve their objectives using online services such as Home Workout and Udemy. [6]

The number of active users of video conferencing software has also increased significantly. Time spent on applications such as Zoom, and House Party has increased by 71.1 percent. The number of active users has increased by 104.1 percent. Other countries in lockdown have had similar experiences. The usage of the Internet had increased in the US, Europe, and china. Companies such as Netflix have since reduced the amount of bandwidth, they use to reduce the stress on the Internet around the world, including in India. [7, 8]

"We continue to witness global decline in monthly fixed broadband speeds, with mobile broadband exhibiting its first month of lower speeds in March 2020," said Ookla in a statement.

Meanwhile, China, which first experienced a shutdown due to the coronavirus pandemic (Covid-19), experienced a comparable rise in Internet traffic. According to the corporation, this has now begun to wear off. There has been a hike in cell phones and download speeds. Fixed broadband download speed has increased by 24.20 percent, while mobile download speed has increased by 76.90 percent. [9]

You all might have probably noticed a technical problem if you have worked from home during the Covid-19 coronavirus pandemic. Perhaps your co-worker's face froze when in a Zoom meeting, or YouTube video also appeared grainier than usual. These may appear to be warning signs that the internet in the US is straining to keep up with an enormous surge in use from millions of home bound computer users. What is really going on is more complicated. [10]

There was a rise in internet traffic during the pandemic weeks, which is attributed in part to more people working from home. Before the epidemic, this was happening on a smaller scale. People started doing everything from home, and a lot of it was online, since the government-imposed lockdowns around the country. People are playing video games online, making video chats, witnessing nerve-wracking news briefings, and, yes, they are working part-time. [11]

Even though the internet was invented in the United States, the country does not have the finest internet in the world. As more people began to use the internet more often during the epidemic, it's understandable that some technical journalists, internet fans, and, to a lesser extent, engineers have been concerned about whether our network infrastructure can handle a massive rise in internet traffic [12]. Even the Facebook CEO Mark Zuckerberg said that his business, which owns WhatsApp and Instagram, is "simply trying to keep the lights on" as on its application usage and platforms reaches all-time highs. However, there is a contrast in between what occurs on the internet and what occurs on platforms like as Facebook, zoom and WhatsApp. [13]

The internet is an extraordinarily robust and resilient network that was particularly designed to respond to massive traffic spikes like the one we're experiencing. However, the platforms and apps that make the internet useful have received less testing. So, the good news is that the internet in America is better prepared for this epidemic than you may expect. The bad news is that Mark Zuckerberg and others are concerned that their platforms will be overwhelmed. Fortunately for you, many specialists believe that everything will be OK.

2 DATASETS

A network of networks makes up the Internet. To transport network traffic between hundreds of other networks or perhaps thousands of them, networks might act as massive traffic hubs, depending on their size and location. In other circumstances, they are poorly connected and close to the consumer at the topology's edge.

To have a comprehensive understanding of the epidemic's effects on the Internet, we monitored it from a variety of angles. They are located on a major Tier-1 Service ISP backbone and peering points, at the centre of the Internet IXP's, and on the Internet's periphery which is a cosmopolitan university network and a mobile operator.

ISP: It is a significant Central European ISP with over fifteen million fixed line users and a transit network called Tier-1. [14]

IXPs: It is a facility for network connections where Networks can join and trade traffic with other participants over the IXP's infrastructure. We concentrate our analysis on three significant IXPs. Central Europe is home to IXP-CE, the largest IXP. It has more than 900 members and more than 9 Tbps at peak traffic. IXP-SE, the second IXP, has more than 170 members and is based in Southern Europe. The US East Coast is home to the third IXP, known as IXP-US, which has more than 250 members. [14]

REDI-Madrid university network: the data is gathered and examined from the REDI-Madrid university academic network, that links sixteen autonomous universities and research institutions in the Madrid area. It provides services to close to 290,000 users, including academic staff, researchers, student residences, WiFi networks, and administrative and support personnel. [15]

Mobile operator: This is a mobile service provider in Europe with more than 40 million clients.

We gather data from each vantage point about traffic flows both before and during the outbreak. This enables, to analyse the pandemic's effects on Internet traffic and talk about corresponding traffic changes. We only evaluate aggregated or anonymised datasets to ensure user privacy. For further information about our measurement process.

3   NETWORK TRAFFIC SHIFTS

During, before as well as after the harshest lock-down times for the spring and fall 2020 waves(both), we firstly checked for the overall modifications to long-established traffic patterns to understand better how traffic changed throughout the pandemic. We normalise the data to make it comparable because all data sources have widely varying traffic levels and characteristics.

The data consumption per day during the 2020 pandemic is shown in Figure 1.

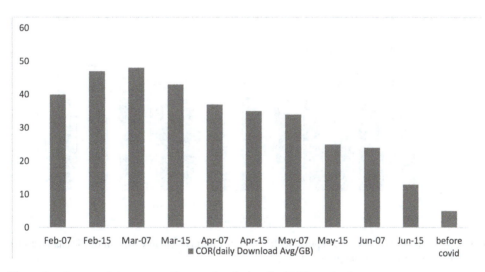

Figure 1.   Average data consumption per day during Covid-19 pandemic.

It is crystal evident from the diagram that internet usage increases significantly during pandemics, just like it will in the month of April 2020. however, in the month of February 2020, it is extremely little.

The average growth of various online material throughout the COVID-19 pandemic is shown in Figure 2. This graph makes it abundantly evident that during the pandemic, digital marketing occupies the mostarea.

We exhibit the normalised total traffic for the years 2018 to 2021 for every month in Figure 3, which focuses on the ISP. Even though the traffic increased by almost 30% from the same month the previous year during the years 2018–2019, things drastically changed after March 2020. To comprehend this, we annotate the rise for each month in the Figure for the years 2019, 2020, and February 2021 above the bars.

The ratio of upstream to downstream traffic is seen in Figure 4. Prior to the COVID-19 pandemic, this ratio typically had values around 9.8 with some obvious variance. This trend

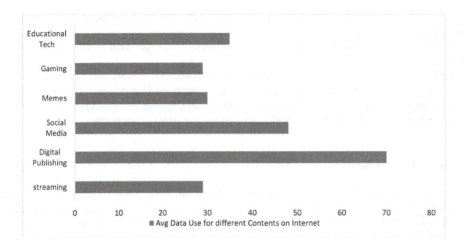

Figure 2. Average growth of different types of Internet content during Covid-19.

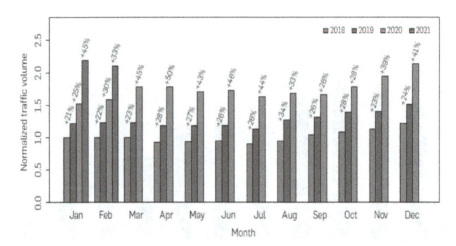

Figure 3. During the COVID-19 epidemic, ISP monthly normalized downstream traffic changed with a percentage increase over the prior year.

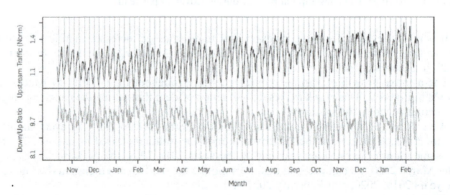

Figure 4. Depicts ISP traffic averaged over 8 hours from Oct 2019 to Feb 2021: upstream traffic normalised increase (top) and downstream v/s upstream traffic ratio (bottom).

alters following the initial shutdown at the starting of March 2020 and lasting through the end of Feb 2021. In fact, the upstream to downstream traffic ratio declines substantially, with normal values around 9 and extremely wide variation. This ratio is as low as 8.1 on workdays. This demonstrates that the proportional growth of upstream traffic is up to eighteen percent greater than that of downstream traffic growth.

4 APPLICATION TRAFFIC VARIATIONS

Now let's focus on the changes in traffic for the various apps of classes that were predicted to be impacted by the corona virus epidemic, such as Web conferencing apps, online gaming, streaming, Video on Demand, and traffic coming from service networks of universities. The following are examples of how traffic load variations are used:

4.1 *Web conferencing*

Applications for web conferencing have experienced a sharp increase during the lockout times.The ISP and IXP-CE saw significant traffic surge in March, immediately following the start of the first shutdown, affecting all hours of the day, but mainly on week days [16]. This trend picks up speed in June & peaks in December and January when both vantage points show increases of more than 300% from the base week. Notably, the high rise continues weekends in December and January. This shows that private social interactions have gone online along with work-related activities.

4.2 *Video on demand*

In both the waves, usage of video streaming applications exhibits rapid development. Strangely, the ISP only sees a little increase during the lockdown in the Ist part of March, and then in the 2nd half of March, the volume drops lower the pre-COVID-19 reference period. This is due, in part, to majorly streaming services in Europe, they are decreasing their streaming resolution by the middle of March for a 30-day period [17]. With regard to the IXP, a comparable although not as strong trend was seen in March. Moreover, there is a noticeable rise in traffic for Video on Demand in the months of January June and December, exceeding 200 percent for IXP & 100 percent for ISP for few days, particularly on week ends.

4.3 *Gaming*

When seen from the IXP perspective, especially during the day, The exponential growth of gaming applications makes more sense. Although the ISP significantly increases in the morning, the Spring wave often sees a decline in the ISP. Keep in mind that this effect is primarily due to the category's extremely high traffic volumes during our baseline week in February 2020[18]. Nowadays, people utilise gaming apps whenever they want, rather than only on the weekends or in the evenings. In June, the tendency begins to flatten; this could be due to individuals taking vacations or spending more time outside. During the fall wave, the ISP observes a rise in gaming-related traffic of up to 300% on all weekdays, with a focus on the morning hours. At the IXP, a comparable trend manifests, albeit with lesser increments. The closure of schools during the fall wave may account for the significant rise seen at both viewpoint sites in the early morning hours [19–21].

4.4 *University networks*

At both points of view, traffic coming from both networks exhibits comparable behaviour, with the ISP exhibiting a more dramatic tendency. Particularly during the fall wave, both vantage points observe a huge spike in traffic of 100% or more [22, 23]. This expansion may be linked to the fact that certain European university networks offer video conferencing services that are now being used by ISP/IXP clients.

5 CONCLUSION

Due to our society's enhanced digitalization and resilience, life went on despite the COVID-19 disruption, because of Internet providing a crucial aid for business, education, entertainment, sales, and social connections. In this study, we examine data on flow of internet from various angles in many industrialised nations. All Together, they give us a clear knowledge of how the COVID-19 waves and the lockout procedures affected Internet traffic. The total traffic volume surged by about 40% a year after the initial lockdown measures were put into place, far beyond the usual yearly growth. Additionally, during lockdowns, the relative gap between workday and weekend traffic patterns has nearly completely vanished. Applications for distance learning and working, like VPN and video conferencing, had traffic spikes of more than 200 %.

Our research highlights the significance of using several distinct lenses to have a full understanding of these events. Our results also focused on the usefulness of tackling traffic engineering with a highlight that goes beyond Huge traffic and favourable traffic classes to take necessary apps for remote working into consideration. Our research work shows that automation, over-provisioning, and the network management that is proactive are crucial for building powerful networks which can survive sudden and unforeseen surges in demand, like those encountered in the era of corona virus. However, because of this pandemic, it is crucial to keep tracking traffic patterns to comprehend usage as most of the companies switched on to work from home.

REFERENCES

[1] Nasajpour, M., Pouriyeh, S., Parizi, R. M., Dorodchi, M., Valero, M., & Arabnia, H. R. (2020). Internet of Things for current COVID-19 and Future Pandemics: An Exploratory Study. *Journal of Healthcare Informatics Research*, 4(4), 325–364.

[2] Bronzino, F., Feamster, N., Liu, S., Saxon, J., & Schmitt, P. (2021, February). Mapping the Digital Divide: Before, During, and After COVID-19. In *TPRC48: The 48th Research Conference on Communication, Information and Internet Policy*.

[3] Purohit, H., Kaur, P., & Choudhary, S. (2022). Mitigating the Effect of COVID Lockdown Period on Channel and Bandwidth Utilization in Mobile Communication Network in North Western Rajasthan (India). *Wireless Personal Communications*, 123(4), 3497–3509.

[4] Chiu, Y. C., Schlinker, B., Radhakrishnan, A. B., Katz-Bassett, E., & Govindan, R. (2015, October). Are we one hop away from a better internet?. In *Proceedings of the 2015 Internet Measurement Conference* (pp. 523–529).

[5] Murata, T. (2021). COVID-19 and Networks. *New Generation Computing*, 39(3), 469–481.

[6] Béraud, G., Kazmercziak, S., Beutels, P., Levy-Bruhl, D., Lenne, X., Mielcarek, N., ... & Dervaux, B. (2015). The French Connection: the First Large Population-based Contact Survey in France Relevant for the Spread of Infectious Diseases. *PloS one*, 10(7), e0133203.

[7] Feldmann, A., Gasser, O., Lichtblau, F., Pujol, E., Poese, I., Dietzel, C., ... & Smaragdakis, G. (2021). A Year in Lockdown: How the Waves of COVID-19 Impact Internet Traffic. *Communications of the ACM*, 64(7), 101–108.

[8] Gasser, O., Lichtblau, F., Pujol, E., Poese, I., Dietzel, C., Wagner, D., ... & Smaragdakis, G. (2021). A Year in Lockdown: How the Waves of COVID-19 Impact Internet Traffic. *Communications of the ACM*, 64(7).

[9] Feldmann, A., Gasser, O., Lichtblau, F., Pujol, E., Poese, I., Dietzel, C., ... & Smaragdakis, G. (2020, October). The lockdown effect: Implications of the COVID-19 Pandemic on Internet Traffic. In *Proceedings of the ACM Internet Measurement Conference* (pp. 1–18).

[10] Feldmann, A., Gasser, O., Lichtblau, F., Pujol, E., Poese, I., Dietzel, C., ... & Smaragdakis, G. (2021). A Year in Lockdown: How the Waves of COVID-19 impact Internet Traffic. *Communications of the ACM*, 64(7), 101–108.

[11] Labovitz, C. (2020, June). Pandemic Impact on Global Internet Traffic. NANOG.

[12] Labovitz, C., Iekel-Johnson, S., McPherson, D., Oberheide, J., & Jahanian, F. (2010). Internet Interdomain Traffic. *ACM SIGCOMM Computer Communication Review*, 40(4), 75–86.

[13] Lakhina, A., Papagiannaki, K., Crovella, M., Diot, C., Kolaczyk, E. D., & Taft, N. (2004, June). Structural Analysis of Network Traffic Flows. In *Proceedings of the Joint International Conference on Measurement and Modeling of Computer Systems* (pp. 61–72).

[14] Leighton, T. Can the Internet keep up with the surge in demand? 2020. https://blogs.akamai.com/2020/04/can-the-internet-keep-up-with-the-surge-in-demand.html.

[15] Lutu, A., Perino, D., Bagnulo, M., Frias-Martinez, E., & Khangosstar, J. (2020, October). A Characterization of the COVID-19 Pandemic Impact on a Mobile Network Operator Traffic. In *Proceedings of the ACM Internet Measurement Conference* (pp. 19–33).

[16] McKeay, M. (2020). Parts of a whole: *Effect of COVID-19 on US Internet Traffic.*

[17] Adam, D. C., Wu, P., Wong, J. Y., Lau, E. H., Tsang, T. K., Cauchemez, S., ... & Cowling, B. J. (2020). Clustering and Superspreading Potential of SARS-CoV-2 Infections in Hong Kong. *Nature Medicine*, 26(11), 1714–1719.2020.

[18] Herrmann, H. A., & Schwartz, J. M. (2020). Why COVID-19 models should incorporate the network of social interactions. *Physical Biology*, 17(6), 065008.

[19] Feldmann, A., Gasser, O., Lichtblau, F., Pujol, E., Poese, I., Dietzel, C., ... & Smaragdakis, G. (2021). A Year in Lockdown: How the Waves of COVID-19 Impact Internet Traffic. *Communications of the ACM*, 64(7), 101–108.

[20] Schilz, B., & Maunier, R. (2020). Experience on Deploying a New Remote PoP During COVID-19 Restriction. *RIPE* 80.

[21] Schlinker, B., Kim, H., Cui, T., Katz-Bassett, E., Madhyastha, H. V., Cunha, I., ... & Zeng, H. (2017, August). Engineering Egress with Edge Fabric: Steering Oceans of Content to the World. In *Proceedings of the Conference of the ACM Special Interest Group on Data Communication* (pp. 418–431).

[22] Feldmann, A., Gasser, O., Lichtblau, F., Pujol, E., Poese, I., Dietzel, C., ... & Smaragdakis, G. (2021). A Year in Lockdown: How the Waves of COVID-19 Impact Internet Traffic. *Communications of the ACM*, 64(7), 101–108.

[23] Wang, Z., Bauch, C. T., Bhattacharyya, S., d'Onofrio, A., Manfredi, P., Perc, M., ... & Zhao, D. (2016). Statistical Physics of Vaccination. *Physics Reports*, 664, 1–113.

Automation and Computation – Vats et al. (Eds)
© 2023 the Author(s), ISBN 978-1-032-36723-1

Understanding immersive technologies for autism detection: A study

Shaurya Gupta, Mitali Chugh & Sonali Vyas*
School of Computer Science, UPES University, Dehradun, Uttarakhand, India

ABSTRACT: The research for early detection of autism spectrum disorder (ASD) has increased rapidly because of the rising autism rate worldwide. The most basic symptoms of autism include the absence of normal activities, not the presence of abnormal ones. A lot of research has been done to detect abnormal behaviors in autistic children, but no one focused on absent normal behaviors in autistic children. This article surveys technology that has been or could be employed to diagnose and assess societal skills, influence and emotion management, communication, managing problem behavior, and synthesizing data. To conclude, the article recommends criteria for getting insights into future technologies that find applications for the treatment and awareness of individuals on the autism spectrum.

Keywords: Autism Spectrum Disorder, Artificial Intelligence (AI), Augmented Reality (AR), Virtual Reality (VR), Internet of Things (IoT), Telehealth, Blockchain

1 INTRODUCTION

Autism spectrum disorder (ASD) is a set of developmental disabilities defined generally with two significant indications of social communication/interaction (SCI) and restricted/repetitive behavior (RRB) [1]. Autism is not caused by any individual known cause; however, the existing facts point to genetic reasons. Autism affects 4.5 times more than females and it exists in all societal clusters and races. Male-to-female ratios range from 2:1 to 5:1, with an estimate of 4:1 in the 2010 Global Burden of Disease study [2]. Researchers have identified autism as a genetic condition through twin studies and have revealed the fact that if one of the monozygotic twins suffers from autism then it is 36-95% likely that the other one will also be detected with ASD. In the case of dizygotic twins, the possibility is reduced to 0-31% that both twins will have ASD [3].

The term Autism was first coined by Psychiatrist Eugen Bleuler in 1911 and described as restricting interactions with humans and with the outside world, expressed as philosophical social isolation. In the last 50 years, ASD has passed through a path of a scarcely defined rare disorder to a well-publicized and lifetime ailment that is rather heterogenous that has now witnessed a spectrum from very moderate to acute. Though, several (but not all) people with ASD need lifetime care for certain forms. To be specific to detect ASD there are no specific biomarkers and it is detected based on behavior. Subject to age and logical aptitudes, people identified with autism have a variable level of interaction deficits. These deficits vary from poor comprehension, pronoun reversal, speech delays, echolalia, or a total absence of speech. In the context of nonverbal communication ASD results in weak interpretation of facial expressions, poor eye contact, etc. A deficit in socio-emotional reciprocity is an additional significant characteristic of persons with ASD. These people to a lesser extent

---

*Corresponding Author: researcher.svyas@gmail.com

DOI: 10.1201/9781003333500-42

begin a discussion, exhibit less concern in cohort interactions, and generally find it hard to change their conduct according to various societal settings. People with ASD usually have comorbid logical incapacity and are susceptible to emotional complexities such as anxiety or depression.

On the regional and world level attempts have been made for early diagnosis and cure of ASD. The World Health Organization (2014) asserts infant and adolescent development monitoring for early diagnosis and cure of ASD are significant for improving the national health system. Hyman *et al.* (2020) in their clinical report provide ASD regulations and practice parameters released in the United States [4]. In addition, several other reports in various countries that address the concerns for the early diagnosis of ASD have been released recently, including New Zealand [5], France [6], and India [7]. Together, they accentuate the significance of practices, guidelines, and processes to enhance the early detection of ASD. Immersive technologies like AI, IoT, AR/VR, etc. are employed to assist and enhance communication for individuals with ASD, irrespective of speech ability. Moreover, these technologies can assist individuals of all ages by enhancing communication, supporting independence, and increasing societal communications [8].

Hence, keeping in view the growing need for early diagnosis and support of ASD, this study presents an account of immersive technologies that assist in the process. The study is structured as section 2 reviews the relevant literature, section 3 presents a descriptive account of immersive technologies that support the need for ASD detection and assistance. Section 4 describes the recent useful applications that are based on assistive technologies and enhance therapeutic efficiency. Section 5 concludes the study and provides guidelines for future research directions.

2 LITERATURE REVIEW

In this section, we review the literature for presenting an understanding of the usage of immersive technologies for early detection and support for ASD. Omar *et al.* (2019) in their study mention that ASD detection through screening is a time-consuming and expensive method. In such a scenario artificial intelligence and machine learning can assist in Autism early-stage prediction. They have proposed an efficient prediction model based on the ML technique and have developed a mobile application for the same. The prediction model uses Random Forest-ID3 (Iterative Dichotomiser and Random Forest-CART (Classification and Regression Trees) and the AQ10 dataset along with a live dataset of 250. The outcomes revealed better accuracy for both datasets [9]. Kong et. al. (2019) constructed a unique brain network for the representation of features and performed classification using a deep neural network (DNN) classifier and investigational outcomes show that the proposed method can attain an accuracy of 90.39% [10]. Mellema *et al.* (2019) have proposed an ML-based diagnosis of Autism over the conventional method of clinical diagnosis that requires a high level of expertise. The researchers have used IMPAC datasets and have selected 900 subjects from them. They compare the shallow learning method and deep learning method for ASD detection. The findings report that the best model was a dense feedforward network and gives added predictive accuracy over classical methods [11].

Banna *et al.* (2020) proposes an AI-based system employing sensor data for patient monitoring based on sentiment and facial representations of the patient and adapting the ML method through stimulating tasks and games as a solution during the COVD19 lockdown. The proposed system is offered as a solution for the caregivers and parents of ASD patients who are not able to approach medical practitioners and are in a situation of fix [12]. Nogay *et al.* (2020) in their study present ML research for ASD detection using functional MRI, structural magnetic resonance image (MRI), and hybrid imaging techniques. They mention ML technology is anticipated to impact substantially the early and rapid diagnosis of ASD and develop to facilitate clinicians [13]. Ke (2020) has used an open-source dataset comprising 1000 MRI scans and has tested using 14 models that included Recurrent neural networks and Convolution neural networks. In their study, the authors have demonstrated the use of deep neural networks as tools for diagnosing and analyzing psychiatric disorders. They mention that their system can support clinicians to ensure a cost-effective and time-efficient diagnosis process [14].

Ghosh *et al.* (2021) mention AI, ML, and the Internet of Things (IoT) are utilized in various medical applications, and autistic people can be aided by the correct usage of automatic systems. They have reviewed the 58 articles and have confirmed the significance of AI, ML, and IoT for ASD detection [15]. In addition the studies in the literature mention use of DL in the detection of brain lesions such as tumors[16], neural imaging ranges from brain MR image segmentation [17], production of artificial structural or functional brain images [18], and diagnosis of the brain functional disorders such as ASD [19].

3 IMMERSIVE TECHNOLOGIES FOR AUTISM DETECTION

1. *Sensor-based technologies [20]*

   Children with ASD are a varied group with changing degrees of functional limits. Several have co-occurred including attention deficit/hyperactivity disorder, seizures, sleep dis-orders, oppositional defiant disorders, and speech delay. Investigators have observed use of sensor-based technologies, including artificial intelligence (AI) algorithms to scan children with ASD. AI is mainly valuable for recognizing patterns in data, which is beneficial when researchers are trying to find markers related with an ASD analysis. Nevertheless, AI algorithms are efficient as the training datasets that feed diagnostic systems. Several AI-related technologies have been developed over the past few years to improve healthcare. In addition, the researchers used sensors to analyze sounds, facial expressions, eye tracking, tactile sensitivity, movement, and interactions with robots yielding promising results in efforts to identify children with ASD. Inappropriately, till date none of such techniques are delicate enough for diagnostic use.

2. *Telehealth-based device enables diagnosis*

   In 2005, Behaviour Imaging Solutions (BIS) was established by Ronald and Sharon Ober Leitner after their 3-year-old child got an ASD diagnosis. Created throughout the span of quite a while and first made available across country in 2018, the framework, called the Naturalistic Observation Diagnostic Assessment (NODA), comprises of 2 parts. Parents use smart Capture, a smartphone-based application, to finish up a formative ques-tionnaire and record and transfer four 10-minute recordings of their kid. Many setups incorporate the youngster play a part alone, the kid performing with others, a family supper time, and a way of behaving of parent's anxiety.

3. *AI for ASD*

   One of the most captivating ways to deal with involving AI in treatment includes making and training robots to communicate with autistic kids. Their motivation is to give autistic youngsters practice with recognizing looks, associating socially, and answering fittingly to expressive gestures. Another review shows that carrying out an artificial knowledge based clinical gadget in primary care settings can assist clinicians with precisely diagnosing autism range jumble in youngsters. The AI application utilizes a progression of activities to help the client quiet down or answer suitably and afterward, contingent upon the temperament of the kid, the model offers activities and afterward learns how the kid answers. AI-based applications are less expensive and simpler to incorporate into ordin-ary homes, schools, and specialists' workplaces than very good quality robots. AI-based diagnostic aids might can possibly help clinicians in primary care settings with ASD finding. [21]

4. AR/VR based ASD

Expanded The truth is a powerful innovation for empowering play, further developing language and vocabulary, verbal communication, nonverbal communication, and expressive gestures, feeling distinguishing proof, and consideration, the learning of new undertakings and for expanding inspiration. AR can likewise be an instrument used to help individuals with autism while they are having a social collaboration. AR considers cooperation with this present reality which makes it more straightforward to sum up genuine circumstances through computerized content. The vivid, visual nature of AR profits by a strength largely held by individuals with ASD and creates greater interest and commitment. Presenting innovation can likewise be profoundly persuading, making a more top to bottom learning experience. Also, AR can be handily adjusted to enhance proof based rehearses, for example, picture inciting and video demonstrating, that are right now being utilized by clinicians.

With AR innovation being generally new, its utilization inside healthcare, and particularly the domain of ASD, is as yet preliminary. All things being equal, concentrates on show that the basic expansion of an AR component to current treatments can expand inspiration, consideration, and concentration. These featured results simply scratch the outer layer of the capability of AR-based intercessions, and the standpoint is promising for future clinical use. [22]

Virtual Reality is an extremely successful innovation to assist the learning with handling overall. In the instance of kids with ASD it is valuable to prepare them to public talking, essential and complex interactive abilities, or to train them to control their tensions. Virtual Reality is particularly powerful in view of its exceptionally vivid perspective. Children can learn new errands or learn to confront distressing circumstances. For instance, for public talking abilities they can utilize a group of people of avatars which trends away the youngster talking doesn't keep eye to eye connection with the avatars. Additionally, they were instructed to check out the crowd and not exclusively to one center point. This game had an excellent reaction from the participants. A game-like treatment is far more proper for a kid. With the utilization of Virtual and Expanded Reality in treatment, it won't just work on the consequences of the treatment, yet it will likewise protect his joy and inspiration. [23]

5. IoT-enabled ASD

IoT can help parents control and oversee conditions. As well as framing part of specific learning programs, it can likewise do something amazing at home, assisting with making a position of quiet where visual and hear-able upgrades don't overpower a youngster. Along these lines, parents of youngsters with ASD are careful to pick fitting kids' toys concerning instructive aims, interests, and obviously, age. There are many toys aimed at youngsters with ASD, a significant number of which are dependent on network association. These can be utilized for pretending, entertainment, or quieting. The IoT can assist with outer highlights like encompassing lighting, loosening up music on request, and the portion of 3D or material walls that kids appreciate contacting and associating with. [24] The Internet of Things can offer kids with ASD more freedom, a more reasonable home climate, and wearable tech to keep them no problem at all.

IoT enabled DEVICE in USE:

- Alert Me Bands-Emergency contact wristband completely adaptable to impart to medical, extraordinary necessities and sensitivity alarms. It furnishes a straightforward solution to speak with the emergency contacts and make awareness of medical requirements or sensitivity sign.
- Amber ready GPS smart-finder-Supportive gadget for autistic patients which synchronizes with an application which helps in observing them at any moment.
- Eye tracking glasses-They identify abnormal look designs for screening autism by estimating x and y directions of look obsession of the patient regarding time.

- Movement trackers-It is utilized for identifying self-oppressive ways of behaving and conducting unsettling influences in advance with the goal that convenient mediations of caretakers is feasible.
- Polysomnography y (Rest Quality Assessment)-It estimates various neurophysiological and cardiorespiratory parameters to give a profound examination of the activity and exercises. It likewise identifies any rest-related issues.

6. *Blockchain for ASD*

The application called Autism Telerehabilitation Application (ATA) utilizes a private blockchain in view of the Ethereum convention to set up history and accounts of all the electronic health records (EHRs) of patients from the smart gadget. Moreover, this ATA would give medical mediations and ongoing patient seeing by sending alarms to the patient and medical subject matter experts. Also, it can get and maintain the record of who has started these activities.

4   APPS FOR ASSISTIVE TECHNOLOGIES FOR AUTISM

Here are the five best apps, based on assistive technologies, in 2021: [25]

- **Aut2Speak:** This is a versatile application keyboard for individuals with autism or other nonverbal circumstances like strokes, who know how to type. There are many highlights in the application, including an adjustable rundown of names, a rundown of sentiments and requirements, a rundown of pronouns, a rundown of word endings, and a novel keyboard. This application works for the two iOS and Android gadgets, and it is valuable for kids ages 6 to 17. It additionally functions admirably for grown-ups with formative or mental incapacities. It costs $1.
- **Autism iHelp:** This is a portable application for Apple gadgets that instructs vocabulary. It was developed initially by a discourse language pathologist whose kid had autism. The parents saw that kids with autism benefit from explicit ways to deal with language inter-cession. The application works like blaze cards or picture cards in conduct treatment, utilizing 24 pictures of genuine things, considering expressive achievements that kids need to reach. These pictures are isolated into three gatherings of eight, and in these gatherings, the learning system feels less overpowering. The app is designed for children who are 4 and older, and it is free.
- **AutismXpress:** While many individuals with milder types of autism can discuss very well with words and expressions, they might in any case battle with looks or conveying their feelings without words. This application assists individuals with autism to perceive and communicate feelings with looks. There are 12 buttons with cartoons addressing feelings like miserable, cheerful, irate, hungry. AutismXpress is intended to assist kids with autism learn, yet it can likewise be helpful for young people or grown-ups. This is a free application to download, and it deals with most cell phones.
- **CommBoards:** This application depends on the Picture Exchange Communication System (PECS), which has for quite some time been utilized on actual boards and through electronic gadgets. Presently, this application makes this strategy for conveying broadly open, without pulling around cards, pieces, or a gadget separate from your PC or telephone. At the point when kids utilize the application, the program expresses the word related to the picture without holding back, empowering the kid to rehash it and learn to articulate the word. You can likewise make your own cards with voice accounts and pictures, to assist your kid with communicating explicit names, places, or items they experience on a regular premise. Unlike some other free applications, CommBoards costs $19.99 to buy. It is planned and supported by clinicians who utilize assistive communication gadgets, so it is one of only a handful of exceptional choices available on cell phones that is explicitly endorsed by individuals who work in autism-related assistive innovation.

- **Leeloo:** This application utilizes an extensive variety of picture choices to assist nonverbal kids and young people with autism use smartphones or tablets to speak with their caregivers. The groundwork of the application is the Picture Exchange Communication System (PECS), and it likewise utilizes augmentative and alternative communication (AAC) standards. A card addresses the words your kid could require over the course of the day to speak with family, instructors, medical professionals, and companions. In the same way as other applications that help individuals with autism, Leeloo is free to download.

5   CONCLUSION

Smart observing and helped living systems for mental wellbeing assessment assume a focal part in assessment of people's medical issue. Autistic youngsters experience the ill effects of certain hardships including interactive abilities, monotonous ways of behaving, discourse, and nonverbal communication, and obliging to the climate all over them. Hence, managing autistic youngsters is a serious general medical condition as it is hard to figure out what they feel with an absence of close to home mental capacity. We have examined various technologies and applications for working on mental capacity and daily living abilities and amplifying the capacity of the autistic youngster to work and participate decidedly locally. Through using vivid technologies like, Artificial Insight (AI) AR/VR, Sensor-based gadgets, and IoT technologies, we can work with the course of variation to the world around the autistic youngsters. Development is being made, and before long parents and suppliers will use technological innovations to treat patients more readily with ASD.

REFERENCES

[1]   Frazier T. W., Youngstrom E. A., Speer Le, *et al* (2012) Validation of Proposed DSM-5 Criteria for Autism Spectrum Disorder. *J Ambient Acad Child Adolesc Psychiatry* 51:28–40

[2]   Brugha T. S., Spiers N., Bankart J., et al (2016) Epidemiology of Autism in Adults Across Age Groups and Ability Levels. *Br J Psychiatry* 209:498–503. https://doi.org/10.1192/bjp.bp.115.174649

[3]   Frazier T. W., Thompson L., Youngstrom E. A., et al (2014) A Twin Study of Heritable and Shared Environmental Contributions to Autism. *J Autism Dev Disord* 44:2013–2025

[4]   Hyman S. L., Levy S. E., Myers S. M. (2020) *Identification*, Evaluation, and Management of Children With Autism Spectrum Disorder

[5]   *Ministries of Health and Education* (2016) New Zealand Autism Spectrum Disorder Guideline

[6]   Haute Autorite de Sante (2018) Autism Spectrum Disorder: Warning Signs, Detection, Diagnosis and Assessment in Children and Adolescents. *HAS/Department Good Prof Pract* 1:1–46

[7]   Dalwai S., Ahmed S., Udani V, et al (2017) Consensus Statement of the Indian Academy of Pediatrics on Evaluation and Management of Autism Spectrum Disorder. *Indian Pediatr* 54:385–393

[8]   Chen Y., Zhou Z., Cao M., et al (2022) *Extended Reality (XR) and Telehealth Interventions for Children or Adolescents with Autism Spectrum Disorder: Systematic Review of Qualitative and Quantitative Studies.* Elsevier Ltd

[9]   Omar K. S., Mondal P., Khan N. S., et al. (2019) A Machine Learning Approach to Predict Autism Spectrum Disorder. *2nd Int Conf Electr Comput Commun Eng ECCE* 2019 7–9. https://doi.org/10.1109/ECACE.2019.8679454

[10] Kong Y., Gao J., Xu Y., et al (2019) Classification of Autism Spectrum Disorder by Combining Brain Connectivity and Deep Neural Network Classifier. *Neurocomputing* 324:63–68. https://doi.org/10.1016/j.neucom.2018.04.080

[11] Mellema C., Treacher A., Nguyen K., Montillo A. (2019) Multiple Deep Learning Architectures Achieve Superior Performance Diagnosing Autism Spectrum Disorder using Features Previously Extracted from Structural and Functional MRI. In: *International Symposium on Biomedical Imaging. IEEE*, pp 1891–1895

[12] Al Banna M. H., Ghosh T., Taher K. A., et al (2020) A Monitoring System for Patients of Autism Spectrum Disorder Using Artificial Intelligence. *Springer International Publishing*

[13]   Nogay H. S., Adeli H. (2020) Machine Learning (ML) for the Diagnosis of autism Spectrum Disorder (ASD) using Brain Imaging. *Rev Neurosci* 31:825–841. https://doi.org/10.1515/revneuro-2020-0043

[14]   Ke F., Choi S., Kang Y. H., et al (2020) Exploring the Structural and Strategic Bases of Autism Spectrum Disorders with Deep Learning. *IEEE Access* 8:153341–153352. https://doi.org/10.1109/ACCESS.2020.3016734

[15]   Ghosh T., Banna M. H. Al, Rahman M. S., et al (2021) Artificial Intelligence and Internet of Things in Screening and Management of Autism Spectrum Disorder. *Sustain Cities Soc* 74:103189. https://doi.org/10.1016/j.scs.2021.103189

[16]   Ghassemi N., Shoeibi A., Rouhani M. (2020) Deep Neural Network with Generative Adversarial Networks Pre-Training for Brain Tumor Classification based MR images. *Biomed Signal Process Control* 57:2–7

[17]   Dolz J., Desrosiers C., Wang L., et al *Deep CNN Ensembles and Suggestive Annotations for Infant Brain MRI Segmentation*

[18]   Delannoy Q., Pham C.-H., Cazorla C., et al (2020) SegSRGAN: Super-resolution and Segmentation using Generative Adversarial Networks-Application to Neonatal Brain MRI. *Comput Biol Med* 10:1–18

[19]   Li X., Dvornek N. C., Papademetris X., et al (2018) 2-Channel Convolutional 3D Deep Neural Network (2CC3D) for fMRI Analysis: ASD Classification and Feature Learning. In: *International Symposium on Biomedical Imaging. IEEE*, pp 1252–1255

[20]   Schuman, A. J. (2021). AI, Telehealth & Sensor-based Technologies Facilitate Autism Diagnosis. *Contemporary Pediatrics*, 38(10), 16–20.

[21]   Sohl, K., Kilian, R., Curran, A. B., Mahurin, M., Nanclares-Nogués, V., Liu-Mayo, S., ... & Taraman, S. (2022). Feasibility and Impact of Integrating an Artificial Intelligence–Based Diagnosis Aid for Autism Into the Extension for Community Health Outcomes Autism Primary Care Model: Protocol for a Prospective Observational Study. *JMIR Research Protocols*, 11(7), e37576.

[22]   Karami, B., Koushki, R., Arabgol, F., Rahmani, M., & Vahabie, A. H. (2021). Effectiveness of Virtual/ Augmented Reality–Based Therapeutic Interventions on Individuals With Autism Spectrum Disorder: A Comprehensive Meta-Analysis. *Frontiers in Psychiatry*, 12, 665326.

[23]   Rega, A., Mennitto, A., Vita, S., & Iovino, L. (2018). New Technologies and Autism: Can Augment Reality (ar) Increase the Motivation in Children with Autism. *INTED2018 Proceedings*, 4904, 4910.

[24]   Sula, A., Spaho, E., Matsuo, K., Barolli, L., Miho, R., & Xhafa, F. (2013, October). An IoT-based System for Supporting Children with an Autism Spectrum Disorder. In 2013 Eighth International Conference on Broadband and Wireless Computing, *Communication and Applications* (pp. 282–289). IEEE.

[25]   Syriopoulou-Delli, C. K., & Gkiolnta, E. (2022). Review of Assistive Technology in the Training of Children with Autism Spectrum Disorders. *International Journal of Developmental Disabilities*, 68(2), 73–85.

Automation and Computation – Vats et al. (Eds)
© 2023 the Author(s), ISBN 978-1-032-36723-1

A cryptographic and consensus-based analysis of the blockchain

Saurabh Jain & Adarsh Kumar
School of Computer Science, University of Petroleum and Energy Studies, Bidholi, Dehradun, India

ABSTRACT: One crypto-heavy technology, blockchain, has been getting much attention lately. Even though blockchain security and privacy have been the subject of many studies in the recent past, the cryptographic building blocks used in blockchain have not been fully explored. Hopefully, it will help cryptographers studying blockchain and financial engineers and managers looking for cryptographic-based solutions for blockchain-assisted and blockchain-based work. This paper primarily focuses on different consensus algorithms and how cryptographic and other evaluation parameters can affect and change blockchain performance.

Keywords: Blockchain, Consensus, Hashing, Cryptography, Proof of work, distributed network.

1 INTRODUCTION

The immutable ledger on a distributed network, which maintains data integrity, is at the core of blockchain technology. In a distributed network, hash values from previous blocks and the ledger, which holds all transaction information, are linked to each data block [1]. This distributed ledger is updated via a consensus technique, which allows the whole network to access the same data blocks [2]. These characteristics are critical for ensuring network data integrity [3]. Recent interest in adopting blockchain technology into Internet of Things (IoT) ecosystems has grown due to its unique characteristics. Multiple IoT devices and things are connected over the Internet. Sensors on these devices, connected to network nodes such as servers or networked computers, collect data and save, send, and analyze it. Every day, enormous quantities of data are created and stored [4,5]. Since of these resource constraints, IoT devices are vulnerable to malicious attacks because they favor less complicated security mechanisms. Because IoT end-node devices rely on hardware with limited resources and battery capacity, they must be energy efficient to enable long-lasting nodes. Smart cities, smart healthcare, smart transportation, and smart industry are just some domains where the Internet of Things is deployed and integrated. Aside from that, blockchain could solve low-constrained device difficulties such as increasing address space and identifying items [6,7]. For these reasons, lightweight blockchain networks are becoming increasingly popular. Many researchers explore ways to use light blockchain technology in their research areas. Specifically, lightweight blockchains provide advantages in the industrial sector (e.g., operation automatically, cost reduction, productivity, etc.) [8].

Further, this paper is divided into sections II to IV, and Section II discusses various consensus algorithms. Section III discussed different evaluation parameters that defined the complexity and execution time for various consensus algorithms in blockchain networks. The last section of this paper establishes some research findings for future scope.

DOI: 10.1201/9781003333500-43

2 BLOCKCHAIN CONSENSUS ALGORITHMS

Reaching consensus on a blockchain network is a vast and challenging undertaking. Once all network nodes have confirmed a new block, new transaction data will add to the blockchain. Once approved, a block can no longer be changed or removed. Blockchain is designed to be legitimate in a network devoid of trust and hostile users. Consensus algorithms are created in different ways. As blockchain technology advances, the number of these algorithms grows daily. However, this section will discuss and analyze the most used consensus methods in blockchain networks [9–13].

2.1 *Proof of Work (PoW)*

PoW is now one of blockchain's most widely used and secure consensus algorithms. Miners must work out challenging mathematical riddles before entering a new block into the ledger. After solving a challenge, the solution is sent to other miners for verification before being added to their ledger copy. Two competing forks will unlikely generate the next block simultaneously, as nodes must adopt a fork that conforms to the PoW algorithm. Using Proof-of-Work (PoW) validation technology, the blockchain's core network prevents double-spending. After verification, the miners authorize the transaction. If someone tries duplicating a transaction, the network will detect it as a forgery and reject it.

2.2 *Proof-of-Stake (PoS)*

PoS is a variant of the PoW algorithm that requires less CPU operations to mine. Despite having the same goal as Proof-of-Work (PoW), the algorithm used in this method is significantly different. A new block is definitively built based on its assets or stake in the proof-of-stake algorithm, while proof-of-work miners are compensated for solving mathematical puzzles and creating new blocks. This indicates that there is no block reward for the PoS method. Therefore, miners earn money from transaction fees. The Point-of-Sale (PoS) system has benefits and drawbacks, and its implementation is somewhat complicated.

2.3 *Byzantine Fault Tolerance (BFT)*

In an academic publication Robert Shostak, Leslie Lamport, and Marshall Pease described the "Byzantine generals' issue," a logical dilemma from which BFT is formed. BFT is employed to address the problem of a malicious or unreliable node. Assume that any community member may send contradictory transaction information to others. In such a scenario, the dependability of the blockchain is compromised, and no central authority can intervene to resolve the issue. PoW provides BFT with sufficient processing power to solve this issue. Alternatively, PoS requires a more definitive solution. Regularly, nodes will vote on which transactions are authentic. Using a PoS variant compatible with BFT seems to be the most practical method for authorizing blockchain transactions.

2.4 *Proof of Elapsed Time (PoET)*

As one of the consensus mechanisms for blockchains, Intel presents the Proof of Elapsed Time method, which, like Proof of Work, requires each miner to calculate the hash problem. Due to block manufacture, each block approver (Miners) is chosen as feasible and based on a reliable function in the shortest time. This election uses the Trusted Execution Environment (TEE) to safeguard its election process. The miner is randomly selected from the network. A specific Intel hardware machine provides TEE.

2.5 Proof-of-Capacity (PoC)

Unique is the algorithmic consensus approach. The cost of storage fluctuates here. The odds of mining the next block and receiving rewards are proportional to the amount of disc space available. The method creates a large number of "plots" that are stored on your hard disc before mining in a PoC. It will have a greater chance of discovering the following block if there are more plots. A substantial amount of storage space will be required to implement this strategy.

2.6 Proof of Game (PoG)

In this consensus technique, PoG participants play a game with new block participants based on the availability of resources. All participants are initially assumed to be honest and willing to provide their computational ability. A recent participant must provide their computational capacity, signature, and block information if interested in adding a block. The resources of the incoming participants will determine how the current blockchain participants start the game. If new participants have enough resources for computation, they are classified as resourceful, and a game requiring complex calculations is then played. If not, the lightweight game is favored, and the player is classified as having few resources. A multiplayer game in the resourceful category requires the new player to confirm the location of random bit/s in a challenge. The hash value's random bit position is demonstrated in the light game. A new player is deemed the winner and permitted to add the block if they successfully verify the challenge. Participants' sincerity is crucial to the success of this consensus method but maintaining a history of fair play is necessary to boost the trustworthiness and security of the current blockchain [13].

3 EVALUATION PARAMETERS FOR BLOCKCHAIN AND ITS CONSENSUS ALGORITHMS

With the rapid growth of blockchain technology and its implementation in various fields, many complex consensus algorithms with specific properties and applications have been developed. The primary objective of this paper is to identify the most influential factors influencing the performance of these consensus algorithms. In [14–21], the authors have conducted an extensive literature review and have identified various parameters for evaluating consensus algorithms:

3.1 Throughput

Throughput can be defined as the rate at which work is completed. The transaction throughput of a blockchain is the rate at which it can process transactions. The transaction throughput of a decentralized protocol is determined by the consensus technology employed by the blockchain platform. Proof-of-work (PoW) blockchains such as Bitcoin tend to have lower throughput than proof-of-stake (PoS) networks such as Cardano. A blockchain's block size, traffic, and transaction complexity are other factors that affect throughput. To increase a blockchain's throughput, developers use a variety of strategies, including rollups, sidechains, state channels, novel consensus procedures, and larger blocks.

3.2 TPS

Many people who talk about cryptocurrencies use the term "transactions per second" to talk about how fast transactions are handled. The number of transactions per second (TPS) shows how fast a computer system can handle requests. The number of transactions that can be done in a certain amount of time is a common way to measure the performance of systems that keep records and handle routine transactions. Think about how fast a platform or network can handle a transaction by how many transactions it can handle per second. The more transactions it can handle at once, the faster a transaction can be started, checked, and finished on a single platform. This rate is usually given in TPS but can also be provided in TPM or TPH (TPH).

3.3 Block latency

Block time or latency refers to the time required to verify and add transactions to a block. The block is then added to the existing chain of blocks within the blockchain. The time between when a value is transmitted to the network and when a decision is made regarding that value is known as latency.

3.4 Block verification time

The time required for a miner or other network validator to verify a block's transactions and add them to a blockchain is called block verification time. If a transaction is requested to be recorded in a blockchain, it will be regarded as a fair exchange between a sender and a recipient. Whenever a user conducts a transaction, his private key is utilized as a digital signature. Authorized transactions are recorded in a block and added to the blockchain database. Once a transaction has been added to a mining block, a single confirmation is generated.

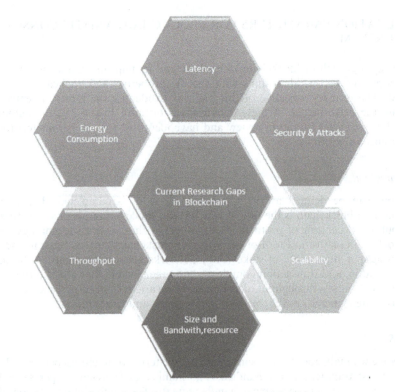

Figure 1. Various research gaps in blockchain and its consensus algorithms.

3.5 Block size

The capacity of a block increases as its size increases. Like other containers, a block has a limited capacity for storing data. The block size limit determines the maximum amount of data stored in a single Blockchain block. Even though crypto transactions keep only a small amount of data, the blockchain's storage capacity is limited relative to current standards. Even though the maximum size of a single Bitcoin block is 1 MB, this relatively small amount of data can still keep track of more than 2000 individual Bitcoin transactions.

3.6 Mining reward

A user that verifies transactions on a blockchain system is rewarded with a block reward. This is a portion of recently minted digital tokens. The people who verify transactions are collectively referred to as validators. However, depending on how the blockchain makes choices, they may be referred to as stakers or miners (proof-of-stake or proof-of-work, respectively.)

4   FUTURE RESEARCH FINDINGS AND CONCLUSION

Blockchain has been a fascinating information and communication technology development in the past 12 years. To fully understand how secure and private blockchain-based systems are, learning more about how they work cryptographically would be helpful. Since blockchain research and uses are still in their early stages, many problems, especially those related to cryptography, have yet to be solved [22]. Some of them are as follows.

(i) *Based on recent research in the literature, the authors have proved that there are plenty of scopes for the researcher to work on cryptographic primitives and evaluate blockchain solutions. Blockchain, a crypto-intensive technology, has been widely recognized as one of the fascinating information and communication technology innovations in recent years. A thorough investigation of the underlying cryptographic primitives would be beneficial for fully understanding the security and privacy of blockchain-based systems.*

(ii) *Cryptocurrencies based on proof of work, like Bitcoin (7 TPS), Ethereum (15 TPS), Litecoin (28 TPS), and Zcash (27 TPS), etc., have a low number of transactions per second. Comparing these cryptocurrencies with PayPal (193 TPS) and VISA (15000 TPS), it has been proved that Blockchain-based Cryptocurrency requires good performance in terms of TPS.*

(iii) *Most popular cryptocurrencies and many Blockchain-based applications use the SHA family algorithm to compute hash values, even PoW, PoS, PoET, etc., using hashing mechanisms for their consensus protocols. The cryptographic hash function and consensus algorithm are essential parameters for Blockchain network performance measurement in a Blockchain network.*

(iv) *Compatibility of lightweight cryptographic techniques with distributed ledgers: This is vital for connecting the Internet of Things (IoT) with blockchain technology With the exception of IOTA, none of the currently existing cryptocurrencies employ lightweight cryptographic techniques. This circumstance offers numerous opportunities for future investigation.*

REFERENCES

[1]   Nakamoto, S. 2008. Bitcoin: A Peer-to-peer Electronic Cash System. *Decentralized Business Review*, 21260.

[2]   Mingxiao, D., Xiaofeng, M., Zhe, Z., Xiangwei, W., & Qijun, C. 2017. A Review on Consensus Algorithm of Blockchain. In 2017 *IEEE International Conference on Systems, Man, and Cybernetics (SMC)* (pp. 2567–2572). IEEE.

[3] Zikratov, I., Kuzmin, A., Akimenko, V., Niculichev, V., & Yalansky, L. 2017. Ensuring Data Integrity Using Blockchain Technology. In 2017 20th *Conference of Open Innovations Association (FRUCT)* (pp. 534–539). IEEE.

[4] Ali, A., Rahouti, M., Latif, S., Kanhere, S., Singh, J., Janjua, U., ... & Crowcroft, J. 2019. Blockchain and the Future of the Internet: A Comprehensive Review. arXiv Preprint arXiv:1904.00733.

[5] Reyna, A., Martín, C., Chen, J., Soler, E., & Díaz, M. 2018. On Blockchain and its Integration with IoT. Challenges and Opportunities. *Future Generation Computer Systems*, 88, 173–190.

[6] Fernández-Caramés, T. M., & Fraga-Lamas, P. 2018. A Review on the Use of Blockchain for the Internet of Things. *Ieee Access*, 6, 32979–33001.

[7] Jain, S., & Kumar, A. 2022. A Security Analysis of Lightweight Consensus Algorithm for Wearable Kidney. *International Journal of Grid and Utility Computing*, 13(5), 505–525.

[8] Seok, B., Park, J., & Park, J. H. 2019. A lightweight hash-based blockchain architecture for industrial IoT. *Applied Sciences*, 9(18), 3740.

[9] BangBit Technologies, "What is Consensus Algorithm in Blockchain & Different Types of Consensus Models," Medium, 14-May-2018. [Online]. Available: https://medium.com/@BangBitTech/what-is-consensus-algorithm-in-blockchain-different-types-of-consensus-models-12cce443fc77. [Accessed: Aug-2022].

[10] Chaudhry, N., & Yousaf, M. M. 2018, December. Consensus Algorithms in Blockchain: Comparative Analysis, Challenges and Opportunities. In 2018 12th *International Conference on Open Source Systems and Technologies (ICOSST)* (pp. 54–63). IEEE.

[11] "Slimcoin," *Slimcoin.info*. [Online]. Available: https://slimcoin.info/. [Accessed: Aug-2022].

[12] "Decred - Secure. Adaptable. Sustainable," *Decred.org*. [Online]. Available: https://www.decred.org/. [Accessed: Aug-2022].

[13] Kumar, A., & Jain, S. 2021. Proof of Game (PoG): A Proof of work (PoW)'s Extended Consensus Algorithm for Healthcare Application. In *International Conference on Innovative Computing and Communications* (pp. 23–36). Springer, Singapore.

[14] Nguyen, G. T., & Kim, K. 2018. A Survey About Consensus Algorithms Used in Blockchain. *Journal of Information Processing Systems*, 14(1), 101–128.

[15] Bano, S., Al-Bassam, M., & Danezis, G. 2017. The Road to Scalable Blockchain Designs. *USENIX; login: magazine*, 42(4), 31–36.

[16] Croman, K., Decker, C., Eyal, I., Gencer, A. E., Juels, A., Kosba, A., & Wattenhofer, R. 2016. On Scaling Decentralized Blockchains. In *International Conference on Financial Cryptography and Data Security* (pp. 106–125). Springer, Berlin, Heidelberg.

[17] "Bitcoin, Litecoin, Ethereum Block Time Chart," BitInfoCharts. [Online]. Available: https://bitinfo-charts.com/comparison/confirmationtime-btc-ltc-eth.html. [Accessed: 15-march-2022].

[18] Xu, X., Sun, G., Luo, L., Cao, H., Yu, H., & Vasilakos, A. V. 2021. Latency Performance Modeling and Analysis for Hyperledger Fabric Blockchain Network. *Information Processing & Management*, 58 (1), 102436.

[19] Denisova, V. 2019. *Blockchain Infrastructure and Growth of Global Power Consumption*. 670216917.

[20] Ciaian, P., Kancs, D. A., & Rajcaniova, M. 2021. *Interdependencies Between Mining Costs, Mining Rewards and Blockchain Security*. arXiv preprint arXiv:2102.08107.

[21] Bamakan, S. M. H., Motavali, A., & Bondarti, A. B. 2020. A Survey of Blockchain Consensus Algorithms Performance Evaluation Criteria. *Expert Systems with Applications*, 113385.

[22] Wang, L., Shen, X., Li, J., Shao, J., & Yang, Y. 2019. Cryptographic Primitives in Blockchains. *Journal of Network and Computer Applications*, 127, 43–58.

Automation and Computation – Vats et al. (Eds)
© 2023 the Author(s), ISBN 978-1-032-36723-1

An analytic study of various machine learning algorithms to predict heart related diseases

Shivam Chawla, Aryan Agarwal, Aryan Ratra & Manisha Aeri
Graphic Era Hill University, Dehradun, Uttarakhand, India

ABSTRACT: Heart Disease, popularly known as cardiovascular disease, is one of the most common causes of deaths worldwide. Some Machine learning algorithms are used here to classify whether a person is suffering from heart disease or not. Some commonly used algorithms used by most of the researchers in this field are KNN Classifier, Support Vector Machine, Decision Tree, etc.

1 INTRODUCTION

A heart attack which is analogous to acute myocardial infarction (AMI) is one of the most serious diseases in the segment of cardiovascular disease. It occurs due to the interruption of blood circulation to muscle of the heart which damages the heart the muscle. Diagnosing heart disease is also a crucial task. The symptoms, physical examination, and understanding of the different signs of this disease are required to diagnose heart disease [17].

With advancements in Science and Technology, we human beings have discovered the cure for many diseases but have also been responsible for the birth of many new diseases. It is the cause of 17.7 million deaths every year. Which is about 31% of all global deaths. "Machine Learning is a way of Manipulating and extraction of implicit, previously known/known and potential useful information about data" [16]. So, I decided to create a model that predicts whether a person's heart is healthy or not. Several Symptoms are associated with heart diseases. With the help of the knowledge of doctors and other experts. Different values are assigned to different attributes and dataset is prepared. So many ML engineers use this knowledge to train a model using this data and predict a person's heart health.

Medical Organization all over the world collect data on various health related concerns. So the data is of large size, which is impossible for a human brain to understand. It is too over-whelming. So, it is processed using Machine learning.

To design such a model, I have implemented some algorithms and managed to obtain accuracy up to 98.5365%. Before these algorithms, I read articles saying that most people do not have enough money to get a proper checkup.

This ignorance is one of the most contributing causes of death of many people. Main motivation behind developing this project is to help an average person check his heart. In recent times, we have seen an exponential increase in the amount of data being generated by electronic devices like PCs, Laptops, Phones and Tablets etc. and other IOT-based devices [1–14]. Machine learning is a highly broad and complex topic, and its application and

DOI: 10.1201/9781003333500-44

breadth are expanding daily. In order to forecast and determine the accuracy of the provided dataset, machine learning incorporates a variety of classifiers from supervised, unsupervised, and ensemble learning. Given that it will benefit many people, we can use that information to our HDPS project. It is the main factor in adult deaths [15].

Nowadays, AI is playing an important role in the field of cardiology, appreciations to massive advancements in equipment, big data, knowledge storage, acquisition, and recovery. Using various data mining techniques, researchers used preprocessing methods on the data to make verdicts using various ML models. In the cataloguing of genetic cardiac illnesses and control subjects, a widespread set of ML algorithms with their variations is used to predict the early stages of heart failure.[18]. KNN, DT, SVC, LR, and RF machine algorithms are examples of heart attack prediction algorithms. Machine learning approaches can be divided into three categories: Supervised ML: task drive, labeled data (classification/regression); Unsupervised ML: data-driven, unlabeled data (clustering); Reinforcement Learning: learning from mistakes (playing games) [19,20].

2 APPROACH AND METHODOLOGY

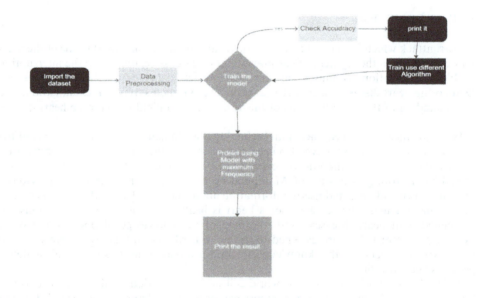

• Algorithms Used

KNN: It is one of the most widely used Supervised learning algorithms in disease prediction by many researchers. It is also used for regression. It is one of the most efficient algorithms for many smaller datasets.

In this algorithm, value of k is decided, by default value of k is 5. Then Euclidean distance is calculated between Neighbours, and out of closest k members count number of neighbors in each category and assign new data points to the category having maximum number of points.

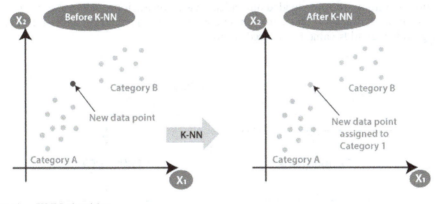

Figure 1. KNN algorithm.

Euclidean distance between neighbors is calculated as:

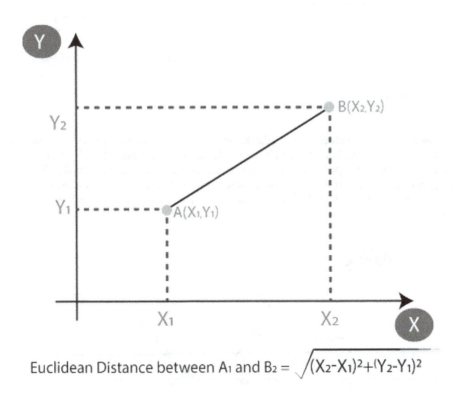

Euclidean Distance between A_1 and $B_2 = \sqrt{(X_2-X_1)^2+(Y_2-Y_1)^2}$

Decision Trees: It is one of the most popular algorithms used for classification and regression purposes. These algorithms can mimic human style of thinking. It is quite easy to understand as it represents a tree–like structure. In case of classification, it decides the class after using a voting system, but in case of regression it calculates the average of decision trees.

In this algorithm, trees are divided into subsets based on attribute values and portioning is done again on the subsets. This partitioning continues in a recursive manner till the partitioning no longer adds value to the predictions.

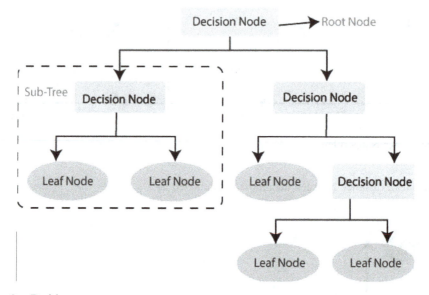

Figure 2.   Decision tree.

Support Vector Machine: One of the most popular classification and regression algorithms. Firstly, a hyperplane is found b/w to differentiate b/w classes. Firstly, points are kept separated and then mapped onto that space and classification is done accordingly.

Hyperplane is linear if class is differentiated b/w two features and nonlinear in case of more than two features. It is difficult to visualize hypervisor's shape in case of more than 3 features.

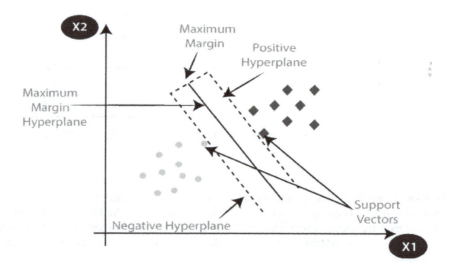

Figure 3.   Support vector machine.

Random Forest Classifier: An algorithm based on the concept of ensemble learning. It combines multiple classifiers to solve a problem. It is possible that prediction of some trees may be wrong, and some may be correct.

But combination of these output will be correct. It takes much less training time as compared to other algorithms and can predict correct output when large proportions of values are missing.

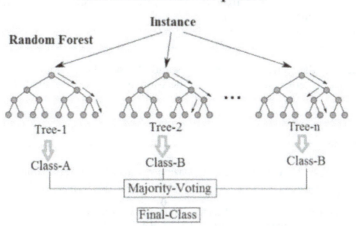

Figure 4.   Random forest classifier.

Logistic Regression: An algorithm used for predicting the probability of dependent variable using independent variables. It basically gives a probabilistic value instead of discrete value such as 0 or 1. It takes 2 assumptions:

->Dependent variables are categorical in nature.
->Independent variable has less multi-collinearity.

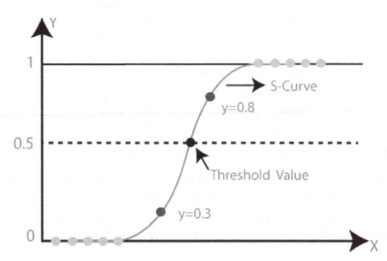

Figure 5.   Logistic regression.

3 RESULTS

The results of different algorithms are shown in this section. The metrics used are confusion matrix and accuracy score.

Algorithm	True Positive	False Positive	True Negative	False Negative	Accuracy_Score
Logistic Regression	24	24	8	5	0.786885
SVM	20	25	12	4	0.803278
KNN	102	100	0	3	0.985365
Decision Tree	102	100	0	3	0.985365
Random Forest	93	98	9	5	0.931707

When considering the metrics, Support KNN and Decision Tree comes out to out to be the best algorithms, in terms of accuracy. So, decision tree is used for prediction as it faster than KNN.

REFERENCES

[1] Vats S., Singh S., Kala G., Tarar R. and Dhawan S. "iDoc-X: An Artificial Intelligence Model for Tuberculosis Diagnosis and Localization," *J. Discret. Math. Sci. Cryptogr.*, vol. 24, no. 5, pp. 1257–1272, 2021.

[2] Vats S., Sagar B. B., Singh K., Ahmadian A., and Pansera B. A., "Performance Evaluation of an Independent Time Optimized Infrastructure for Big Data Analytics that Maintains Symmetry," *Symmetry (Basel).*, vol. 12, no. 8, 2020, doi: 10.3390/SYM12081274.

[3] Vats S. and Sagar B. B., "An Independent Time Optimized Hybrid Infrastructure for Big Data Analytics," *Mod. Phys. Lett. B*, vol. 34, no. 28, p. 2050311, Oct. 2020, doi: 10.1142/S0217984920 50311X.

[4] Vats S. and Sagar B. B., "Performance Evaluation of K-means Cluster- ing on Hadoop Infrastructure," *J. Discret. Math. Sci. Cryptogr.*, vol. 22, no. 8, 2019, doi: 10.1080/09720529.2019.1692444.

[5] Agarwal R., Singh S., and Vats S., "Implementation of an Improved Algorithm for Frequent Itemset Mining Using Hadoop," in 2016 International Conference on Computing, Communication and Automation (ICCCA), 2016, pp. 13–18. doi: 10.1109/CCAA.2016.7813719.

[6] Agarwal R., Singh S., and Vats S., *Review of Parallel Apriori Algorithm on Mapreduce Framework for Performance Enhancement*, vol. 654. 2018. doi: 10.1007/978-981-10-6620-7_38.

[7] Bhati J. P., Tomar D., and Vats S., "Examining *big data management techniques* for Cloud-based IoT Systems," In *Examining Cloud Computing Technologies Through the Internet of Things, IGI Global*, 2018, pp. 164– 191.

[8] Sharma V. et al., "OGAS: Omni-directional Glider Assisted Scheme for Autonomous Deployment of Sensor Nodes in Open Area Wireless Sensor Network," *ISA Trans.*, Aug. 2022, doi: 10.1016/j. isatra.2022.08.001.

[9] Sharma V., Patel R. B., Bhadauria H. S., and Prasad D., "NADS: Neighbor Assisted Deployment Scheme for Optimal Placement of Sensor Nodes to Achieve Blanket Coverage in Wireless Sensor Network," *Wirel. Pers. Commun.*, vol. 90, no. 4, pp. 1903–1933, 2016.

[10] Sharma V., Patel R. B., Bhadauria H. S., and Prasad D., "Policy for Planned Placement of Sensor Nodes in Large Scale Wireless Sensor *Network,"* KSII Trans. Internet Inf. Syst., vol. 10, no. 7, pp. 3213–3230, 2016.

[11] Sharma V., Patel R. B., Bhadauria H. S., and Prasad D., "Deployment Schemes in Wireless Sensor Network to Achieve Blanket Coverage in Large-scale Open Area: A Review," *Egypt. Informatics J.*, vol. 17, no. 1, pp. 45–56, 2016.

[12] Bhatia M., Sharma V., Singh P., and Masud M., "Multi-level P2P Traffic Classification Using Heuristic and Statistical-based Techniques: A Hybrid Approach," *Symmetry (Basel).*, vol. 12, no. 12, p. 2117, 2020.

[13] Vikrant S., Patel R. B., Bhadauria H. S., and Prasad D., "Glider Assisted Schemes to Deploy Sensor Nodes in Wireless Sensor Networks," *Rob. Auton. Syst.*, vol. 100, pp. 1–13, 2018.

[14] Vats S. and Sagar B. B., "Datalake: A Plausible Big Data Science for Business Intelligence," 2019.

[15] Harshit Jindal et al. 2021 *IOP Conf. Ser.: Mater. Sci. Eng.* 1022012072

[16] Soni J., Ansari U., Sharma D. & Soni S. (2011). Predictive Data Mining for Medical Diagnosis: an Overview of Heart Disease Prediction. *International Journal of Computer Applications*. 17(8), 43–8

[17] Nandal N., Goel L., Tanwar R., "Machine Learning-based Heart Attack Prediction: A Symptomatic Heart Attack Prediction Method and Exploratory Analysis" *F1000Research* (2022) 11 1126

[18] Hassan, C. A. ul, Iqbal, J., Irfan, R., Hussain, S., Algarni, A. D., Bukhari, S. S. H., Alturki, N., & Ullah, S. S. (2022). Effectively Predicting the Presence of Coronary Heart Disease Using Machine Learning Classifiers. *Sensors*, 22(19).

[19] Gour, S.; Panwar, P.; Dwivedi, D.; Mali, C. A Machine Learning Approach for Heart Attack Prediction. In *Intelligent Sustainable Systems*; Springer: Singapore, 2022; pp. 741–747.

[20] Available online: https://www.potentiaco.com/what-is-machine- learning-definition-types-applications-and-examples/ (accessed on 10 January 2022).

Automation and Computation – Vats et al. (Eds)
© 2023 the Author(s), ISBN 978-1-032-36723-1

Crop recommendation system: A review

Vikrant Sharma*, Satvik Vats*, Priyanshu Rawat* & Madhvan Bajaj*
Computer Science and Engineering, Graphic Era Hill University, Dehradun, India

ABSTRACT: Crop recommendation system is an emerging system that is being used in the agricultural sector. It is used to provide a better guide to the farmers regarding the method of cultivation of crops, usage of the right fertilizers, and selecting the best crop to cultivate based on the current demand. The researchers have proposed various fascinating for crop recommendation with the help of fuzzy logic, sensors, neural networks, etc.

Keywords: Crop recommendation, farming, Fuzzy logic, Ensemble, sensors

1   INTRODUCTION

Agriculture has been a key occupation for a major part of the world. Even after contributing over 6.4% of world economic production, it still couldn't match the pace with the fastening technology of the world [1]. The degree of improvement that the agriculture sector needed or deserved wasn't provided which led to monotonous cropping methods. Such methods not only affected the income of farmers but also had an adverse impact on the environment.

Lately, the introduction of technology in agriculture has led to many innovations in cropping patterns. Farmers are now much more dependent on technologies for their crop production. Precise agriculture plays a very crucial role and comes at the front part to solve this issue [2,3]. One such advancement is the recommendation system. It suggests the right amount of fertilizers, on what soil type, on which weather condition and temperatures by using various sensors, machine learning techniques, algorithms, IOT, GPS and many more which claims maximum accuracy and effectiveness (in terms of prediction and usability) to predict the precise amount of crop suitable for the farmers to grow.

When it comes to prediction, machine learning (ML) is the best technology to opt for. It is a part of artificial intelligence where a large set of datasets is trained, and an algorithm is prepared on the basis of which the output is generated for various parameters and ultimately the human effort is reduced, Sensors are used to gather data that can be used to train a machine learning model. This data can be used to make predictions or take actions based on the input from the sensors [4].

The use of ML-algorithm such as, K nearest neighbor algorithm not only predicts the suitable crop but also recommends the required soil nutrients and fertilizers as per the need [5].

In the same way with the help of fuzzy logic, the system is developed which provides the detailed report of the required nutrients for the crop at a regular interval as the quality of soil keeps changing from season to season. So, with the help of a fuzzy-logic based system, the required nutrients by taking the inputs like nitrogen and phosphorus levels can be predicted [6]. Crop recommendation system aims to get into use on a wider scale where every farmer can be benefited from it which will result in higher yield and income that may contribute to

---

*Corresponding Authors: vsharma@gehu.ac.in, svats@gehu.ac.in, priyanshurawat152002@gmail.com and bajajmadhvan21@gmail.com

DOI: 10.1201/9781003333500-45

the economy of the nation. In this paper, various models are used for crop recommendation systems, which are studied, classified, and analyzed [7]. The classification of the recommendation system is shown in Figure 1.

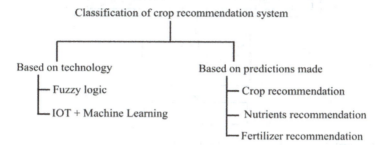

Figure 1.   Classification of crop recommendation system.

2 FUZZY LOGIC

2.1 *Fuzzy Decision Support System (FDSS)*

In agriculture predictive models, information on planting time, crop rotation, crop prediction, harvesting time are provided to the farmers with the help of information provided on the basis of land to be cultivated. The author has proposed a model in which algorithms of Machine learning are used to address the issues such as, Soil electrical conductivity (EC) prediction, soil moisture prediction [8], and crop yield prediction [9]. Soil moisture is predicted with the help of K-means with a support vector machine for regression (SVR). Crop suitability is predicted by the analysis of nutrients of the soil, which is done by using advanced decision tree algorithms. Rice yield is predicted by a vector machine.

Machine learning fails when it has to deal with fundamental issues of leadership, uncertainty, and lack of knowledge. Several methods have been developed by hybridizing machine learning algorithms using coarse and fuzzy sets [10]. The approximate set is generated using Johnson's algorithm and blended with the software set to handle undefined conditions and reduce parameterization issue.

Existing methodologies are not conducive to uncertainty in the data, so we use a combination of fuzzy and sparse sets to deal with boundary conditions. Pre-processing is done through discretization and fuzzification [11]. The output of these two methods is fed into a rough set. It implements an induction algorithm with four rules and analyzes the results of all methods to select the best one.

Implement and analyze various rule inference methods to find the best rule inference method. Rule induction methods are classified into Local and Global. Methods such as CN2, AQ, and LEM2 are used to generate fuzzy rules. (see Figure 2).

2.2 *Mamdani Fuzzy Inference Model (CRS-MFI)*

Recommendation systems give ideas to the users to pick specific things from a substantial pool of things. The motivation for this study is to plan a recommendation system for the farmers for suggesting earlier thought with respect to a yield that is reasonable as indicated by the area of the farmer dependent on the climate state of the earlier months. This system moreover suggests different seeds, pesticides, and machines according to the area and preferences of the farmer when purchasing on the internet [12]. The model uses a cosine comparability measure to predict rice yield in Kharif season in Odisha, India by finding similar users based on farmer's area and fuzzy logic.

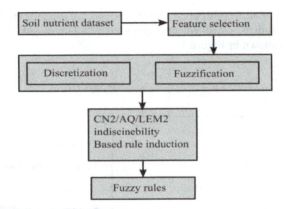

Figure 2.   Proposed system architecture (redrawn from [9]).

The proposed structure is implemented in the Mamdani fuzzy inference model. The outcomes uncover that system gives an earlier thought on the crop before planting of seeds. The author has provided the information about input variable, and output variable and has proposed the experimental design for the model and then decoded it. In input variable description, the system uses three fuzzy variables, and every variable is represented by three different terms, i.e., a1 = Temperature, a2 = Humidity, and a3 = Rainfall.

These can be measured as High, Medium, and Low. In output variables high, Medium, and Low are the terms used for rice crop in the given model. For a fuzzy rule-based system, the rules play a significant part. The rules express the connection between the information and yield factors, furthermore, address the information base of FRBS [13]. There are three inputs and three measures are taken so, the author has considered all the twenty-seven methods in the system. The two examples are:

If a1, a2 and a3 are low then the output would be low. If a1, a2, and a3 are medium then the output would be high.

In experimental design, the framework takes the farmer's state and district then takes the types of season and name of the crop as input and recommends the crop is suitable if the expectation of crop in that season is high [14]. The cosine similarity measure is utilized to discover similar farmers based on area from the data set. The formula is:

$$\cos(e1, e2) = (e1, e2)/|e1| \ |e2| \qquad (1)$$

For good recommendations, the system also provides pesticides, fertilizers, seeds and machines.

3   IOT + MACHINE LEARNING

3.1   Crop recommendation system using Neural Networks (CNN)

In this model, a neural network is used to anticipate the best sort of harvest [15]. There are various sensors that are deployed. To detect the humidity, a humidity sensor is used; a temperature sensor, a soil sensor, and a soil moisture sensor is also used. These are the most significant components of this model. ESP8266 (Wi-fi module) will get the data in digital form from those sensors. Sensors detect the various parameters of the soil. The whole model is going to work on the basis of these sensor readings [16]. The ESP8266 Wi-Fi module will get the readings of different parameters of the soil from the sensors. By using the MQTT (Message Queuing Telemetry Transport) protocol, ESP8266 aggregates the data which were

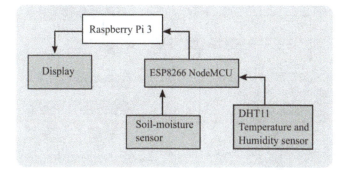

Figure 3.   Soil sensor architecture (Redrawn from [3]).

collected by the sensors to Raspberry Pi (see Figure 3). MQTT is a very light protocol that follows ISO standard messaging protocol and belongs to TCP/IP protocol. It is very beneficial for the recommendation system.

The DHT11 humidity, the soil moisture sensor and the temperature sensor will monitor the different parameters of soil. The data is sent to ESP8266 Node MCU. Then MCU sent the data to Raspberry Pi. Raspberry Pi is the computation-unit of the whole system. It has Bluetooth and wireless LAN as its important features. It performs actions according to the inputs received from ESP8266 [17].

3.2   Intelligent Crop Recommendation System (ICRS)

In this research paper [, the author has used 4 techniques Decision Tree, K-NN, Random Forest, and Neural Network. To ensure that Agro Consultant has the very best possible accuracy, the author has individually implemented the four algorithms mentioned above. Then the performance of the four was compared and therefore the one with the best accuracy was selected for the model. After applying the information to various machine learning algorithms, a trained crop recommendation system model is obtained [18]. Certain crops may be best suited to a given soil and weather. If all farmers in an area use Agro Consultant during the same season, they will surely get the same recommendation. However the author knows that if all the farmers in the area grew the same crops, they would end up losing on their yields. Each crop requires rainfall. If this requirement is not met, the yield will be reduced. Conversely, excessive rainfall can adversely affect crop yields again. Therefore, rainfall can be an important parameter for crop cultivation.

3.3   Sensors driven Agricultural Recommendation Model (AI-ARM)

In the recommendation model, this section discusses the proposed model based on sensor network. The sensor-based data management architecture diagram requires 3 steps of significant importance: acquisition of data, communication using digital means, and processing of data according to our needs. The information collection is done by different sensing devices such as, moisture sensor(soil), pH-sensor, electromagnetic wave (EMW) sensor and salinity sensor. You can also use Wi-Fi, which is used to transmit data in remote farmlands. For superior information processing, the information is then transmitted to the cloud via the web [19]. The cloud engine used here is (AWS) Amazon Web Service, so the data stored is utilized for machine learning so that analysis of data can be done (see Figure 4). Electromagnetic sensors typically measure texture(soil), drainage(internal), water content (available), matter(organic), exchange capacity(cation), saturation and carbonate. The data used in this experiment were collected from several villages in the Tiruvannamalai and

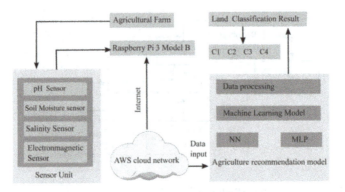

Figure 4.   Architecture diagram of proposed model (redrawn from [13]).

Vellore districts of Tamil Nadu, India. A short list includes various parameters that affect performance, taking into account the opinions of our experts. These datasets contain a characteristic combination of soil, climate and groundwater features. Before training Multilayer Perceptron (MLP) and Neural Network (NN) models, it is important to organize knowledge properly. The actual data received from these sensors is unevenly distributed and the additional information cannot be used properly when training and testing the MLP and NN models. Therefore, for efficient processing, the input features gets normalized and the categorical variables gets converted to numeric data through the use of information tag encoder [20]. The normalized data set then gets divided into independent training and test sets at a ratio of 75:25.

4   CROP RECOMMENDATION

4.1   *Demand based Crop Recommender system for farmers(DCR)*

India being the agricultural land which produces a very large amount of crop but still, its contribution (14%) to the GDP of India is not significant as expected. One of the reasons is the lack of knowledge among the farmers for adequate crop planning. Since we don't have any such system so in this paper the author has given an approach towards building such a system through which one can predict the prices and yield (crop) that a farmer can get in the future through his land, by analyzing the past data patterns [21]. With the help of techniques like sliding-window and non-linear regression, this issue can be resolved and the specific crop can be predicted based on factors affecting the crop production like rainfall, temperature, market price, land area, and past crop production. The analysis is done for the different districts of Tamil Nadu (India). For a few years data is collected and based on that the prediction is done. With the help of data mining techniques, the dataset is classified into three categories that is Excess, Scare, and Neutral on the basis of change in the market price and based on this classification the demand is inferred. With the help of an outlier algorithm, sudden change or misplaced value is found, for example it will always indicate when there is a sudden increase in the crop price. Demand gets calculated by comparing total consumption with total crops cultivated of that same crop type along a specific period of time [22]. For example, if the demand is high, will be recommended as an alternate crop else it won't be recommended. The crop is being recommended on various parameters like land area, consumption and demand grade. It may happen that multiple crops are being recommended then in this case the crop with higher demand grade will be given priority. To help the farmers have better interaction with the interface it is having a text translator which will translate the texts from English to their native language with the help of linguistic rule set by parsing done in the grammar tree. Thus, this system successfully helps in suggesting the farmers about the high demanding crop and helps to cultivate it.

4.2 Crop recommendation technique using Ensemble Technique (CRS-ET)

The agricultural sector has used machine learning (ML) algorithms to create efficient and cost-effective solutions to the challenges faced by farmers. Researchers were able to use PC simulations to conduct initial tests to gauge how well the test might perform when tested with a modified lower atmosphere, to soil composition, climate design and other variables. Ensemble is a technique for building a hunch pattern by combining multiple patterns. The main reason to use an aggregation framework is that it provides a classifier that outperforms individual classifiers [23]. Ensemble uses two frameworks, dependent and independent: in the dependent framework, the output of one classifier is used to develop the next classifier. The second method involves an independent method in which each classifier generates class labels independently (see Figure 5). Random Forest, Naive Bayes, and Linear SVM are three independent primitive learning tools used to build ensemble models. The main advantages of using the Random Forest algorithm are that it has better accuracy, is more stringent on outliers, is faster than bagging and boosting, and provides basic and easy parallelization. Naive Bayes has several applications, such as real-time prediction, probabilistic prediction of different target attribute classes, spam filtering, and when combined with collaborative filtering helps build recommender systems. The main advantage of linear SVM is that it works well with very large data sets and has high accuracy [24]. Linear SVMs also provide better performance when working with multidimensional data. This is the class with the most votes. This method is also called multiple voting. This methodology has mostly been used as an integrative strategy to consider recently proposed strategies.

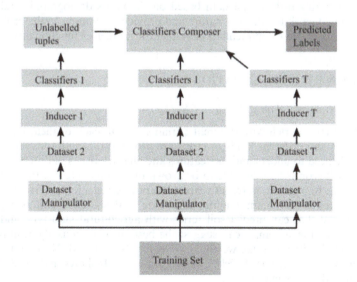

Figure 5.   Ensemble framework (Redrawn from [15]).

4.3   Crop recommendaer system using Weather Forecasts (CRS-WF)

Weather forecasting is the application of science and technology to interesting atmospheric conditions at a particular location. Many industrial and agricultural sectors rely primarily on weather dynamics. Data collected over the last three years are from the UCI dataset, which uses his ANN for accurate weather forecasts, and recommended crops are reasonable based on weather patterns. This article addresses the problem of predicting the next day's cumulative precipitation based on data collected the previous day. The data used in the research work were used for regular online predictions [25]. Various weather conditions make up the weather, such as snow, rain, sunlight, changes in temperature, and wind at a particular time and place. The term climate is used to describe general characteristics of weather over long periods of time. Data preprocessing improves the quality of the input and affects the performance and accuracy of the analysis. A key step in this process is data preparation. This step takes old, useless input and transforms it into new data that can be used in the data mining process. If the supposedly prepared data is not available, the data mining algorithm cannot confirm that the data is correct or indicate a run-time error. The algorithm works at best, but the results are unreliable or inaccurate.

According to the author in this article, the forecasting process is driven by Ann. Neuron selection is done through trial and error based on the input data provided. The latter algorithm is used to realize weather forecast improvements in intelligent farming systems and provide smart city development improvements.

Replicating human intelligence and reasoning has been facilitated by a branch of AI known as (ANN) artificial neural networks [26]. Artificial neural networks can easily recognize patterns such as weather and process and remember data. A given I/O function and its weight determines the output of the ANN. Artificial neural networks can solve important and extremely complex problems typical of existing technologies.

In this research project, we will use RDF triplets to design a CBR case. In this case, the request data provided by the user is used to build RDF triples using the CBR mechanism. RDF triplets are emitted from the CBR engine using the REASONER component, which operates on the CBR precedent ontology. The system may use previous experience recommendations for similar users to provide appropriate recommendations to users who have missed previously obtained interest data based on the user's demographic and preference data. The proposed work is hybrid in nature and provides a solution to the problem of lack of data that is part of the recommendation system for influencer functionality.

5 NUTRITION RECOMMENDATION

5.1 *Crop suggestion based on soil series (CSML)*

The agricultural land is rich with different variants of soil, each of them having different kinds of features and different kinds of crops are grown on it, and not every type can be grown on it. We need to understand the features and characteristics of different soil types to understand which crop is best suitable to be grown on it. To overcome this issue Machine learning can come into action, in recent years it has progressed a lot, and still is an emerging and challenging research domain in agricultural data analysis. Through this paper a model has been proposed that can predict soil series with agricultural land type and as per the prediction given by the system, it can suggest the best crop. Various algorithms have been used for the soil classification like weighted k-Nearest Neighbor (k-NN), Gaussian kernel-based Support Vector Machine (SVM), and Bagged Trees. It shows that the SVM model is better than any other existing models.

The experiment is done on the basis of a data set of 2045 samples of 10 types of soil types of Pune district, India. The experiment is carried out on a soil series of sic Upazila of Khulna district, Bangladesh. Various algorithms have been applied such as Naïve Bayes, JRip, J48, k-means, Random Tree and, Apriori for classification tasks and predicting suitable crops.

The dataset has been collected from Indian Meteorological Department, Statistical Institution, and Agriculture department. K-means clustering has been used to form a table which has been divided into four clusters ranging from 1 to 5 cm from the total no of raining years. The values of those four clusters have different values as minimum, mean and maximum. Then the comparison of the dataset takes place of sowing and average production by applying k-means clustering and MLR techniques.

The proposed method involves two phases training phase and testing phase and two datasets that is crop dataset and soil dataset. Various chemicals attributes are used in the method like pH value, Zinc, Boron, Calcium etc.

5.2 Diagnosis and Recommendation Integrated System (DRIS)

Brazil is a country with a very high rate of cost on cotton. And there is not a good management technique to give a better recommendation for the fertilizers and seeds. So, to bring the cost down on cotton it is necessary to have a good amount of production. Nutrients play a very important role in production, so the nutritional analysis of plants is very important to understand the requirements and influence of nutritional balance on yield. To satisfy this purpose, a Diagnosis and Integrated Recommendation system (DIRS) came into existence.

To solve this problem, the authors of [27] proposed a model called DRIS. This diagnostic system consists of a double ratio of all nutrients (N/P, N/K, K/N, K/Ca, etc.) in diagnosis. However, the use of DRIS requires the development of DRIS norms, and these norms have been developed based on farmland data or experimental units. Most importantly, your data should be from a high-income population.

For each nutrient under diagnosis, provides a DRIS index, which can be either positive or negative, representing the effect of each nutrient on the nutrient balance of the plant. When the DRIS index is close to 0, the nutritional balance is much closer.

For each nutrient, the DRIS index must be interpreted, which depends on the value and sign of the index. A value close to 0 means that the plant is getting enough nutrients.

The purpose of this study is to implement a Diagnostic and Recommendation System (DRIS) to study the nutritional status of cotton crops by determining their nutrient balance and current requirements. DRIS shows that the lower the Nutrient Balance Index (NBI), the higher the yield. It simply means you have to have a low NBI to get high returns. In the study, sulfur, boron and zinc showed higher frequency with the highest phytonutrient requirements. Magnesium and manganese were both required by the plants in small amounts, meaning that both limited high yields due to tissue overabundance.

6 FERTILIZER RECOMMENDATION

6.1 US Corn System (USCS)

For corn crops, nitrogen is a very significant factor. For economic and environmental reasons, nitrogen management became popular. In this study, a model is introduced based on the satellite images which recommend nitrogen fertilization at high resolution.

The sub-field of the executive zones has been depicted by covering the soil review. The given system is used to find the range, growth, and nitrogen uptake of the crop. Based on the target rate, the sidedress rate is calculated (see Figure 6). Nitrogen loss is also calculated by denitrification and the nitrogen left in the soil.

The farmer needs to know the nitrogen management, so they can make the decision on how much nitrogen fertilizers are required. The system is made up on Amazon EC2. First, it is required to find the Nitrogen application rate (see equation 2). The target is dependent on the field productivity and nitrogen credits.

$$T = a_0 + a_1.(U + E) - C \qquad (2)$$

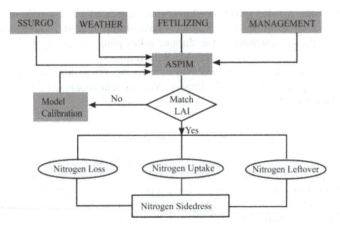

Figure 6. Workflow for generating nitrogen sidedress (Redrawn from [19]).

Where, T = Nitrogen application Rate
 U = Field Average Yield
 E = Adjustment Term
 C = Credit for Soil Nitrogen

From the Soil Survey Geographic database, the soil PH, soil hydraulic properties, and SOM are taken and resampled. The National Climate Data Center provided real-time data such as radiation, precipitation, and temperature. The inputs by users consist of seeding rate, sowing date, and maturity which can be taken from the NASS report if not available.

Now, the next step is model simulation. The APSIM provided the Nitrogen Status in the soil as well as in the crops. The model has worked on the virtual grid and then provided the output. The last step is to find sidedress Nitrogen rate which is denoted by S (see equation 3).

$$S = T + L - U - Le \qquad (3)$$

Where,
 L = Nitrogen loss by denitrification
 U = Cumulative uptake of plant
 Le = leftover inorganic Nitrogen

6.2 *Voting based ensemble classifier*

The authors proposed a joint system for predicting and predicting yield, crop rotation and fertilizer recommendations. This project developed a system that uses a vote-based group classifier to recommend suitable crops. The system also provides the necessary fertilizer to increase crop yield by increasing the nutrients contained in the soil. The is designed to provide a platform for expert harvest advice and fertilizer recommendations. This helps increase your production potential and recommends the best portals to buy fertilizer. It is based on a number of parameters that affect yield, such as arable land area, annual rainfall, and food price index. For example, there are weather parameters that play a key role in predicting things like temperature and relative humidity, and climate-based models that predict things with crop history [28].

It will also help in soil inspection through data mining techniques like it will predict the crop which should be grown based on soil type. It will be mainly useful in defining the soil quality of the agricultural land and then recommending the fertilizers for improving its quality. It is having a framework that is focusing on crops based on the datasets of Rabi and Kharif crop season so that before cultivating they will be able to forecast the crop yield.

It is available with a web application to forecast the impact of climatic variables on crop production for selected plants in some specific areas of Madhya Pradesh [29]. So, in brief, this whole system this application will provide a summary of the probable impact of weather conditions on crop yield.

7   ANALYSIS OF SYTEM BASED ON TECHNOLOGY + PREDICTION

The comparative analysis of the system is shown in the below table (see Table 1). According to the table, most of the systems are using Fuzzy logic and neural networks for prediction like FDSS, CRS-MFI, AI-ARM, and more [30]. These systems are completely dependent on Fuzzy rules which can cause an issue the future as this technology is completely based on human knowledge and it can only give the output on the basis of trained datasets like past year records of crop yield.

Table 1.   Comparative analysis.

SNo.	Research Work	Technology Used	Sensors Used (Yes/No)	Data Source
1	FDSS [9]	Fuzzy Logic	No	Soil Health card
2	CRS-MFI [10]	Fuzzy logic	No	Questionnaires
3	CRS-NN [11]	Neural Network, Sensors, ML	Yes	Sensors
4	ICRS [12]	Neural network, K-NN, random forest, Decision tree	No	Meteorological dataset
5	AI-ARM [13]	Neural Network	Yes	IOT devices
6	DCR [14]	Data mining, NLP	No	_
7	CRS-ET [15]	Ensemble	No	Data repository site of the Government of India
8	CRS-WF [16]	Artificial Neural Networks	No	UCI dataset
9	CSML [17]	Weighted K-NN, Bagged Tree, SVM	No	National Informatics Centre, Tamil Nadu
10	DRIS [18]	DRIS function	No	Commercial fields
11	USCS [19]	Satellite, APSIM	No	66 ha land in USA
12	VBC [20]	Voting based ensemble classifier	No	_

Systems which are using sensors in it for recommendation like CRS-NN and AI-ARM are much efficient to use. It provides the most accurate result and also it collects the data and processes it in the real time so the chances of getting the accurate result as per the current parameters are high.

CRS-ET and VBC are using ensembles for the prediction [31, 32]. The major benefit of using this technique in our model is that it will give the most accurate result as compared to others because the input is passed through various models and the output with the majority voting is chosen [33]. In general, what happens is that when we pass our data through different models then also, we don't achieve as accurate as 95% output, so what we do is we put it into an ensemble by combining it and for each prediction, we take the average of those all four models [34].

The systems which are using the neural networks are having much more benefits over others like fault tolerance, organic learning, and nonlinear data processing and, self-repair.

8 CONCLUSION

Crop recommendation system has been the need of an era for getting the best out of the production of crops and technology is playing a vital role in this field for making it more efficient and optimize to use. It is one of the most emerging systems which is being used in the agricultural sector to provide better guidance to the farmers who are not available with proper information regarding the cultivation of crops, usage of right fertilizers and chemicals, or whether it be selecting the best crop based on the current demand. If these systems are made available to the local public then it will have many benefits like the least amount of loss to the farmers, the government will not have to import the crops from other nations as it will be grown very well in their own country only and as we know that the economy of the most country like Nigeria, India and many more is dependent on agriculture sector only so it will ultimately help in increase in the GDP.

Various research works are being carried out for much more effective prediction with the help of fuzzy logic, sensors, neural networks and many more. Among all the technologies the use of ensemble and neural networks are much effective because in the end we need a model with the most accurate output which will lead to the farmers taking the right decision. And these systems provide the most accurate result.

We can also use sensors in our system for the real time data as it is more cost-effective, it ranges from $5 to $55 being the highest price. So, if we want to have a model prepared in a low budget then it is the best option to opt for. It is much more pocket-friendly and reliable to use.

REFERENCES

[1] Kumaravel, Archana & Saranya, K G. (2020). A Survey on Crop Recommendation System: Techniques, Challenges. Pudumalar S., Ramanujam E., Rajashree R.H., Kavya C., Kiruthika T. and Nisha J., (2017) "Crop Recommendation System for Precision Agriculture," *2016 Eighth International Conference on Advanced Computing (ICoAC)*, Chennai, India, pp. 32–36, doi: 10.1109/ICoAC.2017. 7951740I.

[2] Bharath R, Kempegowda, Balakrishna, Bency A, Siddesha M & Sushmitha R. (2019). Crop Recommendation System for Precision Agriculture. *International Journal of Computer Sciences and Engineering* 7, 1277–1282. 10.26438/ijcse/v7i5.12771282.

[3] Saranya N. & Mythili A. (2020). Classification of Soil and Crop Suggestion using Machine Learning Techniques, *International Journal of Engineering Research & Technology (IJERT)* 9(2).

[4] Katarya R., Raturi A., Mehndiratta A. & Thapper A., "Impact of Machine Learning Techniques in Precision Agriculture," *2020 3rd International Conference on Emerging Technologies in Computer Engineering: Machine Learning and Internet of Things (ICETCE)*, Jaipur, India, 2020, pp. 1–6, doi: 10.1109/ICETCE48199.2020.9091741.

[5] Mariappan A.K., Madhumitha C., Nishitha P. & Nivedhitha S. (Mar. 2020) Crop Recommendation System through Soil Analysis Using Classification in Machine Learning, *IJAST* 29(3), 12738–12747.

[6] Haban J.J.I., Puno J.C.V., Bandala A.A., Kerwin Billones R., Dadios E.P. and *Sybingco* E., "Soil Fertilizer Recommendation System using Fuzzy Logic," *2020 IEEE Region 10 Conference (TENCON)*, Osaka, Japan, 2020, pp. 1171–1175, doi: 10.1109/TENCON50793.2020.9293780.

[7] Attaluri, S., Batcha, N. and Raheem, M. (2020). *Crop Plantation Recommendation using Feature Extraction and Machine Learning Techniques*, vol. 4, pp. 1–4.

[8] Rajeswari A.M., Anushiya A.S., Fathima K.S.A., Priya S.S. and Mathumithaa N. (June 2020) "Fuzzy Decision Support System for Recommendation of Crop Cultivation based on Soil Type," In *Proceedings of the 4th International Conference on Trends in Electronics and Informatics, ICOEI 2020*, pp. 768–773, doi: 10.1109/ICOEI48184.2020.9142899.

[9] Kuanr M., Kesari Rath B. and Nandan Mohanty S. (2018) Crop Recommender System for the Farmers using Mamdani Fuzzy Inference Model, *Int. J. Eng. Technol.*, 7(4.15), 277, doi: 10.14419/ijet. v7i4.15.23006.

[10] Banavlikar T., Mahir A., Budukh M. and Dhodapkar S. (2008) Crop Recommendation System Using Neural Networks, *Int. Res. J. Eng. Technol.*, 9001, 1475, Accessed: Apr. 10, 2021. [Online]. Available: www.irjet.net.

[11] Doshi Z., Nadkarni S., Agrawal R. and Shah N. (Jul. 2018) *"AgroConsultant: Intelligent Crop Recommendation System Using Machine Learning Algorithms,"* doi: 10.1109/ICCUBEA.2018.8697349.

[12] Vincent D.R., Deepa N., Elavarasan D., Srinivasan K., Chauhdary S. H. and Iwendi C. (Aug. 2019) Sensors Driven AI-Based Agriculture Recommendation Model for Assessing Land Suitability, *Sensors*, 19(17), 3667, doi: 10.3390/s19173667.

[13] Raja S.K.S., Rishi R., Sundaresan E. and Srijit V., (Jan. 2018) "Demand based Crop Recommender System for Farmers," In *Proceedings – 2017 IEEE Technological Innovations in ICT for Agriculture and Rural Development, TIAR 2017*, vol. 2018-January, pp. 194–199, doi: 10.1109/TIAR.2017.8273714.

[14] Kulkarni N.H., Srinivasan G.N., Sagar B.M. and Cauvery N.K., "Improving Crop Productivity Through A Crop Recommendation System Using Ensembling Technique," In *Proceedings 2018 3rd International Conference on Computational Systems andInformation Technology for Sustainable Solutions, CSITSS 2018*, Dec. 2018, pp. 114–119, doi: 10.1109/CSITSS.2018.8768790.

[15] KamatchiS. B. and Parvathi R., "Improvement of Crop Production Using Recommender System by Weather Forecasts," In *Procedia Computer Science*, Jan. 2019, vol. 165, pp. 724–732, doi: 10.1016/j. procs.2020.01.023.

[16] Rahman S.A.Z., Mitra K.C. and Islam S.M.M., "Soil Classification Using Machine Learning Methods and Crop Suggestion Based on Soil Series," Jan. 2019, doi: 10.1109/ICCITECHN.2018.8631943

[17] Serra A.P. et al., "Diagnosis and Recommendation Integrated System (DRIS) to Assess the Nutritional State of Cotton Crop in Brazil," *Am. J. Plant Sci.*, vol. 05, no. 04, pp. 508–516, Feb. 2014, doi: 10.4236/ ajps.2014.54065.

[18] Jin Z., Prasad R., Shriver J. and Zhuang Q., "Crop Model- and Satellite Imagery-based Recommendation tool for variable rate N fertilizer application for the US Corn system," *Precis. Agric.*, vol. 18, no. 5, pp. 779–800, Oct. 2017, doi: 10.1007/s11119-016-9488-z.

[19] Archana K. and Saranya K.G., "Crop Yield Prediction, Forecasting and Fertilizer Recommendation using Voting Based Ensemble Classifier," 2020. Accessed: Apr. 10, 2021. *[Online]. Available:* www. international *journal ssrg.org.*

[20] Vats S., Singh S., Kala G., Tarar R. and Dhawan S., "iDoc-X: An Artificial Intelligence Model for Tuberculosis Diagnosis and Localization," *J. Discret. Math. Sci. Cryptogr.*, vol. 24, no. 5, pp. 1257–1272, 2021.

[21] Vats S., Sagar B.B., Singh K., Ahmadian A. and Pansera B.A., "Performance Evaluation of an Independent Time Optimized Infrastructure for Big Data Analytics that Maintains Symmetry," *Symmetry (Basel).*, vol. 12, no. 8, 2020, doi: 10.3390/SYM12081274.

[22] Vats S. and Sagar B.B., "An Independent Time Optimized Hybrid Infrastructure for Big Data Analytics," *Mod. Phys. Lett. B*, vol. 34, no. 28, p. 2050311, Oct. 2020, doi: 10.1142/S02179849205 0311X.

[23] Vats S. and Sagar B.B., "Performance Evaluation of K-means Clustering on Hadoop Infrastructure," *J. Discret. Math. Sci. Cryptogr.*, vol. 22, no. 8, 2019, doi: 10.1080/09720529.2019.1692444.

[24] Agarwal R., Singh S., and Vats S., "Implementation of an Improved Algorithm for Frequent Itemset Mining Using Hadoop," *In 2016 International Conference on Computing, Communication and Automation (ICCCA)*, 2016, pp. 13–18. doi: 10.1109/CCAA.2016.7813719.

[25] Agarwal R., Singh S. and Vats S., Review of Parallel Apriori Algorithm on Mapreduce Framework for Performance Enhancement, vol. 654. 2018. doi: 10.1007/978-981-10-6620-7_38.

[26] Bhati J.P., Tomar D. and Vats S., "Examining Big Data Management Techniques for Cloud-based IoT Systems," *In Examining Cloud Computing Technologies Through the Internet of Things, IGI Global*, 2018, pp. 164–191.

[27] Sharma V. et al., "OGAS: Omni-directional Glider Assisted Scheme for Autonomous Deployment of Sensor Nodes in Open Area Wireless Sensor Network," *ISA Trans.*, Aug. 2022, doi: 10.1016/j.isatra. 2022.08.001.

[28] Sharma V., Patel R.B., Bhadauria H.S. and Prasad D., "NADS: Neighbor Assisted Deployment Scheme for Optimal Placement of Sensor Nodes to Achieve Blanket Coverage in Wireless Sensor Network," *Wirel. Pers. Commun.*, vol. 90, no. 4, pp. 1903–1933, 2016.

[29]   Sharma V., Patel R.B., Bhadauria H.S. and Prasad D., "Policy for Planned Placement of Sensor Nodes in Large Scale Wireless Sensor Network," *KSII Trans. Internet Inf. Syst.*, vol. 10, no. 7, pp. 3213–3230, 2016.

[30]   Sharma V., Patel R. B., Bhadauria H. S. and Prasad D., "Deployment Schemes in Wireless Sensor Network to Achieve Blanket Coverage in Large-scale Open Area: A Review," *Egypt. Informatics J.*, vol. 17, no. 1, pp. 45–56, 2016.

[31]   Bhatia M., Sharma V., Singh P. and Masud M., "Multi-level P2P Traffic Classification using Heuristic and Statistical-based Techniques: A Hybrid Approach," *Symmetry (Basel).*, vol. 12, no. 12, p. 2117, 2020.

[32]   Vikrant S., Patel R.B., Bhadauria H.S. and Prasad D., "Glider Assisted Schemes to Deploy Sensor Nodes in Wireless Sensor Networks," *Rob. Auton. Syst.*, vol. 100, pp. 1–13, 2018.

[33]   Vats, S. and Sagar, B.B., 2019. Data Lake: A plausible Big Data Science for Business Intelligence. In *Communication and Computing Systems* (pp. 442–448). CRC Press.

Automation and Computation – Vats et al. (Eds)
© *2023 the Author(s), ISBN 978-1-032-36723-1*

Big data-based clustering algorithm technique: A review analysis

Kapil Joshi
Uttaranchal Institute of Technology, Uttaranchal University, Dehradun, India

Manisha Khanduja
Uttaranchal School of Computing Sciences, Uttaranchal University, Dehradun, India

Rajesh Kumar
Meerut Institute of Technology, Meerut, India

Umang
Department of Computer Applications, DSB Campus, Kumaun University, Nainital, India

Parul Saxena
Department of Computer Science, Soban Singh Jeena University, Campus Almora, India

Aditi Sharma
School of Engineering and Digital Sciences, Nazarbayev University, Astana, Kazakhstan

ABSTRACT: Clustering is a significant research focus in the science of machine learning, which has shown fast growth in recent years. The clustering technique for tiny data sets has advanced significantly because of extensive and ongoing study, and numerous efficient clustering algorithms have already been examined. When working with huge data sets, it might be challenging to get appropriate results with these algorithms. The primary causes are the high level of computing complexity and the limited capacity to handle large amounts of data. Processing these enormous amounts of data quickly is becoming increasingly crucial. As a result, among the crucial jobs around machine learning is the research into clustering algorithms for large-scale data sets. This article describes the concepts of clustering algorithms or big data, classifies the clustering algorithms that are appropriate for big data, and makes recommendations for future clustering algorithm research related to big data. The upcoming studies are expected to yield some study discoveries.

1 INTRODUCTION

By exploiting the inherent resemblance between the data and the division of the data set into smaller subgroups so the clustering types of an automatic [1] machine learning method is required. Every data subset belongs to a cluster and samples inside a cluster are typically comparable to others' clusters while among clusters, samples are not. Markov distance and Euclidean distance, Jaccard similarity, cosine similarity, Pearson distance, Chebyshev distance, Manhattan distance, and density probability are used to describing how similar two samples are. Real-world applications of clustering algorithms include customer segmentation in business operations, categorization in bioinformatics, gene sequence spam detection on the Internet, and study of electricity for industrial consumption patterns in the market for electricity.

DOI: 10.1201/9781003333500-46

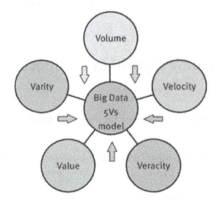

Figure 1.   The 5v Model in big data.

The emergence of the big data era has made data collection methods and processing extremely simple and practical. One after another, large data sets with storage on the GB and even TB scales are developing. Data processing faces enormous [2] difficulties as massive data sets expand in size at an inconceivable rate. As a result, research into clustering for enormous data sets is always developing. Clustering algorithms have so far achieved a significant advance in clustering accuracy for many kinds of medium and small data sets [3]. When dealing with huge data sets, these algorithms always have a lot of issues. High computational complexity, as well as a long computation time, is the key flaws, and this cannot be accepted. The Big data principles are defined and shown in Figure 1: -

Here we explained the classification of big data with 5v's model so our contribution is to examine the complete data on large volume of data and applied some basic concepts of clustering algorithm. Through clustering algorithm, we may also identify the similar group of data and we segregated the clustered part from them.

Overall section of the paper is divided in five sections where first section is about introduction of big data. Overview of big data is classified into second section. Complete study with big data using clustering technique is done in third section. Few steps are performed in forth section. Conclusion of the work is justified in fifth section.

2   OVERVIEW OF BIG DATA

Big data is getting continually more common these days. Big data [4] does not have a single, accepted definition. Large capacity is big data's most fundamental meaning, yet it does have other connotations as well. Additionally, it differs from "extremely huge data" but also "massive data" in a few ways. Big data has been defined in numerous publications, and this paper claims that three concepts are more crucial than others.

The United States National Institute of Standards and Technology defines [5] big data as the "overall capacity of data, the rate of data collecting, and how data are presented, which restrict the analysis conventional relational techniques to process data, as well as all horizontal extension used mechanisms to increase processing effectiveness. Big data can also be further broken down into the [6] big data science and big data frameworks. It is a big data framework software library with a technique for distributed big data analysis and processing between clusters of computer units. Big data science is a research area that covers big data acquisition, adjustment, and assessment technologies [7].

The first to do so was The International Data Center (IDC) investigates whether large data had some influence on big data, according to the attribution definition. Big data technology is described as a new system or technological period required to capture data from detailed, heterogeneous data using rapid discovery, analysis, and capture approaches in IDC's report from 2011. Big data's four key attributes—volume, diversity, velocity, and value—are embodied in this definition. It is known as the "4Vs" definition, and it is frequently used [8].

The definition of big data is "data sets which go beyond the ability Some common database software components can acquire, store, manage, and analyze [9] data" in a 2011 McKinsey study paper. This description does not specifically state the properties of big data, for example, or big data features. This definition presents a developmental perspective on the types of data sets that can be categorized as big data in Figure 2.

This article believes that all definitions of big data are valid and that each one captures a different component of big data. This study contends that big data [10] possesses the qualities listed below after much investigation and analysis. The ability to both the data set is the first and most important consideration when separating big data from conventional data. As an illustration, Facebook reported that every day in 2012, the number of users was 2.7 billion, and left comments. Second, large data can be classified into three categories: both structured and unstructured information. Standard data is frequently organized and simple to tag and keep. The majority of the data produced by users of Twitter, Facebook, and other platforms [11] is presently unstructured, which presents some storage challenges. Third, the rate of processing and analysis of sets of data must keep up with the speed at which big data are obtained [12]. Big data enters the company a stream and requires to be processed as rapidly for applications as possible requiring high requirements for real-time, including include fraud detection with data management for RFID. Finally, significant values can be extracted from huge data with low-value density by using a variety of data mining techniques on big data sets [13]. The organizational structure of big data as well as the essential phases of big data processing is depicted in Figures 1 and 2.

3   STUDY OF BIG DATA SET CLUSTERING ALGORITHM

The formula can be used in either parallel or serial processing. Serial designates the software running on a single actuator sequentially. Simultaneous performance of the programmed on several processors is [14] referred to as parallel. Given the serial and parallel nature of enormous data sets, this research contends greater parallelism suited to the handling of big data sets. As a result, the focus of this study is to outline parallel-compatible clustering methods and provide recommendations for the algorithm's further research. Divide and conquering enormous data is the fundamental [15] idea behind parallel algorithms. Different computing paradigms operate together within the distributed computing architecture to train the clustering algorithm. This study primarily examines the clustering algorithms

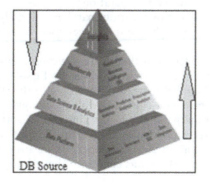

Figure 2.   Pyramid model in big data.

related to the computer framework [16] MapReduce, with Spark computing framework, and the Storm computing framework, including follows. The event and clustered table are defined in Table 1.

Table 1.   Clustered table associated with number of events.

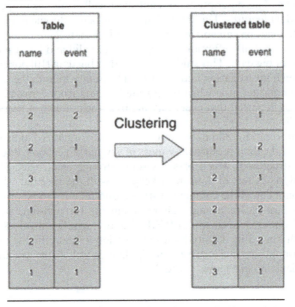

3.1   *Clustering algorithm based on the MapReduce computing framework*

Google has put forth the parallel computing model known as MapReduce. It is focused on massive data collection. The fundamental function of the computing model, specifically the Hadoop distributed file system [17], is to split the data and schedule the processing activities (HDFS, or Hadoop Distributed File System) and divide a big amount of to be processed data into other various each data block is corresponding to a single auto-matically doing a computer job assigns the nodes for computing (such as reduce nodes as well as map nodes) the tasks they need to complete to process the associated data blocks at the same time. The Map node's synchronization control is handled by the execution status, which is tracked. The Map and Reduce operations design as in the MapReduce framework represents the key differences between various parallelization algorithms for [18] the same kind of clustering algorithm. The MapReduce-based k-means algorithm is briefly explained here in Figure 3.

A classic example of a clustering method is the k-means algorithm.2009 saw the realiza-tion of the parallel PK-Means clustering algorithm by First-base MapReduce framework designs including ZHAO (parallel k-means). According to the [19] simulation results, PK-Means can handle Large-scale data clustering with 8 GB issues on a four-node computing cluster. Another based on the parallel k-means clustering algorithm on 10 computing the MapReduce framework-based nodes was created by Jiang Xiaoping in 2011. To avoid iterative correlation, CUI et al. in 2014 developed a k-means clustering technique created using the MapReduce framework. It was confirmed that the algorithm contains 4 296 075

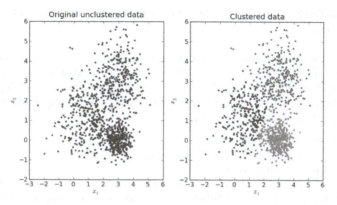

Figure 3.   K-means clustering data in the cluster group.

259 9-dimensional data on such a cluster system with 16 multiple processors. The k-means method, which is based on MapReduce, has already shown effective while processing, resulting in enormous data sets.

3.2   Clustering algorithm based on Spark computing framework

Open-source software is Apache Spark, a memory-based system for cluster computing. Although it is superior to MapReduce, it has some of MapReduce's advantages. There doesn't require to read from or write to HDFS because intermediate output could be kept in memory. As a result, Spark is better suited for iterative machine learning and data mining techniques. Despite its late launch, Spark has gained widespread acceptance as a big data computing engine due to its distinct benefits in solving big data computing issues. Please consult the literature for more information on the development of large data [20] computation within the Spark framework, and analysis. The following is an example of work using the clustering method from the Spark framework: the Spark-based fuzzy c-means technique. Liang Peng is the algorithm's earliest use. This scalable nucleation clustering algorithm scalability KFCM on under the Spark framework was ultimately developed in 2015 by Liang Peng etc. after years of development. The algorithm incorporates the notion of a kernel function [21]. It is fuzzy c-means parallel algorithms clustering [22] performance is considerably enhanced by translating samples of the original material feature space to that same in high dimensions, followed by data division. A paper based on a parallel fuzzy c-means algorithm a particular function of a matrix that takes cache into the Spark framework was sent by Wang Guilan in 2016. The clustering technique created in this research reduces the amount of data that must transit between computing nodes, guaranteeing a Data volume and performance is correlated linearly. Additionally, using With the Spark framework and scalable random sampling is required.

3.3   Clustering algorithm based on Storm computing framework

A massive data processing system with just a strong emphasis on streaming data processing, Apache Storm is a distributed, high-fault, and high-real-time. As a result, in real-time processing, it is also known as Hadoop [23]. The spout component controls the flow of data into both the Storm cluster and sends the data to its bolt. Bolt will send the info to other bolts if it doesn't save it to memory. Data delivered by spout between several bolts are converted by the Storm cluster. There are two different kinds of nodes in a Storm cluster: a master node and several working nodes. Task distribution and execution are the responsibilities [24] of the master node and worker nodes, respectively.

4 MAJOR SUGGESTIONS FOR IMPROVEMENT OF THE CLUSTERING ALGORITHM

According to the research above, the primary approach to solving big data clustering issues is cluster-based distributed computing. This is an example of the "split and conquers" approach to big data. Although the algorithms offer some processing capability for handling big data volumes, they also have glaring flaws. These algorithms require development.

Trilogarithms suggested most steps seen in:

1. The data is not separated properly.
2. There is an excessive reliance on both algorithm's parallelism.
3. There is inadequate automation.
4. There is insufficient timely processing, making it impossible to derive significant value from the data. This makes the following four recommendations as solutions to these issues are mentioned below:

(a) To increase the algorithm's viability, the data division needs careful design. The regularity of the data is discovered by statistical analysis, allowing for a more rational and scientific division of the data.
(b) The algorithm parallelism is replaced by data parallelism. Both the MapReduce and Spark computing frameworks are capable of handling large amounts of data, but each has drawbacks. MapReduce performs iterative calculations poorly. The maximum amount of computer memory is used to calculate Spark's memory. As a result, you may want to consider employing data parallelism rather than algorithm parallelism.
(c) Automating the training of the clustering method. One approach to overcoming huge data clustering issues is to limit human involvement in [25] the clustering process. Because of the variety of importance among the four V's of big data, which correlate to the large amounts of data generated by practical applications.
(d) The capability of real-time data value capture, particularly for data requiring high performance in real-time. In the end, it is brought on by three factors. The real-time accuracy of such data will be much enhanced if the three issues are resolved.

Table 2. Parameters on Big data using clustering algorithm.

Methods	Density Algorithm		Estimation Data	
	Shape	Density	Low Range	High Range
DBSCAN	Yes	No	2.45	6.89
VDBSCAN	Yes	Yes	3.56	8.29
DVBSCAN	Yes	Yes	4.20	7.30
DBCLASD	Yes	Yes	3.38	8.02
Overall	Yes	No	3.58	8.63

5 CONCLUSION

Big data research has been ongoing, and while some results have been obtained in the areas of big data mining, data storage, big data search, etc., they are still inadequate to meet the requirements of the current big data. To fully explore the hidden potential of big data, it is essential to research real-time, extremely reliable, and novel high-efficiency clustering methods.

Many clustering algorithms used in data mining produce results that depend on the parameters being set correctly, making them Intelligent machine learning is far from being a developed and practical field method. A more effective intelligent automatic clustering method needs to be researched and developed for the big data environment. As a result, to keep up with the demands of the most recent big data, the clustering method for big data requires continual study for future work.

REFERENCES

[1] Saeed, M.M., Al Aghbari, Z. and Alsharidah, M., 2020. Big Data Clustering Techniques based on Spark: A Literature Review. *PeerJ Computer Science*, 6, p.e321.

[2] Kumar, K.S., Ramkumar, S., Shankar, K. and Ilayaraja, M., Improving Mapreduce Process By Introducing Aggregator Repartition Data for Big Data Analytics. *Reading*, 10, p.11.

[3] Cebeci, Z. and Yildiz, F., 2016. Efficiency of Random Sampling based Data Size Reduction on Computing Time and Validity of Clustering in Data Mining. *Journal of Agricultural Informatics*, 7(1).

[4] Kumar, S. and Mohbey, K.K., 2019. A Review on Big Data Based Parallel and Distributed Approaches of Pattern Mining. *Journal of King Saud University-Computer and Information Sciences*.

[5] Mahfuz, N.M., Yusoff, M. and Ahmad, Z., 2019. Review of Single Clustering Methods. *IAES International Journal of Artificial Intelligence*, 8(3), p.221.

[6] Mittal, M., Goyal, L.M., Hemanth, D.J. and Sethi, J.K., 2019. Clustering Approaches for High-dimensional Databases: A Review. Wiley Interdisciplinary Reviews: *Data Mining and Knowledge Discovery*, 9(3), p.e1300.

[7] Joshi, K., Kumar, M., Memoria, M., Bhardwaj, P., Chhabra, G. and Baloni, D., 2022. Big Data f5 Load Balancer with ChatBots Framework. In *Rising Threats in Expert Applications and Solutions* (pp. 709–717). Springer, Singapore.

[8] You, G., 2022. Spatiotemporal Data-Adaptive Clustering Algorithm: An Intelligent Computational Technique for City Big Data. *Annals of the American Association of Geographers*, 112(2), pp.602–619.

[9] Sardar, T.H. and Ansari, Z., 2022. Distributed big data clustering using mapreduce-based Fuzzy c-Medoids. *Journal of The Institution of Engineers (India): Series B*, 103(1), pp.73–82.

[10] Pandey, K.K. and Shukla, D., 2022. Stratified Linear Systematic Sampling based Clustering Approach for Detection of Financial Risk Group by Mining of Big Data. *International Journal of System Assurance Engineering and Management*, 13(3), pp.1239–1253.

[11] Parekh, M. and Shukla, M., 2023. Survey of Streaming Clustering Algorithms in Machine Learning on Big Data Architecture. In *Information and Communication Technology for Competitive Strategies (ICTCS 2021)* (pp. 503–514). Springer, Singapore.

[12] Anida, I.N., Norbakyah, J.S., Yussof, W.N.J.H.W., Walker, P. and Salisa, A.R., 2023. Driving Data Analysis for the Development of Kuala Terengganu Driving Cycle. In *International Conference on Mechanical Engineering Research* (pp. 3–14). Springer, Singapore.

[13] Lotfi, C., Srinivasan, S., Ertz, M. and Latrous, I., 2023. A Tool for Study on Impact of Big Data Technologies on Firm Performance. In *Intelligent Communication Technologies and Virtual Mobile Networks* (pp. 501–515). Springer, Singapore.

[14] Khan, S. and Adil, A., 2023. Novel Software Architecture to Calculate Effort Estimation for Industrial Big Data. *In Proceedings of Seventh International Congress on Information and Communication Technology* (pp. 619–626). Springer, Singapore.

[15] Bharany, S., Badotra, S., Sharma, S., Rani, S., Alazab, M., Jhaveri, R.H. and Gadekallu, T.R., 2022. Energy Efficient Fault Tolerance Techniques in Green Cloud Computing: A Systematic Survey and Taxonomy. *Sustainable Energy Technologies and Assessments*, 53, p.102613.

[16] Bharany, S., Sharma, S., Badotra, S., Khalaf, O.I., Alotaibi, Y., Alghamdi, S. and Alassery, F., 2021. *Energy-efficient Clustering Scheme for Flying Ad-hoc Networks Using an Optimized LEACH Protocol. Energies*, 14(19), p.6016.

[17] Lutfi, A., Alrawad, M., Alsyouf, A., Almaiah, M.A., Al-Khasawneh, A., Al-Khasawneh, A.L., Alshira'h, A.F., Alshirah, M.H., Saad, M. and Ibrahim, N., 2023. Drivers and Impact of Big Data Analytic Adoption in the Retail Industry: A Quantitative Investigation Applying Structural Equation Modeling. *Journal of Retailing and Consumer Services*, 70, p.103129.

[18] Bag, S., Dhamija, P., Singh, R.K., Rahman, M.S. and Sreedharan, V.R., 2023. Big Data Analytics and Artificial Intelligence Technologies based Collaborative Platform Empowering Absorptive Capacity in Health Care Supply Chain: An Empirical Study. *Journal of Business Research*, 154, p.113315.

[19] Priyadarshini, R., Barik, R.K., Panigrahi, C., Dubey, H. and Mishra, B.K., 2018. An Investigation into the Efficacy of Deep Learning Tools for Big Data Analysis in Health Care. *International Journal of Grid and High Performance Computing (IJGHPC)*, 10(3), pp.1–13.

[20] Kario, K., Hoshide, S. and Mogi, M., 2022. Digital Hypertension 2023: Concept, Hypothesis, and New Technology. *Hypertension Research*, 45(10), pp.1529–1530.

[21] Diwakar, M., Tripathi, A., Joshi, K., Sharma, A., Singh, P. and Memoria, M., 2021. A Comparative Review: Medical Image Fusion Using SWT and DWT. *Materials Today: Proceedings*, 37, pp.3411–3416.

[22] Winter, J.S. and Davidson, E., 2019. Big Data Governance of Personal Health Information and Challenges To Contextual Integrity. *The Information Society*, 35(1), pp.36–51.

[23] Zhu, K., Liu, H., Ji, Y. and Li, Z., 2021, September. Earthquake Observation data Grading and Storage Research. *In Journal of Physics: Conference Series* (Vol. 2024, No. 1, p. 012038). IOP Publishing.

[24] Marinakis, V., Doukas, H., Tsapelas, J., Mouzakitis, S., Sicilia, Á., Madrazo, L. and Sgouridis, S., 2020. From Big Data to Smart Energy Services: An Application for Intelligent Energy Management. *Future Generation Computer Systems*, 110, pp.572–586.

[25] Kassens-Noor, E., Vertalka, J. and Wilson, M., 2019. Good Games, Bad Host? Using Big Data to Measure Public Attention and Imagery of the *Olympic Games. Cities*, 90, pp.229–236.

Automation and Computation – Vats et al. (Eds)
© 2023 the Author(s), ISBN 978-1-032-36723-1

Design and implementation of sound signal reconstruction algorithm for blue hearing system using wavelet

Parul Saxena*

Assistant Professor, Department of Computer Science, Soban Singh Jeena University, Almora, India

Umang Saini*

Assistant Professor, Department of Computer Application, Kumaun University, Nainital, India

Vinay Saxena*

Professor, Department of Mathematics, Kisan Post Graduate College, Bahraich, UP, India

ABSTRACT: Blue Hearing System is an alternative solution to develop hearing sense in deaf and mute persons. Speech and sound signal reconstruction acquired from outside world is very important part of any hearing device. In the present work initially the speech signal reconstruction algorithm has been developed and then it has been studied using two different wavelet functions wrcoef and waverec. Later this algorithm has been successfully checked for twenty five sound signals from various kind of range. These signals have been studied under the measurement of Mean Asolute Error(MAE), Mean Squared Error(MSE), Signal to Noise Ratio(SNR) and Peak Signal to Noise Ratio (PSNR). Finally it has been concluded that the algorithm works better using wrcoef function in comparison to waverec function. The design and implementation part of this algorithm uses MATLAB interface.

Keywords: Blue Hearing System, Signal Recostruction, Wavelet Tree, Error Measurement

1   INTRODUCTION

Blue Hearing System [1] is the substitute system for deaf and mute and facilitates the user just like the natural hearing system. It uses the device cochlear implant [2] which is the combination of two parts one internal and one external. After major ear surgery the internal part of cochlear implant device is inserted in the inner part of the ear and external part is expected to mount outside the ear. As soon as the speech is obtained from outside world using microphone of cochlear implant device, the speech processor of the device processes it in different aspects to create impulse for the brain stimulation, as a result the patient starts to perceive sound signal and as a conclusion speech development process starts in the patient in case of prelingual deafness. for postlingual deaf patients the sound perception is retrieved back with the help of BHS.

2   WAVELET TRANSFORM FOR INPUT SIGNAL RECONSTRUCTION

Wavelet theory is very important which delivers an integrated structure for different techniques designed separately for different applications of digital signal processing. This theory

*Corresponding Authors: parul_saxena@yahoo.com, anilumang@yahoo.co.in and dr.vinaysaxena@gmail.com

DOI: 10.1201/9781003333500-47

provides generic approach for various applications in the field of signal processing specially for nonstationary signals [2,3]. Wavelet transform dissolves signals as a superposition of simple units from which the original signals can be easily recreated [4,5]. The wavelet packet transform allows for sub-band analysis in the second decomposition without any constraints [6,7]. To send shorter waveforms at high frequencies is more proficient in place of sending pulses of equal period.This is possible by the scaling of a single function nothing but wavelet [8,9].

3 SOUND AND SPEECH RECONSTRUCTION

Sometimes the received sound signal may not be in appropriate form, it may be destroyed due to many problems occurring in the transmission media. The problem of attenuation may be overcome by using amplifiers, the problem of noise may be overcome by using different filters, but for the problem of distortion, the signal reconstruction mechanism plays a very important role. The present work provides a very robust mechanism to reconstruct the signal using wavelet tree decomposition method. Wavelets principle for analyzing the signals according to scale makes it very important for various applications [10–12]. One dimensional wavelet transform may be used very efficiently for sound signal reconstruction method. Here tree command has been used to construct the wavelet tree and it has been decomposed in four levels by using wavelet function wavedec. The branches of wavelet tree are observed for highpass and lowpass coefficients. Each individual branch possesses the chunk of the information of the sound signal.

These coefficients play the major role in signal reconstruction using function wrcoef and waverec.

4 FLOW CHART AND ALGORITHM FOR SIGNAL RECONSTRUCTION

The signal is reconstructed using two different mechanism, in the first mechanism, the signal is reconstructed using waverec function, in which the approximation coefficients and detailed wavelet decomposition coefficients are calculated to reconstruct the tree. In the second mechanism the signal is reconstructed using wrcoef function in which individual branches are constructed from the full vector of the transform coefficients. Figure 1 shows the Flow Chart for Sound Signal Reconstruction Algorithm and the steps of alogritihm are:

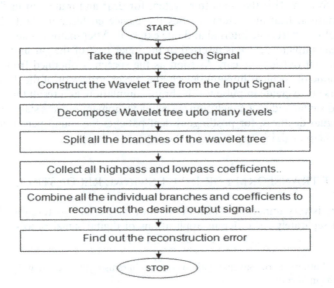

Figure 1.   Flow chart for sound signal reconstruction algorithm.

Step1: Take the input speech from the user or from stored file. This speech signal may also be collected from any audio source

Step 2: Create the wavelet tree for the given input sound signal by using the command tree.

Step 3: Generate various levels from the wavelet tree by decomposing it upto many levels. As higher number of levels produce better accuracy in result.

Step 4: Split all the branches of the wavelet tree individually.

Step 5: Collect all highpass and lowpass coefficients for each individual branch of wavelet tree.

Step 6: Combine all the individual branches and coefficients to reconstruct the desired output signal using waverec or wrcoef function.

Step 7: Find out the reconstruction error in terms of MAE, MSE, PSNR and SNR.

5 RESULTS AND DISCUSSION

The input for the problem may be captured from the microphone of any device or from already existing sound. In the present work the input has been taken from a vast domain of sound signals. One of the example has been shown in Figure 2, then the signal has been decomposed into a four stage wavelet tree as shown in Figure 3. Here we have used one dimentional form of wavelet transform Here in Figure 4 all the lowpass and highpass coefficients in the wavelet decomposition have been represented as d1, d2, d3 and a3. These are approximation and decomposition coefficients. In Figure 5 the individually decomposed branches rd1, rd2, rd3 and ra3 have been represented after decomposition. Figure 6 shows the reconstructed signal using wavrec function which uses wavelet coefficients to reconstruct the signal. Figure 7 shows the reconstructed signal using wrcoef function in which individual branches are used to reconstruct the signal.

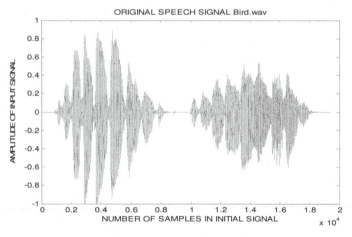

Figure 2.   Input signal bird.wav.

Table 1 represents the results of signal reconstruction using two different MATLAB functions waverec and wrcoef. From the Table 1 it is clear that using function wrcoef the values of MAE and MSE are smaller in comparison to the values of these parameters using function waverec. PSNR is also higher in case of wrcoef. Hence we can conclude that the results of signal reconstruction methods are better for the function wrcoef in comparison to the function waverec.

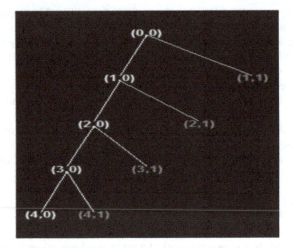

Figure 3. Four stage decomposition of wavelet tree.

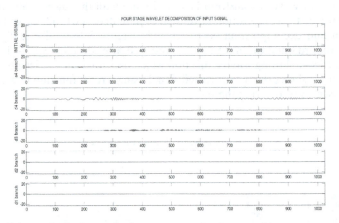

Figure 4. Coefficients of wavelet tree in four stage wavelet decomposition.

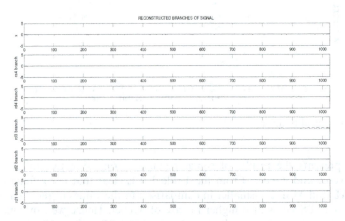

Figure 5. Decomposed branches of four stage wavelet tree.

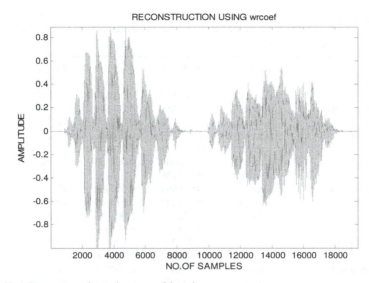

Figure 6.   Signal reconstruction using wrcoef function.

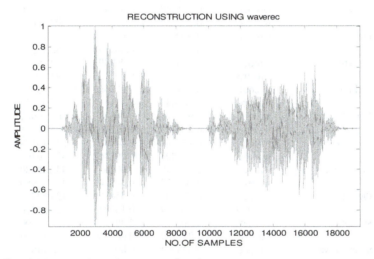

Figure 7.   Signal reconstruction using waverec function.

Table 1.   Various sound signal reconstruction using wrcoef and waverec wavelet function.

S.No.	Input Signal	Function	MAE	MSE	SNR	PSNR
1	Experiment.wav	wrcoef	0.010051	0.000413	50.210238	67.855601
		waverec	0.020334	0.000782	50.210238	65.084479
2	Bearings.wav	wrcoef	0.100876	0.023083	40.036328	48.416182
		waverec	0.161213	0.043138	40.036328	45.700478
3	Bird.wav	wrcoef	0.125271	0.041594	42.890536	46.456906
		waverec	0.214038	0.077600	42.890536	43.748591
4	Bizarro.wav	wrcoef	0.087550	0.014848	44.574580	51.265051

(*continued*)

Table 1. Continued

S.No.	Input Signal	Function	MAE	MSE	SNR	PSNR
		waverec	0.130769	0.028111	44.574580	48.493024
5	Boing.wav	wrcoef	0.060765	0.008872	38.654001	52.263536
		waverec	0.100527	0.016869	38.654001	49.472947
6	Bubble.wav	wrcoef	0.080489	0.014510	44.980761	51.443887
		waverec	0.127118	0.027454	44.980761	48.674495
7	Bubble2.wav	wrcoef	0.142358	0.060303	41.242433	44.503532
		waverec	0.261603	0.114865	41.242433	41.704996
8	Can.wav	wrcoef	0.048400	0.008143	48.096405	54.534326
		waverec	0.092038	0.015431	48.096405	51.758271
9	Chimes.wav	wrcoef	0.056594	0.007734	42.007411	53.582378
		waverec	0.093204	0.014859	42.007411	50.746589
10	Click.wav	wrcoef	0.016584	0.003084	38.553980	56.830247
		waverec	0.051038	0.005912	38.553980	54.004040
11	Clicks.wav	wrcoef	0.036905	0.006635	39.082705	53.621415
		waverec	0.079202	0.012660	39.082705	50.815442
12	Ding.wav	wrcoef	0.034029	0.003365	46.915412	58.157077
		waverec	0.060150	0.006408	46.915412	55.359367
13	Dog.wav	wrcoef	0.095212	0.019183	43.996564	50.039134
		waverec	0.146806	0.036439	43.996564	47.252678
14	Dove.wav	wrcoef	0.071248	0.013405	39.356080	50.627730
		waverec	0.122640	0.025277	39.356080	47.873062
15	Drip.wav	wrcoef	0.010952	0.010952	43.113300	61.081673
		waverec	0.035700	0.002798	43.113300	58.223757
16	Incorrec.wav	wrcoef	0.183851	0.069307	39.979104	43.628917
		waverec	0.281300	0.127253	39.979104	40.989979
17	Jackhmmr.wav	wrcoef	0.174940	0.059466	37.043221	43.631509
		waverec	0.264040	0.111631	37.043221	40.896325
18	Jaillet65.wav	wrcoef	0.064808	0.008129	50.916001	55.036945
		waverec	0.095681	0.015354	50.916001	52.274828
19	Kungfu.wav	wrcoef	0.132489	0.047803	32.340108	43.400321
		waverec	0.235896	0.092367	32.340108	40.539642
20	Marble.wav	wrcoef	0.020370	0.004641	44.726980	56.345335
		waverec	0.064023	0.008743	44.726980	53.594620
21	Pig.wav	wrcoef	0.179702	0.056802	42.658316	45.056425
		waverec	0.260239	0.107745	42.658316	42.876092
22	Smlsplas.wav	wrcoef	0.010952	0.001449	43.113300	61.081673
		waverec	0.035259	0.002756	43.113300	58.289429
23	Sounder.wav	wrcoef	0.038054	0.004422	48.914873	57.332416
		waverec	0.067645	0.008418	48.914873	54.536606
24	Splwater.wav	wrcoef	0.046016	0.010130	43.905112	52.794234
		waverec	0.100012	0.019288	43.905112	49.997342
25	Tms.wav	wrcoef	0.084269	0.015547	51.303499	52.286422
		waverec	0.132565	0.029177	51.303499	51.552473

6 CONCLUSIONS

We have designed and implemented sound signal reconstruction algorithm using MATLAB, which is very robust and applicable for various types of input signals. It is observed that the wavelet function wrcoef works better for the signal reconstruction in comparison to wavelet function waverec. Hence in BHS the signal reconstruction method using the function wrcoef may be adopted for maximizing the benefit to the user.

REFERENCES

[1]   Saxena P., 2010. A Note on Blue Hearing System (BHS) to Develop Speech in Deaf and Mute People, *Recent Research in Science and Technology*, 2(9): pp.04–07.

[2]   Saxena P. and Mehta A., 2016. Refinement of input speech by suppressing the unwanted amplitudes for blue hearing system, *2nd International Conference on Advances in Computing, Communication, & Automation (ICACCA) (Fall)*, pp. 1–4.

[3]   Ioul O. and Vetterli M., October 1991. Wavelet and Signal Processing, *IEEE Magazine*,1053–5888. pp.14–38.

[4]   Chen Jie, May 14, 1993. Application of Wavelet Transform to SignalAnalysis, Honors Project paper 5. http://digitalcommons.iwu.edu/math-honproj/5

[5]   Gulden Kokturk, 2012. Wavelet Based Speech Strategy in Cochlear Implant, *Cochlear Implant Research Updates*, Dr. Cila Umat (Ed.), 2012 ISBN: 978-953-51-0582-4, InTech, http://www.intecho-pen.com/books/cochlear-implantresearch-updates/wavelet-based-speech-strategy-incochlear-implant

[6]   Rezaee A. and Khaleqhi A.N.,2012. Application of Convolutionary Algorithm for Wavelet Filter Design, *International Journal of Information and Electronics Engineering*, Vol.2 No. 4.

[7]   Qazi O.U.R., Dijk B.V., Moonen M., and Wouters J., Jan 2012. Speech Understanding Performance of Cochlear Implant Subjects using Time-Frequency Masking based Noise Reduction, *Transactions On Biomedical Engineering*, Technology Centre Belgium.

[8]   Daubechies, 1992. Ten Lectures on Wavelets, *of* CBMS-NSF Regional Conference Series in Applied Mathematics SIAM Philadelphia, vol. 61.

[9]   Daubechies, September 1990. The Wavelet Transform Time-Frequency Localization and Signal Analysis, *IEEE Transaction on Information Theory*, vol. 36, pp. 961–1005.

[10] Chen Jie, May 14, 1993. Applications of the Wavelet Transform to Signal Analysis, *Illinois Wesleyan University*.

[11] Saxena, P., and Saxena, V, 2022. Comparative Study of White Gaussian Noise Reduction for Different Signals using Wavelet. *International Journal of Research - GRANTHAALAYAH*, 10(7), 112–123. doi:10.29121/granthaalayah.v10.i7.2022.4711

[12] Razza, S., Zaccone, M., Meli, A., and Cristofari, E., 2017. Evaluation of Speech Reception Threshold in Noise in Young Cochlear Nucleus System 6 Implant Recipients using Two Different Digital Remote Microphone Technologies and A Speech Enhancement Sound Processing Algorithm, *International Journal of Pediatric Otorhinolaryngology, Elsevier* 103,71–75. https://doi.org/10.1016/j.ijporl.2017. 10.002

Automation and Computation – Vats et al. (Eds)
© 2023 the Author(s), ISBN 978-1-032-36723-1

A deep learning model for classification of mental illnesses like depression and anxiety via social network

Vaibhav Sharma*

Research Scholar, School of CA & IT, SGRR University, Dehradun, Uttarakhand, India
ORCID ID: 0000-0002-1404-2012

Parul Goyal*

Professor & Dean, School of CA & IT, SGRR University, Dehradun, Uttarakhand, India
ORCID ID: 0000-0001-9729-1155

ABSTRACT: Social networking sites are a common way for people to express their feelings in the modern world. These kinds of emotions are frequently examined in order to forecast user behaviour. These feelings are categorized here using a DL model to predict the mental disorder of users. The study is performed using social networking site, and CNN and RNN are employed to create the implementing deep learning model. In this study, multiclass classification is used to distinguish between mental problems or illness like anxiety and depression. Correctness, precision, recall, and F1 score are the performance metrics for assess the models. With a class prediction accuracy of more than 93 percent, the suggested ensemble model outperformed the other models when utilised for multiclass classification.

Keywords: Mental Illness, Depression, Anxiety, Deep Learning Model

1   INTRODUCTION

It is quite common for people to express themselves through social media. Multiple social media channels or online discussion boards for social health are frequently used to anonymously share mental health difficulties or illnesses [1]. Joining an online health community will allow you to express sympathy for those who exhibit comparable symptoms. From above, many users seek facts about their trait on social media in order to diagnose himself or herself. A number of academics have studied mental illnesses and emotional states through social media, including depression, anxiety, and schizophrenia. A recent study collected tweets from people with depression, according to the study. The Linguistic Inquiry and Word Count (LIWC) method of analyzing and tracking Twitter social interaction was used to describe and track the evolution of social interaction during a session. In order to greater comprehension of the dangers among users of social media for postpartum depression, In one investigation, the prevalence of postpartum depression was compared between the prenatal and postnatal phases using specific psychometric measures. In their study, author [12] used picture analysis to forecast depression among social network users. Face identification and colorimetric analysis were applied to Instagram images. One study used language modeling, N-grams, and vector embedding to predict who would acquire anxiety disorders [2].Previous research has shown that social media data may be used to monitor or identify emotions or mental health issues. The goal of the project is to develop the model using Deep learning identify different mental health issues in person via social media data on mental health condition that include

---

*Corresponding Authors: vsdeveloper10@gmil.com and profdrparul@gmail.com

DOI: 10.1201/9781003333500-48

depression, anxiety etc. Users of social Networking sitewere asked to provide data via groups like depressed peoples group. In order to identify depression or anxiety in a person, we collected information from some groups which is related with mental health issues, like autism, schizophrenia, bipolar, psychiatric disorders [3].A semi supervised method and an expert assessment procedure were used to identify mental health-related groups among the most popular groups. All groups related to depression, anxiety are fall into this category.

We were able to determine whether individual posts on Social networking site could be divided into relevant forms of mental issues by looking at its mental health groups. Since bipolar disorder is difficult to diagnose, patients may forget to recognize the signs of bipolar disorder and depression due to their similar symptoms. People typically talk about their mental health in general health-related channels on Social networking site, but neglect to identify the specific health problems that are talked about there. At first sight, we assumed people were asking for assistance when they used phrases like "mental health," "mental disease," or "mental condition" to describe themselves." To identify users with mental health problems on social media, we are looking for data on their behavior [4]. The following research question will be addressed in this study.

Millions of people of all ages suffer from depression, according to the WHO's Mental Disorders Fact Sheet compiled in 2020 [5]. As COVID-19 spreads, mental health disorders have also increased. There will soon be more people receiving depression treatment. The lack of resources for individuals with mental illnesses makes suicide one of the leading causes of death among those aged 14 to 32. As a result of this lack of medical care, more than eighty five thousand people die each year. Within the past decade, the number of adolescents suffering from major depression has steadily increased. According to projections [6], mental illness will affect the global economy with a $5-6 trillion economic impact. Intervention and prevention strategies seem to be in high demand.

People are increasingly expressing themselves and connecting with others through social media sites like Twitter and Facebook, according to research. The result is the generation of massive amounts of social data that reveals the preferences, moods, and activities of people. Social media is expected to be used by 59 percent of the globe's population by 2022, according to survey. A global increase of more than 15 percent in unique users has occurred over the past year. As medical decision support systems which can examine signs of mental problems, online screening tools are very useful for diagnosis of mental diseases since conventional methods are largely dependent based on polls and interactions. Consequently, social media data can be used to develop new screening tools for mental illness. In recent studies, ML techniques are used to distinguish mental health states, for instance, based on language patterns.

This process can be broken down into three steps: creating a database of mental illnesses, assessing estimating accuracy and building a framework based on the information available. In the absence of sufficient datasets for mental health research, there has been a slowdown in the field. An effective solution to this problem can enhance crisis informatics, the rapid diagnosis and action of crises, and the effectiveness of treatment. These gaps in time hinder the creation of successful intervention programs because identifying mental disease risk factors is frequently involves taking action quickly. Researchers may also find that mentally ill individuals are less cooperative. There are many different types of social data that can be analyzed for psychological signals that, while textual elements prevail in the evaluations (especially tweets), examine numerous additional types [7]. There has been relatively little attention paid to visual qualities, for example. Although social media platforms generate enormous amounts of content, certain studies indicate that photographs are most popular. As a relatively new domain of study, mental illness social data poses many concerns.

2   RESEARCH REVIEW

In social media-based mental illness prediction, data may be collected either directly from users using tools like surveys or APIs (application programming interfaces), or indirectly through visitors of social networks with their consent and data gathering tools. A crowd-sourcing platform or data donation website, was used by researchers to publish study details and invite people to fill out questionnaires, consent to gather their social information. Depression is frequently assessed using the Beck Depression Inventory (BDI) and the Center for Epidemiologic Studies Depression Scale (CESD). The Suicide Probability Scale and the Life Satisfaction Scale are also available that can be used to monitor the risk of suicidal ideation, the second method gathers relevant content from social media networks by employing keywords connected with such phrases. In post retrieval, self-reported diagnoses can include the following regular expressions: "I was diagnosed with (disorder)". In order to use APIs for further research, posts must be analyzed first. It is considered irrelevant to include quotes or articles that denial suicide ideation, discuss other people's suicides, or report/news about suicide. A self-report diagnosis is only valid when its contents include hypothetical statements, quotations etc.

Since we discussed the issues in the introduction, many researchers suggest human annotators as a method for validating regular expressions. There are some predefined datasets that can also be gathered using other methods of gathering data, such as those produced by the eRisk workshops, myPersonality project, and Language modeling and Clinical Psychology (CLPsych) [8,9]. Preprocessing of data consists of deleting stop words, hashtags, URLs, and tweets and lowercase alphabets before analysis [13]. The data was also analyzed using ASCII-converted emojis. In addition to reducing training time and stimulating interpretation, selecting the most relevant features will increase the possibility that the model will generalize & avoid overfitting [14], and speed up the learning process. Support vector machines (SVMs), linear, decision trees, logistic regression, naive Bayes etc. just a few of the machine learning methods that are well recognized [15]. A deep learning approach has also been used to study depression and suicide-related psychiatric stresses [16]. A reputation for the classification model's accuracy is assessed by using the evaluation mechanism, despite most earlier research did not provide sufficient detail on feature normalisation and parameter tuning. Several metrics and visualisation tools,like accuracy classification call, F1 score, confusion matrix, precision etc., can be used to assess the performance of a model (ROC).

Some studies have shown that machine learning techniques can be used to identify relevant specifications. Rather than developing learning methods, this research focusing on finding undiscovered links and patterns in the data, This allows them to find hidden patterns by training a classifier. Textual content and linguistic patterns are used extensively in mental illness research. Depression, bipolar, and seasonal affective disorder can be discovered through NLP methods developed by Coppersmith et al. [17]. 1st pronouns, aggressive language, and other offensive words tendencies are also linked to mental disorders. Several factors contributed to these findings, including the LIWC. Various psychological dictionaries were used by psychologists to build this database manually. Using this technique, you can extract positive/negative emotions and personal pronouns from the text, for instance. A number of research papers used sentiment analysis tools to measure textual sentiment in addition: SentiStrength [18] and OpinionFinder. In numerous attempts to identify user posts' unused topics, latent dirichlet allocation (LDA) has been used [19]. In spite of this, few studies use image analysis techniques, with the majority focusing on textual features. In order to interpret the emotional content of Twitter photographs, author utilised SIFT descriptors and colour compositions. According to author [12]. A recent study in paper [11] suggests that predicting depression can also be done utilising the image classifier VGG-Net [10]. Instagram users can be predicted to have depression based on the hue, saturation, and brightness of an image. An analysis of multiple convolutional neural networks in the above discussed paper took a substantial amount of time. It is therefore based on the extraction of features by convolutional neural networks, followed by classification by recurrent neural networks.

3 PROPOSED METHODOLOGY

The mental illness especially depression and anxiety are focused in this study. Among these diseases that were classified is anxiety versus non-anxiety, depression versus non-depression. Proposed deep learning models are used to classify these diseases. It is unique to use an ensemble-based approach to classify different types of medical disorders using LSTM and convolutional neural networks. Let's take a closer look at the proposed methodology.

3.1 *Dataset*

The dataset from Social networking site was taken out using the NLTK Python package. The dataset contains sufficient records that contrast anxiety, bipolar, depression and psychotic. Here our focus for classification among anxiety and depression. Data cleaning and lemmatization will be used to process the dataset. About 40 million adults aged 18 and older suffer from mental health disorders in America. Data from the research paper is mostly taken from the dataset.

3.2 *DL models*

In this study, a CNN and two RNN (bidirectional LSTM) are used. A first stage involves preprocessing the dataset, followed by passing it through various DL models such as CNN and RNN. A flowchart illustrating the proposed work is shown in Figure 1, as well as a detailed description of the proposed methodology.

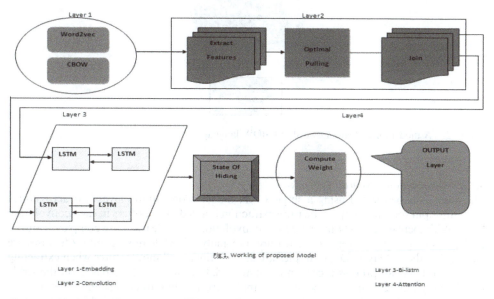

Fig.1. Working of proposed Model

Layer 1-Embedding

Layer 2-Convolution

Layer 3-Bi-listm

Layer 4-Attention

Figure 1. Working of proposed model.

3.2.1 *Continuous bag of words model for word embedding*
Processing of natural language has benefited greatly from Word2vec. This has resulted in its widespread acceptance and use. In order to convert the text of the dataset into vector format, an embedding technique called Word2vec is used. The vectors' size vary with the size of each word in your dataset. CBOW models take context into account and use it as an input.

Following that, it forecasts relevant words. Here is an illustration of this. The term "depression" is given to the neural network, which then analyses the text. This exercise is directed towards creating a neural network that will predict the word "depression." One-hot encoding will be used for input words, and prediction error rates will be measured for the encoded target word. By determining the term with the lowest error, we can predict the result. This diagram illustrates the architecture of the CBOW model. Models are able to anticipate target words by recognizing the context of adjacent words. Here is a good example, "Help with declination, depression as well as suicide problem". The (context word, target word) approach is used to divide this sentence into word pairs. Window size must be customized by the user. A 2-word context word might mean (help, with], rejection), (help, with], rejection], depression), and (rejection depression], suicide). These word pairs are used to make predictions using contextual terms. Four 1XW input vectors will make up if we use four word embeddings to anticipate a given target word, the input layer. The buried layer multiplies these input vectors using a WXN matrix. As shown in Figure 2, the final activation and output are determined by summarizing the 1XN hidden layer outputs.

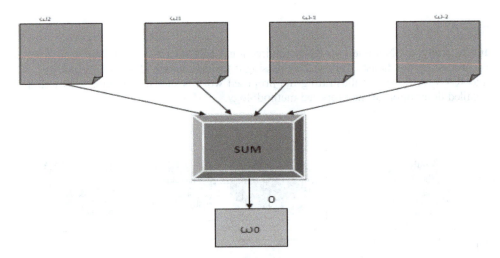

Figure 2.   A model for embedding word in CBOW language.

3.2.2 *Convolutional Neural Network*
The analysis is performed using a single CNN model. Convolution layers extract features from the input training samples. Feature extraction is aided by the filters in each convolution layer. With increasing CNN model depth, convolution layers learn more complex features. In the first convolution layer, simple features are analyzed, while more complex features are analyzed in the last layer. Data samples are convolutionalized into features when extracting features. Filters can process a certain amount of data at a given time based on the stride length and padding value. It is possible to apply zero padding to data samples before convolution. Convolutional output is further processed using ReLUs (rectified linear units). A nonlinear transformation of data is performed by this unit. Negative convolution output clips ReLU output to zero. It is impossible to use sigmoid units as activation units because of the vanishing gradient problem. Thus, the output of the network varies slightly as a result. In other words, converging is either slow or nonexistent. For such a situation to be avoided, ReLU is preferred.

Once the ReLU has completed its work, the output is sent to a pooling layer. Any redundant features are removed during convolution by a pooling layer. The result is a

416

smaller data sample. An image is pooled if adjacent pixels have similar values. Pooling consists of averaging, averaging the minimum, and averaging the maximum values of four adjacent pixels. Input images are halved by using a 2 * 2 filter. There may be no padding applied to the input data before pooling. Convolution and pooling layers are repeated in the CNN model. For educational purposes, it is repeated twice or four times. Convolution and pooling layers are processed using multilayer neural networks. An individual neuron's feature map functions as an example of a feature map. The dropout layer reduces overfitting of the CNN model by making it noise-resistant. An already fully connected neural network typically adds layers such as these between two layers. Due to this, two systems connected by data were temporarily disrupted. To classify accurately under noisy conditions, the model must be taught to classify in this manner. Therefore, an overfitted model is less likely to classify incorrectly. A CNN model's output is computed using the SoftMax function. The probability of the output for each class is provided by SoftMax instead of just * 0.5 for sigmoid output. Output accuracy is increased when SoftMax is used to find output results.

3.2.3 *Recurrent Neural Network*

We primarily use attention-based Bi-LSTM architectures in our classification model. CNNs reduce input features so that they can be used for prediction, but each word affects its final classification in a unique way. In this study, we intend to utilize CNNs and Bi-LSTMs to their full potential. Consequently, Bi-LSTMs enable Long-Distance word dependencies to be encoded more effectively. Bi-LSTM is depicted in Figure 1. A new layer of features is generated based on the features from the CNN stage. Due to the fact that BiLSTM provides accessibility towards both previous and following contextual data, Its data is presented in two separate textual forms. As part of the Bi-LSTM model, the Convolutional Neural Network features are used to create a representation of a sequence. After selecting the most correlated features, the attention layer uses them for classification. In response to "I'm angry". The phrase should be categorised as depressive in order, attention mechanism places more weight on "angry" than "self". In this case, it shows that the "angry" attitude remains unwavering despite the suffering of the illness. In addition to improving prediction accuracy, the attention mechanism reduces learnable weights.

4 ANALYSIS AND DISCUSSION

Python programming is used for implementation of proposed work. Correctness, precision, recall, and F1 parameters are used to evaluate the models' performance.

Proposed model is compared to an existing model in Table 1. A CNN model is inferior to the proposed model, as shown in the table.

As shown in Figure 3, this work has been compared to the existing work in terms of accuracy or correctness. By utilizing this work, it is more accurate to classify anxiety vs depression than other mental illnesses.

Table 1. For comparison of DL models.

	Convolutional Neural Network				Proposed DL model			
Class	Correctness (%)	CNNPrecision	Recall	F1 Score	Correctness (%)	Precision	Recall	F1 Score
Anxiety	78.81	88.44	42.34	58.25	82.34	90.63	55.14	64.30
Depression	76.15	88.1	72.75	79.9	83.45	92.46	86.17	80.72

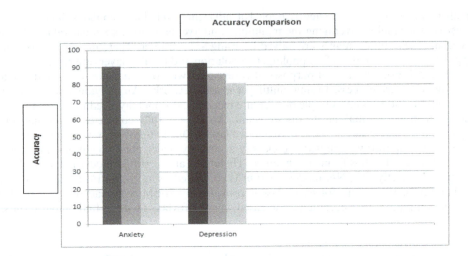

Figure 3.　Accuracy comparison of proposed work.

5　CONCLUSION

In this Research Paper, deep learning is used to classify mental illnesses. For organizing mental illnesses like anxiety and depression, this DL model combines CNN with RNN. According to parameters such as accuracy, precision, recall, and F1 score, this DL model performed superior to current models. It is possible to construct convolutional networks using images to classify mental illness. Here NLTK python library was used to collect the dataset from social networking site.

REFERENCES

[1]　Smith-Merry J., Goggin G., Campbell A., McKenzie K., Ridout B. and Baylosis C., "Social connection and onlineengagement: Insights from Interviews with Users of a Mental Health Online Forum," *JMIR Mental Health*, vol. 6, no. 3,Article ID e11084, 2019.

[2]　Roark B., Saraclar M. and Collins M., "Discriminative N-gram Language Modeling," *Computer Speech & Language*, vol. 21,no. 2, pp. 373–392, 2007.

[3]　Kim J., Lee J., Park E. and Han J., "A Deep Learning Model for Detecting Mental Illness from User Content on Social Media," *Scientific Reports*, vol. 10, no. 1, 2020.

[4]　Ma Z., Zhao J., Li Y. et al. "Mental Health Problems and Correlates Among 746 217 College Students during theCoronavirus Disease 2019 Outbreak in China," *Epidemiology and Psychiatric Sciences*, vol. 29, 2020.

[5]　"2020 Alzheimer's Disease Facts and Figures," *Alzheimer's and Dementia*, vol. 16, no. 3, 2020.

[6]　Knapp M. and Wong G., "Economics and Mental Health: the Current Scenario," *World Psychiatry*, vol. 19, no. 1, 2020.

[7]　Tsapatsoulis N. and Djouvas C., "Opinion Mining from Social Media Short Texts: Does Collective Intelligence Beat Deep Learning?" *Frontiers Robotics AI*, vol. 6, 2019.

[8]　Howard D., Maslej M.M., Lee J., Ritchie J., Woollard G. and French L., "Transfer Learning for Risk Classification of Socialmedia Posts: Model Evaluation Study," *Journal of Medical Internet Research*, vol. 22, no. 5, 2020.

[9]　Masood R., Ramiandrisoa F. and Aker A., "UDE at Erisk 2019: Early Risk Prediction on the Internet," *CEUR WorkshopProceedings*, vol. 2380, 2019.

[10]　Nair R., Vishwakarma S., Soni M., Patel T. and Joshi S., "Detection of COVID-19 Cases through X-ray Images Using Hybrid Deep Neural Network," *World Journal of Engineering*, vol. 19, 2021.

[11]   Guntuku S.C., Lin W., Carpenter J., Ng W.K., Ungar L.H. and Preotiuc-Pietro D., "Studying Personality through the Content of Posted and Liked Images on Twitter," In *Proceedings of the 2017 ACM on Web Science Conference*, New York, NY, USA, June 2017.

[12]   Reece A.G. and Danforth C.M., "Instagram Photos Reveal Predictive Markers of Depression," *EPJ Data Science*, vol. 6,no. 1, 2017.

[13]   Naseem U., Razzak I., Khushi M., Eklund P.W. and Kim J., "COVIDSenti: A Large-Scale Benchmark Twitter Data Set forCOVID-19 Sentiment Analysis," *IEEE Transactions on Computational Social Systems*, vol. 8, no. 4, 2021.

[14]   Nair R. and Bhagat A., "Feature Selection Method to Improve the Accuracy of Classification Algorithm," *International Journal of Innovative Technology and Exploring Engineering*, vol. 8, 2019.

[15]   Byun S., Kim A.Y., Jang E.H. et al., "Detection of Major Depressive Disorder from Linear and Nonlinear Heart Ratevariability Features During Mental Task Protocol," *Computers in Biology and Medicine*, vol. 112, 2019.

[16]   Nair R. and Bhagat A., "Genes Expression Classification Using Improved Deep Learning Method," *International Journal onEmerging Technologies*, vol. 10, 2019.

[17]   Khaira U., Johanda R., Utomo P. E. P., and Suratno T., "Sentiment Analysis of Cyberbullying on Twitter Using SentiStrength,"*Indonesian Journal of Artificial Intelligence and Data Mining*, vol. 3, no. 1, 2020.

[18]   Feldhege J., Moessner M. and Bauer S., "Who Says What? Content and Participation Characteristics in an Online Depressioncommunity," *Journal of Affective Disorders*, vol. 263,2020.

Automation and Computation – Vats et al. (Eds)
© 2023 the Author(s), ISBN 978-1-032-36723-1

A novel design of a drain cleaning robotic vehicle

Tushar Awasthi, Abhishek Purohit & Utkarsh Shah
Student, Department of Electronics & Communication Engineering, JSS Academy of Technical Education, Noida, UP, India

Gayatri Sakya*
Department of Electronics & Communication Engineering, JSS Academy of Technical Education, Noida, UP, India

ABSTRACT: Since the majority of health problems are caused by unhealthy environments, it's important to maintain a clean environment as well as underground sewage pipelines. Since all domestic and commercial building waste is collected through underground sewers, the water inside of them will always be contaminated, and when it is released back into the environment, it will contaminate the wells and groundwater in your neighborhood. For this, a human worker goes inside these sewage pipelines, putting their lives in danger due to the various hazardous things present. Therefore, the goal of our project is to create a robotic vehicle that can clean the sewage which will replace human scavenging and pipe inspection. This is a very significant issue in the context of the Indian scenario and needs to be handled with the utmost urgency. For this, we'd build a robotic vehicle with an IR camera installed in it to relay real-time images of the sewage to the user-controlled mobile application. The user would then select whether or not to act on it by instructing the robotic arm attached to the vehicle.

1   INTRODUCTION

In today's modern world we are developing rapidly in terms of technology, but this is also accompanied by an increase in issues related to human lifestyle, one of which is the ongoing growth of sewage under the pipelines. Cleaning this up is our first responsibility, but doing so is inefficient if done manually because it could endanger that person's life, as such procedures put him into contact with things which are extremely risky for their life. Therefore there is a great need for techniques or solutions that can stop human connection with sewage, so we are developing a robotic vehicle that will clean sewage and will prevent people from entering sewage pipelines for cleaning purposes. The scope of the solution proposed by this project is as follows:

(a)   To develop a robot able to reach the desired location.
(b)   To enable an IR camera on the robot so that it can take pictures of the area in which the robot is currently located and will send them to the user.
(c)   To develop a mobile application with a user interface that enables user interaction with the robot using real-time images received by the IR camera.

2   LITERATURE REVIEW

Feng Chen et al. [1] Provided a force/velocity control method for the arm wrestling robot (AWR) system after analyzing the variations in the force of interaction between the arm wrestling entertainment system and the human hand. S.P. Singh *et al.* [2] proposed the

---

*Corresponding Author

                                                   DOI: 10.1201/9781003333500-49

design and creation of embedded-controlled robotic sewer inspection equipment. It employs ultrasonic sensors, sewer blockage detection systems, embedded systems, and other technologies. The embedded system collects data, preprocesses it, and provides information about the sewer line blockage. M. Lee et al. [3] Proposed Underwater robots without having to remove hundreds of tonnes of water, might be utilised to clean the bottoms of industrial reservoirs. An ongoing effort at the Pohang Institute of Intelligent Robotics is described in this paper, an underwater cleaning robot called PIRO-U3. S. Samaimak et al. [4] demonstrates the use of parallel reading in a wireless receiver with eight channels that can be read both sequentially and simultaneously for semi-automatic robot control using an Arduino. The outcome demonstrates that in this experiment, reading time was utilised more quickly when reading in parallel across all eight channels than when reading in sequence. It has an impact on the enhancement of remote robot control sensitivity. B. P. Aniruddha Prabhu et al. [5] Unarmed ground vehicle (SUGV) is a robot built using a toy car as a base, an Arduino Uno controller, batteries, Bluetooth module, and a smart phone mounted on it. The goal of the project is to create a robot that can relay footage to the military to help them in close-quarters combat while still being reasonably priced. S.Sharmili, et al. [6] focused on the Design of an Intelligent Drainage Cleaner Using Arm Robot. For effective cleaning and disposal of the drainage wastes, they employed an Arduino UNO and GSM modem with an RF receiver mounted on the street poles, together with a transmitter-enclosed RF transmitter and diode logic to the receiver. P. Atre et al. [7] focused on how to build an Efficient and Feasible Gesture Controlled Robotic Arm. This robot is controlled by a microcontroller and it is claimed to solve the major issue of cleaning the pipelines which can prevent damages leading to defects in pipelines. A. Salunke et al. [8] Focused on the cleaning of sewage present inside the pipeline. This robot is controlled by a microcontroller and it is claiming to solve the major issue of cleaning the pipelines, it can prevent damages leading to defects in pipelines.M. S. Ahsan et al. [9] suggested using a mobile robot that can be controlled manually to clear the drain. The gadget that can operate the vehicle over Bluetooth is that one. The drains will be cleaned in this case by a grabber system powered by chain sprockets. The methodology in this paper includes the Chain-Sprocket Mechanism, Design Procedure, Design Block, Controlling App and many more. The limitation or the research gaps in the research paper is focusing on the fact that the grabber system can only grab big and lightweight particles such as water bottles, plastic wrappers etc. Pon, et al. [10] proposes the development of the Robotic Arm Control using a Bluetooth device with an Android Application. This project mainly deals with the idea of how to use a mobile android application to see the location of the robot and to give instructions to the robotic arm.

R. Asma et al. [11] focuses on the implementation of a completely contactless robotic hand controlled by a mobile application. It protects cleaning workers from health-related issues such as bronchitis, skin cancer, and respiratory problems the worst being the loss of life. Jinendran Chandrasekhar et al. [12] proposed the development of a modular In-Pipe Robot for Inspection and Cleaning. It uses a camera to see the location of the robot and a jet nozzle from which very high-pressure water is jet on the blockage to clear it. This project mainly deals with the problem of blockage in sewer lines and how to clear that blockage for which they use the camera to detect the location of the blockage and the robot's jet spray to clear it. S. Khrueangsakun et al. [13] concentrated on creating a Cyber-Physical System for Real-Time Web-based Visualization and Control of a Robot Arm.They used ESP32 WROOM with a Wi-Fi module to control the robot and a WebSocket protocol through which the virtual model was displayed and controlled via the web browser to analyse the robot's position and use the 4-DoF robotic arm to clear the blockage. H. He et al. [14] suggested the creation of a mobile robotic arm with gesture recognition. The STM32 NUCLEO-F411RE development board is being used by the controller to analyse and evaluate the gesture data that was received over Bluetooth. The gesture data, which was captured via a gesture sensor, was used to operate the robotic arm and the automobile.

The gesture data was also analysed using a computer. Sulthana et al. [15]focuses on the creation and modeling of a robot that cleans drains. They created a robot that can drive itself through pipelines to look for obstructions using an IR camera and ultrasonic sensor. In order to prevent it from escaping and to be retrieved after the block has been cleared, it is tied via a data wire. G. Singh et al. [16] focuses on developing a device that makes use of the aforementioned technologies, a robotic arm installed on top of a rover for selecting and putting tasks in perilous environments. The system's brain is the NodeMCU, and the complete arrangement is controlled by a smartphone.

3 PROPOSED MODEL AND ITS NOVELTY

The proposed model is a Bandicoot Robot that can clean sewage and go to the defined region. It would use an Arduino and robotic arm to accomplish the objectives, and an IR camera would be enabled in the Robot so that according to camera inputs, the user interface will automate the robotic vehicle, and the user can control it via a mobile application. This model differs from the others as it includes a Bandicoot Robot equipped with an IR camera. Furthermore, this model will allow users to control the robot using a mobile application to complete all tasks.

3.1 *Robotic vehicle*

The Circuit Diagram for the Robotic Vehicle is shown below: In this circuit diagram we are using L298 Motor driver, HC-05 Bluetooth module, Arduino UNO, 5V DC to DC Buck Converter. In this , L298 Motor driver is responsible for the movement of the DC motors in the desirable direction. The mobile application and robot are in two-way communication (as transmitter and receiver) which is done by using the HC-05 Bluetooth module.

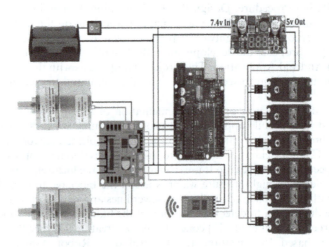

Figure 1.   Circuit diagram of robotic vehicle.

Arduino UNO is responsible for controlling all the functionality of the other components. Here the program is uploaded according to which the whole system is working.5V DC to DC buck converter is also used for converting any input voltage to 5 Volts of DC Voltage. In this case a battery of 7.4V (3.7Vx2) emf is also being used.

Figure 2. Robotic vehicle design.

The Figure 2 shows the snapshot of the robotic vehicle that has so far been created.

The Figure below shows the snapshot of an Mobile application that has been created so far that is responsible for controlling the robotic vehicle and would also be responsible for controlling the Robotic Arm Movements.

Figure 3. Snapshot of user interface.

3.2 *Robotic arm*

The proposed robotic arm for the drain cleaning is designed as shown in Figure 4.

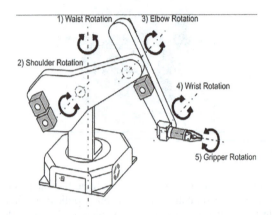

Figure 4.   Robotic arm for drain cleaning.

Servo Motors are used here in the robotic arm in order to rotate it from the required junction. We will be using 3–4 servo motors for the movement and grabbing of the waste/ sewage, which will be controlled through a mobile application. There will be two types of movement in the arm which are the vertical and rotation within the arm. we have used an Arduino UNO to implement the robotic vehicle and we are controlling this robotic vehicle using an Android Application that receives signals from the Bluetooth module connected by maintaining two-way communication between the Robot and the Mobile Application, the same Bluetooth module receives commands from the Mobile Application and transmits them to the Robot. The next step of the work is:

a. Implementation of a Robotic Arm on the Vehicle to pick and drop the Waste Material.
b. Installation of an IR Camera and building of an app to provide user interface which will
    be responsible for automating the robotic arm according to the camera inputs.

4   CONCLUSION

In recent times, there are so many studies carried out in order to automate the process of cleaning the drainage. Our work also stands with the same ideology but has major improvements in it. Many studies have been studied in great depth. We focus more on making the system completely contactless and remotely accessible from anywhere.

a. The robotic vehicle can reach the desired location and can clean sewage.
b. The mobile application's user interface will be responsible for automating the robotic
    vehicle.
c. The robotic vehicle according to the camera inputs and the user can take actions via
    mobile application.

So the work is extremely important in view of social relevance and society will be benefited if government approved such designs for drainage cleaning robots. We will complete the design future work and will come up with fully fledged prototype model of robot.

REFERENCES

[1] Chen F., Yu Y., Song Q. and Ge Y., *"Force Analysis of Arm Wrestling Entertainment System and its Application to Control Design,"* 2006 International Conference on Mechatronics and Automation, 2006

[2] Singh S.P., Verma A. and Shrivastava A.K., *"Design and Development of Robotic Sewer Inspection Equipment Controlled by Embedded Systems,"* 2008 First International Conference on Emerging Trends in Engineering and Technology, 2008

[3] Lee M. et al., *"An Underwater Cleaning Robot for Industrial Reservoirs,"* 2012 IEEE International Conference on Automation Science and Engineering (CASE), 2012

[4] Samaimak S., Tipsuwanporn V., Numsomran A. and Harnnarong S., *"The Improved Sensitivity of Remote Robot Control via Parallel Method,"* 2015 15th International Conference on Control, Automation and Systems (ICCAS), 2015

[5] Aniruddha Prabhu B.P. and Hebbal S., *"Small Unarmed Robot for Defense and Security:A Cost Saving method by the help of Arduino Uno,"* 2017 2nd International Conference On Emerging Computation and Information Technologies (ICECIT), 2017

[6] Sharmili, S., Saarika, T. and Malar, R., 2017. Intelligence Drainage Cleaning Using Arm Robot. *Int. J. Emerg. Technol. Comput. Sci. Electron. (IJETCSE)*

[7] Atre P., Bhagat S., Pooniwala N. and Shah P., *"Efficient and Feasible Gesture Controlled Robotic Arm,"* 2018 Second International Conference on Intelligent Computing and Control Systems (ICICCS), 2018

[8] Salunke A., Ramani S.,Shirodkar S., Vas O. and Acharya K., worked on *"Pipe Cleaning Robot,"* in 2019 International Conference on Nascent Technologies in Engineering (ICNTE), in 2019

[9] Ahsan M.S., Munna M.S. and Nayeem A., *"Design and Implementation of a Drain Cleaner Robotic Vehicle,"* 2019 3rd International Conference on Electrical, Computer & Telecommunication Engineering (ICECTE), 2019

[10] Pon, M. & Natarajan, Shivaanivarsha. (2019). *Robotic Arm Control using Bluetooth Device with an Android Application.*

[11] Asma R. and Matilda S., *"Keluthi Roomba – A Robotic Hand for Cleaning Sewage,"* 2020 International Conference on System, Computation, Automation and Networking (ICSCAN), 2020

[12] Jinendran, Chandrasekhar. (2020). *Development of Modular In-Pipe Robot for Inspection and Cleaning.*

[13] Khrueangsakun S., Nuratch S. and Boonpramuk P., *"Design and Development of Cyber Physical System for Real-Time Web-based Visualization and Control of Robot Arm,"* 2020 5th International Conference on Control and Robotics Engineering (ICCRE), 2020

[14] He H. and Dan Y., *"The Research and Design of Smart Mobile Robotic Arm based on Gesture Controlled,"* 2020 International Conference on Advanced Mechatronic Systems (ICAMechS), 2020

[15] Sulthana, S.F., Vibha, K., Kumar, S., Mathur, S. and Mohile, T.A., 2020, August. Modelling and design of a drain cleaning robot. In *IOP Conference Series:* Materials Science and Engineering (Vol. 912, No. 2, p. 022049). IOP Publishing.

[16] Singh G., Singh A.K., Yadav A., Bhardwaj I. and Chauhan U., *"IoT developed Wi-Fi Controlled Rover With Robotic Arm Using NodeMCU,"* 2020 2nd International Conference on Advances in Computing, Communication Control and Networking (ICACCCN), 2020.

[17] Patidar, V., Mishra, A., & Tiwari, R. (2018). *Robotic Gripper Arm System with Effective Working Envelope.* 2018 Second International Conference on Intelligent Computing and Control Systems (ICICCS).

[18] Cristian Moldovan, Valentin Ciupe, Ion Crastiu, Valer Dolga. Model Free Control of a 2DOF Robotic Arm using Video Feedback. *2019 6th International Symposium on Electrical and Electronics Engineering (ISEEE).*

[19] Le, N. T., & Ngo, T. Q. (2019). Proposal of a Sewerage Cleaning Robot to Collect Garbage Applying for Ho Chi Minh City. *2019 IEEE/SICE International Symposium on System Integration (SII).*

[20] Varun Kumar Ojha, Paramartha Dutta, Hiranmay Saha and Sugato Ghosh (2012). Linear Regression Based Statistical Approach for Detecting Proportion of Component Gases in Manhole Gas Mixture. *2012 1st International Symposium on Physics and Technology of Sensors (ISPTS -1).*

[21] Kasat, N. N., Gawande, P. D., & Gawande, A. D. (2019). *Smart City Solutions On Drainage, Unused Well And Garbage Alerting System For Human Safety.* 2019 9th International Conference on Emerging Trends in Engineering and Technology – Signal and Information Processing (ICETET-SIP-19).

[22] Rana, A., Venkateshwar, & Joshi, G. S. (2010). Manhole Cleaning Robotic System (MCRS). INTERACT-2010.

[23]   Joshi, G., Rana, A., & Venkateshwar, R. (2011). *Robotic System for Cleaning Manholes (RSCM)*. 2011 IEEE Student Conference on Research and Development.

[24]   Deng, S., Xu, X., Li, C., & Zhang, X. (2010). Research on the Oil Tank Sludge Cleaning Robot System. *2010 International Conference on Mechanic Automation and Control Engineering.*

[25]   Le, N. T., & Ngo, T. Q. (2019). Proposal of a Sewerage Cleaning Robot to Collect Garbage Applying for Ho Chi Minh City. 2019 *IEEE/SICE International Symposium on System Integration (SII).*

[26]   Haswani, N. G., & Deore, P. J. (2018). Web-Based Realtime Underground Drainage or Sewage Monitoring System Using Wireless Sensor Networks. *2018 Fourth International Conference on Computing Communication Control and Automation (ICCUBEA).*

Automation and Computation – Vats et al. (Eds)
© 2023 the Author(s), ISBN 978-1-032-36723-1

Intelligent surveillance system: Review

Lipsa Das*
Amity University, Greater Noida, India

Laxmi Ahuja*
Amity University, AIIT, Noida, India

Priyansha Tiwari*
MRIIRS University, Faridabad, Haryana, India

Vishal Jain*
Sharda University, Greater Noida, India

Khushi Dadhich* & Bhuvi Sharma*
Amity University, Greater Noida, India

ABSTRACT: ISS expanded as "The Intelligent Surveillance System" is a monitoring structure with the ability to intelligently analyse monitoring data automatically and carry out crucial activities like producing an alarm or warning. With or without human assistance, The ISS is equipped to automatically evaluating collected information, comprising audio, picture, and video. Computer vision and electronics (sensing equipment), AI (machine learning), communication (IoT), and other fields are all included in the interdisciplinary field of ISS. By eliminating and limiting human involvement in the surveillance system, the proposed work intends to establish an ISS. The major goal of this monitoring system is to keep an eye on the property using wireless devices and CCTV cameras. CCTV cameras are used in the surveillance system to record video, which is then converted and processed using picture processing and ML methods. The system also provides a live data feed to the user. If a suspect motion is identified along the frame that records the motion, the owner will receive the pushed notification. The surveillance system also aids facial recognition-based automatic door access systems. A security warning is delivered automatically by being sent to the user's cell phone. The owner has access to the alarm system and uses a mobile device to remotely operate the door.

Keywords: Intelligent Surveillance System, Internet of Things, Computer Vision, Machine Learning

1 INTRODUCTION

For security applications such as computer vision, database systems, multimedia disciplinary studies such as computer animation, computer networks, digital imagery processing, as well as programming and algorithms, monitoring has grown in importance. In contrast to conventional visual surveillance techniques, ISS allows for automatic and remote monitoring of security risk factors in homes, banks, casino, airports, and public spaces. Public safety and

*Corresponding Authors: lipsaentc9@gmail.com, lahuja@.amity.edu,
Priyansha.tiwari191@gmail.com, drvishaljain83@gmail.com, Khushidadhich08@.amity.edu and
Bhuvisharma86@gmail.com

DOI: 10.1201/9781003333500-50

the issue of surveillance security are intertwined. Protection of property is a constant priority for all organizations, and governments take significant steps to mitigate the risks of terrorist attacks. Transportation applications, such as marine environments, airports, railways, highways, and subways; Open spaces, such as banks, homes, supermarkets, parking lots, and retail stores; Remote monitoring of human activities, such as many industrial processes, as well as military applications to accomplish some quality checks. A control center is equipped with calibrated cameras, that are a basic component needed for intelligent monitoring. The work of the security staff and the number of parents who require supervision can help to assist the advantages of monitoring the larger districts.

The recession's costs and the quick shrinking of video cameras have led to widespread usage of Critical infrastructure, including military posts, airports, power plants, banks, colleges, etc., is being monitored by a vast array of surveillance cameras and other sensors. Manual supervision by operators is an unsuitable or indeed unrealistic method since human resources are pricey and have narrow skill sets [1]. An intelligent surveillance system (ISS) is designed to autonomously monitor the framework and surrounding with less to no human involvement. Automatically finding and tracking items (such persons or vehicles) are a part of these monitoring operations, which are then accompanied by additional study and action. Such intelligent devices are made possible by techniques for signal processing, image processing, and AI (machine learning) [2]. The most frequent modalities (devices) for surveillance systems are visible cameras, like CCTV. It has been used to supervise people, events, and settings for long period of time.

This review paper is based on a thorough investigation for automatically interpreting audio and video data. This paper's primary goals are to present a comprehensive outline of intelligent surveillance systems and to review the recent methods to every of the steps in their processing.

2 NEED FOR INTELLIGENT SURVEILLANCE SYSTEM

This industry uses new technology to fulfil the following end users" increasing demands:

- Improved picture quality
- Simpler installation and upkeep
- More trustworthy and secure technology
- Increased video recording retention
- Decrease in expenses
- Scalability and size
- Remote surveillance capabilities
- Combination with different systems

The bulk of monitoring systems entail manual supervision by human operators, which is an ineffective or even unrealistic answer given the cost and skill limitations of human resource. Due to this significant flaw, surveillance systems must adhere to the following limitations [10]:

- The monitoring personnel must keep an eye on several displays for an extended amount of time, which causes them to lose focus and miss the ideal moment to make a choice.
- Due to limitations imposed by the physical layout and spatial distance, the monitoring personnel are required to continuously monitor the screens from a specific spot;
- Traditional monitoring systems capture the picture instantly without being able to manage and classify the graphical data, that makes retrieving and tracing a challenge in the future. increased system intelligence inbuilt.
- Hardware dependence is a problem when installing a traditional non-network monitoring system. The establishment of a control booth is necessary due to the fixed site and distance limitations on the linked camcorders.

Commercial video solutions are numerous as a result of the growth of monitoring devices, but they all employ a "post-event" forensic analysis tool. To provide actual monitoring capabilities, this has been a key motivator for machine vision designers to create smart monitoring solutions. Given the wide range of capabilities of such items, it is useful to group them according to the superior classification given as:

First generation: conventional analogue CCTV systems with digitized video camera recording capabilities.

Second generation: huge IP networks and very capable "video management systems" (cameras with encoders can be digital or classic). A variety of basic image processing tools are present in these systems (For example, abandoned object recognition, perimeter intrusion detection, etc.)

Third generation: systems of intelligent surveillance that are truly multi-view capable and have significant data extraction capabilities.

The majority of commercial solutions are based on second-generation technology and frequently exhibit significant false-alarm rates. Such kinds of monitoring tools have been used to keep an eye on people, places, and activities for long time. Numerous researches have been done to automatically interpret information from security camera (pictures or video) [10].

3 INTELLIGENT SURVEILLANCE SYSTEM (ISS) OVERVIEW

With a enormous number of safety cams, an ISS is a tracking technique with the ability to intelligently evaluate collected information and take appropriate actions, including creating an alarm or warning. Machine vision and patterns identification, ai (machine learning), networking, communications, and other fields are also involved in the interdisciplinary study of the ISS. Intelligent surveillance systems have the potential to be used in many different settings and contexts. The following is a list of some application areas:

- Both intruder detection and home security [3,4].
- Safety and home care [5].
- Tracking of traffic [6].
- Analysis and management of the crown [7].
- Autonomous vehicles and pedestrian detection [8].
- Public xspaces including campuses, colleges, and governmental structures.
- Remote border monitoring, power plant and enterprise perimeter surveillance, etc.

VSAM expanded as "Video surveillance and monitoring" [12], W4 [13], PRISMATICA (pro-active integrated systems for security management by technological institutional and communication assistance) [14], and ADVISOR (annotated digital video for intelligent surveillance and optimised collection) [15] are a few examples of the various monitoring systems that have earlier been researched or created to have automation or smart abilities. A summary of the PRISMATICA system, which has been suggested to enhance passenger safety and security in the public transportation system is shown in Figure 1. It has a number of significant parts, including a camera network (current CCTV), a smart system of cameras, transmission, an monitoring of audio system, an operator, and a main server (MIFSA).

DARPA ARGUS-IS (autonomous real-time ground ubiquitous surveillance imaging system) is another outstanding surveillance system [20]. With a 1.8 Gigapixel video system, ARGUS-IS can use a single platform to automatically track any moving item in an area of 40 square kilometers (the size of a small city). DETEC AS (www.detec.no) and DETER (detection of events for threat evaluation and recognition) are two examples of commercially accessible ISS solutions [21].

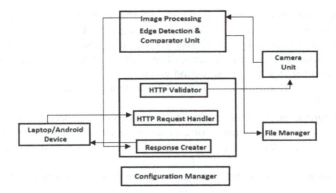

Figure 1.   General surveillance system framework [9].

Best video surveillance advantages

Intelligent surveillance systems may be very important for both public and private safety and security. Given its practical utility, the following benefits can be listed:

Use for security - In the past, malls and shopping centres were the only places where surveillance techniques were used. Closed-circuit televisions are now present practically everywhere, from tiny shops to residences and religious institutions. They therefore provide increased public security at a lower cost.

Real-time observation: Large enterprises have traditionally benefited from video surveillance operated by security specialists. In the past, events that were caught on camera were utilized to reveal crucial information and serve as evidence after the fact. However, users of contemporary technologies can check and respond to alarms right away [9][11].

Performance - including system precision.

Robustness: The system ought to be resistant to issues encountered in the actual environment, such as shifting lighting, clutter, occlusion, shifting weather, camouflage, and cybernetic attack [10].

Efficient in terms of cost: A cost-effective framework is needed for realistic implementation in various range of actual-world monitoring practices, such as indoor intrusion recognition and outside monitoring of significant constructions.

4   ARCHITECTURE OF GENERAL VIDEO SERVILLANCE SYSTEM

The intelligent system for video surveillance has a three-tier design that consists of a dbms, an application server, and clients. The GSM modem-equipped server device is what makes up the application server. Only the browser is required on the client side device. It will use the server's IP address and port number to connect. The http server module and the image analysis module make up the majority of the server side device. The http protocol is used to connect two devices. So, a http server is needed. It will come with a built-in server-side device. This's primary functions are to manage incoming requests, validate them, and produce a response. The module for image processing's task is to find intrusions.

5   TECHNIQUES FOR COMPUTER VISION

An ISS's goal is to develop "smart places" that function automatically and synergistically on a network made of geographically scattered distributed cameras. Static and dynamic pan-tilt-zoom (PTZ) recording devices can be used for this. Such a high standard is composed of a number of separate layers:

1. Detection of objects;
2. Detecting and tracking objects;
3. Camera control;
4. The Use of Sensors
5. Layers for interaction and visualization;
6. Behavioral Assessment.

Detection of objects:

The key feature of ISS is the ability to identify and classify various object like animal, vehicle or any human being. Because of varying face appearance and body structure as well as different kind of background or lighting with any type of different clothes, it is difficult for a computer like machine to detect a human [16]. There are number of methodologies on visual camera, have been proposed by many no of researchers to identify a human.

- The research article of Enzweiler *et al* [17] paper shows high resolution of image and the advantage of HOG/linSVM [18] with a low processing speed.
- It is found that FDDW [22] shows highest accomplishment in the research work of Dollar *et al* [19]. However the cost of computing is not a parameter then MULTIFTR +MOTION [23] will be the best choice.
- Spinello *et al* [24] stated the detection of people by bottom-up and top-down detector depends on lidar data, A layered person model is learned from the back of specified classifiers having different people's height levels those continuously vote in space by bottom-up model. The features of computed under the voxel of boosted tessellation are used to classify the candidate by top-down model.
- Whereas, the 3D lidar point data is mapped into a depth image by Banedek *et al.* [25], who also perform 2D people detection.
- A people detection method based on RGBDepth sensors that provide both range & image information was presented by Spinello *et al.* in [24].

Many of the suggested methods for categorising and identifying objects focused mainly on a few item kinds, such people and cars. Various animals and other factors that can compromise security or safety are just a few of the many considerations that must be made in the real world.

Object recognition and tracking:

The object is typically tracked in the spatiotemporal domain by surveillance systems after device recognition. Object tracking in realistic situations is a challenging task because of lighting fluctuations, obstacles, clutter, sensor motion, and other issues. The algorithm related to Visual tracking (based on visible cameras) have been proposed quite a bit recently. Considering the different methods that is based on visual camera can be differentiated into five groups such as:

- model-based method
- appearance-based
- contour- and mesh based
- feature-based
- hybrid methods.

When accuracy and robustness are taken into account, the discriminative scale space tracker (DSST) used in the VOT2014 (Visual Object Tracking Challenge) yields the best results [27]. Smeulders *et al.* [26] conducted an experimental survey based on Amsterdam Library of Ordinary Videos (ALOV) for 19 online trackers. With robust scale estimation, this tracker improved upon the MOSSE i.e, minimum output sum of squared errors tracker [28].

The technologies of radar, lidar, and other types are currently used to track people. Stereo range data was used by Mitzel *et al.* [31] to track multiple people in real-time. Along with 2D images, they looked at range information from stereo cameras. An automatic target classifier is created by Javed *et al.* [30] using ground surveillance radar (such as pedestrians and vehicles).Ultra-wideband (UWB) radars were used by Kocur *et al.* for their robot surveillance system in [29]. Spinello *et al.* [24] proposed 3D people tracing using a multi-target, multi-hypothesis tracking approach based on lidar data, where they used a bottom-up top-down detector to Spinello *et al.* [24] used a bottom-up top-down sensor to identify people and offered a multi-target, multi-hypothesis monitoring strategy. Banedek *et al.* [25] used deterministic background modelling, multi people monitoring, and online re-identification in their approach to real-time 3D people surveillance. The tracker module was also evaluated in actual outdoor settings with plenty of occlusions and lots of persons reappearing over the course of the observation.

Camera control: For smart camera like CCTV systems to achieve greater as compared to simple array of CCTV cameras, cams management is crucial. In these circumstances, manually calibration and operating of cameras involves time taking work that needs to be done by a professional specialists. In these circumstances, modifying cams physically is laborious job which needs to be done by a qualified expert. Understanding the relationships between monitoring cameras and between video monitoring and the CCTV operating software is necessary for the interconnection and data transmission in between images acquired in a worldwide computing devices or systems. A room needs to be recalibrated if it is moved, taken out, or added. This turns into a major barrier to the effective application of automatic monitoring systems in systems with multiple cameras. The quantity of manual input necessary is being eliminated or at least reduced by developing automatic calibration and cameras control technology.

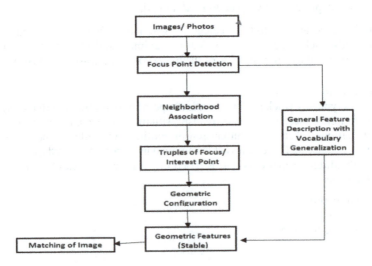

Figure 2.   Camera management structure.

Systems for managing surveillance cameras track and detect network modifications and update calibration data as necessary. An essential component of automatic calibration is the capacity to recognize correlation between overlapping images. Cameras are rarely placed in a monitoring system so as to enhance the range of the region. Searching the relationship among the photographs taken is exceedingly difficult as a result. To accomplish these goals, reliable extraction and matching approaches have been created. Even the most sophisticated matching techniques can fall short of offering sufficient precise correspondence to begin calibration. This might be as a result of a scene's lack of distinguishing elements, low quality of photo, or a high percentage of inaccurate comparisons/match ups. There exists just some techniques rest on Hessian Affine region sensors (WiDense method) [32] improves correspondences, decline false match ups, and provide a significant amount of extremely correct point correspondence. The de-noise, deinterlace, sharpening, scaling, and colour processing procedures perform better because to this sophisticated video processing logic. This collection of software and hardware innovations combines cutting-edge video techniques to reduce jitter and provide sharper images, making it easier to read license plates and recognize faces.

- **Visible camera**

  Common technology modalities for surveillance systems include visible (video) cameras. It has been employed for a very longer time to track individuals, occasions, and environments. This technique is used by the vast majority of available commercially monitoring sensors, including low-cost IP cameras and expensive professional Cameras. Surveillance cameras have been installed everywhere, including on streets, in public buildings, and at international borders. There has been a lot of study done on visual or video surveillance systems. For surveillance systems, various types of visible cameras, including colour (or RGB) cameras, monocular, stereo, omnidirectional cameras, etc., have been studied. According to Valera *et al.* [34], there are three generations of visual monitoring systems: the first is analogue CCTV systems, the second is automated optical surveillance using computer vision technology, and an automation process for broad monitoring is the third (3rd).

- **Infrared (IR) and thermal camera**

  Noticeable cameras only function effectively in environments with sufficient lighting or light intensity, such as during the day. Visible cameras are not very successful at capturing scenes in low light conditions or at night. Two choices are available in this situation: a thermal camera or an infrared sensor. Images captured by thermal or infrared cameras make objects (like people) that have different temperatures from their surroundings more easier to differentiate than images captured by visible cameras.

  The words "infrared camera" and "thermal camera" are sometimes used interchangeably since both cameras record infrared light that is not visible to the humans. To improve visibility, an infrared camera often refers to one that records NIR (near-infrared) or SWIR expanded as "short-wavelength infrared" emissions. Low illumination situations are a good fit for infrared sensors. Far-infrared radiation abbreviated as FIR or Long-wave is given out by entities, and a thermal based cams are cameras that can record this radiation. If the environment is absolutely black, a thermal camera can be useful. There are two categories of thermal cameras: cooled and uncooled. Higher resolution and picture quality are provided by cooled thermal cameras, although they are typically more expensive and power-hungry.

- **Radar and lidar**

  Distance recognition is an intriguing detector modality for the reason of its precision, broad range of vision, and resilience to changes in lighting. Radar (radio detection and ranging) and lidar are examples of this type of range sensing (light detection and ranging). Lidar employs light or a laser while radar uses radio signals for sensing. By removing all information outside the tracking area, changes in the backdrop in range data may be easily filtered out. Range data often provides less information than sight information for recognizing people or objects, which is a downside.

Numerous sensor modalities, including audio, ultrasonic, passive infrared (PIR) pressure sensor, and others, have been investigated to enhance or support surveillance systems. Environmental sounds such as glass breaking, dogs barking, screaming people, fire alarms, gunfire, and other similar sounds can help determine if a situation is "safe" or "insecure." A noticeable camera (or another sensor) is engaged for more accurate object recognition once certain sensors are employed for alerting. An imbedded home security system based on several ultrasonic sensors was developed by Bai et al. [3]. They used pressure sensors and pyroelectric ir sensor (PIR) as an alarm system to conserve electricity in other work [35].

Integration of Sensor:

It seems sense that combining several sensors will yield more precise data about the intended object. There are two types of many sensors: heterogeneous (multiple modality, like so many cameras) and homogeneous (different modalities). Since they collect data from a 2D perspective, several sensor modalities, like visual, infrared, and thermal cameras, are seemingly closed and complimentary. Similar to each other, lidar and radar both record data in the range domain (two dimensional or three dimensional). Data fusion is an example of a basic level of sensor fusion. Decision fusion is an example of a high degree of sensor fusion. Every sensor transmits the first measure to fusion centre in data fusion, which then decides whether to consider the event to have occurred. Every detector takes a separate choice on the basis on its particular measurements in decision fusion, and the fusion centre then makes an overall judgement according to every singular choice (like using majority voting). Every fusion situation offers unique benefits and disadvantages. Issues faced in sensor fusion include dealing with various information modality (optical, audio, radio signals, etc.), data flaws, conflicting data, sensor topologies, and other problems [36].

Numerous cameras have been used in extensive work on visual surveillance systems. Recent advancements in multi-camera smart surveillance footage, such as multiple camera modification, calculating camera network topology, multiple camera observation, entity re-recognition, and analysis of activities of multiple camera, are covered in Reference [3]. On outdoors people recognition for moving platforms, Robertson et al. [37] use detectable, IR, and thermal based cams technology (vehicle). Premebida et al. [38] suggested pedestrian recognition using dense LIDAR data and RGB camera data.

Visualization and interaction layer:

In a simulated viewing environment, like a comprehensive view or a 3D replica of the operational environment, viewer layer controllers can engage with the common video elements of an image. The fact that huge portions of structures typically share a common layout and that man-made surroundings frequently contain multiple flat surfaces is the foundation for many of these methods. This layer offers resources for looking up events, things, and individuals of interest as well as alerts to pertinent staff so they may decide on a course of action and carry it out. The ISS is built to automate those chores, but high-level choices would still be made by a human operator. The following characteristics are present in this kind of automation: The soft-biometric module can utilise motion detection and tracking algorithmsEvery camera stream operates in timely manner or very near to real time and is accessible over an IP network. They have the ability to operate a network of cams of any capacity.

Behavioral analysis:

Automated monitoring scene analysis is gaining popularity because it analyses events at a higher level than only objects, including object detection and tracking. There are few areas where particular attention is required. These areas are crown analysis [41], incident analysis, group behaviour [40] and automated human behaviour analysis [39]. Human behavior analysis can remarkably enhance safety by shortening the time it takes to prevent unwanted incidents and seeing them at the starting level of scepticism [42]. Analyzing human behavior is difficult despite its importance. An essential component of studying human behavior is classification. Numerous categories have been used to classify human behaviour. The classification is expanded upon by Foroughi *et al* [44] into normal, exceptional, and abnormal. Positive, neutral, and negative are the earlier classisfication of the activities by Park and Aggarwal [45]. Humans can be seen alone, in small groups, or in vast crowds. Bottlenecks, fountainheads, tracks, arch, and blocks are five crowd behaviours that Solmaz *et al* [47] developed a method for recognizing in visual settings. Bremond *et al.* [46] proposed an activity monitoring framework for recognising behaviours involving either lone individuals, groups of persons, or crowds in the domain of visual monitoring of metro scenes using several cameras. Simple classification was suggested by Kiryati *et al.* [43] normal and abnormal.

6 PROBLEMS WITH CURRENT SYSTEM

- Although surveillance cameras are inexpensive and widely accessible, the cost of the labour needed to monitor them is high.
- As a result, the footage from such devices is typically only occasionally or never examined; instead, it is frequently utilized as a recorder to go over an occurrence that had occurred.
- It also includes the labor-intensive access control monitoring.
- It's becoming more difficult to control people and preserve personal information in an increasingly digital society.

How to guarantee that personnel and guests can enter the appropriate locations at the appropriate moments.

7 THE PROPOSED ARCHITECTURE

Figure 3 depicts the overall design of the proposed ISS, which is equipped with video surveillance cameras and a video processing unit to enable high-level data collection. This processing unit can communicate with the network's control and alert handling center. The video database and retrieval tool, which allow for searching and storing of the chosen video objects,

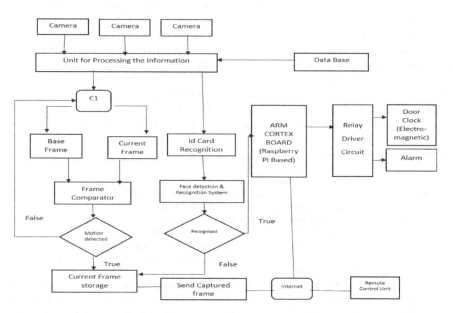

Figure 3.   General framework of intelligent surveillance system architecture.

sections, and associated content, are also crucial components. The processing unit receives the information from multiple cameras and uses background subtraction to check for motion. The Detection and Recognition algorithm for faces is also used to recognize ID cards. The adafruit.io and firebase server, which handles IoT tasks, is connected to everything. The edge devices that assist with alarm and door access are then connected to the servers.

Following that, the ISS will capture the camera footage, manipulate it, and process it using Open CV on the chosen local processing unit (Raspberry Pi or Intel-powered CPU). Additionally, the user will get live data sent by the ISS. If suspicious movement is discovered, the owner will get a push notification or alert along with the image. The face recognition automatic door access system will benefit from the ISS as well. The automated notification was accomplished by texting the user's phone a security alert. The owner will be in charge of the alarm system and will be able to operate the door remotely using a mobile device.

8   BENEFITS OF PROPOSED SYSTEM

The above proposed system will enables users to watch videos even when they are in distant locations. Due to the use of the http protocol, the application offers online video streaming capabilities so that users can view the videos from web browsers, including those on computers and Android devices. For image matching and intrusion detection, we don't require the use of any extra hardware. Because of the image matching technique used by our system, the results are more precise and accurate. Using this architecture, the entirety of smart surveillance can be conducted remotely. The user can even operate the system from a distance. He is able to issue commands to turn the system camera on or off. As soon as an intrusion is discovered, the user is informed. As a result, the user can act appropriately right away. Intelligent video movement detection analysis systems are integrated with SMS notification systems for smart video surveillance.

9   CONCLUSIONS

Although significant advancements in computer vision and other fields have been made, still there exists significant scientific obstacles to vanquish prior the aspiration of trustable automatic monitoring can be realized. Practical factors like poor weather and lighting robustness, smart processing of video for occasion identification, and effectiveness in respect of actual-time operation and amount add to these technical difficulties. Because they are restricted to a specific circumstance, the majority of supervision systems executes through singular modality and deficient robustness. Thermal infrared, audio, and other forms of media provide supplemental and auxiliary surveillance data of the environment. The precise fusing of data from diverse media streams can improve object recognition, surveillance, and activity recognition in the real world, based on a review of the research on multi-modal data fusion.

REFERENCES

[1]   Sulman N., Sanocki T., Goldgof D. and Kasturi R., How Effective is Human Video Surveillance Performance?, *19th International Conference on Pattern Recognition, Tampa, FL*, 2008, pp.1–3R.

[2]   Ibrahim S.W. A Comprehensive Review on Intelligent Surveillance Systems. *Communications in Science and Technology*. 1(1) (May 2016). DOI:https://doi.org/10.21924/cst.1.1.2016.7.

[3]   Bai Y.W., Li-Sih Shen and Zong-HanLi, Design and Implementation of an Embedded Homesurveillance System by use of Mu ltiple Ultrasonic Sensors, *IEEE Transactions onConsumer Electronics*, 5(6) (2010) 119–124.

[4]   Castro J.L., Delgado M., Medinaand J. and Ruiz-Lozano M.D., Intelligent Surveillance System with Integration of Heterogeneous Information for Intrusion Detection, *Expert Systems with Applications*, 38 (2011) 11182–11192.

[5] Prati A., Cucchiara and R. Vezzani, *A Multi-camera Vision System for fall Detection and Alarm Generation,Expert Systems*, 24(5) 334–345.

[6] Chen Y.L. Bing-Fei Wu, Hao-Yu Huang and Chung-Jui Fan, A RealTime Vision System for Nighttime Vehicle Detection and Traffic Surveillance, *IEEE Transactions on Industrial Electronics* 5 (8) (2014) 2030–2044.

[7] Junior J.C.S.J., Musse S.R. and Jung C.R., Crowd Analysis Using Computer Vision Techniques:A Survey, *IEEE Signal Processing Magazine* 27 (2010) 66–77

[8] Dollar P., Wojek C., Schiele B. and Perona P., Pedestrian Detection: An Evaluation of the State ofthe Art, *IEEE Transactions on Pattern Analysis and Machine Intelligence* 34 (2012) 743–761.

[9] Khot Harish S., Gote Swati R., Khatal Sonali B. and Sangmesh P., Smart Video Surveillance, *International Journal of Emerging Engineering Research and Technology* 3(1), January 2015, PP 1 ISSN 23494395 (Print) & ISSN 234909112 4409 (Online)

[10] Adrian A.I., Ismet P. and Petru P., "An Overview of Intelligent Surveillance Systems Development," *2018 International Symposium on Electronics and Telecommunications (ISETC)*, 2018, pp. 1–6, doi:10.1109/ISETC.2018.8584003.

[11] Amer A. and Regazzoni C. Editorial: Introduction to the Special Issue on Video Object Processing for Surveillance Applications. *Real-Time Imaging* Vol. 11, pp: 167–171, 2005.

[12] Collins R., Lipton A., Kanade T., Fujiyoshi H., Duggins D., Tsin Y. Tolliver D., Enomoto N. and Hasegawa O., *A System for Video Surveillance and Monitoring, Tech.Rep. CMU-RI-TR- 00-12, Robotics Institute*, Carnegie Mellon University (2000).

[13] Haritaoglu I., Harwood D. and Davis L.S., W4: *Real-time Surveillance of People and their Activities, Patter Analysis and Machine Intelligence*, 2(2) (2000) 809–830.

[14] Benny Ping Lai Lo, Jie Sun and S A.Velastin, Fusing Visual and Audio Information in a Distributed Intelligent Surveillance System for Public Transport Systems,*Acta Automatica Sinica*, 29 (2003) 393–407.

[15] Siebel N.T and Maybank N.T., The Advisor Visual Surveillance System, Proc. ECCV2004 Workshop Applications of Computer Vision,*Prague, Czech Republic*, 2004, pp. 103–111

[16] Paul M., Haque S.M. and Chakraborty S., *Human Detection in Surveillance Videos and its Applications-A Review*, EURASIP Journal onar Advances in Signal Processing 2013 (2013) 1–16.

[17] Enzweiler M. and Gavrila D.M., Monocular Pedestrian Detection: Survey and Experiments, IEEETrans. *Pattern Analysis and Machine Intelligence* 31 (2009) 2179–2195.

[18] Dalal N.and Triggs B., Histograms of Oriented Gradients for Human Detection, *IEEE Int'l Conf. Computer Vision and Pattern Recognition*, San Diego, CA, USA, 2005, pp. 886–893

[19] Dollar P., Wojek C., Schiele B. and Perona P., Pedestrian Detection: An Evaluation of the State ofthe Art, *IEEE Transactions on Pattern Analysis and Machine Intelligence* 34 (2012) 743–761.

[20] Leininger B., et al. Autonomous Real-time Ground Ubiquitous Surveillance-imaging System (ARGUS-IS). *Proc. SPIE* 6981 (2008) 69810H.

[21] Morellas V., Pavlidis I. and Tsiamyrtzis P., DETER:Detection of Events for Threat Evaluation and Recognition, *Machine Vision and Applications*, 15 (2003) 29–46.

[22] Dollar P., Belongie S. and Perona P., The Fastest Pedestrian Detectorin the West, *Proc. British Machine Vision Conf.*, 2010, pp. 68.1–68.11.

[23] Walk S., Majer N., Schindler K. and Schiele B., New Features and Insights for Pedestrian Detection, *IEEE Conf. Computer Vision and PatternRecognition*, San Francisco,CA, USA (2010), 1030–1037.

[24] Spinello L., Luber M. and Arras K.O., Tracking People in 3 D using a bottom-up top-down Detector, 2011 *IEEE International Conference on Robotics and Automation (ICRA)*, Shanghai (2011), pp.1304–1310.

[25] Csaba Benedek, 3D People Surveillanceon Rangedata SeqUences of a Rotating Lidar,Pattern Recognition Letters 50 (2014) 149–158.

[26] Smeulders A.W.M., Dung M. Chu, R. Cucchiara, S. Calderara, A. Dehghan and M. Shah, Visual Tracking: An Experimental Survey, *IEEE Transaction on Pattern Analysis and Machine Intelligence* 3 (6) (2014) 1442–1468.

[27] Danelljan M., Hager G., Khan F.S. and Felsberg M., Accurate Scale Estimation Forrobust Visual Tracking, *Proceedings of the British Machine Vision Conference BMVC*, Nottingham,UK (2014).

[28] Bolme D.S., Beveridge J.R., Draper B.A. and Y.M. Lui, Visual Object Tracking using Adaptive Correlation Filters, *Comp. Vis. Patt. Recognition* (2010), San Francisco, CA, USA, (2010), pp.2544–2550.

[29] Kocur D., Kazimir P., Fortes J., Novak D., Drutarovsky M., Galajda P. and Zetik R., Short-range UWB radar: Surveillance robot equipment of the future, *IEEE International Conference on Systems, Man and Cybernetics (SMC)*, San Diego, CA, USA (2014), pp.3767–3772.

[30] Javed A., Ejaz A., Liaqat S., Ashraf A. and Ihsan M.B., Automatic Target Classifierfor a Ground Surveillance Radar using Linear Discriminant Analysis and Logistic Regression, *Radar Conference (EuRAD), 2012 9th European*, Amsterdam (2012), pp.302–305.

[31] Mitzel D. and Leibe B., Real-time Multi-person Tracking with Detector Assisted Structure Propagation, *Computer Vision Workshops (ICCV Workshops), 2011 IEEE International Conference on*, Barcelona (2011), pp. 974–981.

[32] "Dense Correspondence Extraction in Difficult Uncalibrated Scenarios,"in Proc. Digital Image Computing: *Techniques and Applications*, Melbourne, Australia (2009).

[33] Siebel N.T. and Maybank S., The ADVISOR visual surveillance system, *Proc. ECCV2004 Workshop Applications of Computer Vision*, Prague, Czech Republic, 2004, pp.103–111.

[34] Valera M. and Velastin S. A., Intelligent Distributed Surveillance Systems: A Review, *IEEE Proceedings-Vision, Image and Signal Processing*, vol. 152 (2005) 192–204.

[35] Bai Y.W., Xie Z. l. and Li Z. H., Design and Implementation of a Home Embedded Surveillance System with Ultra-low Alert Power, *IEEE Transactions on Consumer Electronics*, 57 (2011) 153–159.

[36] Khaleghi B., Alaa Khamis, Fakhreddine O. Karray and Saiedeh N. Razavi, Multisensor data fusion: A review of the state-of-the-art, *Information Fusion* 14 (2013) 28–44.

[37] Robertson N.M. and Letham J., Contextual Person Detection in Multimodal Outdoor Surveillance, Signal Processing Conference (EUSIPCO), 2012. *Proceedings of the20thEuropean, Bucharest, Romania* (2012), pp. 1930–1934.

[38] Premebida C., Carreira J., Batista J. and Nunes U., Pedestrian detection combining rgb and dense lidar data, *International Conference on Intelligent Robots and Systems*, Chicago, IL, USA (2014), pp. 4112–4117

[39] Rougier C., Meunier J., St-Arnaud A. and Rousseau J., Robust Video Surveillance for Fall Detection based on Human Shape Deformation, *IEEE TransCircuits Syst Video Technol* 21 (2011) 611–622.

[40] Lin W., Sun M.T., Poovendran R. and Zhang Z., Group Event Detection with a Varying Number of Group Members for Video Surveillance. *IEEE Trans Circuits Syst Video Technol* 20(8) (2010) 1057–1067.

[41] Idrees H., Nolan Warner andMubarak Shah, Tracking in dense crowds Using Prominence and Neighborhoodmotion Concurrence, *Image and Vision Computing* 32 (2014) 14–26.

[42] Gowshikaa H., Abirami S and Baskaran R, Automated HumanBehaviour Analysis from Surveillance videos: A survey, *Artificial Intelligence Review* 42 (2014) 747–765

[43] Kiryati N., Riklin T.R., Ivanchenko Y. and Rochel S., Real-time Abnormal Motion Detection in Surveillance Video, *IEEE 19th International Conference on Pattern Recognition, Tampa, Florida, USA* (2008), pp.1–4

[44] Foroughi H., Yazdi H.S., Pourreza H. and Javidi M., An Eigen Space Based Approach for Human Fall Detection using Integrated Tim Emotion Image and Multi-class Support Vector Machine, *IEEE 4th international Conference on Intelligent Computer Communication and Processing*, Cluj Napoca (2008), pp. 83–90.

[45] Park S. and Aggarwal J. K., Semantic-level Understanding of Human Actions and Interactions Using Event Hierarchy, *IEEE Computer Society Conference on Computer Vision and Pattern Recognition Workshops*, Washington, DC, USA, 2004, pp 12.

[46] Bremond F., Thonnat M. and Zuniga M., Video Understanding Framework for Automatic Behavior Recognition, *Behav Res Methods* 38 (2005) 416–426

[47] Solmaz Brian B., Moore E. and Mubarak Shah, Identifying Behavioursin Crowd Scenes Using Stability Analysis for Dynamical Systems, *IEEE Transactions on Pattern Analysis and Machine Intelligence* 3(4) (2012) 2064–2070.

Automation and Computation – Vats et al. (Eds)
© *2023 the Author(s), ISBN 978-1-032-36723-1*

Adam optimizer based deep learning approach for improving efficiency in license plate recognition

Pushpendra Kumar Rajput*
School of Computer Science, University of Petroleum and Energy Studies, Dehradun, Uttarakhand, India

Kiran Kumar Ravulakollu
School of Technology, Woxsen University, Hyderabad, Telangana, India

Naveen Jagadam
CEO, N2force Technologies Inc. Naperville, IL, USA

Pardeep Singh
School of Computer Science, Graphic Era Hill University, Dehradun, Uttarakhand, India

ABSTRACT: Image segmentation has been a significant phase of digital image processing. Being a critical phase, it is necessary to ensure accuracy in object detection is achieved at initial stage for maximum efficiency unlike post processing optimization. On contrary to traditional image processing, learning based image processing has been very effective due to recent advancements delivered by deep learning based neural network models. Optimization in neural network architecture has always been an intrinsic part that can address computational cost especially due to their vast learning time. In this research optimization of learning time is challenged with the association of Adam optimizer specific to effectively balancing the momentum with learning rate. The paper uses image dataset containing car number plates captured in multiple traffic stops for validation of optimizer using YOLO-V3 architecture approach for quantification of success. Evaluation of the proposed experiment is measured based on the mAP score and the convergence time while training the model.

1 INTRODUCTION

With the growth in the number of cars and transport infrastructure over the last three decades, traffic monitoring has become more important for the optimal use of road networks. In addition, human based monitoring has become more challenging and costly in such scenarios (Prabhu *et al.* 2017). License plate identification is a technology associated with digital image processing that is widely used to identify vehicles in vehicle monitoring systems (Sharma 2018). Numerous recognition methodologies have been proposed, and number plate identification systems are now widely utilized in a variety of traffic and security applications, including parking, border control, and the monitoring of stolen vehicles (Yao *et al.* 2019).

Researchers from academics and commercial businesses have adopted a number of license plate identification approaches, including texture-based (Yang 2013), color-based (Davix 2017), template-based matching (Thidarat *et al.* 2018), and edge detection-based (Zhao & Gu 2012) in order to enhance the performance of license plate position and character identification. In recent years, deep learning has shown a remarkable character in image

---

*Corresponding Author

DOI: 10.1201/9781003333500-51

processing field. Researchers have proposed a number of deep learning networks that employs different machine learning algorithms. Convolution Neural Networks (CNN) is one of the most acceptable networks in this field.

In this research, an optimizer-based learning criteria is proposed for which the flow of content is organized as follows. In Section 2 a discussion on the preliminary's knowledge about learning approach and optimization method used in completing the research is provided. In Section 3 various approaches are discussed adopted earlier by different researchers to solve the problem. Section 4 outlines our proposed methodology in terms of system architecture, operation, and algorithm adopted. The results of the experiments are presented in Section 5 along with a discussion on findings. Finally, the conclusion of the study is presented in Section 6.

2 LEARNING AND OPTIMIZATION

2.1 Overview

In a way, learning is often treated as a novel way of optimization that intensifies the way of processing at a logical level. Learning in computational methodology is a way of achieving generalization through approximation. Which means, when a machine learning method is applied, it tries to solve an approximation problem through datasets. Hence, we can, effectively claim optimization has a core component for learning.

During learning, optimization is not always necessary to be applied in modelling stage. When it comes to any machine learning approach, approximations are typically surrounded by heavy optimizations. Starting from data preparation till hyperparameter turning during data modelling, optimization is applicable.

Most of the optimization approaches uses continuous iteration criterion for achieving it. This is often tedious, computationally heavy and includes heavy complexities. Hence, different learning approaches are derived that can support various learning strategies. From those deep learning model one widely used and valued model is YOLO.

2.2 YOLO net

YOLO employs the concept of spatially separated bounding boxes and defines object identification as a regression application (Redmon et $al.$ 2016). A single evaluation estimates bounding boxes and probabilities of label from a full picture using one neural network.

The object is detected through prediction of bounding boxes. After dividing image into different grids, bounding boxes and confidence are predicted with the system. Each bounding boxes comprises five predictions (a, b, w, h), where (a, b) represents centre of the box. YOLO utilizes sum-squared error to be optimized. The loss function for the same is defined as Equation 1.

$$\lambda_{Bco} \sum_{i=0}^{S^2} \sum_{j=0}^{B} \mathbb{1}_{i,j}^{obj} [(a_i - \hat{a}_i)^2 + (b_i - \hat{b}_i)^2]$$

$$+ \lambda_{Bco} \sum_{i=0}^{S^2} \sum_{j=0}^{B} \mathbb{1}_{i,j}^{obj} [(\sqrt{w_i} - \sqrt{\hat{w}_i})^2 + (\sqrt{h_i} - \sqrt{\hat{h}_i})^2]$$

$$+ \sum_{i=0}^{S^2} \sum_{j=0}^{B} \mathbb{1}_{i,j}^{obj} (C_i - \hat{C}_i)^2 + \lambda_{Confno} \sum_{i=0}^{S^2} \sum_{j=0}^{B} \mathbb{1}_{i,j}^{obj} (C_i - \hat{C}_i)^2$$

$$+ \sum_{i=0}^{S^2} \mathbb{1}_{i}^{obj} \sum_{c \in classes} \left(p_i(c) - \hat{p}_i(c) \right)^2$$

(1)

where B is the number of bounding boxes predicted by a grid cell, S represents dimension of $S \times S$ grid, $\mathbb{1}_{i}^{obj}$ shows presence of object in cell i, $\mathbb{1}_{i,j}^{obj}$ represents j^{th} box in i^{th} cell,

C denotes class probability, λ_{Bco} and λ_{Confno} are two parameters used for increasing the loss through bounding box prediction and decreasing the loss from confidence.

2.3 Adam optimizer

Diederik and Jimmy (2014) proposed Adam, a stochastic optimization method. It implied advantages of AdaGrad and RMSProp, the two methods that work efficiently in sparse gradients and in non-stationary environment. Adam differentiates itself by making use of the squared gradients. It helps to increase the learning rate. It also employed advantage of momentum for which moving average of the gradient is utilized as comparative to gradient itself. The example of the approach is SGD with momentum. Adam works in an adaptive manner where it computes separate learning rates for various parameters.

Adam optimizer considers an objective function $f(x)$, which is scaler and can be differentiated with respect to parameter x. It minimizes the expected value of $f(x)$. The following steps demonstrates working of Adam optimizer.

1. *Initialization:* Assign initial values to step size α, exponential decay rates $(r1, r2 \in [0, 1))$, parameter vector x_0, first and second moment vectors $mv1_0$ and $mv2_0$, and timestamp t.
2. *Update:* update the following attributes in each iteration until parameter vector is not converged.
    (a) Calculate gradients, which is a partial derivative of $f(x)$ at current timestamp with respect to x as shown in Equation 2.

$$g_t = x f_t(x_{t-1}) \tag{2}$$

   where f_t represents realization of $f(x)$ at timestamp, t.
    (b) Update 1st and 2nd moments with bias using Equation 3 and Equation 4.

$$mv1_t = r_1 . mv1_{t-1} + (1 - r_1) \quad . \quad g_t \tag{3}$$

$$mv2_t = r_2 . mv2_{t-1} + (1 - r_2) \quad . \quad g_t^2 \tag{4}$$

   (c) Calculate estimation of bias corrected moments as demonstrated in Equation 5 and Equation 6.

$$\widehat{mv1}_t = mv1_t / (1 - r_1{}^t) \tag{5}$$

$$\widehat{mv2}_t = mv2_t / (1 - r_2{}^t) \tag{6}$$

   (d) Update the parameter vector by applying Adam update rule as defined with Equation 7.

$$x_t = x_{t-1} - \alpha . \widehat{mv1}_t / \left(\sqrt{mv2_t} + \varepsilon \right) \tag{7}$$

3. *Return:* resulting parameter vector x_t (after t iteration)

3 LITERATURE REVIEW

License plate recognition is an application of object detection approach in image processing area. At first, the object detection is appraised as a classification problem in learning field. Various approaches exist which are developed with the fusion of machine learning and image processing for license plate recognition (Pathak *et al.* 2018). Availability of strong learning tools motivated researchers to study deeper features and perform object detection by employing them.

Fast R-CNN, introduced by Ross Girshick, is a method of object detection which utilizes window extraction procedure in place of sliding window algorithm. The method employs distinct training for feature extraction and classification (Ding & Zhao 2018). This combination of feature isolation and categorization reduces training time of fast R-CNN nine times comparative to R-CNN (Ren *et al.* 2016). Kim et al (Kim *et al.* 2018) proposed a framework that uses background subtraction with CNN. They have applied their approach to detect the moving thing using CCTV.

Joseph Redmon *et al.* (Redmon *et al.* 2016) introduced a novel architecture for the detection of objects named as You Only Look Once (YOLO). It utilizes a regression model to solve object identification. An enhanced variant of YOLO was presented. Fully connected layer was removed and in place of that anchor boxes was used to predict bounding box (Redmon & Farhadi 2017). An Incremental Improvement in which some challenges have been addressed regarding the localization and classification process was presented as YOLOv3 (Redmon & Farhadi 2018). In place of batch normalization, residual block has been used and at the outer layer, softmax is removed and logistic classifier has been used. Tanvir Ahmed *et al.* (2020) presented a modified network which utilizes YOLOv1 with optimization of loss function. It introduces a new inception method and perform better.

Wei Liu *et al.* (2016) developed a novel approach for object detection. The method utilizes only one deep NN. The method is named as Single Shot MultiBox Detector (SSD). It uses multiscale convolutional bounding box, attached to several feature map. Recently, Chen et al (Chen *et al.* 2021) used anchor boxes and specific regression loss function. The method has been verified with face recognition application. The method is based on YOLOv3 and performed better than previous YOLO versions.

Alex Krizhevsky *et al.* (2012) constructed an ImageNet Classification tool with Deep Convolutional Neural Network. Narendra Singh Tomar *et al.* (2018) developed a method for recognizing optical character from segmented number plate. Gajendra Sharma (2018) introduced a novel method that perform character recognition through template matching process. The system was tested on 90 patterns using two different methods that are phase correlation and normalized cross correlation where normalized cross correlation has given the maximum accuracy of 67.98% over phase correlation which has given the accuracy of 63.46%. A smart vehicle number plate detection technique with improved segmentation method is proposed which is also employing the template matching process (Balaji & Rajesh 2017).

Sarbjit Singh Kaur (2014) proposed a morphological operation, thresholding and sobel edge detection baased method for number plate identification and character recognition. The test was conducted on 40 vehicle images where 35 of them were successfully extracted with a success rate of 90%. However, the character recognition accuracy was recorded as 96.6%. B. Sachin Prabhu *et al.* (2017) presented a paper on identification of Indian number plate from live stream videos in which OpenALPR, K-NN and Convolutional Neural Network were used for the recognition of digits in the localized number plate.

Tejas *et al.* (2019) utilizes bounding box segmentation, Sobel edge detection, and neural networks techniques for detecting Indian license plate detection. However, the experiment shows that bad environmental conditions were not considered during testing. Arafat at al. (2019) presented a systematic review on various number plate recognition approaches. Salau *et al.* (2021) presented a work in this dimension and focused on number plate region position and achieved an remarkable accuracy of 99.8%. However, the authors left character recognition task a future consideration of proposed work.

In this proposed work the authors have designed an Indian vehicle license plate identification and recognition using an optimized learning network. The authors addresses varying illumination, bright and dark objects, live capturing of images in noisy environment, images from cross angle, and skewness in the images.

4 METHODOLOGY

In this proposed work, we used tiny-yolov2 architecture where the input images are considered of size 416 x 416 x 3. The methodology comprises four important phases including dataset formation, pre-processing, number plate detection, and license number recognition. Each of these phases are discussed in following subsections.

4.1 Dataset formation

To perform the proposed research, a dataset of 2125 images is created. The cases are developed through capturing live images. An organization entry gate and parking area are targeted for capturing input images or videos in their original form.

Two different positions of camera are discussed to capture the live images for testing data. The following Figure 2 depicts real scenario of organization entry gate along with distances to be covered by the camera. The scenario depicts that a vehicle coming to the organization is passing three speed breakers. The distances from boom barrier to first, second, and third speed breaker are 5 meter, 6 meter, and 15 meters respectively.

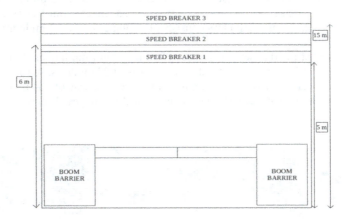

Figure 2.   Top view of the scenario with distance measurement.

The positions of the camera considered for investigation are back of the boom barrier and corner to the boom barrier. Figure 3 and 4 depict the real positioning of camera along with distances covered, termed as field of view.

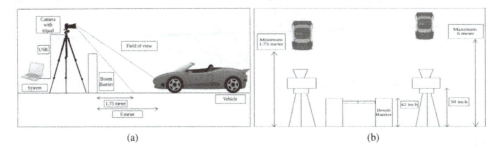

(a)   (b)

Figure 3.   Camera positioned behind the boom barrier: (a) Front view, (b) Camera height and field of view.

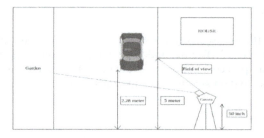

Figure 4.   Camera placed in the right most corner to get the different field of view.

4.2   *Pre-processing*

The dataset is partitioned into two parts. The first part of the dataset is used to check the number of success case of detection of number plates when comparing optimized tiny-yolov2 with the original tiny-yolov2. The second part consists of a greater number of images then the first part and is used to check the performance of the optimized architecture.

The first part of the dataset consists of 297 images in total having 5 different classes. First class consists of 65 images, second class consists of 53 images, third class consists of 61 images and fourth class, and fifth class consists of 59 images each. Sample images for each class is show in Figure 5.

The second part of the dataset consists of 1828 images in total having 7 different classes. First class consists of 288 images, second class consists of 227 images, third class consists of 230 images, fourth class consists of 265 images, fifth class consists of 234 images, sixth class consists of 280 images and seventh class consist of 304 images. Figure 6 shows sample images. class.

Figure 5.   Sample images of dataset consisting of 5 classes.

Figure 6.   Sample images of dataset consisting of 7 classes.

4.3   *Number plate detection*

4.3.1   *System architecture*

An optimized tiny-YOLOv2 net is used for number plate recognition. Figure 7 depicts the layered architecture of tiny-YOLOv2. The first layer of the architecture is responsible for

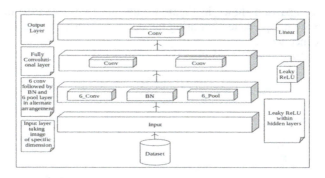

Figure 7.   Architecture used for detection of number plate.

receiving an input image. The second layer in the architecture signifies that there are 6 convolutional layers attached with batch normalization and then followed by the overlapping pooling layer for noise reduction stacked one after another. Then comes the fully convolutional layer, which are used in place of fully connected layers as the proposed task is based on localization plus classification. At the last is the output layer which is also the convolutional layer. The convolutional layer, which acts as hidden layers, consists of leaky ReLU activation function whereas output layer consists of linear activation function. It did not used softmax or sigmoid activation function even consisting of multiple classes just because to reduce the problem of vanishing gradient.

4.3.2   Overlapping pooling
Overlapping pooling as a part of pooling layer always uses less number of filters than convolutional layer filters. The overlapping pooling helps in the reduction of overfitting situation when the model is performing extraction job (Krizhevsky *et al.* 2012). Therefore, this paper includes overlapping pooling. To match the eligibility threshold the size of filter is kept 2 pixels while the size of filter at convolutional layer is set to 3. The proposed values satisfied the properly of overlapping pooling as $3 > 2$.

4.3.3   Leaky ReLu activation
'Dying ReLu' issue is one of the most challenging problems in convolutional neural network. Leaky ReLU is one step to fix this issue. Rather than the value of function being zero when $x < 0$, a leaky ReLU will instead have a trivial negative incline. It computes $f(x) = 1(x < 0)(ax) + 1(x >= 0)(x) f(x) = 1(x < 0)(ax) + 1(x >= 0)(x)$ where a is a constant which is very small.

4.3.4   Bounding box
The network employs anchor boxes as dimension clusters for predicting bounding boxes. The structure estimates 4 different directions $(d_x, \quad d_y, \quad d_w, \quad d_h)$ for each bounding box. In the event that the cell is stabilized from the upper left corner of the image by offset $(O_x, \quad O_y)$ and the bounding box predicted earlier has original width and height $(A_w, \quad A_h)$ then, at that point, the predictions compare to:

$$E_x = \quad \sigma(d_x) + \quad O_x, E_y = \quad \sigma(d_y) + \quad O_y, E_w = \quad A_w^{e^{d_w}}, E_h = \quad A_h^{e^{d_h}}$$

5   RESULTS AND DISCUSSION

In first phase, we used the existing network that is tiny version of yolov2 to conduct the experiment with a dataset of 297 images in total. After that, we optimized the network by using Adam optimizer and a learning rate of 0.0001 (at initial it used 0.001 with momentum optimizer). The comparison shown in Figure 8 indicates that, using the original network, class3 and class5 are not predicted but using the optimized version, all classes are predicted with some confidence score.

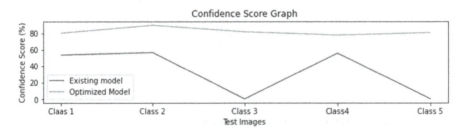

Figure 8. Comparison based on success rate of detection of vehicle number plate.

In second phase, the experiment has been scaled up to larger dataset where each class has 200 – 300 images. It is shown in Figure 9 that for all the 7 classes confidence score for test images are increased. A higher confidence score of approx 90% is recorded in majority of the test cases when using optimized network. Figure 10 depicts that proposed optimized network also offer a remarkable decrease in error rate which is at its lower level below 10% for majority of test cases.

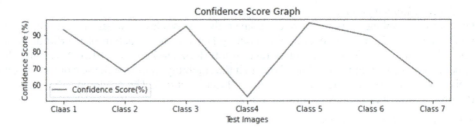

Figure 9. Confidence score graph on test images.

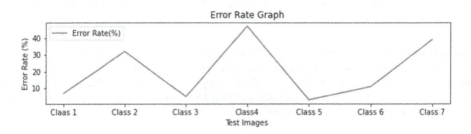

Figure 10. Error rate graph on test images.

After performance check of the model on a larger dataset, some other tests are performed to verify the feasibility of the optimized model in real scenarios. The target is to detect the name plate of a moving vehicle through capturing the live images or videos. The following subsections details about these tests performed at three different timing to capture the impact of light.

5.1 At night

Due to some constraint, the moving vehicle part was not conducted at night and thus standing vehicle detection was commenced. Three different cases at various light impact shows the performance of model, shown in Figure 11. Figure 11(a) consists of total 453 frames where the detection starts from 1st frame. A total of 343 images are used for training at night. Result depicted in Figure 11(b) is recoreded from a video consisting of total 732 frames where the detection of the vehicle starts from 1st frame. Same training data was used as with the previous frame.

Figure 11(c), a case of extreme dark, shows one of the frame from the video which we have processed for classification. There were total of 1075 frames in captured video and the detection started from 1st frame. For training this class, data consists of total 265 images including both day and night. The results show that confidence level is high for bright object.

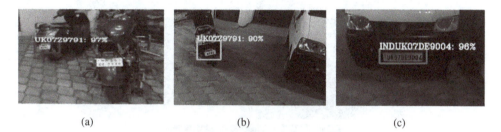

| (a) | (b) | (c) |

Figure 11. Test on vehicle at night which is in (a) light, (b) half in light and half in dark, (c) extreme dark.

5.2 At evening

In evening, the video was taken of moving vehicles. The evening dataset is of 294 training images in total. The detection was done during evening time at around 6:30 pm. Figure 12(a) and Figure 12(b) are the frames captured with an average speed. The detection started from 498th frame out of 561 frames in total.

A high speed vehicle is also considered and a video is captured, the frames are depicted in Figure 12(c). The detection started from 171th frame out of 232 frames in total. The impact of movement is clearly visible in the performance of optimized network. For an average speed confidence score is around 50% at the same time the confidence score is recorded around 40% as the vehicle is passing at high speed. The effect of camera distance is also significant and can be seen in the variation of confidence score as 8%.

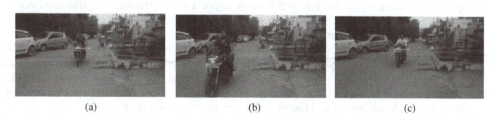

| (a) | (b) | (c) |

Figure 12. Test on vehicle at evening which is moving with (a) & (b) an average speed, (c) a high speed.

5.3 At morning

In a bright day light the test is conducted at same entry gate's boom barrier of a moving vehicle where the dataset is captured. A total of 363 images of this category were used for training purpose. At a slow speed the vehicle is passing the barrier. Figure 13 shows 5 different frames captured during movement. First frame is captured before 2nd speed breaker. It has been clearly observed that the robust performance has started from 2nd speed breaker. It can be seen with confidence score acquired. However, distance can vary on the position and the quality of the camera. Various frames shown in Figure 13 depicts that as the vehicle is coming closer to the camera the confidence score is approaching 100%.

Figure 13.   Figure consists of 5 frames in total taken from different videos.

6   CONCLUSION

Learning based image processing has been prominent and is proven to be effective for many a real-time application. Deep learning on their hand had proven to be more effective for most of the problem-solving aspect of many a image processing applications. However, optimization in the context of network architecture has always been effective in achieving computational cost. Adaptation of adam optimizer has significantly improved the classification, due to effective balancing of momentum. Adam optimizer based yolo-v3 is found effective in recognizing the object from segmented region. After extensive validation, the model is proven to reflect a greater success over the mAP score of above 98% and convergence time is much shorter to only 1/6 of the frame rate. Thereby in real-time, the model is able to detect with greater precision from a distance of 1.75 – 2.28 meters i.e., only when the object is 4/5th distance away from the stationed camera. As of now, the model is influential towards the light effects such as day and night times, which can be worked out in the next version of this paper.

REFERENCES

Ahmad T, Ma Y, Yahya M, Ahmad B, Nazir S (2020) Object Detection Through Modified YOLO Neural Network. *Scientific Programming*, 2020.

Arafat MY, Khairuddin ASM, Khairuddin U, Paramesran R (2019) Systematic Review on Vehicular Licence Plate Recognition Framework in Intelligent Transport Systems. *IET Intelligent Transport Systems*, 13 (5):745–755.

Balaji GN, Rajesh D (2017) *Smart Vehicle Number Plate Detection System for Different Countries Using an Improved Segmentation Method.* vol, 3, 7.

Chen W, Huang H, Peng S, Zhou C, Zhang C (2021) YOLO-face: A Real-time Face Detector. *The Visual Computer*, 37(4): 805–813.

Davix XA, Christopher CS, Christine SS (2017) License Plate Detection Using Channel Scale Space and Color Based Detection Method. In: *2017 IEEE International Conference on Circuits and Systems (ICCS)*, IEEE, pp. 82–86.

Ding S, Zhao K (2018) Research on Daily Objects Detection based on Deep Neural Network. In: *IOP Conference Series: Materials Science and Engineering*, IOP Publishing, Vol. 322, No. 6, p. 062024.

Kaur S, Kaur S (2014) An Efficient Method of Number Plate Extraction from Indian Vehicles Image. *International Journal of Computer Applications*, 88(4).

Kim C, Lee J, Han T, Kim YM (2018) A Hybrid Framework Combining Background Subtraction and Deep Neural Networks for Rapid Person Detection. *Journal of Big Data*, 5(1):1–24.

Kingma DP, Ba J (2014) Adam: A Method for Stochastic Optimization. arXiv Preprint arXiv: 1412.6980.

Krizhevsky A, Sutskever I, Hinton GE (2012) Imagenet Classification with Deep Convolutional *Neural Networks. Advances in Neural Information Processing Systems*, 25:1097–1105.

Liu W, Anguelov D, Erhan D, Szegedy C, Reed S, Fu CY, Berg AC (2016) Ssd: Single Shot Multibox Detector. In *European Conference on Computer Vision*, Springer, Cham, pp. 21–37.

Mondal M, Mondal P, Saha N, Chattopadhyay P (2017) Automatic Number Plate Recognition Using CNN based Self Synthesized Feature Learning. In: *2017 IEEE Calcutta Conference (CALCON)*, IEEE, pp. 378–381.

Pathak AR, Pandey M, Rautaray S (2018) Application of Deep Learning for Object Detection. *Procedia Computer Science*, 132:1706–1717.

Prabhu BS, Kalambur S, Sitaram D (2017) Recognition of Indian License Plate Number From Live Stream Videos. In: *2017 International Conference on Advances in Computing, Communications and Informatics (ICACCI)* (pp. 2359–2365). IEEE.

Redmon J, Farhadi A (2017) YOLO9000: Better, Faster, Stronger. In: *Proceedings of the IEEE Conference on Computer Vision and Pattern Recognition*, pp. 7263–7271.

Redmon J, Farhadi A (2018) Yolov3: An Incremental Improvement. arXiv preprint arXiv:1804.02767.

Redmon J, Divvala S, Girshick R, Farhadi A (2016) You only look once: Unified, real-time object detection. In: *Proceedings of the IEEE Conference on Computer Vision and Pattern Recognition*, pp. 779 – 788.

Ren S, He K, Girshick R, Sun J (2016) Faster R-CNN: Towards Real-time Object Detection with Region Proposal Networks. *IEEE Transactions on Pattern Analysis and Machine Intelligence*, 39(6):1137–1149.

Salau AO, Yesufu TK, Ogundare BS (2021) Vehicle plate number localization using a modified GrabCut algorithm. *Journal of King Saud University-Computer and Information Sciences*, 33(4):399–407.

Sharma G (2018) Performance analysis of vehicle number plate recognition system using template matching techniques. *Journal of Information Technology & Software Engineering*, 8(2):1–9.

Tejas K, Reddy KA, Reddy DP, Bharath KP, Karthik R, Rajesh Kumar M (2019) Efficient License Plate Recognition System with Smarter Interpretation through IoT. In: *Soft Computing for Problem Solving, Springer, Singapore*, pp. 207–220.

Thidarat P, Worawut Y, Narumol C, Mahasak K (2018) License Plate Tracking Based on Template Matching Technique. In: *18th International Symposium on Communications and Information Technologies (ISCIT)*, Bangkok, THAILAND, pp. 299–303.

Tomar NS, Sachan P, Mittal P, Agarwal S (2018) Vehicle Number Plate Detection Using MATLAB. *International Research Journal of Engineering and Technology*.

Yang X (2013) Self-adaptive Model of Texture-based Target Location for Intelligent Transportation System Applications. *Optik*, 124(19):3974–3982.

Yao L, Zhao Y, Fan J, Liu M, Jiang J, Wan Y (2019) Research and Application of License Plate Recognition Technology Based on Deep Learning. In: *Journal of Physics: Conference Series*, IOP Publishing, Vol. 1237, No. 2, p. 022155.

Zhao Y, Gu X (2012) Vehicle License Plate Localization and License Number Recognition using Unit-linking Pulse Coupled Neural Network. In: *International Conference on Neural Information Processing*, Springer, Berlin, Heidelberg. pp. 100–108.

Automation and Computation – Vats et al. (Eds)
© *2023 the Author(s), ISBN 978-1-032-36723-1*

Interdependence of SGPA of preceding and upcoming semester for computer science undergraduates

Piyush Chauhan*

School of Computer Science, University of Petroleum and Energy Studies, Dehradun, India

ABSTRACT: Identifying interdependence between the semester-wise performance of undergraduate students in the computer science stream of university pupils is very crucial to assist weak students. However, for any mathematical model to work efficiently we need to find out the statistical importance of various SGPA (Semester Grade Point Average) spread across all 8 semesters of undergraduate students. Regression is one method of statistics that can be used to find out semester Grade Point Average interdependence. Verifying the interdependence of a SGPA of various semesters on each other will act as the foundation for various mathematical models to predict the grades of students. In this research work, we have tried to investigate this interdependence using a regression for computer science undergraduate students at one university. Computer science students are further subdivided into various specializations.

1 INTRODUCTION AND LITERATURE REVIEW

To identify the interdependence (Romero C. & Ventura S.2010.) of SGPA across eight semesters of computer science undergraduate students' author has used simple linear regression. the equation for understanding Simple linear regression is as follows:

$$\text{Dependent Variable} = \text{Slope of Line}(\text{Independent Variable})$$

$$+ \text{Dependent Variable axis Intercept.} \qquad (1)$$

A statistical measure used to determine whether a link exists between a dependent and an independent variable is the P-value. We examine whether the coefficient's actual value is equal to 0 which means no relationship. Hypothesis testing is the term for this statistical test. A minimum P-value (<0.05, which we used in this study) indicates that there is a good chance the coefficient will not equal zero. If the P-value is large (> 0.05), we cannot draw the conclusion that the independent/explanatory variable influences the dependent variable. An insignificant P-value is another name for a P-value higher than .05.

Another Statistical measure is taken into consideration in this research work which is R-Squared. A statistical measure in a regression model called R-Squared (sometimes called the coefficient of determination) estimates the percentage of variance in the dependent variable (SGPA of later semester in our research work) that can be explained by the independent variable (SGPA of the previous semester).

Now that we have understood these basic statistical terms used in the research work, we will highlight a few important and very efficient research articles which worked on evaluating the performance of students on various parameters (Romero C. & Ventura S.2010).

---

*Corresponding Author: pchauhan@ddn.upes.ac.in

  DOI: 10.1201/9781003333500-52

In (T. D. Gedeon T. D. & Turner H. S. 1993) For predicting student success in a huge undergraduate Computer Science course at New South Wales University, the authors have utilized a feed-forward neural network that is back-propagation trained. This network's goal is to enable students to forecast their anticipated final grades based on their performance so far. Additionally, if the anticipated grade falls short of their expectations, it enhances their performance. The ability to extract meaningful educational knowledge from a neural network trained on student grades to forecast grades has been proven by authors.

Another Similar work has been done in (Vasileva E. E. (eds).2019.). Universities are social systems that necessitate the use of nonlinear mathematical models and the discovery of hidden links. Using data on student's academic performance in their second year, this study uses a neural network approach to answer the problem of early grade point average marks prediction for a university graduate. According to the research's findings, the neural network logically evaluated students' chances of achieving a particular average score. The created artificial neural network shows that, even with little input data, it is still possible to forecast the average grade of university graduates early on. The conclusions obtained have significant practical importance for managing the university educational process. However, many social factors also impact grade of students. In (Kiu C. -C. 2018) authors have tried to identify impact of these factors.

Authors in (Kiu C. -C. 2018) study conducted an analysis to determine the significance and impact of a student's background, social activities, and coursework accomplishments in predicting academic performance. In order to predict mathematical achievement in secondary school, supervised educational data mining techniques such as Random Forest, Naive Bayesian, Decision Tree J48, and Multilayer Perceptron were utilized. The authors of this research paper have shown how important it is to use student background information and social activities to predict student performance early on and to identify students who may be at risk.

Another fascinating work is done in (Sweeney M.*et al.* 2015). According to (Sweeney M *et al.* 2015.). Student retention for successful graduation is a persistent problem in higher education. In a traditional university context, authors created a system for the task of projecting students' course grades for the following enrolment term. Students sign up for a set number of courses each term, earning grades ranging from A to F for each course. The primary goal of this research project is to forecast each student's grades in the classes they will enroll in during the upcoming term using past grade data. The difficulty of predicting a student's grade for the following term now resembles a rating prediction quite a bit. The most advanced technique is to use the factorization machine (FM), a general-purpose matrix factorization (MF) algorithm suitable for this task. According to the authors' experiments, FMs obtain the lowest prediction error. In general, as authors gather more grade information, their forecasts get better. The system, however, appears to take a while to adjust to changing characteristics in the learners' population as a whole. The authors' approach performs badly in predicting failing grades, which are a crucial part of accurate grade prediction.

Finally, we are highlighting the education analytics work done on one of Indian University in research article (Krishna Kishore K. V. & Alekhya S(eds).2014).In (Krishna Kishore K. V. & Alekhya S(eds).2014). Grade Point Average (GPA) of university students is recommended to be predicted using a prediction tool which is based on Multilayer Perceptron and is using a student's prior academic performance. The architecture of the author's proposed application is as follows:

It all starts with a database of graduate students. Firstly, we do data collection from the database of graduate students. Next comes the data pre-processing and cleaning step post-data collection from the database. After data pre-processing it comes to a very crucial step of feature selection. Author's application Split data into training data and test data. Training data is utilized to build models using the knowledge base and finally provided to the classification algorithm. Test data is also directly provided to the classifier algorithm. Finally, the author's application does the prediction of accuracy utilizing test data and training data on the classifier algorithm. This multilayer perceptron beats existing classification techniques like j48 and Naive Bayes, achieving 97.3 percent accuracy on the test data set. All these techniques use latest advancement in field of Machine Learning and Artificial Intelligence. However, for these algorithms to work efficiently we need to correctly identify the parameters which assist us in predicting the future performance (Yagsi A. & Cevik M. 2019) of students (Al-Sheeb Bothaina A. & Abdella Galal M. (eds).2019) perfectly.

2 PROPOSED ALGORITHM

SGPA is obviously the final parameter that sums up the performance in various subjects for an entire semester of any pupil. Also, if we keep into consideration only SGPA then we need not worry about the intricacies of how an SGPA is calculated. Various variations of subjects are used in different universities across world to calculate SGPA for different semesters of computer science undergraduate students belonging to various specializations. Below figure 1 shows the algorithm used to identify significance of relationship between previous and upcoming semesters of undergraduate program of computer science students from various specializations.

Input: N^{th} semester SGPA as Independent Variable, $(N+1)^{th}$ SGPA as dependent variable
START
1) Select: N^{th} semester SGPA
2) Select $(N+1)^{th}$ SGPA
3) Evaluate Ordinary least square (OLS) by applying simple linear regression in (1).
4) Evaluate P value
5) Evaluate R-squared
6) Test the significance level (SL) of both p value and R-squared value
7) If (SL is high)
8) Return relationship as successful relationship
9) Else
10) Return relationship as unsuccessful relationship
END
Output: relationship significance

Figure 1. Simple linear regression to predict significance of relationship between Nth and (N+1)th semester SGPA.

In the given figure we have given to imports. The first input for our algorithm is the Nth semester SGPA marks for an undergraduate student. The second input for our algorithm is (N + 1)th semester SGPA for the very same undergraduate student. We select these two inputs and calculate using the ordinary least square method to evaluate simple linear regression shown in equation (1). We get the P value and R squared value. Now we calculate the significance level of both the P value and R squared value. If the significance level is very high then we return the relationship between Nth semester SGPA and (n + 1) th semester SGPA as a successful relationship. I will return the relationship between the nth and (n + 1)th semester SGPA as an unsuccessful relationship. However, if the significance level is low then we will return the relationship in SGPA of two consecutive semesters as an unsuccessful relationship (Meijer E & Heiser W.J. (eds). 2019) (Lu et al. 2018). Results are shown in next section for the eight semester's SGPA and their relationship significance.

3 RESULTS AND DISCUSSIONS

As shown in figure 2 we have a very small p-value when we run simple linear regression between semester 1 and semester 2 where semester 1 is an independent variable and semester 2 is a dependent variable. The adjusted r-squared value is also shown in Figure 2. as is Visible that for these 6 semesters we always have a very low P value when there is a simple linear regression model for any two consecutive semesters. This can be very helpful in the further application of the SGPA marks of various semesters applied in advanced statistical, regression, and AI-based algorithms for the prediction of marks for an undergraduate student of computer science across all specialization courses. The below figures 3,4,5,6,7 give actual and fitted SGPA(N+1) th vs SGPA(N) th semester marks and this shows that there is a significant level of interdependence between the marks of any two consecutive semesters.

Nth Semester	P value	R Squared Vale	(N+1)th Semester
1	8.3E-131	0.493132	2
2	6E-132	0.498565	3
3	1.47E-68	0.297677	4
4	2.7E-122	0.466807	5
5	1.1E-193	0.632901	6

Figure 2. p value and r squared value for nth semester and (n+1)th semester of undergraduate students marks across six semesters in computer science.

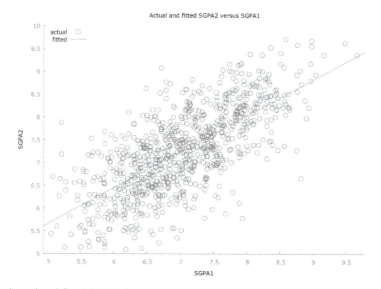

Figure 3. Actual and fitted SGPA for second semester versus SGPA of first semester.

453

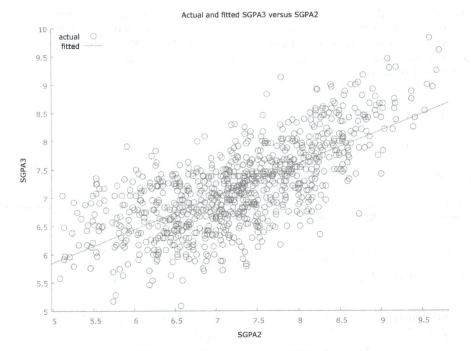

Figure 4.   Actual and fitted SGPA for third semester versus SGPA of second semester.

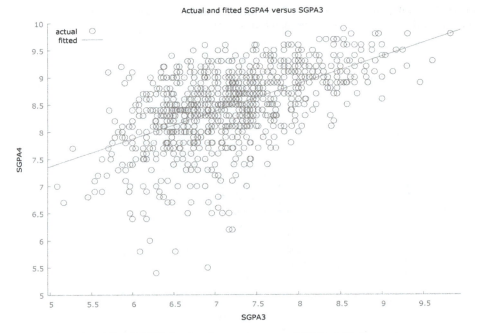

Figure 5.   Actual and fitted SGPA for fourth semester versus SGPA of third semester.

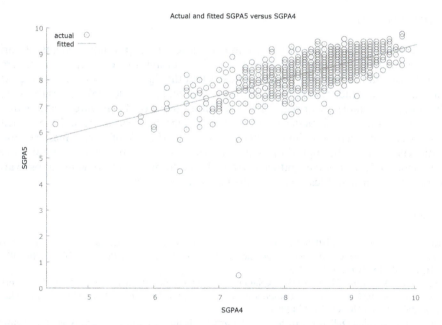

Figure 6. Actual and fitted SGPA for fifth semester versus SGPA of fourth semester.

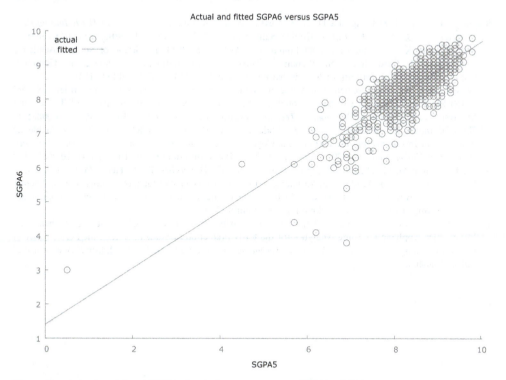

Figure 7. Actual and fitted SGPA for sixth semester versus SGPA of fifth semester.

4 CONCLUSION

We can conclude that in the case of computer science undergraduate students of various specializations there is a relationship between SGPA of any two consecutive semesters for the first three years. A very low P value indicates that the significance level is very high when we take as the independent variable the Nth semester and as the dependent variable the (n + 1) th semester in simple linear regression. The future scope of this research work is that we can make a mathematical model using multiple independent variables and then calculate their impact in predicting a single dependent variable (SGPA of next semester). Further Logistic regression can be applied to identify a weak student as a dependent variable utilizing various independent variables gathered from the performance of students across various quizzes, assignments, and tests which happened in an entire semester.

REFERENCES

Romero C., & Ventura S. 2010. Educational Data Mining: A Review of the State of the Art. *In IEEE Transactions on Systems, Man, and Cybernetics, Part C (Applications and Reviews)*, 2010, 40(6).601–618. DOI: 10.1109/TSMCC.2010.2053532

Gedeon T. D. & Turner H. S. 1993. Explaining Student Grades Predicted by a Neural Network. *In Proceedings of 1993 International Conference on Neural Networks (IJCNN-93)*:609-612 vol.1, -Nagoya, Japan doi: 10.1109/IJCNN.1993.713989.

Vasileva E. E., Kurushin D. S. &Vlasov S. S. 2019. Early Prediction of the Grade Point Average of University Students Diploma: *Neural Network Approach. In 2019 XXII International Conference on Soft Computing and Measurements (SCM))* 259-262. St. Petersburg, Russia. doi: 10.1109/SCM.2019.8903629.

Kiu C.-C. 2018. Data Mining Analysis on Student's Academic Performance through Exploration of Student's Background and Social Activities. *In Fourth International Conference on Advances in Computing, Communication & Automation (ICACCA)*. 1-5. Subang Jaya, Malaysia. doi: 10.1109/ICACCAF.2018. 8776809.

Sweeney M., Lester J. & Rangwala H. 2015. Next-term Student Grade Prediction. *In IEEE International Conference on Big Data (Big Data)*. 970-975. Santa Clara, CA, USA doi: 10.1109/BigData.2015.7363847.

Krishna Kishore K. V., Venkatramaphanikumar S & Alekhya S. 2014. Prediction of Student Academic Progression: A Case Study on Vignan University. In International Conference on Computer Communication and Informatics.1-6. Coimbatore, India. doi: 10.1109/ICCCI.2014.6921731.

Yagsi A. & Cevik M. 2019. Prediction of Academic Achievements of Vocational and Technical High School (VTS) Students in Science Courses Through Artificial Neural Networks (Comparison of Turkey and Malaysia). *In Education and Information Technologies*24(5):2147-2167. DOI:10.1007/s10639-019-09885-4

Al-Sheeb Bothaina A., Hamouda A.M. & Abdella Galal M. 2019. Modeling of Student Academic Achievement In Engineering Education using Cognitive and Non-cognitive Factors. *In Journal of Applied Research in Higher Education*. 2019. Vol. 11 No. 2.178-198. https://doi.org/10.1108/JARHE-10-2017-0120

Meijer E., Cleiren M.P.H.D., Dusseldorp E., Buurman V.J.C., Hogervorst R.M., Heiser W.J. 2019. Cross-Validated Prediction of Academic Performance of First-Year University Students: Identifying Risk Factors in a NonselectiveEnvironment (Article). In Educational Measurement: *Issues and Practice*. Volume 38. Issue 1. Spring 2019.36–47. https://doi.o rg/10.1111/emip.12204

Lu, Owen H. T., Anna Y. Q. Huang, Jeff C.H. Huang, Albert J. Q. Lin, Hiroaki Ogata, & Stephen J. H. Yang. 2018. Applying Learning Analytics for the Early Prediction of Students' Academic Performance in Blended Learning. *In Journal of Educational Technology & Society* 21, no. 2. 220–32. http://www.jstor.org/stable/26388400.

Automation and Computation – Vats et al. (Eds)
© 2023 the Author(s), ISBN 978-1-032-36723-1

Survey on learning based super resolution techniques for enhancement of digital images

Kavya T.M.

Research Scholar, Department of P.G. Studies and Research in Computer Science, Kuvempu University, JnanaSahyadri, Shankaraghatta, Shivamogga, Karnataka, India

Yogish Naik G.R.

Assistant Professor, Department of P.G. Studies and Research in Computer Science, Kuvempu University, JnanaSahyadri, Shankaraghatta, Shivamogga, Karnataka, India

ABSTRACT: The process of enhancing low resolution photos to high resolution photographs with minimal loss of image quality appears to be known as super resolution. The most effective method of splendid decision is primarily relies on interpolation, that anticipates the high resolution pixels depending on their surroundings. This technique has a reduced computational cost and a great real-time performance, Bilinear or bicubic interpolation methods comes under simple interpolation methods, for example, to produce smooth images with ringing and jagged artefacts, which leads to poor iterative reconstruction quality. When there are few or no accessible input images, and whenever the needed magnification power is high, the quality of the reconstructed image quickly declines. The result in these circumstances could be overly straightforward and omit crucial high frequency data. In this article, we describe how a learning-based super resolution reconstruction method can utilize the prior knowledge in low resolution images and the prior information collected through learning. Even with a high magnification factor, excellent reconstruction results can be produced. Learning-based super resolution has thus emerged as among the most efficient methods for a variety of real-world applications.

Keywords: Bicubic, Bilinear, Super Resolution, Reconstruction, Learning based

1 INTRODUCTION

One of the key characteristics of an image acquisition device that affects how well a picture looks is resolution. In general, the degree of detail present in an image can be referred to as image resolution. Thus, images with higher resolution provide more pictorial information of a scene. Aliasing, blurring, and noise are three major limitations of digital images captured by low-resolution cameras. Aliasing can happen as a result of lacking in image sensor elements, that causes in an under sampled spatial resolution, resulting in a substantial loss of high frequency information such as corners and patterns. Camera motion, jitter, out of focus, and other factors can cause image blur. Apart from blur, numerous noises can be introduced into an image during the imaging process, which diminishes the image quality.

The primary objective of super resolution is to transform a low-resolution image into a high-resolution image. Image super resolution is a low-cost method of boosting image resolution while maintaining high detail. In computer vision applications, high resolution is extremely important to better pattern recognition and image analysis. High-resolution images are usually unavailable. Expensive imaging technology is required for a high-resolution image. Because it is expensive and requires a lot of disc space, such imaging

DOI: 10.1201/9781003333500-53

technologies are exceedingly difficult to use in common applications like video surveillance and remote sensing. In order to avoid upgrading the hardware, methods for enhancing resolution are being explored. To improve performance, signal processing approaches are applied. There are several methods for creating a high-resolution image from an image. These high-resolution techniques don't require additional hardware to be used.

Increasing the area of interest is beneficial for satellite imagery, science, and surveillance. Increasing the area of interest is beneficial for satellite imagery, science, and surveillance. It can be occasionally required to enhance details in the scene for surveillance reasons, such as a criminal's face or a vehicle's license plate. When using medical imaging techniques like CT and MRI, high quality images are essential for making an accurate diagnosis. in High resolution images of the Earth or the Moon can get more extensive geographic features, comprehending tasks like as segmentation [1], detection [2–4], and recognition [5]. Images with high resolution are essential. Super resolution images can also completely illustrate their benefits in a variety of different situations. Using high resolution technology for standard shooting instruments and adverse weather can address numerous difficulties.

Depending on the quantity of input images, image super resolution is divided into two categories: single image super resolution and multiframe image super resolution. The method of estimating the high resolution image by integrating various low resolution images of the same scene is known as multiple picture super resolution. A single degraded lower resolution image is referred to as a single image super resolution when it is used to approximate an image. Single picture superresolution is not only more efficient in implementation than multi frame superresolution, but it also has a larger application scenario in reality.

2 REVIEW OF LITERATURE

The three categories of spatial domain-based methods comprise interpolation-based techniques, reconstruction-based approaches, and learning-based approaches. Interpolation-based SR image method: Image interpolation is one of the most fundamental methods used in image processing. Another method involves transforming the function to discrete samples in order to determine the potential values at a place between its phases. The HR picture from the LR input is changed via interpolation-based SR techniques employing a smooth kernel function. The size of LR is upscaled using parametric or nonparametric techniques in SR approaches based on interpolation. This approach approximates all pixels inside the HR grids using a base1function to produce a better resolution image. Non-adaptive and adaptive approaches are the two categories into which this method fits.

Non-adaptive Interpolation: Instead of evaluating the picture contents of features, pixels are directly modified in this approach. This approach is shown by Nearest Neighbor[12], Bilinear[13], and Bicubic[14]. The term "nearest neighbour interpolation" refers to the translation1of known pixel values. A good outcome is produced when1the image has1high quality pixels. Some1information is lost at1the edges as a result of this. An estimated value of adjacent pixels on either side is utilised in bilinear interpolation. This is easy to put it into practise. Our method performs better than closest neighbour interpolation and computes faster than bicubic interpolation. Bicubic1interpolation is the translation1of a mean value of the1nearest pixels. Although it takes longer to compute, this approach yields superior results. When there is no time limitation, this strategy outperforms all other non-adaptive strategies. The interpolant in spline interpolation is a specific sort of piecewise polynomial. However, despite their low computing complexity, These methods fail to maintain image spatial characteristics, resulting in poor image quality.

An edge-directed interpolation technique with low complexity for natural images was introduced by Li [15]. Because low-resolution and high-resolution covariances have a geometric duality, After estimating the covariance coefficients with an LR image, the interpolation at a higher resolution is modified using these estimations. To minimize total computation complexities, an hybrid method of switching between covariance-based adaptive interpolation and bilinear interpolation is described. With the method suggested, a 320x240 flower image is super-resolved to create a 1280x960 image. Choosing a window size of 3 improves the efficacy of covariance-based adaptation. However, because isolated dots are viewed as tiny edges, this method fails to maintain them.

To retain edge sharpness and eliminate ringing effects, Zhang [16] developed an original edge-guided nonlinear interpolation technique derived from data fusion and a directed filter. Zhang [16] developed an original edge-guided nonlinear interpolation technique derived from data fusion and a directed filter. The linear minimal mean square-error estimation (LMMSE) method is used to combine these directional values, into a more accurate estimate, which are depicted as separate noisy readings of the missing pixel. The study further shows a streamlined variation of the LMMSE-based interpolation method to lessen computing complexity, the proposed approach is put to the test. The PSNR for these photos is 29.28, 29.28, 28.11, 29.30, and 25.18, which is more than 2dB higher than the PSNR obtained using traditional methods A Markov random field (MRF) model-based edge-directed interpolation method was proposed by Li (2008) [17], This results in sharp edges across edges and smooth edges along edges. The orientations of edges, which are suggested by statistically based techniques, are indirectly estimated by the length-16 weighting vectors.

By employing a single-pass method, the computational complexity of MRF is minimised. On a realistic picture, the suggested approach is evaluated for neighbourhood structure sizes of 3x3, 5x5, 7x7, and 9x9. The PSNR values found for 3x3, 5x5, 7x7, and 9x9 are 34.03, 34.25, 34.37, and 34.28, respectively. The times in seconds for 3x3, 5x5, 7x7, and 9x9 are 89.63, 205.42, 535.32, and 811.55, respectively.

There are two types of reconstruction-based algorithms: deterministic algorithms and stochastic algorithms. Priors are used in the deterministic procedure to store expectations for the appearance of the high resolution picture, and the limited least squares method is used to regularise the result. On top of a least-squares optimization, a smoothness prior is commonly imposed via regularisation. The regularisation term ensures that the optimization function is symmetric and distinguishable. Most often, regularisation is utilised to add a smoothness prior to a least-squares optimization. A regularisation term ensures the symmetry and distinctness of the optimization function. Stochastic approaches that treat SR restoration as a statistical estimator has piqued the interest of researchers because they provide a sound foundation for the incorporation of a-priori1 constraints required for the effective resolution of the ill-posed SR inverse1 problem. Prior information and noise are explicitly handled by statistical approaches. Using a stochastic technique, including previous information is generally more natural. The strength of this approach it can be easily combined by other typical image processing tasks like as de-noising, de-convolution, augmentation, and so on. Irani and Peleg [21] present an iterative back-projection (IBP) approach that the difference among observed and simulated low-resolution images is used periodically to forecast the quality image.

Joshi[22] described a method for obtaining the super-resolution brightness field from a series of magnified samples. A very closely cropped image determines the overall scene resolution, which only depicts a small portion of it. The model parameters were supplied by the observation with the highest zoom level, and the high resolution picture could be modelled as an MRF or a SAR. The super-resolved image is then recovered using a MAP estimate for the MRF model and an appropriate regularisation method for the SAR model. In Marquina's[23] unique time-dependent super-resolution convolution model, which is based on a limited variational model, the total variation of the signal acted as a regularising functional. Lijun [24] suggested a super resolution technique based on a wavelet-domain Hidden Markov Tree (HMT) model and a Morphological Component Analysis (MCA) model. The author decomposes the image's texture and smooth element first, then uses interpolation to improve the smooth section.

Xu [25] proposes a single image super resolution method depending on specific fractal analysis. Fractal dimension invariance is used to sharpen the image, and the picture intensity is represented as an approximation of a set of fractals. The gradient of an image is a measurement of the fractal set that makes up a natural image's pixels. Fractal dimension scale invariance is used to determine the regularisation term. With RMSEs of 22.14, 12.94, and 17.38 on photos of a kid, a koala, and a Kodak camera, respectively, this method was evaluated. In this study, the picture gradient was the only fractal metric used, which may not be sufficient for such complex image contents. However, the Total Variation (TV) model is preferred over earlier models due of its capacity to preserve edge information. But on "Foreman" video sequence at 352x288 pixels and the "Bulletin" video sequence at 640x480 pixels, the method is examined by varying the regularisation value. At a regularisation value of 0.512, the author produced 37.663dB.

Qiang [27] suggested a spatially weighted TV-based SR method. To restrict the SR approach, information spread over distinct sections of the picture is added. This approach minimises artefacts in fat parts of the picture while preserving edge information. Aerial photographs with a resolution of 200x200 pixels, spot-5 images with a resolution of 256x256 pixels, and cameraman images with a resolution of 200x200 pixels are used to assess the approach. These images have PSNR values of 33.413, 32.636, and 28.65210, respectively. To save texture information, Ren created a regularisation method for super-resolution based on fractional order total variation (TV). To retain discontinuities and visual structures, the variational formulation includes regularisation, picture integrity, and TV regularisation.

Deshpande [29] [30], [31], A framework based on total variation and Gaussian process regression (GPR) was provided for super-resolve long-range recorded iris pictures. The motion vectors utilised in total variation are calculated using the diamond search technique. The pictures are super resolved using the GPR linear kernel covariance function. Furthermore, the super-resolved images are utilised to identify the individual.

The motion vectors utilised in total variation are calculated using the diamond search technique. The pictures are super resolved using the GPR linear kernel covariance function. Additionally, the person's identity is determined by the super-resolved images. The initial stage is to record images in two domains: spatial and range. The LR images' point spread function (PSF) is then specified with a Gaussian function, followed from an approximation based on the maximum a posteriori (MAP) using a blind de-convolution method. There are two terms in the MAP cost function: a regularised part and a data fidelity part. The Huber-Markov prior is used in the regularized part, that can also reduce noise and artefacts while keeping picture edges, whereas the L2 norm is used in the data fidelity term.

Shi [33] suggested a new total variation hybrid approach for image retrieval in wireless sensor networks(WSN) that is both local and nonlocal. An image smoothing approach divides the deteriorated image into two parts before it is preprocessed. By the local TV term, one comprises edges and flat sections. Shi [33] proposed a new total variation hybrid method that is both local and nonlocal towards image recovery in wireless sensor networks (WSN), The feature function is then optimized using the alternate direction method of multipliers (ADMM), and two critical parameters are provided with enhanced quality.

Liu [34] developed a strategy for total variation (TV) regularisation. Using the high-resolution (HR) SAR image, At each stage, the TV regularization parameter adaption is done. For a fair evaluation of SAR picture super-resolution reconstruction performance, SAR pictures are subjected to several assessment criteria. This computationally effective method for HR picture estimation is noise-resistant in SAR environments. This highly effective method for HR image recognition is noise-resistant in SAR settings. The suggested split Bregman super-resolution technique has an excellent noise suppression effect and can successfully avoid the speckle noise produced by some odd textures. This operationally important factor is noise-resistant in SAR situations for HR image estimation. The proposed split Bregman super-resolution method effectively prevents noise present caused by some uncommon patterns and also has a good noise-suppression impact, while also efficiently retaining the SAR picture information.

Super resolution is a technique to increase the resolution of an image or a sequence of images beyond what imaging technology is capable of. Precision has always been essential in imaging. Medical imaging, satellite imaging, multimedia applications, the restoration of old historic photographs, tracking, licence plate identification systems, and other applications have all demonstrated that super resolution image reconstruction may be successful.

Weisheng Dong *et al.* [2011] developed a new scant representation-based method for image deblurring and single picture super resolution using adaptive sparse domain selection (ASDS) and adaptive regularisation. This approach selects the pre-learned dictionaries for each local patch adaptively from a dataset of high-quality sample patches. Using ASDS increases the effectiveness of sparse modelling, which considerably enhances the outcomes of picture restoration. The training dataset is used to create a collection of autoregressive (AR) models, which are then used to smooth out the local smoothness of the image. To benefit from image non-local redundancy, a new regularisation term called image non-local similarity is introduced. The suggested ASDS approach with adaptive regularisation is accomplished using an iterated contraction mechanism. [7]

Mittu George. P*et al.* [2015] suggested a deblurring approach that combines adaptive sparse domain selection with a smoothness regularisation parameter. This paper proposes a method for allocating a sub dictionary to each tiny section of a picture in an adaptable manner. This sub dictionary was selected from a set of sub dictionaries that had been trained on high-quality picture patches, and a smoothing regularisation parameter was used to improve the edge reconstruction. [8]

3 OBJECTIVES

The proposed methodology focuses on using learning-based super resolution techniques to produce a high-resolution image from a low-resolution image.

- To study and investigate the recent and existing techniques for image super resolution.
- To propose a method for upscale a picture using a learning-based super resolution methodology.
- To propose a method for improving digital image clarity utilising a learning-based super resolution technology.
- To improve existing learning-based algorithms for picture super resolution.

4 PROPOSED FRAMEWORK

There are three types of picture super resolution techniques [18]: interpolation-based approaches, reconstruction-based approaches, and learning-based approaches. Interpolation is the simplest way to achieve super resolution, which estimates high resolution pixels based on their surroundings. Simple interpolation techniques, like bilinear or bicubic interpolation, frequently result in images with ringing and too smooth aberrations, but this approach has a low computational cost and great real-time performance, As a result, the image reconstruction is of low quality. When the desired resolution is exceeded, reconstruction-based techniques are routinely employed to build high-resolution images from low-quality images, but the clarity of the restored image gradually decreases. The number of available images for input is restricted, or the magnifying factor is large. In certain circumstances, the outcome could be overly smooth and devoid of high frequency material. A learning-based super resolution reconstruction method may fully utilize both low resolution image prior information and learning-based prior information. Even with a high higher magnification, superior reconstruction results can be achieved. Learning-based super resolution has consequently become one of the most promising technologies for a variety of real-world uses.

Approaches for transforming a single low-resolution image to a high-quality image are used in single-image super resolution methods. The suggested method uses a single low-resolution image as the input image, which is then separated into patches of the same size.

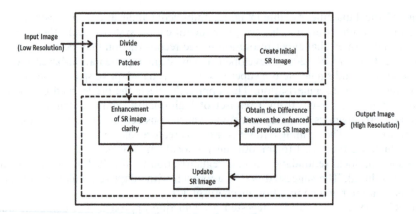

Figure 1.　General block diagram of learning based super resolution.

Following patch generation. To produce a first estimate of the super resolution image, the input image is up sampled. In the following stage, progressively and iteratively restore the missing information to the super resolution image. Each cycle compares the final version of the super resolution image to the improved super resolution image. An iterative algorithm is used to update the super resolution picture estimation. New details are computed in each iteration. The current super resolution image is then enhanced with these attributes, providing a more accurate approximation of the super resolution image.

　Figure 1 illustrates the overall work flow of learning-based super resolution (SR). This technique uses a single low-resolution image as input image, which is then divided into

Table 1.　Comparison of super resolution techniques.

Methods	Algorithms	Processing	Advantages	Disadvantages
Interpolation based method	Nearest neighbor interpolation Bicubic interpolation Bilinear interpolation	Estimate new pixel from neighbor pixel, interpolate missing pixel value into image to create HR image	simple and easy to implement,fast computation	Interpolation based SR methods perform well in low-frequency areas but poorly in edge (high frequency) areas in that they are prone to blurring and jaggy artifacts along edges
Reconstruction based method	Maximum a posteriori Markov random field Kernel steering Total Variation	Contains three steps 1. image registration 2. Image fusion 3. Post processing	Image overly change detection	Small discontinuity and loss of pixels near the edges
Example based method	Markov random field Sparse coding Gaussian Process Regression	Estimates HR image by comparing images patches in external/ internal image database	High magnification factor	Requires huge image database Increase in execution time as the image patcher of test image needs to be searched in image database

patches. Following patch fabrication, a preliminary estimate of the super-high-resolution image is made, along with high resolution and low resolution dictionaries. The following stage is, an iterative process is used to update the super resolution image estimate. These features are then incorporated to the current super resolution image, resulting in a more accurate super resolution image estimation.

5 CONCLUSION

For the enhancement of low resolution images, the Learning based method has been proposed. A learning-based super resolution reconstruction method may fully utilize both low resolution image prior information and learning-based prior information. Even with a high higher magnification, superior reconstruction results can be achieved. As a result, learning-based super resolution has emerged as one of the most promising technologies for a wide range of practical applications. When comparing multiple ways for improving image resolution, learning-based strategies may produce a better outcome than other techniques. Future work will concentrate on improving SR methods through learning time optimization and developing image enhancing algorithms for applications.

REFERENCES

[1] Lei S., Shi Z., Wu X., Pan B., Xu X., and Hao H., "Simultaneous Super-resolution and Segmentation for remote Sensing Images," *In IEEE International Geoscience and Remote Sensing Symposium (IGARSS)*, 2019, pp. 3121–3124.

[2] Haris M., Shakhnarovich G., and Ukita N., "Task-driven Super Resolution: Object Detection in Low-Resolution Images," arXiv Preprint arXiv:1803.11316, 2018.

[3] Pang Y., Cao J., Wang J., and Han J., "JCS-Net: Joint Classification and Superresolution Network for Small-scale Pedestrian Detection in Surveillance Images," *IEEE Transactions on Information Forensics and Security*, vol. 14, no. 12, pp. 3322–3331, 2019.

[4] Zhang Y., Bai Y., Ding M., Xu S., and Ghanem B., "KGSnet: Keypoint-guided Super-resolution Network for Pedestrian Detection in the Wild," *IEEE Transactions on Neural Networks an Learning Systems*, 2020.

[5] Yang X., Wu W., Liu K., Kim P. W., Sangaiah A. K., and Jeon G., "Long-distance Object Recognition with Image Super Resolution: A Comparative Study," *IEEE Access*, vol. 6, pp. 13 429–13 438, 2018.

[6] Jianchao Yang, John Wright, Thomas Huang, and Yi Ma, *"Image Super-Resolution via Sparse Representation Published in IEEE Transactions on Image Processing*,2010,vol:19, DOI: 10.1109/TIP.2010.2050625.

[7] WeishengDong,Lei Zhang, Guangming Shi,Xiaolin Wu "Image Deblurring and Super-Resolution by Adaptive Sparse Domain Selection and Adaptive Regularization *"IEEE Transactions Image Processing January* 2011, DOI: 10.1109/TIP.2011.2108306.

[8] Mittu george p, vivek m, "Image Restoration in Sparse Domain Using Adaptive Dictionary with Regularization" *International Journal of Advances in Electronics and Computer Science, ISSN*: 2393-2835 Volume-2, Issue-5, May-2015.

[9] Dong C., Loy, C. C., He, K., et al. Image Super-resolution Using Deep Convolutional Networks. *IEEE Transactions on Pattern Analysis and Machine Intelligence*,2015, 38(2), 295–307.

[10] Saeideh Sarmadi and Zari Shamsa, *"A New Approach In Single Image Super Resolution"*, 6th International Conference on Computer and Knowledge Engineering (ICCKE 2016), IEEE, Ferdowsi University of Mashhad.

[11] Chao Lai, Fangzhao Li, Bao Li, ShiyaoJin," *Single Image Super-Resolution via Classified Sparse Representation"*, 13th International Conference on Embedded Software and Systems,2016,DOI 10.1109/ICESS.2016.26.

[12] JianweiZhao, TiantianSn, FeilongCao "Image Super-resolution Via Adaptive Sparse Representation and Self-learning", *Published in IET Journal, IET Compute. Vis.*, 2018, Vol. 12 Iss. 5, pp. 753–761 DOI: 10.1049/iet-cvi.2017.0153.

[13]   Yasser K. Badran, Gouda I. Salama, Tarek A. Mahmoud, Aiman Mousa "Single Image Super Resolution Based on Learning" 2019 *IEEE International Conference on Innovative Trends in Computer Engineering (ITCE'2019)*, Aswan, Egypt.

[14]   Rasoul AsgarianDehkordi, Hossein Khosravi, Alireza Ahmadyfard "Single Image Super-resolution based on Sparse Representation Using Dictionaries Trained with Input Image Patches" *Published in IET journal, IET Image Process*, 2020, Vol. 14 Iss. 8, pp. 1587–1593, DOI: 10.1049/iet-ipr.2019.0129.

[15]   Tong, T., Li, G., Liu, X., et al. Image Super-resolution using Dense Skip Connections. *In Proceedings of the IEEE International Conference on Computer Vision2017.* (pp. 4799–4807).

[16]   Zhang, Y., Tian, Y., Kong, Y. Residual Dense Network for Image Superresolution(2018). *In Proceedings of the IEEE Conference on Computer Vision and Pattern Recognition* (pp. 2472–2481).

[17]   Ledig C, Theis L., Husźar, F. Photo-realistic Single Image Super-resolution using A Generative Adversarial Network, 2017. *In Proceedings of the IEEE Conference on Computer Vision and Pattern Recogniti* on (pp. 4681–4690).

Automation and Computation – Vats et al. (Eds)
© 2023 the Author(s), ISBN 978-1-032-36723-1

Integrated pathway analysis for metabolic engineering and development of better pharmaceutical products

Devender Arora
Bioinformatics Core, Purdue University, West Lafayette, USA

Vikrant Sharma & Satvik Vats
Computer Science Department, Graphic Era Hill University, Dehradun, India

ABSTRACT: Metabolic pathways are the central machinery systems that help us understand biological processes governed by a group of molecules, ranging from proteins, genes, and secondary metabolites that are functionally related. These pathways trigger a chain of events and perform biological functions that ultimately result in functional events and reflect the overall importance of building up the necessary process. Understanding this chain of events is essential for various applications, ranging from drug discovery to the development and optimization of pharmaceutical products. Metabolic engineering plays a critical role in optimizing and developing important drugs. Enhancing the yield of certain drugs requires an understanding of pathways and the host environment, which is a challenging process. To help support and overcome these challenges, computational analysis has been proven to speed up and accurately assist in developing better strategies. In this investigation, we have performed an analysis to demonstrate how computational analysis could help optimize the end product and extract the most from secondary products from such analysis.

Keywords: Metabolic Engineering, secondary metabolites, Pharmaceutical, Drug discovery

1 INTRODUCTION

Metabolic engineering is the process of modifying the metabolic pathways of an organism to produce the desired product or to improve its performance [1]. This can be achieved by introducing new genes into the organism, removing or disabling existing genes, or modifying the expression levels of existing genes. The aim of metabolic engineering is to optimize the organism's metabolism to make it more efficient at producing the desired product, such as a drug, chemical, or biofuel [2]. This involves understanding the organism's metabolic pathways and identifying bottlenecks or inefficiencies that can be improved. Metabolic engineering can be applied to a wide range of organisms, including bacteria, yeasts, and plants [3, 4].

Secondary metabolites, such as Malate, are industrially important products used in multiple sectors as flavoring agents and pH regulators by the processed food industry. They also increase the shelf life of food and are used by the beauty and personal care industry as active ingredients in skin lotions and conditioners [11]. Similarly, Lactate is a pharmaceutical important product used in various medical applications, such as in control drug release systems, dialysis, and surgical sutures. The use of lactic acid in the cosmetic industry is an important integral for hygiene and aesthetic product manufacture segments as well as producing products for better skin care [5].

DOI: 10.1201/9781003333500-54

Multi-omics analysis opens up new frontiers to address research problems in an accurate and more efficient manner. It involves the study of different types of omics data, generated from various sequencing platforms such as genomics, transcriptomics, proteomics, and metabolomics [6, 7]. Multi-omics analysis can be used in a variety of fields, including medicine, ecology, and agriculture. These approaches are used to gain a more comprehensive understanding of a biological system and its interactions and can help identify patterns and connections that might not be apparent when studying a single type of omics data in isolation [8]. At the genome level, we understand that these genes influence the pathways, which can alter the metabolic behavior of the organism. These pathways work in a tight control manner, and even a single metabolite change can change the overall metabolic activity of the cell.

With the emergence of computational approaches and a better understanding of the cells and associated causes, we are demonstrating how multiple pathways that interact together can be studied in one platform and help in multiple applications. In this work, we will be performing an analysis on a convergence pathway from Glycolysis, Glutaminolysis, and the Electron Transport Chain reaction, which produces the described pharmaceutical yield and presents a novel approach to enhance these secondary metabolites in a more efficient manner.

2 MATERIAL AND METHODS

The pathway was constructed using existing knowledge as it's essential to know the association of connected metabolic pathways [9] Thus we exploited the existing network using a literature search that works together and has been studied in multiple studies for a different biological reason to better understand the metabolism-related product of interest, identification of drug targets, and study protein-protein interaction.

2.1 Construction of biological network

System Biology Graphical Notation (SBGN) was adopted and modified from [10] and represented in Figure 1. Here, cell designer was used which comprised of various components of different reactions like state transition that occurred during the process [11]. Cell designer that enabled us to describe interfaces using a well-defined graphical notation [12]. This presents a structured (Extensible Markup Language (XML)) format data that can be easily reduced into the discrete event simulation framework which was further stimulated by SBML Ordinary differential equation in cell designer Figure 2 [13].

2.2 Kinetics analysis

The goal of this study is to model and understand the behavior of secondary metabolites at a specific concentration. By developing such a model and identifying the right target, we can optimize our strategies to yield high-value secondary metabolites with minimal input required. This is important because the fast growth in various industries increases the demand for such chemical compounds, and it is necessary to make their production economically viable and available to all. To achieve this, we use systems biology to identify the target molecule for regulating the whole mechanism. By using simulation results, we can propose new experiments and identify targets for different biological applications, thus helping the metabolic engineering discipline achieve goals in an economical way and assisting the pharmaceutical industry in obtaining the desired product in an efficient manner.

Kinetics analysis is a key step in this study. After developing the model, we assigned kinetic equations to each molecule based on Michaelis-Menten, mass action, Hill etc, using SBML squeezer plugin. SBML squeezer is a java-based plugin used to assign kinetics to each

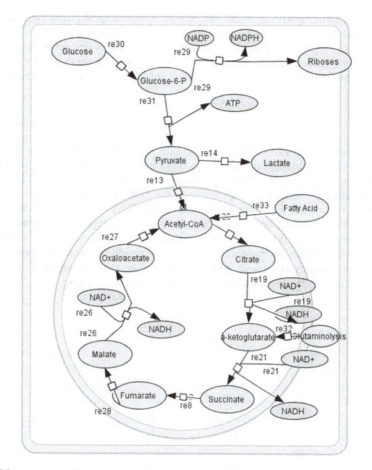

Figure 1.   Map representation of the model prepared in CellDesigner ver.4.4. The graphical representation of convergence of three pathway in accordance with system biology graphical notation (SBGN).

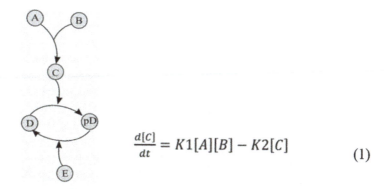

$$\frac{d[C]}{dt} = K1[A][B] - K2[C] \tag{1}$$

Figure 2.   Ordinary differential Equation representation as in Equation. (1) here product C is formed from substrate A and B.

molecule in the network with different rate laws for biochemical pathways, each including activation, inhibition, and reversibility or irreversibility as studied earlier, representing different pathways [13] [14]. The kinetics equation for the pathway was set manually as the real concentration, experimental data was not investigated for such analysis. We presumed and set each molecule in the model equal to a unit of amount, the value ranged from 0.3 to 0.5.

3 RESULTS

The goal of this study was to demonstrate how computational analysis could help in enhancing the yield of industrial important compounds varies from drugs to the cosmetic industry. We presented the importance of integrating multiple pathways in one model and simulated the same by mimicking the biological function and reporting the output.

This study explains the outcome response from studying multiple pathways Glycolysis, Glutaminolysis, and integrating with the Electron transport chain that expression of various secondary metabolites in ideal conditions. During the simulation, we observed the expression of industrially important metabolites in an ideal condition such as Figure 3 and 4. We have seen these changes at various level (Table 1). The effect is visible for the arbitrary concentration and can be optimize with real world experiment set up.

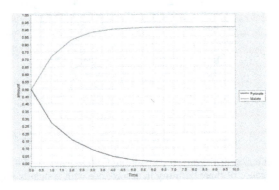

Figure 3. Response curve showing Pyruvate and Malate expression during the biological process.

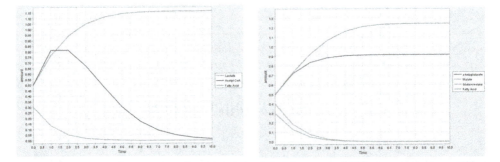

Figure 4. Response curve showing Lactate produce vs AcetylCoA and Malate expression with Fatty acid consumption during the biological process.

Table 1. Change of expression of different molecules in computational environment.

S. no.	Molecules	Initial concentration	Simulation effect/ time
1	Fatty acid	0.4	0/4
2	Glutamynolysis	0.4	0/10
3	Malate	0.5	0.81/2–0/10
4	Oxalacetate	0.4	0.56/4.2
5	A-ketoglutarate	0.5	0.67/4

4 CONCLUSION

The primary objective of this study is to analyze and understand the metabolic pathways that govern the production of secondary metabolites. By modeling the behavior of these compounds at specific concentrations, we aim to optimize our strategies for achieving high yields of these valuable chemicals.

Additionally, we employed the use of kinetics analysis to further understand the behavior of these metabolic pathways. We used the SBML squeezer plugin, a java-based tool, to assign kinetic equations to each molecule in the network based on different rate laws such as Michaelis-Menten, mass action, and Hill. This allowed us to study the activation, inhibition, and reversibility or irreversibility of different pathways as previously studied [13], [14], [15]. The kinetics equation for the pathway was set manually as experimental data was not available for such analysis. We assumed and set each molecule in the model equal to a unit of amount, with values ranging from 0.3 to 0.5. This in-depth analysis of the kinetics of the metabolic pathways will provide valuable insights into the behavior of these systems and aid in the optimization of our strategies for achieving high yields of secondary metabolites.

The rapid growth of various industries has led to an increased demand for these compounds, making it crucial to develop efficient methods for their production. Systems biology approaches have been helpful in identifying key molecules and regulatory mechanisms that can be targeted for metabolic engineering. In this study, we demonstrate the importance of fatty acids and Acetyl-CoA in the pathways leading to secondary metabolite production. By simulating different scenarios and identifying potential targets, we propose new experiments and strategies for achieving our goals in an economically viable manner. This approach has the potential to benefit the pharmaceutical industry by improving the efficiency and yield of valuable secondary metabolites.

REFERENCES

[1] Pickens L.B., Tang Y. and Chooi Y.H., 2011. Metabolic Engineering for the Production of Natural Products. *Annual Review of Chemical and Biomolecular Engineering*, 2, p. 211.

[2] Chubukov V., Mukhopadhyay A., Petzold C.J., Keasling J.D. and Martín H.G., 2016. Synthetic and Systems Biology for Microbial Production of Commodity Chemicals. *NPJ Systems Biology and Applications*, 2(1), pp. 1–11.

[3] Pham J.V., Yilma M.A., Feliz A., Majid M.T., Maffetone N., Walker J.R., Kim E., Cho H.J., Reynolds J.M., Song M.C. and Park S.R., 2019. A Review of the Microbial Production of Bioactive Natural Products and Biologics. *Frontiers in Microbiology*, 10, p. 1404.

[4] Sheng H., Sun X., Yan Y., Yuan Q., Wang J. and Shen X., 2020. Metabolic Engineering of Microorganisms for the Production of Flavonoids. *Frontiers in Bioengineering and Biotechnology*, 8, p. 589069.

[5] Huang H.C., Lee I.J., Huang C. and Chang T.M., 2020. Lactic Acid Bacteria and Lactic Acid for Skin Health and Melanogenesis Inhibition. *Current Pharmaceutical Biotechnology*, 21(7), pp. 566–577.

[6]   Arora D., Srikanth K., Lee J., Lee D., Park N., Wy S., Kim H., Park J.E., Chai H.H., Lim D. and Cho I.C., 2021. Integration of Multi-omics Approaches for Functional Characterization of Muscle Related Selective Sweep Genes in Nanchukmacdon. *Scientific Reports*, 11(1), p. 7219.

[7]   Arora D., Park J.E., Lim D., Cho I.C., Kang K.S., Kim T.H. and Park W., 2022. Multi-omics Approaches for Comprehensive Analysis and Understanding of the Immune Response in the Miniature Pig Breed. *Plos One*, 17(5), p. e0263035.

[8]   Jaiswar A., Arora D., Malhotra M., Shukla A. and Rai N., 2022. Broad Applications of Network Embeddings in Computational Biology, Genomics, Medicine, and Health. *Bioinformatics and Medical Applications: Big Data Using Deep Learning Algorithms*, pp.73–98.

[9]   O'Hara L., Livigni A., Theo T., Boyer B., Angus T., Wright D., Chen S.H., Raza S., Barnett M.W., Digard P. and Smith L.B., 2016. Modelling the Structure and Dynamics of Biological Pathways. *PLoS Biology*, 14(8), p. e1002530.

[10]   Foster D.A., Salloum D., Menon D. and Frias M.A., 2014. Phospholipase D and the Maintenance of Phosphatidic Acid Levels for Regulation of Mammalian Target of Rapamycin (mTOR). *Journal of Biological Chemistry*, 289(33), pp. 22583–22588.

[11]   Jatain I., Yadav K., Nitharwal R.G., Arora D. and Dubey K.K., 2022. A System Biology Approach for Engineering Non-oxidative Glycolysis Pathway in Streptomyces Toxytricini for High Lipstatin Biosynthesis. *Bioresource Technology Reports*, 19, p. 101188.

[12]   Funahashi A., Morohashi M., Matsuoka Y., Jouraku A. and Kitano H., 2007. CellDesigner: A Graphical Biological Network Editor and Workbench Interfacing Simulator. *In Introduction to Systems Biology*. Humana Press, pp. 422–434.

[13]   Drager A., Zielinski D.C., Keller R., Rall M., Eichner J., Palsson B.O. and Zell A., 2015. SBMLsqueezer 2: Context-Sensitive Creation of Kinetic Equations in Biochemical Networks. *BMC Systems Biology*, 9(1), pp. 1–17.

[14]   Arora D. and Singh A., 2018. Systems Biology Approach Deciphering the Biochemical Signaling Pathway and Pharmacokinetic Study of PI3K/mTOR/p53-Mdm2 Module Involved in Neoplastic Transformation. *Network Modeling Analysis in Health Informatics and Bioinformatics* 7(1), pp. 1–11.

[15]   Arora, D., Chaudhary R. and Singh A. System Biology Approach to Identify Potential Receptor for Targeting Cancer and Biomolecular Interaction Studies of Indole [2, 1-a] isoquinoline Derivative as Anticancerous Drug Candidate Against it. *Interdisciplinary Sciences: Computational Life Sciences* 11(1), 125–134, 2019.

Automation and Computation – Vats et al. (Eds)
© 2023 the Author(s), ISBN 978-1-032-36723-1

Performance analysis of a Rotated Hexagonal Sierpinski Gasket (RHSG) fractal antenna for wireless communications

D. Tiwari* & A. Shukla
Graphic Era Hill University, Dehradun, Uttarakhand, India

N.L. Taranath
Alliance College of Engineering and Design, Alliance University, Bengaluru, India

ABSTRACT: In this article a high gain, multiband fractal microstrip antenna has been developed for wireless applications. Here a Sierpinski gasket model is developed on a hexagonal patch. By applying the fractal concepts the radiated patch size is reduced by 43.75% and patch perimeter is increased by 131.8%. This RHSG antenna is analysed by Ansoft HFSS v15 simulation tool. The proposed design presents the Quad-band at 7.49, 11.16, 17.11, 26.16 (GHz) resonant frequencies having Peak Gain (dB) of 7.76, 6.93, 9.56 and 8.73 dB respectively and efficiency of almost 80%. The proposed RHSG fractal antenna have high gain, high efficiency and multiband performance in X/Ku and K band application from 6 to 30 GHz. So the RHSG fractal antenna design is useful or wireless applications like Milletry and RADAR application, Setellite and DTH application 5G/6G wireless applications etc.

Keywords: Hexagona Sierpinski gasket, Multiband antenna, wireless communications, Fractal microstrip antenna, 5G/6G wireless applications

1 INTRODUCTION

A Simple Microstrip antenna is a basic component of any wireless devices and act as an electronic eye and ear to the world [1]. Currently miniaturized fractal microstrip antennas (MSAs) [2,3] are commercially avalable due to its charcteristics of light weight, high gain, Low profile, easy integration and conformal [4,5]. Due to its charecteristcs it is useful for high gain, high directive, multiband and UWB perpose for 5G/6G & IoT applications [6–9]. Fractal miniaturization Technique can overcome the limitations of MSA like low Gain, Low bandwidth [9] etc.

The word 'Fractal' was introduced by famous mathematician B. Mandelbrot in his book "Fractal geometry in nature" in 1975. Self-similarity and space-filling are the two main charecteristics of any fractal geometry [10–20]. The antenna using miniaturized fractal structure is known as fractal antenna. Sierpinski is the name of scientist who developed two basic fractal geometry in 1916 using triangle and square strecture. The fractal antanna using these structure are known as Sierpinski Gasket [17], [20–24] and Sierpinski Carpet fractal [19] antenna and usefull for multiband applications.

In this design, a hexagonal slotted Sierpinski gasket structures are analysed for wireess communications. In this design, a Hexagonl Sierpinski Gasket (HSSG) fractal microstrip antenna have been analysed for wireless communication [25–29]. Using four different

---

*Corresponding Author: deveshcdpm@gmail.com

DOI: 10.1201/9781003333500-55

iterations the patch size is reduced by 43.75% and patch perimeter is increased by 131.8 %. The proposed design is analysed using HFSS v15 simulation software which produce the high gain, multiband performance upto 30 GHz frequency. It is found that RHSG design fractal structure are better than Sierpinski gasket antenna.

Here, this paper is arranged continuously in five sub-sections. Section 1, briefly covere the introduction and literature survey of different fractal microstrip antenna. While the mathematical analysis and design requirements of the RHSG antenna are discussed respectively in Sections 2 and section 3. The result of proposed design is discussed in section 4. At the end, conclusion of paper is discussed briefly in section 5.

2 MATHEMATICAL DESIGN OF RHSG FRACTAL ANTENNA

A microstrip antenna (MSA) contains two metallic (Copper) sheets on top and bottom of the dielitric substrate having permittivity ($\varepsilon_r \leq 10$). The top and bottom layer of metallic sheet is known as radiating patch and ground plane respectively. The patch can be fed by line or coaxial feed in contacting method and shown in Figure 1 (a) and (b). Generally low permitivety material is used for better radiation and broadband and high permitivety material is used for miniaturization perpuses for multiband and high gain applications. Here we have used FR-4 substrate having dielectric substrate of permitivety 4.4. For method of analysis of antenna cavity model is applied.

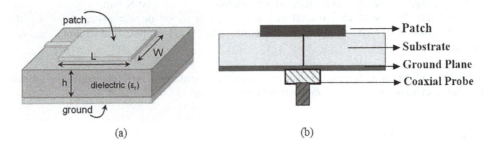

Figure 1.   Microstrip antenna: (a) top view, (b) front view.

Here

$$R_1 = Qr/\omega_r C_1 \tag{1}$$

$$L_1 = 1/\omega_r^2 C_1 \tag{2}$$

$$C_1 = \frac{L W \varepsilon_0 \varepsilon_e}{2h} \cos^2(\pi x_0/L) \tag{3}$$

in which h = hight of dielectric substrate, L = Length of patch, W = Width of Patch, x_0 = coaxial feed

Where Qr and ε_e can be calculated as-

$$Q_r = \frac{c\sqrt{\varepsilon_e}}{fh} \tag{4}$$

$$\varepsilon_e = \frac{\varepsilon_r + 1}{2} + \frac{\varepsilon_r - 1}{2}\left(1 + \frac{12h}{W}\right)^{-1/2} \tag{5}$$

where f = desired frequency of antenna, ε_r = relative permittivity dielectric substrate, c = light velocity and ε_e = effective permittivity. Equivlant total impedance of proposed antenna is calculated and connected in parallel with impedance of basic microstrip antenna as shown in Figure 2.

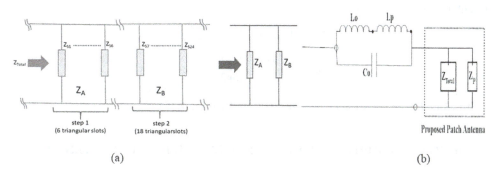

(a)   (b)

Figure 2.   (a) total input impedance (b) designed fractal antenna.

Here, $\mathbf{Z}_{\text{Total}}$ is the total equivalent impedance hexagonal fractal patch.

$$\mathbf{Z}_{\text{Total}} = \frac{1}{\frac{1}{Z_A} + \frac{1}{Z_B}} \tag{6}$$

where Z_A and Z_B is 6 and 18 triangular slot of equal dimentions and can be written as-

$$\mathbf{Z}_A = \frac{1}{\frac{1}{Z_{s1}} + \ldots + \frac{1}{Z_{s6}}} = \frac{Z_{sA}}{6}; \ \mathbf{Z}_{sA} = Z_{s1} = Z_{s2} \ldots\ldots\ldots = Z_{s6}$$

$$\mathbf{Z}_B = \frac{1}{\frac{1}{Z_{sB1}} + \ldots + \frac{1}{Z_{sB18}}} = \frac{Z_{sB}}{18}; \ \mathbf{Z}_{sB} = Z_{s7} = Z_{s8} \ldots\ldots = Z_{s24}$$

Here various antenna parameter are used to analyse the behavior of antenna can be formulated as-

$$\text{Reflection co} - \text{efficient,} \quad \Gamma = \left(\frac{\mathbf{Z}_{Total} - \mathbf{Z}_o}{\mathbf{Z}_{Total} + \mathbf{Z}_o}\right) \tag{7}$$

$$\text{ReturnLoss, RL} = 20 \log |\Gamma| \tag{8}$$

$$\text{VSWR} = \frac{1 + \Gamma}{1 - \Gamma} \tag{9}$$

Bandwidth (BW) of antenna can be written using highest frequency (f_h) and lowest frequency (f_l) as

$$\text{Bandwidth, } BW = \frac{(f_h - f_l)}{(f_c)} \times 100\% \tag{10}$$

$$\text{where, central frequency } (f_c) = \frac{(f_h + f_l)}{2} \tag{11}$$

By using fractal concept of similarity dimension D_S and Hausdorff dimension D_H can be calculated as:-

$$D_H = \log_6 18 = \log_2 3 = 1.584 \tag{12}$$

$$D_S = \frac{\log 18}{\log 6} = \frac{\log 3}{\log 2} = 1.584 \tag{13}$$

It is found from equation (12–13) that D_H and D_S are 1.584.

3 DESIGN DESCRIPTION OF RHSG FRACTAL MICROSTRIP ANTENNA

Here, a FR-4 dielectric substrate with a height of 1.6 mm and a relative permittivity (ε_r) of 4.4 is applied to separate a hexagonal metallic patch and a square ground plane. The RHSG fractal patch of side length 21.47mm and both ground plane and FR4 substrate size of 50 × 50 mm². Table 1 summarizes the dimensional analysis of the proposed model. According to Figure 3, the proposed RHSG fractal antenna has six Sierpinski Gasket strectures coupled in a hexagonal manner. Here in step 1, the hexagonal patch of side length 18.6 mm is used.

Table 1. Design specification of proposed RHSG fractal antenna.

Symbol	Parameter	Dimension (mm)
L_p	Length of RHSG patch (L_p = 2H)	37.20
W_P	Width of RHSG patch (W_P = 2R)	42.95
A_1	Step 1, triangular side length	18.6
A_2	Step 2 triangular side length	9
A_3	Step 3 triangular side length	4.5
$L_s = W_s = A_s$	Length and Width of square substrate	50
$L_g = W_g = A_g$	Length and Width of square ground plane	50
(x_o, y_o)	Coaxial fed point	(–5.20,–9)
h	FR-4 Substrate thickness	1.6

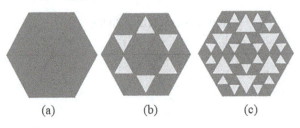

(a) (b) (c)

Figure 3. RHSG fractal antenna geometry: (a) step 1 (b) step 2 (c) step 3.

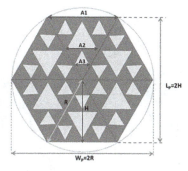

Figure 4. Designed RHSG fractal antenna structure.

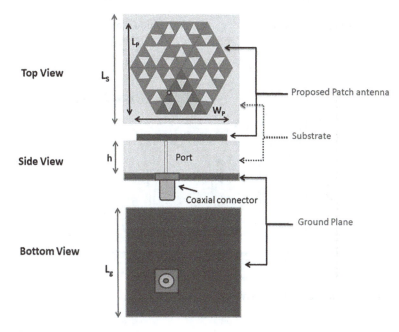

Figure 5. The detailed RHSG fractal antenna structure.

Then 6 and 18 slots are loaded in step 2 and step 3 continuously. The scaling factor in each step of sierpinski gasket design is 0.5 and shown in Table 1 and Figure 4. In RHSG fractal microstrip antenna, the side arm of each triangle is A1, A2, A3 and is 21.7, 9, 4.5 respectively as shown in Figures 4,5 and summarized in Table 1.

4 RESULTS AND DISCUSSION

The designed antenna is analyzed from 6 to 30 GHz using HFSS v-15 simialtion software and verified with fabricated antenna results. Magnitude and vector E-field Current distribution plot is shown in Figure 6. Figure 7 shows the S11 parameter versus frequency plot and Figure 8 shows the VSWR versus frequency plot and summarized in Table 2. The S_{11} parameter (Return Loss) of analyzed antenna are -21.32, -31.92, -27.52 and -38.36 dB while VSWR 1.16, 1.03, 1.08, 1.01 at frequency 07.49, 11.16, 17.11, and 25.16 GHz respectively.

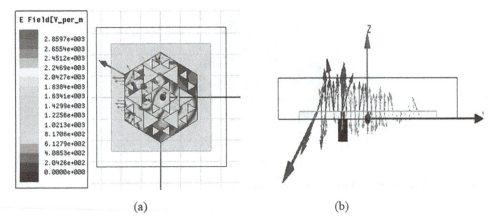

(a)   (b)

Figure 6.   The E-field magnitude and vector current distribution plot.

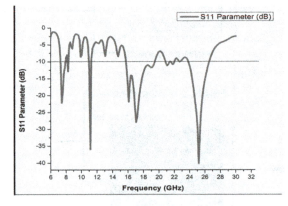

Figure 7. S_{11} parameter versus frequency plot.

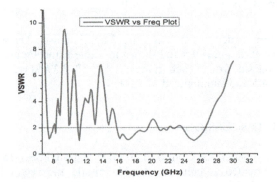

Figure 8.   VSWR versus frequency plot.

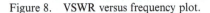

Table 2. The S_{11} parameter (dB) and VSWR value oh HSSG antenna at different frequency (GHz).

Band	f_r (GHz)	S_{11} (dB)	VSWR
1	07.49	−21.32	1.16
2	11.16	−31.92	1.03
3	17.11	−27.51	1.08
4	25.16	−38.36	1.01

Frequency unit in GHz.

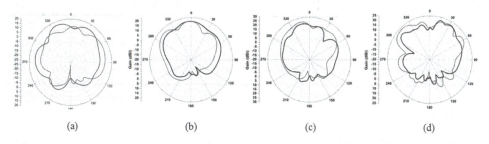

| (a) | (b) | (c) | (d) |

Figure 9. RHSG fractal antenna radiation pattern at (a) 7.49 (b) 11.16 (c) 17.11 (d) 25.16 frequency (GHz).

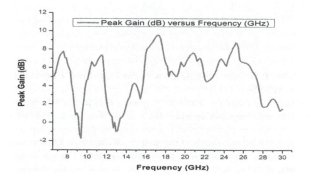

Figure 10. Peak gain (dB) versus frequency plot.

The analysys shows that the multiband behavior is obtained with better matching and radiation pattern. Figure 9 shows the radiation patern of designed RHSG fractal antenna at 0.7.49, 11.16, 17.11, and 25.16 GHz frequency respectively. The omni-directional radiation pattern are achieved at quad-band.

The achieved peak gain (dB) versus frequency (GHz) plot is presented Figure 10 and also summrized in Table 3. It has been observed that the peak gain of antenna are 7.76, 6.93, 9.56 8.73 (dB) are 7.49, 11.16, 17.11, and 25.16 GHz frequency respectively with radiation efficiency of almost 80%. The comparision Table is shown in Table 4 From literature survey presents the Gain of different fractal antenna geometry for wireless applications.

Table 3. Peak gain of designed multiband antenna.

S. No.	Resonant Frequency	Gain
1	07.49 GHz	7.76 dB
2	11.16 GHz	6.93 dB
3	17.11 GHz	9.56 dB
4	25.16 GHz	8.73 dB

Table 4. Comparison between available literature with proposed antenna.

Reference	Size (mm^2) [L×W]	Iterations	Gain(dB)
[28]	22×33.4	4	1.15–3.2
[29]	28×31	3	2–5.8
[12]	31×45	4	0–7
Proposed design	50×50	4	6.9–9.6

5 CONCLUSION

The designed HSSG fractal antenna is very much useful for modern wireless applications. The multiband performance is obtained at 07.49, 11.16, 17.11, and 25.16 GHz resonant frequency with return Loss of –21.32, –31.92, –27.52 and –38.36 dB and VSWR 1.16, 1.03, 1.08, 1.01 respictively. That works well for microwave X/Ku and K frequency band respectively. Due its charecteristics the proposed model is useful for wireless applications as 5G/6G communication, satellite and DTH application, militry and radar applications etc.

REFERENCES

[1] Carver K. R., Mink J. W. Microstrip Antenna Technology. *IEEE Trans Antennas Propag* 1981;29:2–24. doi:10.1109/TAP.1981.1142523.

[2] Anguera J., Puente C., Borja C., Montero R., Soler J. Small and High-directivity Bow-tie Patch Antenna Based on the Sierpinski Fractal. *Microw Opt Technol Lett* 2001;31:239–41. doi:10.1002/mop.1407.

[3] Anguera J., Andújar A., Benavente S., Jayasinghe J., Kahng S. High-directivity Microstrip Antenna with Mandelbrot Fractal Boundary. *IET Microwaves, Antennas Propag* 2018;12:569–75. doi:10.1049/iet-map.2017.0649.

[4] Borja C., Puente C., Medina A. Iterative Network Model to Predict the Behaviour of a Sierpinski Fractal Network. *Electron Lett* 1998;34:1443–5. doi:10.1049/el:19981020.

[5] Khandelwal M. K., Kanaujia B. K., Dwari S., Kumar S., Gautam A. K. Analysis and Design of Dual Band Compact Stacked Microstrip Patch Antenna with Defected Ground Structure for WLAN/WiMAX Applications. *AEU - Int J Electron Commun* 2015;69:39–47. doi:10.1016/j.aeue.2014.07.018.

[6] Chaudhary A. K. and Manohar M., "A Modified SWB Hexagonal Fractal Spatial Diversity Antenna With High Isolation Using Meander Line Approach," *in IEEE Access*, vol. 10, pp. 10238–10250, 2022, doi:10.1109/ACCESS.2022.3144850.

[7] Yu Z., Yu J., Ran X., Zhu C. A Novel Koch and Sierpinski Combined Fractal Antenna for 2G/3G/4G/5G/WLAN/navigation Applications. *Microw Opt Technol Lett* 2017;59:2147–55. doi:10.1002/mop.30698.

[8] Hong W., Jiang Z. H., Yu C., Zhou J., Chen P., Yu Z., *et al.* Multibeam Antenna Technologies for 5G Wireless Communications. *IEEE Trans Antennas Propag* 2017;65:6231–49. doi:10.1109/TAP.2017.2712819.

[9] Ribeiro D. and Carrijo G., "Smart Fractal Antennas in Transition Projects to 5G," *in IEEE Latin America Transactions*, vol. 20, no. 12, pp. 2463–2470, Dec. 2022, doi:10.1109/TLA.2022.9905615.

[10] Azari A. *UWB Fractal Microstrip Antenna Design* 2008;2:7–12.

[11] Electron I. J., Aeü C., Bangi I. S., Sivia J. S. Minkowski and Hilbert Curves based Hybrid Fractal Antenna for Wireless Applications. *Int J Electron Commun* 2018;85:159–68. doi:10.1016/j.aeue.2018.01.005.

[12] Devesh Tiwari, Mohd. Gulman Siddiqui, A. K. Saroj, J. A. Ansari NA and MK. Analysis of Modified Swastika Shaped Slotted (MSSS) Microstrip Antenna for Multiband and Ultrawideband Applications. *Lect Notes Electr Eng* 2020;587:189–98.

[13] Anguera J., Daniel J. Metallized foam for Antenna Design : Application to Fractal-shaped Sierpinski-Carpet Monopole. *Prog Electromagn Res PIER* 2010;104:239–51.

[14] Anguera J., Martínez-Ortigosa E., Puente C., Borja C., Soler J. Broadband Triple-frequency Microstrip Patch Radiator Combining a Dual-band Modified Sierpinski Fractal and a Monoband Antenna. *IEEE Trans Antennas Propag* 2006;54:3367–73. doi:10.1109/TAP.2006.884209.

[15] Behera B. R. Engineering Science and Technology, an International Journal Sierpinski Bow-Tie Antenna with Genetic Algorithm. *Eng Sci Technol an Int J* 2017;20:775–82. doi:10.1016/j.jestch.2016.10.017.

[16] Werner D. H., Gangul S. An Overview' of Fractal Antenna Engineering Research 2003;45.

[17] Devesh Tiwari, Jamshed Aslam Ansari, Abhishek Kr. Saroj, Mukesh Kumar, Analysis of a Miniaturized Hexagonal Sierpinski Gasket Fractal Microstrip Antenna for Modern Wireless Communications, *AEU – Int. J. Electron. Commun.* 123 (2020), 153288, https://doi.org/10.1016/j.aeue.2020.153288. ISSN 1434-8411.

[18] Varamini G., Keshtkar A., Naser-moghadasi M. International Journal of Electronics and Communications (AEÜ) Compact and Miniaturized Microstrip Antenna based on Fractal and Metamaterial Loads with Reconfigurable Qualification. *AEUE - Int J Electron Commun* 2018;83:213–21. doi:10.1016/j.aeue.2017.08.057.

[19] Kaur N., Kaur A. A Compact Plus Shaped Carpet Fractal Antenna with an I-Shaped DGS for C-band/X-band/UWB/WIBAN applications. *Wirel Pers Commun* 2019. doi:10.1007/s11277-019-06645-y.

[20] Baliarda C. P., Borau C. B., Rodero M. N., Robert J. R. An Iterative Model for Fractal Antennas: Application to the Sierpinski Gasket Antenna. *IEEE Trans Antennas Propag* 2000;48:713–9. doi:10.1109/8.855489.

[21] Ghatak R., Poddar D. R., Mishra R. K. Design of Sierpinski Gasket Fractal Microstrip Antenna using Real Coded Genetic Algorithm. *IET Microwaves, Antennas Propag* 2009;3:1133–40. doi:10.1049/iet-map.2008.0257.

[22] Puente-baliarda C., Romeu J. *On the Behavior of the Sierpinski Multiband Fractal Antenna* 1998; 46:517–24.

[23] Devesh, M. G. Siddiqui, A. K. Saroj and J. A. Ansari, "Performance Analysis of Rotated Square Sierpinski Gasket (RSSG) Fractal Antenna for Wireless Communication," *2018 5th IEEE Uttar Pradesh Section International Conference on Electrical, Electronics and Computer Engineering (UPCON)*, 2018, pp. 1-4, doi:10.1109/UPCON.2018.8597125.

[24] Devesh, Ansari J. A., Siddiqui M. G., Saroj A. K. Analysis of Modified Square Sierpinski Gasket Fractal Microstrip Antenna for Wireless communications. *AEU - Int J Electron Commun* 2018. doi:10.1016/j.aeue.2018.07.027.

[25] Mondal T., Samanta S., Ghatak R., Chaudhuri SRB. A Novel Tri-Band Hexagonal Microstrip Patch Antenna Using Modified Sierpinski Fractal for Vehicular Communication 2015;57:25–34.

[26] Singhal S., Singh A. K. CPW-fed hexagonal Sierpinski Super Wideband Fractal Antenna. *IET Microwaves, Antennas Propag* 2016;10:1701–7. doi:10.1049/iet-map.2016.0154.

[27] Shahu B. L., Pal S., Chattoraj N. A Compact Super Wideband Monopole Antenna Design using Fractal Geometries. *Microw Rev* 2014;20:20–4.

[28] Subbarao A., Raghavan S. Coplanar Waveguide-fed Ultra-wideband Planar Antenna with WLAN-band Rejection. *J Microwaves, Optoelectron Electromagn Appl* 2013;12:50–9. doi:10.1590/s2179-10742013000100005.

[29] Darimireddy N. K., Reddy R. R., Prasad A. M. A Miniaturized Hexagonal-Triangular Fractal Antenna for Wide-Band Applications [Antenna Applications Corner]. *IEEE Antennas Propag Mag* 2018;60:104–10. doi:10.1109/MAP.2018.2796441.

Automation and Computation – Vats et al. (Eds)
© *2023 the Author(s), ISBN 978-1-032-36723-1*

Data security for digital environment

Bhishm Dutt* & Karan Singh*
JNU New Delhi, India

Satvik Vats*
Graphic Era Hill University, Dehradun, India

Manisha Manjul*
G B Pant DESU Campus1, New Delhi, India

Vikrant Sharma*
Graphic Era Hill University, Dehradun, India

Ahmad Aziz*
University of Benha, Benha, Egypt

ABSTRACT: The IoT environment and its highly interconnected technologies are now evolving quickly. Academic study, scientific research, industry, banking & finance, business & marketing, and IoT devices and services are involving in digitation. The large amount of digital data is generated during the data publishing, sharing and transferring. Social networks play an important role in today's interconnect development like LinkedIn, Facebook, Twitter etc. The data security is crucial for maintaining privacy rights; thus, it must be safeguarded against unauthorized access and dissemination. There are several cryptographic techniques and different privacy strategies involved in protection, however "Data anonymization" is just another option or a straightforward strategy. According to literature work there are many data anonymity tools on data security, presently all of anonymity tools using the concept of Differential Privacy (DP). It adds the noise by using the Approximate differentiation with Laplace distribution and Gaussian distribution. It provides more security in data hiding and protection against new privacy threads. In the use of assessment analysis and outcome analysis, the real-world dataset repositories used.

1 INTRODUCTION

Today's increasingly interconnected technologies produce a lot of digital data on online platforms, especially in the fields of academia, science, the IT sector, finance, and business [1] [2]. Considering the extraordinary growth of data publication, there is currently an excessive amount of data that is gathered through social network services like LinkedIn, Twitter, and Facebook and made publicly available. Users can create their personal accounts and control the links to those profiles on social networking sites. According to Facebook data, there are 1.15 billion daily users and 1.94 billion monthly users. There are also 1.94 billion new members every month. As it has gotten simpler to acquire and gather this data, numerous publishers have started publishing the data for those who are interested in study. On digital platforms, publishing and sharing of data is necessary to prevent it from unauthorized public

---

*Corresponding Authors: bhishm2009@hotmail.com, karan@mail.jnu.ac.in, svats@gehu.ac.in, manishamanjul@gbpec.edu.in, vsharma@gehu.ac.in and ahmad.aziz@fci.bu.edu.eg

DOI: 10.1201/9781003333500-56

access. Therefore, protecting privacy has become a key consideration in further research into internet technologies. To protect the data security from unauthorized access, there may be a variety of problems or difficulties. There are several ways to provide data privacy. Cryptographic methods have a large processing overhead. Data anonymity is a cycle or one type of solution to ensure the protection of sensitive information on open networks, inter-personal networks, and external apps. To put it another way, anonymity is a technique for protecting confidential and sensitive data that involves encrypting it and removing the information that identifies a person from the storage of information. The distinct entity employs, in addition to the anonymity model, tactics that deal with success in a given level of data anonymization. A number of entirely original alterations to the datasets are made under the direction of the anonymization technique.

1.1 Anonymization

Anonymization is a procedure to protecting the personal information and sensitive infor-mation through the encryption and erasing the identifiable information, which is form of individual's stored data.

The Figure 1, is an overview of the social user's data anonymization framework.

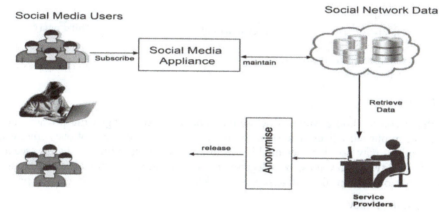

Figure 1.   Anonymization framework.

Chapter organization: The book chapter one is introduced or provide overview of the present privacy preserving technology, including their properties, uses, and limitation. Finally, research context is discussed followed by the significance of the research, motiva-tion, objective, scope, and research organization. There are Six sections in this book chapter, the Section 1 describes the overview on the privacy, Section 2 explanation of data privacy, Section 3 is related to Comparative study, Section 4 detailed the anonymity tools and tech-niques, Section 5, expresses the conclusion and future work, and the last Section 6 detailed the references.

2   DATA PRIVACY

Presently, there are profoundly associated advancements which produces a lot of compu-terized information on internet-based stage, like the field of Academic, Scientific examina-tion, IT industry, Finance and Business Academic study, all are currently going through deeply related advancements that produce a lot of computerized data on an online platform. It may be difficult or problematic to safeguard information security from illegal access and publicly distribution. Cryptographic methods are just one of numerous methods for achieving information security.

The Figure 2, demonstrate the structure of whole data publishing ways on digital environment, first it collected from data owners using of social site through service providers with third parties, from data generation to data publishing.

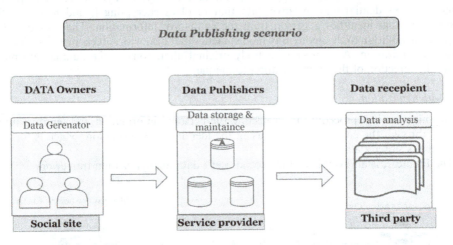

Figure 2.   Example of data anonymity.

2.1   *Privacy*

With the option to choose who to disclose our thoughts and feelings with, privacy plays a crucial role in our daily lives. The right to privacy gives a person control over how, when, and why others may use their personal information. To maintain human dignity, security, and individuality, privacy must be safeguarded. An individual's personality can be developed on their own. Four alternative privacy models have been used to classify the privacy,

- Information privacy
- Bodily privacy
- Communication privacy
- Territorial privacy

2.2   *Privacy*

Identity discloser: In the context of differential privacy, this is commonly known as distinctive identification. This is a significant attack strategy and a well-known privacy thread within the PPDP. When an adversary sides with a person in a privacy-protected printed record, the Identity discloser occurs [9,10].

Attribute discloser: In attribute discloser, the attacker takes advantage of the delicate attributes. As soon as each sensitive characteristic is combined with knowledge, the attribute discloser occurs. Unbalanced datasets will automatically cause this type of thread to spawn [9,10].

Membership discloser: This is frequently regarded as linking knowledge that enables a potential attacker to determine whether or not information on a specific individual is present in a dataset [9,10].

2.3 Anonymity classification

According to the survey study and assessment, the data anonymity has been categorized into three different historically approaches, k-anonymity, L-diversity, and t-closeness, and it has also been categorized into a variety of other forms.

- **k- Anonymity**: Anonymizing relational or tabular data is commonly done using the well-known k-anonymity synaptic privacy model. Each individual in a generalized block is identical to at least k-1 additional individuals, where k is the total number of parameters. The number "k" indicates the degree of anonymity.
- **L-diversity**: To overcome the shortcomings of k-anonymization, the L-diversity privacy was created. This type of gradual information loss is used to save security informative data and is a form of combination base anonymization. This model states that an equivalence class only satisfies the L-diversity property when the value "L" accurately represents the sensitive feature. The intragroup diversity for price-sensitive in anonymity method is promoted by L-diversity.
- **t-closeness**: t-closeness is a group-based data anonymization that is used to maintain data sets' secrecy by gradually reducing the distance between them. It is a further improvement over l-diversity. It closes to quality dispersion in the overall table and necessitates the conveyance of delicate property in any comparability class. The shortest distance between two datasets is called the earth movers' distance. The work is simply divided by the weighted average of the work.

2.4 Relational data

The expression "relational data" refers to information that has a relationship between two tables and is organized by row and column. The actual thing that a table represents in the real world is included in relational data. The data is represented by each row and tuple of records in various attribute kinds. Most databases, including IBM DB2, Oracle, MS Access, MySQL, and MS SQL Server, store data in a relational model and tabular format. Various organizations save their data in various database formats, encrypting or retrieving that data to make it anonymous before storing it [12,13].

2.5 Graph data

There are various corporations, corporate agencies, computing, and cloud communication systems that create and store specific information or unique data in "Graph structure" [7]. There are many nodes and edges in the graph. Examples of systems that employ the graph structure to store user records include social networks, business application systems, message and mail delivery systems, tracking network systems, and others. Typically, this type of data can be used to store and represent "Graph data" [7]. The data is typically published, transferred, and shared across the entire digital platform for public access for research purposes, data mining tasks, network mining, government or commercial uses. So, it is crucial to safeguard individuals' privacy when sharing, transferring, and publishing [7].

2.6 Relational data and graph data anonymization

Essentially, the goal of these anonymization strategies is to frustrate the initial graph building process while keeping as much predicted data utility as possible to ensure user security [7,8].

The Figure 3, An example of anonymization on relational and graph data. This Figure 3 contains two relational tables, one of which describes the four original data records and the other of which displays the anonymized data records.

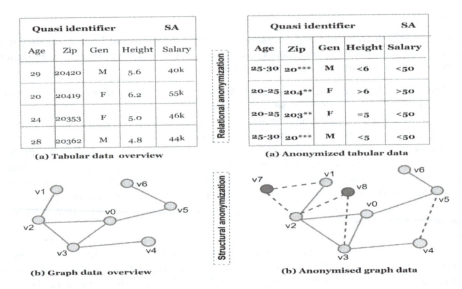

Quasi identifier				SA
Age	Zip	Gen	Height	Salary
29	20420	M	5.6	40k
20	20419	F	6.2	55k
24	20353	F	5.0	46k
28	20362	M	4.8	44k

(a) Tabular data overview

Quasi identifier				SA
Age	Zip	Gen	Height	Salary
25-30	20***	M	<6	<50
20-25	204**	F	>6	>50
20-25	203**	F	=5	<50
25-30	20***	M	<5	<50

(a) Anonymized tabular data

(b) Graph data overview

(b) Anonymised graph data

Figure 3. Rational and graph data anonymizations.

2.7 Graph anonymity techniques

Whether publishing or distributing structural data or graph data, there are several general procedures or approaches for anonymizing the data that take into account the security of the individual's personal information [4].

- Graph Modification
- Graph generalization
- DP based technique

2.8 Graph modification techniques

- **k-Degree anonymity**: Degree-based anonymity in a naive anonymity graph is more dependent on prior information of the target's target to identify its victims. In order to maintain anonymity, k-degree based technology is also utilized. This modifies the network to a degree of k by adding fictional vertices rather than new links.
- **k-neighborhood anonymity**: In this scenario, the enemy attacked the neighborhood using the suspect's education and prior knowledge of the friendship as justification, with the neighbor's anonymity acting as a shield.
- **k-Automorphism**: Adversary presented a method to counter sub-graph-based privacy attacks by identifying the target from a published graph using its prior knowledge of sub-graphs. According to the concept of K-automorphism, the adversary can recreate the original network using k-identical subgraphs in which each node is an automorphism with the other k-1 nodes.

2.9 Graph generalization

The clustering-based approach was called by generalizing the graph. In order to include a merge statement of sub-networks from a super vertex, generalization refers to grouping or linking into a group. The local structural data of the graph can be examined or verified using this method. the clustering of vertices with comparable structural characteristics, as determined by the graph normalization approach, where the total number of vertices in each cluster must be more than k (>=k). In this proposed vertex normalization approach, the vertices of a network are organized as a disjoint set.

Methods: The method of grouping vertices with comparable structural properties is used when the number of vertices in each cluster must be greater than or equal to k *(>=k)*.

2.10 *DP based technique*

The guarantee offered by DP-based tactics is that the attacker will be unable to gather any sensitive information about his victims. DP can be interactive or non-interactive, but in any case, the user's personal data is protected by adding noise to query results or actual data. LaPlace distribution is a popular method for adding noise. The two separate datasets are operated on by DP, the difference can only be one.

2.11 *Attack models*

- **Homogeneity attack**: This concurrent attack, even though the data is k-anonymized, takes use of the fact that a set of k records with the same sensitive value may be acceptable.
- **Linkage attack**: The linkage attack attempts to re-identify a specific person in the data collection by fusing anonymized data with background knowledge. By employing quasi-identifiers like zip codes or postcodes, salaries, gender, ages, and other characteristics that are present in both collections, linkage can create an identifiable connection.

2.12 *Relational data attack*

- **Re-identification**: Re-identification poses a serious threat to relational data since it refers to the practice of cross-matching anonymous data with supplementary information or supporting data to identify the individual with whom the data is related. The practice of comparing anonymous personal data to its original value is known as re-identification. In order to protect the security interests of customers, personal identifiers, such as name, age, number, or Aadhar number (government's unique identification), are typically abstracted from data sets containing important data.

2.13 *Graph data attack*

- **Mutual friends attack**: Each edge connecting two distinct vertices here is connected to one of their buddies. The number of common friends of the victims can be discovered by the attacker in order to confirm the mutual friend assault. Once the attacker has the necessary background information, they can launch an attack against published data to re-identify connections between two victims.
- **Node identity discloser (Identity discloser)**: When an attacker tries to associate and map to re-identify a person from the published anonymized graph, node identity disclosure occurs.
- **Relationship/connection discloser (Edge discloser)**: The edge discloser displays the users' or nodes' revelation. For instance, the relationship between a teacher and student might be very private, thus it needs to be secure. The edge discloser can happen with the assumption that the student may be knowledgeable about research if the teacher is a research specialist.
- **Vertex/edge label discloser (Content discloser)**: When a private attribute connected to a vertex or edge is made public through an anonymized graph, this is known as vertex/edge

label disclosure. And from the original or unaltered graph, this given information can be used to pinpoint a specific person.

- **Affiliation Link discloser**: A link is considered to be direct when it can be directly connected to B and is declared with confidence to be between a user V and an affiliation group H that is greater than or equal to T. A node V that is either a member of the affiliation group H or is not a member of it is an affiliation link discloser.

Summary: This section briefly describes the whole background knowledge about data privacy. we have discussed in depth about the topic's information, and whole review on literature work. This section teaches, what is the privacy and how to secure private or personal data while publishing or sharing it on digital platforms.

3   COMPARATIVE FIELD

3.1   *k-Anonymization*

Anonymization was required since related datasets had to be shared while protecting the privacy of the entities they contained. If a row in a dataset cannot be identified from at least k-1 other rows, it is said to be k-anonymous [10]. Lataney Sweeney serves as the pylon in the anonymity analysis, re-identifying a governor as a case history by comparing it to a list of citizens that is readily available to the public. The data fly formula was developed by Lataney Sweeney. This algorithm ensures that a dataset has been sufficiently anonymized, along with the widely used anonymous dataset megahertz Shetty's k-anonymity construct. In addition to defining defined words like quasi-identifier and attribute mathematically and describing k-anonymity, Lataney Sweeney discusses how information is re-identified [8,9].

Figure 5, shows a relational data anonymization example utilizing the k-anonymity algorithm, where k is the number of the suppressed value. The anonymous database contains two groups, each of which has the value k 2.

1- Original Database			2-	Anonymous Database		
Birthdate	**Gender**	**Zip-code**		**Birthday**	**Gender**	**Zip-code**
			Group 1	**/*1/79	human	5****
21-01-1979	M	53715		**/*1/79	human	5****
10-01-1979	F	55410				
01-10-1954	F	90210	Suppress			
21-02-1983	M	2274				
19-04-1982	M	2237	Group 2	**/*/8*	M	220**
				**/*/8*	M	220**

Figure 5.   k-Anonymity on relational data table.

3.2   *Differential privacy*

Differential privacy is a well-known model for preserving privacy. Differential privacy functions by adding useless data to the original data without speculating about future

486

invaders. The semantics class of privacy models includes the DP privacy. In the literature, differential privacy is referred to be a mathematically demonstrated assurance of private defense against a variety of privacy assaults, such as dereferencing, linking, and re-construction attacks [3], [6].

Differential notions refer to the dataset D and its nearby dataset $\bar{D}$. between the two databases, there are just one or two records, described as:

$$DP = \|D\| - \|\bar{D}\| = 1 \tag{1}$$

$$Pr[f(x) \in S] \leq exp(\varepsilon) \, pr[f(y) \in S] + \delta \tag{2}$$

Here is the random function f with the range S, which define as $S \subseteq Range(f)$.

The Figure 6 shows, the differential privacy model's actual computation, that demonstrating there is always at least one difference between the output and the input data.

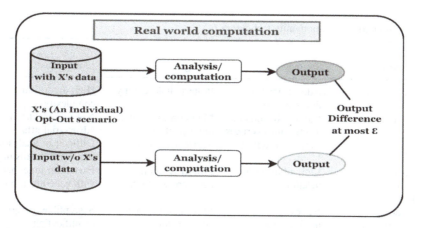

Figure 6.   Differential privacy model.

3.3   K2-NMF

The 2-NMF method is used to connect the K number of triangles in the graph. Additionally, SPL (shortest path length) is used by this technique to join the two vertices. There are at least three triangles present, ensuring the introduction of an adequate number of triangles. We need to massively anonymize the vertices in order to reduce the 2-nmf of the newly added edges to zero so that social networks can obey the law of force [5,6]. There is a certainty that the number of triangles will be presented if there are at least three triangles [6,7].

3.4   K-NDDP algorithm

The k-NDDP technique for undirect graph degree anonymity is used. This algorithm simply transforms the graph G (V, L) into an anonymous graph G based on the input parameter k. The insert operation is used by this algorithm to modify the network, adding the fewest number of links possible while minimizing the cost of anonymization and keeping the structural characteristics of both the original and anonymized graph [5], [6], [8].

3.5   UK-NDDP algorithm

The Updated k Node Degree Differential Privacy "UK-NDDP" that is degree anonymization model and extension of k-anonymity with differential privacy. The concept of this

UK-NDDP model is used on approximate differential privacy to prevent the information disclosing. Here, the Laplace mechanism (Laplace distribution) and Gaussian mechanism (Gaussian distribution) are utilized to roughly satisfy the differential privacy. It is expected that approximate differentiation, Laplace distribution, and Gaussian distribution will offer more security and concealment of data against emerging privacy threads. The UK-NDDP aims to propose a reliable data anonymity model, through the improvement of existing models (k-NMF, k-NDDP).

The above comparison Table 1 shows the overview on comparison on various anonymity techniques as the methodologies or strategies, strength and their drawback or limitations.

Summary: The section-3, consider as whole comparatively study or literature review on data anonymization, A writer should often help readers understand the background of their study problem and compare their current work with previous work. This also describes in detail the advantages and disadvantages of exiting research work, which provide the foundation for the various proposed models in this study.

Table 1. Comparison on anonymity techniques.

Ref. No.	Technique	Strategy	Strength	Limitations
[2]	k-anonymization	Add or deletes links/vertices	Protects link identity	High algorithm complexity
[8]	Rand add/del	Randomly adds or deletes links/vertices	Maintain the actual number of links	Not as considered adversarial attack
[6]	Vertex Clustering	Vertex clustering	Allow structural queries	Difficult to analyses local structural details in graph
[7]	Edge Differential Privacy	Link addition or deletion	Protect relation between two vertices	Execution time increases with generalization
[5]	Node Differential Privacy	Vertex modification	Protect vertex and adjacent links	Susceptible to vertex identification
[3]	Edge Clustering	Vertex generalization	Balances utility and privacy	N/A
[4]	k-degree anonymity	Addition of fake links/vertices	Conserve much of the characteristics of the graph.	Unsecured neighbor connections
[9]	k-NMF	Fake links addition	Protect sensitive links between vertices	Unsecure neighbor connections and Increased runtime
[5]	k-automorphism	Vertex modification	Susceptible to	Susceptible to identity disclosure
[1]	Rand switch	Switch old links with new links	Preserve spectral characteristics of the graph	Not considered any adversarial attack
[12]	k-neighborhood anonymity	Fake vertices and links addition	Re-identification attack protection	Utility loss (extreme change)

4 ANONYMITY TOOLS

There are many data anonymity tools available, some of them are free (Open-source) and some are paid application tools. Below description as some examples of anonymity tools.

1. ***ARX***: The ARX is the open-source toll, or its work under the licences of Apache 2.0. The functionally it has classified in four different categories and all phase works uniquely steps by steps and perform anonymity operation.

   - Configuration perspective: In this section, data is imported and some data transformation rules are developed. It is also deciding which privacy and quality model will be used for the data.
   - Exploration perspective: A solution space is drawn here among the possible transformation for the input data. For a certain level it is necessary to assess the risk threshold and quality of the data.
   - Utility analysis perspective: This perspective determines, how appropriate a particular change is in a particular situation in the main task. Using the output data and descriptive statistics, categorization models can be builds.
   - Risk analysis perspective: Multiple statistical models are used to calculate different indicators related to privacy issues.

2. ***Clover DX's data anonymity***: Clover DX's data anonymization tool converts our crucial production-level data into a set of anonymized data. The production level data is anonymized to remove the sensitive elements while keeping the essential details. The following are a few key benefits of using this anonymity tool.

   - It provides a highly trustworthy data collection thanks to its comprehensive method to anonymizing the data.
   - Stable private protections, such as the ability to insert randomized data and control how anonymous the output should be in comparison to previous data.
   - After setting up ones Automated Anonymization System, anyone can produce anonymized data whenever they wish.

3. ***Amnesia anonymity tool***: Amnesia's strong encryption methods ensure that the original data cannot be linked to any anonymised data. Additionally, a ReST API makes it simple for developers to integrate this data anonymization engine into their workflow.

   - Amnesia allows separated text files with large data sets. Individuals will indeed be prompted to select a specific type of data.
   - Customers can design procedures to swap out real data for an imaginary value.
   - It offers many alternatives, each of which includes one different set of rearranged numbers and statistics upon that accuracy of the data.

4. ***BizDataX***: The anonymity tools from BizDataX are skilled with processing a variety of data formats. However, if you think of a particularly unique circumstance, anyone can utilize a straightforward plugin where you can add your own logic.

   - The method consists of three simple steps: removing data from the BizDataX app cache, processing the data to make it anonymous, and entering the information into database. They typically write roughly 1 billion items every hr of knowledge.
   - It is common for the integrity of the main database to be hampered when data masking activities are underway. BizDataX maintains a watch out for any potential problems.

5. ***g9 Anonymizer***: The utility g9Anonymizer has a highly customised anonymization method. It can be replicated whenever needed and is therefore easy to deploy across a range of database. The database supports a wide range of functions, such as masking data with random values, combining data records, adding random noise to randomize the data, and more. Mastering the precise operation of the g9 Anonymizer phases are,

   - Make a copy of your production data first.
   - The data would need to be imported into the g9 Anonymizer.
   - Establish the guidelines for anonymization that will be used to imported data.

Summary: This section-4 contains the explanation analysis the various tools or models that works on real-time simulation and demonstration the result analysis using the different scenario. There are six anonymity tools discussed as some of them are open-source and some other are paid. All anonymity tools uniquely perform.

5 CONCLUSION AND FUTURE WORK

5.1 *Conclusion*

The conclusion states that this chapter establishes the background knowledge or fundamental framework of the data anonymity models, which enables us to understand the fundamentals of data anonymization. Personal information can take in many different forms, such as relationship or graph data, but it is always dangerous to allow unauthorized access or release. In today's digital age, data security is crucial for data publishing or sharing. In this section we have concluded the whole degree anonymity method, which is based on the degree of sequence and degree partition scheme. During the noise addition, we used the approximate differential privacy to prevent private information disclosing. Here for satisfy the approximate differential privacy the Laplace mechanism (Laplace distribution) and Gaussian mechanism (Gaussian mechanism) are used. Our main contribution is to provides the best protection against linkage attacks. Another advantage of updated k-NDDP model is its support for large datasets. It also provides effectiveness and feasibility in maintain privacy on digital platforms.

5.2 *Future work*

While the study efforts have demonstrated their efficacy through the establishment of proposed models, research on data security is not over. In the near future, various direction remains to be investigated. we plan to study the relationship between data anonymity techniques and anti-discrimination, considering indirect discrimination. furthermore, we should plan to formulate different perspectives on data anonymity to protect the original data records against the both privacy and discrimination threats.

REFERENCES

[1] Wieringa J., Kannan P.K., Ma X., Reutterer T., Risselada H. and Skiera B., "Data Analytics in a Privacy-concerned World," *J. Bus. Res.*, vol. 122, pp. 915–925, Jan. 2021.

[2] Narayanan, A. and Shmatikov, V. "Robust De-anonymization of Large Sparse Datasets," In *Security and Privacy, IEEE Symposium on Security and Privacy*, SP 2008, pp. 111–125, May 2008.

[3] Hassan M. U., Rehmani M. H. and Chen J., "Differential Privacy Techniques for Cyber Physical Systems: A Survey," *In IEEE Communications Surveys & Tutorials*, vol. 22, no. 1, pp. 746–789, First quarter 2020.

[4] Mach J. et al. "*l-Diversity: Privacy on the far side k-Anonymity.*" vol. 2. Id-421–57. IEEE. Access 2009.

[5] Ali M., Malik S. U. and Khan S. U., "DaSCE: Data Security for Cloud Environment with Semi-trusted Third Party," *IEEE Transactions on Cloud Computing*, vol. 5, no. 4, pp. 642–655, 2015.

[6] Sankar L., Rajagopalan S. R. and Poor H. V., "Utility-privacy Tradeoffs in Databases: An Information-Theoretic Approach," *IEEE Trans. Inf. Forensics Security*, vol. 8, no. 6, pp. 838–852, Jun. 2013.

[7] Shao Y., Liu J., Shi S., Zhang Y. and Cui B., "Fast De-anonymization of Social Networks with Structural Information," *Data Sci. Eng.*, vol. 4, no. 1, pp. 76–92, Mar. 2019.

[8] Yağar F., "Growing Concern During the COVID-19 Pandemic: Data Privacy," *Turkiye Klinikleri J. Health Sci.*, vol. 6, no. 2, pp. 387–392, 2021.

[9] Vats S., Singh S., Kala G., Tarar R. and Dhawan S., "iDoc-X: An Artificial Intelligence Model for Tuberculosis Diagnosis and Localization," *J. Discret. Math. Sci. Cryptogr.*, vol. 24, no. 5, pp. 1257–1272, 2021.

[10] Vats S., Sagar B. B., Singh K., Ahmadian A., and Pansera B. A., "Performance Evaluation of an Independent Time Optimized Infrastructure for Big Data Analytics that Maintains Symmetry," *Symmetry (Basel).*, vol. 12, no. 8, 2020, doi: 10.3390/SYM12081274.

[11] Vats S. and Sagar B. B., "An Independent Time Optimized Hybrid Infrastructure for Big Data Analytics," *Mod. Phys. Lett. B*, vol. 34, no. 28, p. 2050311, Oct. 2020, doi: 10.1142/ S021798492050311X.

[12] Vats S. and Sagar B. B., "Performance evaluation of K-means Clustering on Hadoop Infrastructure," *J. Discret. Math. Sci. Cryptogr.*, vol. 22, no. 8, 2019, doi: 10.1080/09720529.2019.1692444.

[13] Agarwal R., Singh S. and Vats S., "Implementation of an Improved Algorithm for Frequent Itemset Mining Using Hadoop," in *2016 International Conference on Computing, Communication and Automation (ICCCA)*, 2016, pp. 13–18. doi: 10.1109/CCAA.2016.7813719.

[14] Casas-Roma J., Salas J., Malliaros F. D. and Vazirgiannis M., "K-degree Anonymity on Directed Networks," *Knowl. Inf. Syst.*, vol. 61, no. 3, pp. 1743–1768, Dec. 2019.

[15] Agarwal R., Singh S. and Vats S., *Review of Parallel Apriori Algorithm on Mapreduce Framework for Performance Enhancement*, vol. 654. 2018. doi:10.1007/978-981-10-6620-7_38.

[16] Bhati J. P., Tomar D. and Vats S., "Examining Big Data Management Techniques for Cloud-based IoT Systems," In Examining Cloud Computing Technologies Through the Internet of Things, IGI Global, pp. 164–191, 2018.

[17] Sharma V. et al., "OGAS: Omni-directional Glider Assisted Scheme for Autonomous Deployment of Sensor Nodes in Open Area Wireless Sensor Network," ISA Trans., Aug. 2022, doi: 10.1016/j. isatra.2022.08.001.

[18] Sharma V., Patel R. B., Bhadauria H. S., and Prasad D., "NADS: Neighbor assisted deployment scheme for optimal placement of sensor nodes to achieve blanket coverage in wireless sensor network," *Wirel. Pers. Commun.*, vol. 90, no. 4, pp. 1903–1933, 2016.

[19] Sharma V., Patel R. B., Bhadauria H. S., and Prasad D., "Policy for Planned Placement of Sensor Nodes in Large Scale Wireless Sensor Network," *KSII Trans. Internet Inf. Syst.*, vol. 10, no. 7, pp. 3213–3230, 2016.

[20] Bhatia M., Sharma V., Singh P., and Masud M., "Multi-level P2P Traffic Classification Using Heuristic and Statistical-based Techniques: A Hybrid Approach," *Symmetry (Basel).*, vol. 12, no. 12, p. 2117, 2020.

[21] Vikrant S., Patel R. B., Bhadauria H. S., and Prasad D., "Glider Assisted Schemes to Deploy Sensor Nodes in Wireless Sensor Networks," *Rob. Auton. Syst.*, vol. 100, pp. 1–13, 2018.

[22] Vats S. and Sagar B. B., "Data Lake: A Plausible Big Data Science for Business Intelligence," 2019.

[23] Dohare, Indu, and Karan Singh. "Green Communication in Sensor Enabled IoT: Integrated Physics Inspired Meta-heuristic Optimization-based Approach." *Wireless Networks* 26, no. 5 (2020): 3331–3348.

Automation and Computation – Vats et al. (Eds)
© 2023 the Editor(s), ISBN 978-1-032-36723-1

Author index